CLINICAL CHEMISTRY:

Interpretation and Techniques

CLINICAL CHEMISTRY:

Interpretation and Techniques

ALEX KAPLAN, Ph.D.

Head, Clinical Chemistry Division
Professor Emeritus, Department of Laboratory
Medicine,
University of Washington, Seattle, Washington

and

LAVERNE L. SZABO, M.S.

Assistant Professor, Clinical Chemistry and
Medical Technology Divisions,
Department of Laboratory Medicine,
University of Washington,
Seattle, Washington

Second Edition

LEA & FEBIGER Philadelphia

Lea & Febiger
600 Washington Square
Philadelphia, Pa. 19106
U.S.A.

First Edition, 1979
Reprinted, 1981
Second Edition, 1983

Library of Congress Cataloging in Publication Data

Kaplan, Alex, 1910–
  Clinical chemistry.

  Includes index.
  1. Chemistry, Clinical. I. Szabo, Laverne L.
II. Title. [DNLM: 1. Chemistry, Clinical. QY 90 K17c]
RB40.K38 1983   616.07'56   82-17249
ISBN 0-8121-0873-6

Printed in the United States of America

Print No.   4   3

*To those dedicated medical technologists
who choose to serve humanity by
working assiduously and accurately
in clinical chemistry laboratories*

PREFACE

Our basic aim in this second edition is to improve the first edition and make it a better teaching manual for those who work, supervise, or teach in clinical chemistry. We have made some changes, enlarged the content and treatment of some chapters, reordered the material, and made many revisions in order to achieve those goals. Methods that have become obsolete are omitted. We have profited by many of the suggestions and thoughtful comments that were made by users and reviewers of the book.

Most of the methods that are described in detail are the ones that we use at the University Hospital, University of Washington. The selection reflects both our experience and prejudice. In a few situations, several methods are given for a single constituent. For example, an enzymatic and a chemical method are described for cholesterol, since both are widely used; this will enable students to become familiar with the principles of both and to function in a wider range of laboratories. In a book of this size, it is impossible to outline the advantages and disadvantages of the various methods and to justify the selection of each method.

In general, the laboratory tests are grouped according to the function or organ system being tested. The chemical principles of the chosen method are discussed, as are the physiologic and biochemical changes for particular constituents that occur in normal and disease states. The material bearing on clinical interpretation serves as a motivating link between the laboratory worker and the physician as their joint effort is directed toward the diagnosis and treatment of disease. In general, work performance in the laboratory is better when technologists have an understanding of the application of their results, because it helps to give them a feeling of participation in the total medical effort rather than of being robots who crank out numbers of unknown significance.

Chapter 2, "Basic Principles," includes a section in which general principles of chemical analysis are presented for the purpose of review and to encourage accurate, precise, and intelligent work in the laboratory. Another section describes common laboratory instruments and techniques and discusses their basic principles. Several new sections have been added, one of which explains the principles of different types of automated analyzers (continuous flow, discrete, centrifugal, thin-film). Only the basic approaches are mentioned since technologists will have to learn the operation of the particular instruments used by their laboratories. The other new sections deal with radiochemistry and laboratory computers. This chapter also includes a section on laboratory quality control and statistics that emphasizes the need for recognition of random, systematic, or procedural errors in the laboratory. It is necessary not only to recognize error but to determine its magnitude in order to assess the validity of test results or to evaluate the usefulness of procedural changes. Finally, a section dealing with laboratory safety is presented to acquaint students with the potential hazards of working in a clinical chemistry laboratory and the means for minimizing the risks.

All of the succeeding chapters have been updated and modified to some extent. In particular, the chapter on enzymes includes a section on enzyme kinetics; the lipid chapter deals more extensively with lipoproteins; the chapters on endocrinology and on immunochemical techniques have been revised extensively; and the toxicology chapter has been expanded, especially in the sections dealing with therapeutic drug monitoring and pharmacokinetics.

A hospital is a complex institution, and there are many links in the chain before a sample is analyzed and the report received by the physician. When a physician requests a blood test, a nurse or clerk transcribes the request upon a laboratory request form, a member of a blood-collecting team usually obtains the specimen that is brought to the laboratory, given an acquisition number, processed and analyzed. After the analysis has been completed, the result is entered upon an appropriate laboratory form, which goes to a nursing station and then to the ordering physician. Computers may or may not be used in the data handling system. The chain is long and complex and subject to error at a number of points, many of which are beyond the control of the laboratory. It is the laboratory's function to perform the test as accurately and as speedily as possible and to see that the results are properly and correctly entered into the system for delivery to the physician. A laboratory result, no matter how accurately and swiftly performed, becomes useless unless it reaches the attention of the attending physician. Conscientious laboratory workers must be prepared to check various aspects of the chain of communication to insure that the results reach their destination without undue delay. Moreover, the specimen itself must be handled adequately and properly from the time of drawing the blood until the analysis is carried out in the laboratory.

In this age of automation and mechanization, there is a tendency to place too much emphasis upon machine capability and too little upon the capabilities, training, and judgment of the technologists and technicians who operate the instruments. Automated instruments can carry out a number of operations in a repetitive fashion and can make possible the performance of a large number of tests, but most of these instruments require adjustment, calibration, adequate maintenance, and constant surveillance to make sure that the results they generate are both precise and accurate. This requires supervision and control by people who understand the instruments and know what they are doing; in the clinical chemistry laboratory this means that the technologists must have a good basic understanding of clinical chemistry and analytical chemistry.

Manual methods are described in detail in the text because the chemical principles are the same in both manual and automated tests; many teaching laboratories require the students to perform tests manually as well as by automation; and a number of laboratories perform some tests manually or use these tests as a back-up for an automated system.

In addition to technical competence, good laboratory workers must have the feeling at all times that they are members of a medical team dealing with sick people. The work that they do is extremely important because the modern health care expert relies heavily upon the results of chemical measurements of constituents in body fluids and tissues. This sense of concern, when accompanied by the initiative to make sure that the results are dependable and that they reach their proper destination, makes the difference between a good and a mediocre laboratory.

Alex Kaplan, Ph.D.

Seattle, Washington   Laverne L. Szabo, M.S.

ACKNOWLEDGMENTS

Many of our departmental colleagues and Senior Fellows in the Postdoctoral Training Program assisted in the preparation of the second edition by offering helpful suggestions and constructive criticism. We appreciate the assistance of Professors K. J. Clayson, J. S. Fine, K. E. Opheim, V. A. Raisys, and E. K. Smith of the Department of Laboratory Medicine. We are also grateful to the following Senior Fellows: Drs. V. M. Brautigan, K. B. Cox, L. F. Ferreri, W. G. Haigh, and J. B. A. Ross.

A. K.
L. S.

CONTENTS

Chapter 1

INTRODUCTION TO CLINICAL CHEMISTRY

The clinical chemistry laboratory is an important contributor to the medical team involved in the diagnosis and treatment of disease because it performs approximately 50% of the total laboratory requests. Physicians rely heavily upon laboratory test results before making decisions.

It is obvious that laboratory results must be reliable for proper decision-making. Good technique and an understanding of the chemical processes are prerequisites for the production of high-quality analytical work. Clinical chemistry, however, involves more than analysis of body fluids and tissues.

Clinical chemistry has been defined as the study of the chemical aspects of human life in health and illness and the application of chemical laboratory methods to diagnosis, control of treatment, and prevention of disease.[1] Thus, clinical chemistry is a fundamental science when it seeks to understand the physiologic and biochemical processes operant both in the normal state and in disease; it is an applied science when analyses are performed on body fluids or tissue specimens to provide useful information for the diagnosis or treatment of disorders. In some countries, the term "clinical biochemistry" is reserved for the fundamental science and "clinical chemistry" for the applied science, but in the United States, the latter term usually covers both categories.

Some elementary biology is presented for student orientation because a knowledge of the working of the human body under healthy, normal conditions is essential for the understanding of changes that may occur in abnormal or pathologic states. An introductory text in biochemistry[2,3] or biology[4] is recommended for those who have not had courses in biochemistry or molecular biology.

CELLS

The basic unit of the body is the cell, the place where most of the body's chemical reactions occur. The cell has an outer membrane that separates it from other cells and from the tissue fluids that bathe them. A mammalian cell is illustrated in Figure 1.1.

Composition

Mammalian cells have a well-defined nucleus that contains the genetic material (DNA) distributed among chromosomes, and many organized structures or compartments (organelles) in the cytoplasm, in which many different functions and chemical reactions take place. Several of these (mitochondria, vacuoles, endoplasmic reticulum) are illustrated in Figure 1.1. The main oxidative

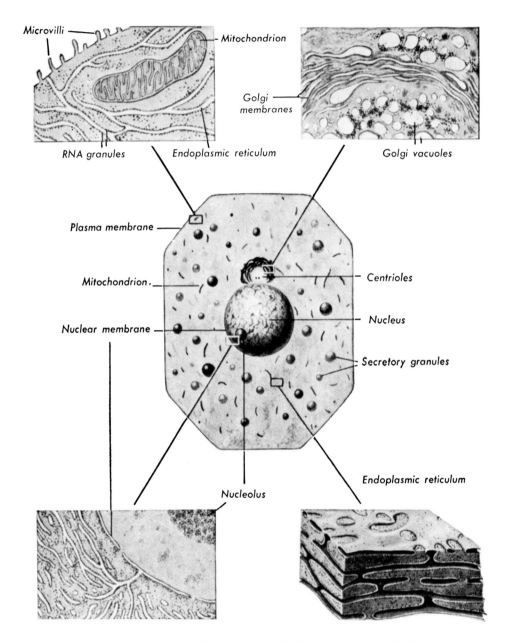

Fig. 1.1. Schematic representation of a cell as seen with the light microscope and enlargements showing the fine structure of some of the cell constituents as revealed by electron microscopy. (From Gray's Anatomy, 29th ed. Philadelphia, Lea & Febiger, 1973.)

reactions for the production of energy occur in the mitochondria; protein synthesis takes place on ribosomes; and many enzymes are located on the endoplasmic reticulum.

The outer cell membrane is composed of a mosaic of a double layer of lipid molecules interspersed with protein molecules and constitutes a means for selective permeability. Some ions and small molecules are able to pass easily into or out of the cell; others require special transport mechanisms for entry or exit while some cannot cross the membrane of the living cell. Special molecules (glycoproteins) on the membrane act as recognition points or receptors for the attachment of particular substances, such as hormones or antibodies, that are present in the circulating blood plasma.

The chemical constituents of cells are simple and complex proteins, nucleoproteins, carbohydrates, lipids, intermediates of these compounds, inorganic salts, and water. The organization within specific cells may vary depending upon the cell's structure and function. The formation, growth, or functioning of cells requires the presence of appropriate raw materials and enzymes and a readily available supply of energy.

Malfunction

Malfunction of a cell may be caused by a variety of factors, such as (1) destruction by trauma or by invasive agents, which include pathogenic microorganisms, viruses, toxins; (2) genetic deficiency of a vital enzyme; (3) insufficient supply of one or more essential nutrients (amino acids, vitamins, or minerals, for example); (4) insufficient blood supply; (5) insufficient oxygen supply; (6) malignancy (uncontrolled tissue growth); (7) accumulation of waste products; (8) failure of a control system; or (9) by a defect in the cellular recognition of certain signals.

ROLE OF THE CLINICAL CHEMISTRY LABORATORY

The major efforts of the clinical chemistry laboratory are devoted to measuring the concentration of various constituents in body fluids (primarily serum but also urine, cerebrospinal fluid, or other fluids). However, it is not easy to predict what is happening at the cellular level when the concentration of a serum constituent is abnormally high or low; additional information is needed. An elevated serum constituent concentration could be caused by excessive intake, excessive body synthesis, excessive breakdown of cells, deficient utilization, deficient excretion, or severe dehydration. The reverse is true for a low concentration. Information derived from the physical examination and the patient's history helps to elucidate the problem, but additional selected tests may be necessary. Gross abnormalities always indicate that something is wrong. It is difficult to make a diagnosis early in a disease, when symptoms are obscure or absent and changes in blood constituent concentrations are minimal. When concentration changes are small, the validity of the test result may be questioned. This is why precision and quality control are so necessary; your contributions are essential for good patient care.

The material in the following chapters provides information on both phases of clinical chemistry: (1) the basic information needed for performing accurate and precise analyses and (2) an explanation of the physiologic and biochemical processes occurring in the body for the major constituents that we analyze. The latter will help in understanding why certain tests are ordered and in the interpretation of results.

REFERENCES

1. Sanz, M.C., and Lous, P.: Suggested definition of "clinical chemistry." Newsletter, Int. Fed. Clin. Chem. 6, 1971.
2. Lehninger, A.L.: Biochemistry: The Molecular Basis of Cell Structure and Function, 2nd ed., New York, Worth Publishers, 1975.
3. Stryer, L.: Biochemistry, San Francisco, W.H. Freeman, 1981.
4. Luria, S.E.: Life: The Unfinished Experiment, New York, Charles Scribner's Sons, 1973.

Chapter 2

BASIC PRINCIPLES

The clinical chemistry laboratory measures chemical changes in the body in the interest of diagnosis, therapy, and prognosis of disease. The main work of its technologists consists in the assay of various chemical constituents in blood, urine, and other fluids or tissues. Normally, the concentrations of these constituents are relatively constant, but in disease states their levels become altered, the magnitude of change usually paralleling the degree of disease. In advanced disease, the abnormalities are usually large and easily detected and present no analytical challenge to the laboratory. Early detection of organ dysfunction is much more difficult, however, because the chemical changes are usually slight and have to be distinguished from possible errors in the performance of the test. Hence, test accuracy is a prerequisite for the proper interpretation of a laboratory test.

The technologist must always keep in mind that each laboratory specimen is taken from a person with a real or potential health problem, and that a technologist is an essential part of a highly skilled team that contributes to an assessment of the patient's condition.

GENERAL LABORATORY INFORMATION

Most test procedures are quantitative and require careful, precise measurements. An understanding of the principles of the testing method and some knowledge of the medical uses of the determinations provide the necessary background knowledge both for performing the tests and for understanding their rationale. A technologist, like any skilled worker, must understand thoroughly the workings of the various instruments and equipment in use in the laboratory. Accordingly, this chapter will summarize some of the background information and the basic principles that are common to most chemistry laboratory methods and essential to good laboratory practice and technique.

Also included are a review of common laboratory calculations and an introduction to quality control. A brief description of medical usage and an interpretation of laboratory tests appear in the chapters dealing with the tests themselves.

Units of Measure

All quantitative measurements must be expressed in clearly defined units that are accepted and understood by all scientists. The metric system is employed in scientific measurements; hence, the gram, meter, liter, and second are employed as the basic units for the expression of weight, length, volume, and time, respectively.

Identical prefixes are used with the basic units to denote larger or smaller size. Thus, one *millimeter* (mm) refers to one-thousandth part of a meter (10^{-3} meters), while *milligram* (mg), *milliliter* (ml), and *millisecond* (ms) are used to describe 10^{-3} g, 10^{-3} L, and 10^{-3} seconds, respectively.

Table 2.1 summarizes the more commonly used prefixes and their abbreviations. The terms marked with asterisks are correct but are seldom used in the clinical laboratory.

Despite the fact that scientists used the metric system, discrepancies in the expression of many units began to appear in different disciplines, different countries, or even different laboratories. Wavelength was designated in angstrom units (10^{-10}m) in some countries or laboratories and as millimicrons (10^{-9}m) in others; the concentrations of blood constituents were expressed as g/L in some countries and as mg/dl in others. Some authors expressed the concentrations of reagents in g/100 ml, but others worked with molar concentrations. This led to confusion and to misunderstanding. A start in the standardized presentation of clinical chemical laboratory data was made in 1967 when the 1966 Recommendation of the Commission on Clinical Chemistry of the International Union of Pure and Applied Chemistry and of the International Federation for Clinical Chemistry[1] was published. The recommended units and mode of expression are known as the Système Internationale or the SI. In SI units, wavelength is designated as nanometers (10^{-9}m) and abbreviated nm.

The units of length, weight, and volume in the metric system were designed to be interrelated with each other. The liter was originally defined as the volume occupied by one kilogram of water at the temperature of its greatest density (4° C), but in 1964 the definition was changed by international agreement so that the liter is now defined as being exactly equal to a cubic decimeter. The old liter was 1.000028 times as large as the new one, a negligible difference. Although the terms *cubic centimeter* (cc) and *milliliter* (ml) are synonymous, only the latter term, ml, is acceptable to the Système Internationale.

Temperature

The temperature scale most commonly used in the clinical laboratory is the centigrade scale (Celsius or ° C), which places the freezing point of water at 0° C, and the boiling point of water at 100° C.

TABLE 2.1
Some Basic Units and Their Prefixes Denoting Multiple or Decimal Factors

PREFIX Basic Unit	ABBREVIATION	MULTIPLE OF BASIC UNIT $10^0 = 1$	WEIGHT g (gram)	LENGTH m (meter)	VOLUME L (liter)
kilo-	k	10^3	kg		
deci-	d	10^{-1}	dg*	dm*	dl
centi-	c	10^{-2}	cg*	cm	cl*
milli-	m	10^{-3}	mg	mm	ml
micro-	μ	10^{-6}	μg	μm	μl
nano-	n	10^{-9}	ng	nm	nl
pico-	p	10^{-12}	pg	pm*	pl*

*Denotes terms seldom used in a clinical chemistry laboratory.

For calculations involving temperature, a scale based on absolute zero is needed. The Kelvin scale (° K) employs units identical to those in the centigrade system, but its zero point corresponds to absolute zero (0° K), which is equivalent to $-273.15°$ C. Thus, for conversion from centigrade to Kelvin temperatures, we can use the following formula:

$$°\,K = °\,C + 273$$

In the SI, temperatures are expressed as ° K, but clinical laboratories still use ° C.

Volumetric Equipment

Clinical chemistry procedures require accurate measurements of specimens and reagents for valid, useful results; it is essential that all equipment and instruments be accurate and reliable. High grade volumetric glassware should be purchased, and calibrations should be verified.

Pipets

Several types of pipets are available, each designed for a specific purpose, but a few rules for correct pipetting technique are common to all of them. The pipet must be held in a vertical position while setting the liquid level to the calibration line and during delivery. The lowest point of the meniscus should be level with the calibration line on the pipet when it is sighted at eye level. The flow of the liquid should be unrestricted when using volumetric pipets; flow may have to be slightly slowed with the finger when using graduated pipets for fractional delivery.

One rule should *always* be observed in the laboratory: Never pipet anything by mouth. Always use a safety bulb or other pipet filler.

TRANSFER PIPETS. The *volumetric*, or transfer, pipet is used when the greatest accuracy and precision are required. It consists of a long, narrow tube with an elongated or rounded bulb near the middle and is designed to deliver a known volume of water at a specified drainage time. The tip is tapered to slow the rate of delivery, and there is a single calibration ring etched into the tubing above the bulb. Class A pipets, conforming to narrow limits specified by the National Bureau of Standards, are necessary for the greatest accuracy. In use, the pipet tips are touched to the inclined surface of the receiving vessel until the fluid has ceased to flow and for 2 seconds thereafter. The tip is then withdrawn horizontally from contact with the receiver. A volumetric pipet with a broken or chipped tip is inaccurate and must be discarded. Pipets are now available in which the glass has been chemically tempered to increase strength. Because the pipets are especially resistant to chipping and breaking, they are economical in the long run.

OSTWALD-FOLIN PIPET. This specialized version of the volumetric pipet was designed for transferring viscous fluids in order to minimize drainage errors. The pipet is shorter, the bulb is near the delivery tip, and the opening is slightly larger for faster drainage. As soon as drainage has stopped, the last drop is blown out. All blowout pipets are marked with an etched single or double ring around the mouthpiece.

MEASURING PIPETS. Formerly called Mohr pipets, measuring pipets consist of long, straight tubes with graduated markings to indicate the volume delivered from the pipet. The calibrations are made between two marks etched on the straight portion of the tube. The tapered tips are not part of the calibrated portion.

SEROLOGIC PIPETS. These are like measuring pipets but are calibrated to the tip. These are blowout pipets and are marked with the etched ring on the mouthpiece; to deliver the total volume, the last drop must be blown out. Serologic pipets may be used for "point-to-point" measurements in the straight part of the tube, just as measuring pipets are used, or for total volume measurements, blowing out the last drop. For good precision when dispensing aliquots of a reagent (for example, 2-ml aliquots from a 10-ml serologic pipet), only the calibrations on the straight tubing should be used, not the tapered section. Thus, only four 2-ml aliquots can be obtained by using a 10-ml serologic pipet.

All of the above pipets are designed "to deliver" (TD) the stated volumes and are marked with "TD" near the upper end of the tube. There are other pipets, designed for measuring small quantities of fluids, that are calibrated "to contain" (TC) rather than to deliver the specified volume. These pipets are described below.

MICROPIPETS. The trend in medical practice is to order more and more chemistry tests on individual patients. To conserve the volume of blood required for analytical purposes, microtechniques that use small amounts of specimen have been incorporated into most laboratory procedures. Some may require as little as 5 microliters (μl) or up to 250 μl of patient specimen. When measuring such small volumes, variations in the liquid film adhering to the pipet wall become significant. The problem of residual liquid in micropipets calibrated to contain a given volume of liquid is eliminated by rinsing the pipet several times with the dilution mixture into which its contents have been dispensed; the last drop is then blown out. There are also micropipets that are calibrated to deliver a measured volume.

The apparatus used to measure the volumes of specimen samples and of standard solutions must be accurate. Class A pipets are accurate enough for routine clinical chemistry work and need no further checking. Other pipets, including automatic pipets, should be carefully calibrated to determine the true volume.

AUTOMATIC PIPETS. Automatic pipets are designed to pick up and to dispense a pre-set volume of solution as a plunger is smoothly released and then depressed; these are available commercially (Fig. 2.1). In some models, disposable plastic tips are used to eliminate carry-over from sample to sample. Another type employs a Teflon-tipped plunger that sweeps the capillary delivery tip clean as the contents are expelled. Automatic pipets deliver reproducible volumes, but the accuracy may be diminished at low volumes. It is advisable to check the accuracy of any particular pipet to be used for measuring small volumes of sample. The attainment of good precision with automatic pipets requires close adherence to the directions for usage, smooth manual action, and practice.

Automatic micropipets may be calibrated by weighing a solution of known density delivered by the pipet. The actual delivery volume in ml is calculated

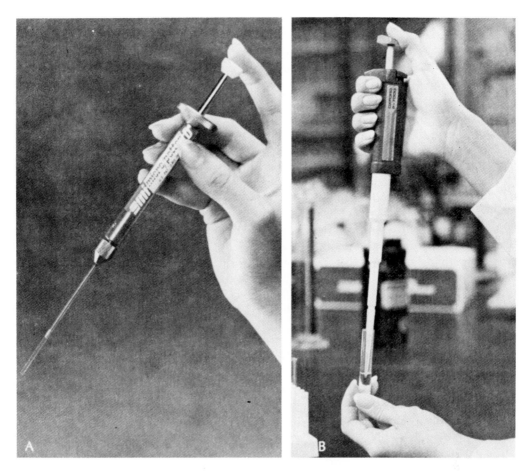

Fig. 2.1. Automatic pipets: A, with reusable glass tip. (Courtesy of Scientific Manufacturing Industries, Emeryville, CA 94608.) B, with disposable plastic tip. (Courtesy of Oxford Laboratories, San Mateo, CA 94401.)

by dividing the weight in g by the density. The calibrated pipets are used as standards for calibrating other pipets by a simpler colorimetric procedure.*

PLUNGER TYPE DISPENSERS. Reagent supply bottles may be fitted with a dispensing apparatus built into the screw top lids. These can be adjusted in the laboratory to repeatedly deliver the desired volume of reagent with good precision. Such dispenser bottles save time when repetitive volumes of a reagent are dispensed.

In large laboratories there are often procedures involving great numbers of samples. For handling work loads of this magnitude, *electrically operated dispensers* and combination *sampler-dilutors* are most practical.

---

*For the colorimetric procedure, 10 ml water is dispensed into tubes (10 tubes for each pipet to be calibrated). An aliquot of $K_2Cr_2O_7$ (4.00 g/dl) is delivered into each tube and the absorbance is read at 450 nm against water. The actual volume of each pipet is calculated from the absorbance and dilution factor compared to that of the calibrated pipet.

Volumetric Flasks

Volumetric flasks that meet Class A requirements should be used for all standard solutions and many laboratory reagents. Since changes in temperature cause variations in the volume of solutions, it is important that glassware and reagent solutions be at room temperature at the time of final dilution in the volumetric flask. Solid reagents must be completely dissolved, and the solutions must be well mixed before final dilution to the calibration mark.

Graduated cylinders are relatively inaccurate measuring devices and are used for situations in which accuracy of a high order is not required. They are convenient for the measuring of 24-hour urine volumes of 500 to 2000 ml, where a low order of accuracy is sufficient. They may also be used for the preparation of noncritical solutions.

Beakers, flasks, and test tubes sometimes have graduated markings along one side indicating approximate volumes. These are only rough approximations, convenient though they may be, and are not appropriate when accurate measurements are required.

Burets

Burets may be used in the laboratory for titrations or for dispensing aliquots of a solution. They are available in a wide range of sizes, most commonly from 1 ml to 100 ml capacity. Microburets include those with a total volume of 10 ml or less. Some microburets are fitted with fine metal tips that deliver drops of approximately 0.01 ml. For delivery and measurement of small volumes (microliters), there are syringe microburets that use the forward motion of a plunger to displace the fluid that is dispensed. The plunger movement is measured accurately by a micrometer. Since the plunger's area of cross section is known, the volume of fluid dispensed is directly proportional to the distance the plunger moves.

All burets must be thoroughly clean so that a uniform film of liquid coats the walls as the liquid is dispensed. When reading a buret, the meniscus of the solution should be at eye level, so that parallax will not introduce serious errors into the reading. Also, the major calibration marks that encircle the buret can help prevent parallax errors. For ease in reading, a white card with a heavy black line should be held behind the buret so the black marking is just below the meniscus. The meniscus will appear darkly outlined against the white background.

Burets may be purchased with either Teflon or ground glass stopcocks. The major advantage of the Teflon stopcock is that it does not require lubrication; the Teflon plug is also highly resistant to alkali and will not "freeze" to the buret when exposed to such solutions. The glass stopcocks require a thin coating of lubricant that must be applied sparingly so that an excess does not plug the stopcock channels or buret tip. Excess lubricant may also adhere to buret walls, making subsequent measurements inaccurate. The grease must be frequently renewed when exposed to alkaline solutions. According to some manufacturers, "freezing" of glass stopcocks in alkali may be prevented by using graphite as an additional stopcock lubricant. The ground glass surface of the stopcock is cleaned, dried, and rubbed with a soft pencil lead to produce a thin black layer before greasing. This prevents "freezing" as long as the black layer can be seen readily.

Burets are indispensable for certain determinations and for standardization of some reagents, but are often the despair of the laboratory. If heavily greased, they become plugged, or dirty; if sparingly greased, the stopcocks are easily pulled so they leak, and the grease still seems to creep and contaminate the buret walls. If burets are used only occasionally and are filled with water for storage between use, they need frequent cleaning.

The mechanical piston buret eliminates many of the above problems (Fig. 2.2). Set up as a closed system by direct attachment to an aspirator bottle protected by a soda lime tube, it is not exposed to dust or to evaporation. It needs no greasing and need not be emptied or cleaned during the intervals it is not in use. There is finger tip control, making it easy to deliver small drops, and the scale is easily read, with no possibility of parallax.

Cleaning Volumetric Glassware

It is essential that glassware used in the laboratory be clean. One quick test of surface cleanliness is to observe the glass surface as the final rinse water drains off. The water should move with a "sheeting" action, leaving a thin film

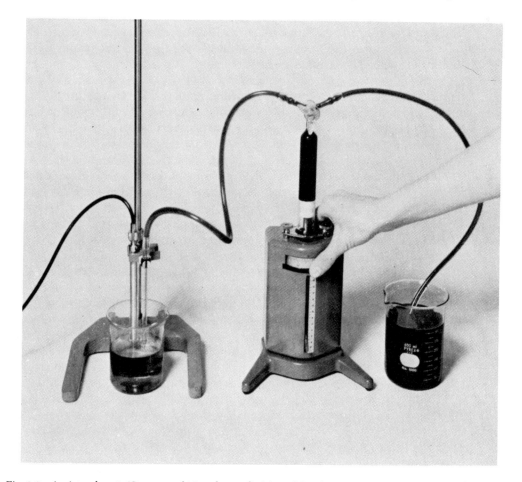

Fig. 2.2. A piston buret. (Courtesy of Metrohm, a division of Brinkmann Instruments, Inc., Westbury, N.Y. 11590.)

over the whole surface. If the film breaks up into droplets, or if the surface is unevenly wet, the piece is not clean.

The subsequent cleaning of laboratory glassware is greatly simplified if the various pieces are emptied after use, rinsed with water, and soaked in a strong solution of detergent. Then they may be cleaned readily by some mechanical dishwashers that employ high pressure jets of hot detergent solution, followed by tap water and deionized water rinses.

For certain critical tests, such as radioimmunoassays, or with some enzymes, machine washing may not be sufficiently reliable; traces of contaminants may ruin a test. For sensitive tests, disposable tubes are used as much as possible, and the glassware for the preparation of solutions is washed by hand after employing an efficient cleaning agent (special formulations of detergents, many of which are found on the market today).

Glassware requiring special treatment is cleaned by brush and detergent, rinsed, and then soaked in the special cleansing solution for 1 to 16 hours. It is then thoroughly rinsed with tap water, then deionized water, and then drained and allowed to dry.

The Analytical Balance

Reagents for the preparation of standard solutions must be weighed accurately; the weighing error should be less than 0.1%. Good single pan analytical balances with internal weights are convenient to use and are accurate to at least ± 0.1 mg.

Operating instructions for analytical balances may vary greatly for different models, and directions supplied by each manufacturer should be followed. A few general rules, however, apply to all analytical balances.

1.   The balance must be level and vibration-free; marble balance tables greatly reduce vibration, but where they are not practical, vibration can be substantially reduced by inserting a cushioned marble slab between the analytical balance and the laboratory bench.
2.   To protect the delicate knife-edge of the balance, weights must be added or removed gently, and only while the balance beam is raised off the knife-edge. This is true whether external weights are added manually to the pan or internal weights are lifted by turning a knob.
3.   Air currents during the final step of weighing must be avoided by closing the doors of the weighing compartment. Since temperature differences also cause convection currents, the samples must be at room temperature.
4.   The balance must be kept scrupulously clean. Weights, pans, and internal parts of the balance must never by touched with the fingers; chemicals left on the balance, either spilled or in fingerprints, may corrode the metal and impair its accuracy. Any spills should be carefully brushed off the pans and removed from the balance compartment. Like all precision instruments, an analytical balance requires periodic maintenance and cleaning for trouble-free performance.

Distilled and Deionized Water

Tap water is not pure enough for use in chemistry procedures: therefore, *distilled or deionized water is required for the preparation of all reagents, standards, and other solutions.* Water that has been prepared by a single distillation is not pure; it may be contaminated by dissolved material from the distillation apparatus itself, dissolved gases, airborne particles, and nonvolatile substances that may be carried over as an aerosol with the steam distillate. The usual contaminants originating from a still are copper and salts leached from the glass; the most likely aerosols include sulfate and carbonate salts of sodium, potassium, and magnesium.

Dissolved gases may be removed by boiling the water for a short time, and airborne materials, such as dust, may be effectively reduced by protecting the condensing and collection portions of the apparatus from the air. Distilled water is not sterile and often contains nitrogenous materials that can support the growth of bacteria. A second distillation, from an alkaline potassium permanganate solution, produces a more nearly pure water by removing most of the contaminants.

Deionized water is prepared by running tap water through a bed of mixed resins to remove ions from the water. A mixed-bed resin removes both cations and anions and thus produces a water low in conductivity, a criterion for estimating water purity. This water may still contain bacteria and organic compounds. If deionized water is also passed through a charcoal bed, contaminants are adsorbed, and water of high purity is produced. The charcoal bed may also be used in conjunction with distillation, replacing the second, or permanganate, distillation.

As water is distilled or deionized, its ability to conduct an electric current is vastly reduced. A conductivity cell placed in the water measures the resistance to a current passing between two spaced electrodes. Water with a resistance of over 1 megaohm/cm is satisfactory for reagent use.

Either distilled or deionized water may become contaminated by the container in which it is stored; soda lime glass, especially, may be a source of metal ions in the water. For this reason, water should not be stored for long periods before it is used, and any reagents that are to be kept for a long time should be stored in borosilicate glassware.

In the procedures described in the following chapters, high purity distilled or deionized water is to be used in the preparation of all aqueous solutions.

Laboratory Chemicals

All chemicals used in clinical chemistry procedures are called reagents and must be prepared from high quality chemicals. For most chemicals, this is the grade known as *Reagent Grade, Analytical Reagent Grade,* or *ACS grade* (meets American Chemical Society specifications). These chemicals are highly purified compounds, guaranteed to meet certain stated standards that are printed on the label. The label may show the actual assay for the chemical and a number of specified contaminants or state that the contaminants are below certain specified concentrations.

For some laboratory procedures, such as fluorometry, gas chromatography, and trace metal analysis, reagents of exceptionally high purity are needed. Until recently, it was necessary for the analyst to further purify each batch of the purchased chemical or at least to check each lot against the laboratory's special requirements. To eliminate this inconvenience, manufacturers now offer selected chemicals that have been subjected to additional purification steps to further reduce the content of all the impurities present. These reagents may be known by such titles as *Ultra pure* or *Ultrex*. Other reagents may be treated to remove impurities interfering in one type of laboratory procedure; they are sold as *enzyme quality*, *Spectrophotometric* grade or Nanograde reagents. The latter two reagents are designed for sensitive photometric procedures such as fluorometry or for high performance liquid chromatographic assays.

On the other hand, manufacturers label some chemicals with only descriptive terms such as *purified* or *highest purity*. There is no standard for such labeling, and the quality of the product varies from one brand to another. Organic compounds are often graded in this way, with the melting point range given as a further indication of purity. These chemicals are not recommended for the clinical chemistry laboratory, but they may have to be used when no better quality is available. It may be necessary to further purify these compounds before using them or to include a reagent blank with each batch of tests.

Technical or *commercial* grade chemicals are not suitable for the laboratory except for such purposes as cleaning solutions.

A USP rating on a reagent bottle indicates that the contents conform to the specifications of the *United States Pharmacopeia*. Since these specifications allow no impurities at a concentration level that would be injurious to health, pharmacists can safely use them for medications. Some of these chemicals are pure enough for laboratory purposes; others are not. Chemical purity as such is not the primary goal of this classification, and such purity must not be assumed by the user of USP grade reagents.

For some of the assays described in this book, a source for a reagent or other material is given in parentheses (Elvanol 70–05, DuPont). Appendix 10 gives the complete names and addresses of manufacturers so mentioned.

Primary Standards

A *primary standard* is a substance that can be accurately weighed or measured to produce a solution of an exactly known concentration. To qualify as a primary standard, a substance must be essentially free of impurities. The IUPAC (International Union of Pure and Applied Chemistry) has set up criteria for primary standards, which require that the compound be at least 99.98% pure (for working standards, 99.95% pure).[2] This level of purification is impossible to attain with most biological materials and is not necessary for routine clinical laboratory tests.

It is difficult to obtain biologic materials of acceptable purity, or even of *known* purity. To help alleviate this problem, the National Bureau of Standards (NBS) has a program of producing a series of Standard Reference Materials for Clinical Chemistry, most of which are at least 99% pure.[3,4] These include cholesterol and bilirubin, compounds that are difficult to purify but that are needed as standards for laboratory procedures. The NBS standards are much too expensive for daily use in the laboratory, but they serve as excellent reference standards. With the aid of the reference standard, a laboratory may accurately assay its commercially obtained reagent, which may then serve as a routine (secondary) standard.

Laboratory Calculations

Weight of Solute per Volume of Solution

The simplest reagents are made by transferring a weighed amount of the desired chemical to a volumetric flask, dissolving it in solvent, and filling the flask exactly to the calibration mark. With pure chemicals and good technique, the concentration of the resulting solution is an exact quantity that may be expressed, in terms of mass per volume of solution, in several ways. As an illustration, 5 g of Na_2SO_4 in a liter of solution is equivalent to 5 g/L, 0.5 g/dl, 500 mg/dl, or 5 mg/ml.

Such expressions of mass of solute per unit volume do not indicate the functional concentration of a solution—that is, the relative number of molecules available to react with other molecules. Molar concentration, which is discussed later, is based on the latter concept, but the mass per volume designation is often used in the laboratory as a convenience in identifying and making simple reagents.

Per Cent Solutions

Another convention consists of expressing the concentration of a solute as "per cent" (per hundred parts of the total solution). The three variations of per cent solutions are illustrated below:

1. *Weight/volume (w/v)*: A solution containing 5 g of Na_2SO_4 dissolved in water and diluted to a final volume of 100 ml of solution can be designated as 5% (w/v) solution.
2. *Volume/volume (v/v)*: This designation is convenient for solutions composed of two liquids. Thus, 5 ml of glacial acetic acid diluted with water to a total volume of 100 ml can be described as a 5% (v/v) acetic acid solution.
3. *Weight/weight (w/w)*: The percentage of solute can also refer to grams of solute per 100 g of final solution. For example, 5 g Na_2SO_4 dissolved in 95 g of water (approx. 95 ml) would result in a concentration of 5% (w/w): 5 g solute in a total weight of 100 g (solute plus solvent). This designation differs significantly from the 5% (w/v) Na_2SO_4 solution described above and is seldom used in clinical chemistry.

Per cent solutions that do not designate weight or volume for solute and solution are ambiguous at best and are open to error and confusion.

Another term, *milligrams per cent* (mg%), has traditionally been used to represent milligrams per 100 ml of solution. This is nondefinable as percentage because mg/dl is one part in 100,000. *Milligrams per deciliter* (mg/dl) is the correct term implied by the above expression and should always be used in place of it.

The SI recognizes moles per liter (mol/L) as the only mass per volume designation; the use of % or mg/dl is discouraged.

Moles; Molarity

In the previous section, the concentrations of solutions were expressed in terms of mass of solute in a given volume of solution. Except when actually weighing out the reagent, this designation is not particularly useful. A more meaningful expression is the relative number of molecules available in the solution to react with other molecules. To express this relationship, the SI[1] recommends that the liter be used as the unit of volume and that the solute concentration be expressed as mol/L whenever the molecular structure is unequivocably known. The use of mol/L allows direct comparisons of the functional concentrations of solutions.

A mole of any pure compound is the molecular weight of that compound expressed in grams. It consists of the same number of molecules (6.02×10^{23}) as one mole of any other compound. "Moles per liter" is sometimes expressed as *molarity*, a one-molar (1 M) solution containing one mole of solute per liter of solution. The SI does not recommend use of the symbol M to represent molar concentration, but it is widely used, and technologists must be familiar with it.

To calculate the number of moles in a given weight of solute, divide the number of grams by the molecular weight.

Example: 5 g Na_2SO_4 is equivalent to how many moles? The molecular weight of Na_2SO_4 is 142, so there are 5/142 or 0.035 moles. If the 5 g Na_2SO_4 were dissolved in water to make 1 L of solution, the concentration would be 0.035 mol/L. To avoid cumbersome decimal fractions, small concentrations may be expressed with an appropriate prefix to the mol/L. Thus, 0.035 mol/L may also be written as 35 mmol/L (millimoles per liter). Micromoles (μmol) and nanomoles (nmol) representing 10^{-6} and 10^{-9} moles, respectively, may also be used where appropriate.

Example: What is the concentration in SI units of a solution containing 1.20 g of Na_2CO_3 in 200 ml of solution?

$$\text{mol/L} = \frac{1.20 \text{ g}}{106 \text{ g/mol}} \times \frac{1000 \text{ ml/L}}{200 \text{ ml}} = 0.0566 \text{ mol/L}$$

$$\left(\frac{1.20g}{106g}\right)\left(\frac{1\,mol}{}\right)\left(\frac{1000\,ml}{}\right)\left(\frac{}{200\,ml}\right) \qquad = 56.6 \text{ mmol/L}$$

Gram Equivalent Weight; Normality

The gram equivalent weight of a substance may be defined as the weight of that substance that can combine with, or displace, 1.008 g of hydrogen.

Example: What is the gram equivalent weight of calcium (atomic weight = 40.08)?

Comparison of the formulas, $CaCO_3$ and H_2CO_3, shows that 40.08 g calcium (1 mole) displaces 2.016 g hydrogen. Setting up the ratio gives:

$$\frac{40.08}{\text{eq wt Ca}} = \frac{2.016}{1.008} = \frac{2}{1}$$

$$\text{eq wt Ca} = \frac{40.08}{2} = 20.04$$

The above example illustrates that the equivalent weight of an element is equal to the atomic weight divided by the number of atoms of hydrogen with which one atom of that element can combine or that it can displace. For aqueous solutions of electrolytes (acids, bases, and salts), the gram equivalent weight may be determined by dividing the gram formula weight by the total positive or negative charge.

Examples: (1) in Na_2O ($2Na^{+1}$, O^{-2}) the equivalent weight of oxygen is 16.00/2 = 8.00, and the equivalent weight of the compound Na_2O is 62/2 = 31.

(2) The equivalent weight of $CaCO_3$ (Ca^{+2}, CO_3^{-2}) is equal to the molecular weight (100.1) divided by 2, or 50.0.

Note: The *gram formula weight* is defined as the weight in grams of the entity represented by a formula. The gram formula weight then is equal to the gram molecular weight of a compound, to the gram atomic weight of an element, or to the weight in grams expressed by the formula for an ion, such as SO_4^{--}.

As discussed in the preceding section, a *molar* solution is a solution that contains one mole of solute per liter of solution. Correspondingly, a normal solution contains one gram equivalent of solute per liter of solution. *Normality* is defined as the number of equivalents of solute that are present per liter of solution; a 3 normal (3N) solution of $CaCO_3$ contains three gram equivalents, or 150.1 g of $CaCO_3$ per liter.

Example: What is the normality of a solution containing 5 g Na_2SO_4 per liter? The molecular weight is 142 and the equivalent weight is 142/2 = 71, since the two atoms of Na can be replaced by two atoms of hydrogen, as in H_2SO_4. The normality of the solution is 0.070 N (5 g/L ÷ 71 g/eq).

The terms *normality* and *equivalents* are not recommended in the SI, but both terms are widely used in clinical laboratories and must be understood by technologists. These units truly represent the functional or stoichiometric quantities of substances; one equivalent of any substance will react with exactly one equivalent of another substance, whether the reaction is a simple combination or displacement reaction or a more complex oxidation-reduction reaction.

It may be convenient, but it is by no means necessary, to use the concepts of "normality" and "equivalence" to calculate the quantities of reactants that react stoichiometrically. The calculations can readily be made using balanced equations containing molecular formulas. An acid-base titration is a simple example of this type of problem.

Example: How many ml of NaOH (1.5 mol/L) will be required to neutralize 10 ml of H_2SO_4 (2.0 mol/L)?

Solution A. $2\ NaOH\ +\ H_2SO_4 \rightarrow Na_2SO_4\ +\ 2\ H_2O$. From this equation we can see that 2 moles of NaOH are required to neutralize 1 mole of H_2SO_4.

$$\text{mmol } H_2SO_4 = \text{Volume (ml)} \times \text{Concentration} \left(\frac{\text{mmol}}{\text{ml}}\right) = 10 \times 2 = 20$$

The amount of NaOH required = 2×20 mmol = 40 mmol

$$\text{ml NaOH} = 40 \text{ mmol}/1.5 \text{ mmol/ml} = 26.67 \text{ ml}$$

Solution B. The problem may also be solved by using the normalities of the solutions:

$$1.5 \text{ M NaOH} = 1.5 \text{ N NaOH}$$

$$2.0 \text{ M } H_2SO_4 = 4.0 \text{ N } H_2SO_4$$

Since we are dealing with equivalents, the equation becomes $V_1C_1 = V_2C_2$.

$$1.5\ V_1 = 10 \times 4$$

$$V_1 = 26.67\ \text{ml of NaOH}$$

Either method of solving the problem is acceptable; use the one that is easier for you.

Hydrated Salts

When salts are crystallized, water molecules sometimes form an integral part of the crystal structure as water of hydration. Some salts have several different hydrated forms that contain differing amounts of water in the crystal lattice: these are always indicated in the molecular formula and contribute to the molecular weight.

For example, sodium sulfate crystals may appear in three forms:
1. Anhydrous (Na_2SO_4); Mol wt. 142
2. Heptahydrate ($Na_2SO_4 \cdot 7H_2O$); Mol wt. 268
3. Decahydrate ($Na_2SO_4 \cdot 10H_2O$); Mol wt. 322

A 1 mol/L solution of each of the above salt forms would contain 142, 268, and 322 g/L, respectively; each solution would contain exactly the same concentration of sodium sulfate as each of the others.

When making up reagents, the technologist must always be aware of the hydration state of any crystalline compound to be measured. For example, if directions for a reagent specify 8.0 g of Na_2SO_4 (anhydrous), but the available salt is $Na_2SO_4 \cdot 10\ H_2O$, a simple ratio using the molecular weights of the two salt forms will adjust the instructions.

$$\frac{X\ g}{8\ g} = \frac{322\ g/mol}{142\ g/mol}$$

$$X = 18.1\ g$$

Dilutions

In the daily routine of the laboratory, a technologist frequently makes dilutions of samples and solutions. This may be done as part of a test procedure, as when a protein-free filtrate is made from serum or when a concentrated stock standard is diluted to make a less stable daily working standard. Dilution of patient specimens (serum, urine) may be required when some constituent is too concentrated to be accurately measured in the routine procedure—a high blood glucose in a diabetic patient, for example.

In either case, the new concentration may be calculated by a simple formula, which is based on the fact that the diluted sample contains the same total amount of the constituent as the original concentrated sample. If 1 ml of the diabetic patient's serum is diluted to 5 ml, the total amount of glucose in the specimen before and after dilution is the same. Assuming an original serum concentration of 500 mg/dl, the glucose concentration in the diluted solution is calculated as follows:

Equation 1: $V_1C_1 = V_2C_2$

$$1 \text{ ml} \times 500 \text{ mg/dl} = 5 \text{ ml} \times C_2$$
$$C_2 = 500 \text{ mg/dl} \times \frac{1}{5}$$
$$C_2 = 100 \text{ mg/dl}$$

Dilutions are generally expressed as a *ratio of the original volume to the total final volume*; the above example is described as a 1:5 dilution of the patient's serum. The formula $V_1C_1 = V_2C_2$ is valid as long as the same units of volume and concentration are used for both the original and the final solutions.

Example: What is the resulting concentration of Na^+ if a serum that contains 140 mmol/L is diluted 1:100?

$$1 \times 140 \text{ mmol/L} = 100 \; C_2$$
$$C_2 = 1.4 \text{ mmol/L}$$

Example: To what volume should 2 L of a stock HCl solution, 4 mol/L, be diluted to provide a solution of 0.8 mol/L?

$$2 \text{ L} \times 4 \text{ mol/L} = V_2 \times 0.8 \text{ mol/L}$$
$$V_2 = \frac{2 \times 4}{0.8} = 10 \text{ L}$$

It is sometimes necessary to make a large dilution, such as 1:1000, of a solution when some stock standard solutions are diluted to working standards. A 1:1000 dilution may be performed in steps as follows: 1 ml of the original solution is diluted to 100 ml (1:100), and 1 ml of the product is further diluted to 10 ml (1:10). The concentration of the final solution would be 1/100 × 1/10, or 1/1000 of the original concentration.

When a patient specimen is diluted for a laboratory test procedure, the dilution factor must be included in the calculation of the final results. In the example of the diabetic patient whose serum was diluted 1:5, the test result on the diluted serum would indicate a glucose concentration of 100 mg/dl. Before the result is reported, the concentration of the *original* serum must be obtained by multiplying by the dilution factor:

$$100 \text{ mg/dl} \times 5 = 500 \text{ mg/dl}$$

Example: 3 ml of urine were diluted with 2 ml of water, and an aliquot was assayed for creatinine. To calculate the creatinine concentration of the original patient urine:

The dilution factor = final volume/original volume = 5/3
Creatinine concentration = concentration in diluted aliquot × 5/3

Conversion of Units

Concentrations of solutions may be expressed in many ways, and a technologist must be able to convert units of concentration from one form to another.

The concentration of some blood constituents such as glucose has been traditionally reported as milligrams per deciliter (mg/dl), and others such as chloride in milliequivalents per liter. For some blood constituents, however, there is no one traditional method of reporting, so that the units may vary from laboratory to laboratory. Magnesium and calcium concentrations are still reported in meq/L or mg/dl even though the SI recommends mmol/L. In utilizing data from another laboratory, it may be necessary to convert the data from one set of units to another.

Such a conversion is a straightforward task of multiplying by a series of factors, with each factor designed to convert one dimension from the given unit to the desired one. For example, one factor may change the volume measurement from deciliters to liters, while another converts the weight or mass from milligrams to milliequivalents.

Example: A serum calcium concentration is 10 mg/dl. Express this concentration in meq/L. (Calcium: atomic weight 40, and equivalent weight 20.)

$$10 \text{ mg/dl} \times \frac{10 \text{ dl}}{1 \text{ L}} \times \frac{1 \text{ meq}}{20 \text{ mg}} = 5 \text{ meq/L}$$

Example: The same principle applies in converting a magnesium concentration of 2 meq/L to mg/dl. (Mg: atomic weight 24, equivalent weight 12.)

$$2 \text{ meq/L} \times \frac{1 \text{ L}}{10 \text{ dl}} \times \frac{12 \text{ mg}}{1 \text{ meq}} = 2.4 \text{ mg/dl}$$

When the conversion factors are all set up correctly, the original units of dimension are canceled, leaving only the units that are desired for the final result.

As SI units become more widely used, technologists must be able to convert concentrations from other units to mol or mmol per liter and vice versa.

Example: The normal range for serum glucose levels in a laboratory is 70 to 105 mg/dl. What is the same normal range expressed in SI units? (The molecular weight of glucose, $C_6H_{12}O_6$, is 180.)

For the lower limit:

$$\text{mol/L} = 70 \text{ mg/dl} \times \frac{10 \text{ dl}}{1 \text{ L}} \times \frac{1 \text{ g}}{1000 \text{ mg}} \times \frac{1 \text{ mol}}{180 \text{ g}}$$

$$= 70 \times \frac{1}{100} \times \frac{1}{180} \text{ mol/L}$$

$$= 0.00388 \text{ mol/L} = 3.88 \text{ mmol/L}$$

For the higher limit:

$$\text{mmol/L} = 105 \text{ mg/dl} \times \frac{10 \text{ dl}}{1 \text{ L}} \times \frac{1 \text{ mmol}}{180 \text{ mg}} \times 5.83 \text{ mmol/L}$$

Thus, the normal range for glucose is 3.88 to 5.83 mmol/L.

Example: In SI units, a serum uric acid concentration is 0.75 mmol/L. Express this concentration in mg/dl.

The uric acid molecule has the formula $C_5H_4O_3N_4$, Mol wt 168.

$$0.75 \text{ mmol/L} \times \frac{1 \text{ L}}{10 \text{ dl}} \times \frac{168 \text{ mg}}{1 \text{ mmol}} = 12.6 \text{ mg/dl}$$

For some serum constituents, molecular weights cannot be unequivocably known. Serum proteins are quantitated as a family of compounds consisting of molecules that vary greatly in size. No single molecular weight can represent all protein molecules; an "average" molecular weight becomes meaningless as the proportions of different classes shift with individuals and with disease states. Protein concentration is reported in g/dl in conventional systems and g/L in the SI.

Molarity of Concentrated Reagents

The common acids and ammonium hydroxide are supplied as concentrated solutions. Each lot of a reagent is assayed by the manufacturer, and data are provided concerning the specific gravity and percentage by weight of the reagent; from these data the concentration in mol/L of the reagent is readily calculated.

Example: A concentrated H_2SO_4 solution has the following composition:

Specific gravity   1.84

H_2SO_4 95% by weight

To express the concentration in mol/L

1. Determine density in g/L:
   1.84 g/ml $\times$ 1000 ml/L = 1840g/L
2. Since 95% of the total weight is H_2SO_4:
   1840 g/L $\times$ 0.95 = 1748 g H_2SO_4/L
3. Convert to molar concentration:
   (H_2SO_4 has a molar weight of 98.1 g)

$$1748 \text{ g/L} \times \frac{1 \text{ mol}}{98.1 \text{ g}} = 17.8 \text{ mol/L, or } 17.8 \text{ M}$$

Although each lot of the reagent may differ slightly in composition from the others, the range of concentrations for each individual acid or base is quite narrow; a concentrated solution of sulfuric acid, for example, always contains about 18 moles of H_2SO_4 per liter. Approximate concentrations of the common concentrated solutions are shown in Appendix 3.

Hydrogen Ion Concentration

The following definitions are used in the discussion of acids and buffers:

1. *Acid:* Any substance that can dissociate to form protons (hydrogen ions, H^+). For example,

$$H_2CO_3 \rightleftharpoons H^+ + HCO_3^-$$

An acid is termed a *strong acid* if its aqueous solutions are highly dissociated (e.g., HCl, H_2SO_4, HNO_3). It is termed a *weak acid* if the degree of dissociation in aqueous solution is low. Stated another way, weak acids have small dissociation constants (less than 10^{-4}). Carbonic acid, H_2CO_3, is a very weak acid, with a dissociation constant of 4.3×10^{-7} for the first H^+.

2. *Base:* Any substance that can accept H^+. For example,

$$NH_3 + H^+ \rightleftharpoons NH_4^+$$

3. *pH:* Hydrogen ion concentration of plasma is about 0.00000004 mol/L (4×10^{-8}). People working with biologic fluids needed a simpler method for expressing the H^+ concentration and adopted the method of expressing it in terms of pH, where $pH = \log 1/[H^+] = -\log[H^+]$. Thus, the hydrogen ion concentration $[H^+]$ of pure water is 10^{-7} mol/L; the pH = 7.0.

Each increase of 1 pH unit represents a tenfold decrease in the concentration of hydrogen ion. If adding a base to a solution causes a pH change from 4 to 7 there is a shift of 3 pH units and the $[H^+]$ is decreased from 10^{-4} to 10^{-7} mol/L, a 1000-fold dilution.

The calculation of $[H^+]$ when the pH is not a whole number requires the use of a logarithm table. The process requires that the mantissa, or decimal part of the logarithm, must always be converted to a positive number before it can be found in a table of logarithms. Also, multiplication is carried out by the process of adding logarithms. For example, to find the concentration of hydrogen ion in a solution of pH 5.2:

$$pH = -\log[H^+] = 5.2, \text{ or } \log[H^+] = -5.2$$

$$[H^+] = \text{antilog } (-5.2) = \text{antilog } (-6 + 0.8).$$

From log tables, the antilog of 0.800 is 6.31, and the antilog of -6 is 10^{-6}, so

$$[H^+] = 6.3 \times 10^{-6}$$

If the hydrogen ion concentration of a solution is known, the pH may be calculated. For a solution in which

$$[H^+] = 0.0005 \text{ mol/L} = 5 \times 10^{-4}$$

$$pH = -\log (5 \times 10^{-4}) = -\log 5 - (-4) = 4 - \log 5.$$

From log tables, $\log 5 = 0.699$, so

$$pH = 4 - 0.699 = 3.3$$

BUFFER SOLUTIONS. The rates of many chemical reactions, particularly those involving enzymes, depend upon the hydrogen ion concentration. Most enzymes have a narrow pH range for optimal action, and activity falls off rapidly on either side of the optimum. Buffer salts stabilize the pH of solutions by accepting or donating protons to prevent a fall or rise in the pH.

A buffer is a mixture of a weak acid with its salt of a strong base, or a weak base with its salt of a strong acid. Buffers resist a change in pH when acid or alkali is added by forming weak acids or bases that consist mostly of undissociated molecules; hence, they are only slightly ionized. Adding HCl to an acetic acid-sodium acetate buffer results in the formation of more of the weak acetic acid and neutral NaCl. Adding NaOH to the same buffer generates sodium acetate and water. The regulation of the pH of body fluids to within narrow limits is made possible only by the presence of various buffer systems.

The equation for the reversible dissociation of a weak acid, HA, may be written:

$$HA \rightleftharpoons H^+ + A^-$$

and expressed mathematically as:

$$Equation\ 2:\ K = \frac{[H^+]\ [A^-]}{[HA]}$$

where K is the symbol for the dissociation constant of the acid HA at equilibrium. If $K = 10^{-5}$ for the acid HA, and concentration of the acid is 0.1 mol/L,* then

$$10^{-5} = \frac{[H^+]\ [A^-]}{10^{-1}}$$

$$10^{-6} = [H^+]\ [A^-]$$

$$[H^+] = [A^-]\ so\ [H^+]^2 = 10^{-6}$$

$$[H^+] = 10^{-3}\ mol\ L$$

The pH of the 0.1 mol/L solution of the weak acid HA is 3 and the $[H^+] = 10^{-3}$ mol/L. Since the concentration of HA is 0.1 mol/L, it is 1% ionized ($10^{-3} \div 0.1 \times 100 = 1\%$).

According to equation 2, the dissociation constant K determines the product of the concentrations of the hydrogen ion and the anion (A^-). Since K is a constant for the acid HA, the hydrogen ion concentration may be varied by changing the concentration of the anion—that is, by adding a salt of the acid to the solution. This principle, known as the "common-ion effect," is the basis for many widely used buffers.

To illustrate, equation 2 may be rearranged to:

$$Equation\ 3:\ [H^+] = K \frac{[HA]}{[A^-]}$$

If 0.1 mol/L of the sodium salt of HA is added to the 0.1 mol/L of HA, $[A^-]$ is increased, resulting in a change in $[H^+]$. The total concentration of A^- consists of the $[A^-]$ provided by the completely ionized salt plus the $[A^-]$ produced by the dissociation of HA. The dissociation of HA is usually so slight that the final concentration of A^- is, for practical purposes, the same as the initial concentration of NaA.

---

*$[HA]$ is actually equal to 0.1 mol/L minus the concentration of the dissociated acid or $[H^+]$. In this instance, $[H^+] = 10^{-3}$ and $[HA] = 0.1 - 0.001$ or 0.099 mol/L, a difference of 1%. In the preparation of a buffer, the normality of the acid would not be determined by titration; a 1% variation is within the limits of error and is considered negligible.

By the same reasoning, [HA] at equilibrium is equal to the initial concentration of acid, minus the amount that dissociates into H^+ and A^-. Since our hypothetical acid is a weak acid ($K = 10^{-5}$), the loss of [HA] from dissociation is extremely small compared to the initial concentration of HA, and the concentration of HA at equilibrium is essentially equal to the initial concentration of HA.

From equation 3:

$$[H^+] = (10^{-5}) \frac{10^{-1}}{10^{-1}} = 10^{-5}$$

and pH = 5

Thus, it was possible to reduce the hydrogen ion concentration of HA one hundredfold from 10^{-3} to 10^{-5} mol/L by the addition of the salt.

Conversely, the addition of a strong acid (HCl) drives the reaction to the left $HA \rightleftharpoons H^+ + A^-$ as the hydrogen ions from the HCl combine with the buffer salt anions to form a weak acid. Cl ion replaces A ion in solution. A strong base, such as NaOH, would shift the equilibrium to the right, releasing more hydrogen ions to neutralize the hydroxyl ions. Again this would result in the formation of the buffer salt, plus water. Thus, adding either a strong acid or a strong base to a buffer solution results in the formation of a weak acid or a weak base with a relatively small change of hydrogen ion concentration.

Henderson-Hasselbalch Equation: From equation 3, a useful relationship between pH and the ionization constant may be developed:

$$[H^+] = K \frac{[HA]}{[A^-]}$$

$$\log[H^+] = \log K + \log \frac{[HA]}{[A^-]}$$

$$-\log[H^+] = -\log K - \log \frac{[HA]}{[A^-]}$$

but $pH = -\log[H+]$ and $pK = -\log K$.

Substituting pH and pK in the above equation:

$$pH = pK - \log \frac{[HA]}{[A^-]}$$

A negative logarithm is equal to the log of the reciprocal of the number, so the above expression can be written in the usual form of the Henderson-Hasselbalch equation:

$$\text{Equation 4: } pH = pK + \log \frac{[A^-]}{[HA]}$$

In a mixture of a weak acid and its sodium salt, the anion concentration depends upon the concentration of the salt, which is completely ionized. The equation may be transformed to:

$$\text{Equation 5: pH} = \text{pK} + \log \frac{[\text{salt}]}{[\text{acid}]}$$

The following examples illustrate the effects of varying the acid and salt concentrations of HA whose pK = 5.0.

With concentrations of both HA and NaA at 0.1 mol/L,

$$\text{pH} = 5 + \log \frac{0.1}{0.1} = 5 + \log 1 = 5 + 0 = 5$$

Thus, when the salt concentration and the acid concentration are equal, the pH is equal to the pK, and the buffering capacity of the solution is at its maximum.

When 0.1 mol/L salt and 0.01 mol/L acid concentrations are used:

$$\text{pH} = 5 + \log \frac{0.1}{0.01} = 5 + \log 10 = 5 + 1 = 6$$

With 0.01 M salt and 0.1 M acid:

$$\text{pH} = 5 + \log \frac{0.01}{0.1} = 5 + \log 10^{-1} = 5 - 1 = 4$$

The Henderson-Hasselbalch equation may be used to calculate the proportions of salt and acid necessary to make a buffer of the desired pH. In selecting a buffer system, the technologist should choose one with a pK value close to the desired pH. If a buffer with a pH of 5.0 is desired, an acetic acid-sodium acetate (HAc-NaAc) buffer may be selected. The pK of this system is 4.7. This means that when the salt and the acid concentrations are equal the solution has a pH of 4.7:

$$\text{pH} = \text{pK} + \log \frac{[\text{Ac}^-]}{[\text{HAc}]}$$

$$\text{pH} = 4.7 + \log \frac{[0.1]}{[0.1]} = 4.7 + 0 = 4.7$$

To make a buffer of pH 5.0, the Henderson-Hasselbalch equation is used:

$$5.0 = 4.7 + \log \frac{[\text{Ac}^-]}{[\text{HAc}]}$$

$$\log \frac{[\text{Ac}^-]}{[\text{HAc}]} = 0.3$$

$$\frac{[\text{Ac}^-]}{[\text{HAc}]} = \text{antilog } 0.3 = 2$$

Whenever the sodium acetate is twice as concentrated as the acetic acid in the buffer solution, the pH is 5.0. The concentrations that are finally chosen depend upon other requirements of the system: the buffering capacity needed, the ionic strength, and solubility limitations. Usually, the buffer for an enzyme reaction system is chosen by selecting the one that gives the highest enzyme activity when varying the pH and ionic strength.

Some common laboratory buffers are listed in Appendixes 5 through 7 with directions for their preparation.

LABORATORY INSTRUMENTS

The clinical chemistry laboratory today could not operate without the wide-spread use of instrumentation. Most laboratory instruments measure changes in (1) the absorbance of light (spectrophotometers, atomic absorption spectrophotometers); (2) the intensity of emitted light (fluorometers, emission flame photometers); (3) electric potential or current flow (pH meter, ion-selective electrodes); (4) the elapsed time for passage of a constant current (coulometer); or (5) the volume of solution needed to reach an end point (titrometer). A discussion follows concerning some basic types of instruments and the principles of their operation. In some cases, specific instrument models may be mentioned while illustrating a principle; this does not imply an endorsement of the instrument but serves as an example for teaching purposes.

Photometry

Photometry means the "measurement of light." A camera light meter is an example of a photometer that quantitates the total light intensity that strikes the photocell. In the laboratory, relative light intensity is usually measured after passage of the light through a solution. The light absorbed by the solution is limited to a narrow portion of the spectrum.

A review of some of the basic characteristics of light may make this section more understandable. Light is a form of electromagnetic energy that travels in waves. The wavelength, or distance between the peaks of a light wave, is a function of its energy. High energy gamma rays that we associate with nuclear reactions have short wavelengths, in the order of 0.1 nm. At the other end of the scale are the long, low energy radiowaves, 25 cm or more in length. Visible light, light that can be seen by the human eye, constitutes a small portion of the electromagnetic radiation and is limited to the region of 380 to 750 nm (see Table 2.2). A very good spectrophotometer may measure light from 180 or 200 nm in the near ultraviolet region to about 1000 nm in the near infrared region.

The color of light is a function of its wavelength (Table 2.3). As the wavelength is changed within the visible range, an alteration in color is detected. For example, when a strip of filter paper is placed in a spectrophotometer cuvet so that the paper interrupts the light beam, the spectral colors change as the wavelength decreases from about 800 nm to below 400 nm. The first light that appears is red; it gradually becomes orange and then moves through yellow, green, blue, and finally to violet before entering the ultraviolet range at 380 nm.

Light from the sun (or from incandescent light bulbs) contains the entire visible spectrum; this continuum of light appears "white" or colorless. Objects that appear colored absorb light at particular wavelengths and reflect the other

TABLE 2.2

Wavelengths of Various Types of Radiation

	TYPE OF RADIATION	APPROXIMATE WAVELENGTH (nm)
	Gamma	<0.1
	X-rays	0.1–10
E ν λ	Ultraviolet	<380
	Visible	380–750
	Infrared	>750
	Radiowaves	over 25×10^7

A listing of types of electromagnetic radiation in the order of the energy involved. E = energy; ν = frequency; λ = wavelength. The vertical arrows indicate the direction of increase in magnitude.

TABLE 2.3

The Visible Spectrum

APPROXIMATE WAVELENGTH (nm)	COLOR OF LIGHT ABSORBED	COLOR OF LIGHT REFLECTED
400–435	violet	green-yellow
435–500	blue	yellow
500–570	green	red
570–600	yellow	blue
600–630	orange	green-blue
630–700	red	green

The wavelengths are approximate. The colors of the light absorbed and reflected change gradually from one color to the next with no clear line of demarcation. The sum of the colors of the reflected light forms the apparent color of the object to the viewer.

parts of the visible spectrum, thus giving many "shadings" of color. For example, a substance that absorbs violet light at 400 nm reflects all other light and appears as yellow-green, and a substance absorbing yellow light at 590 nm is seen as blue which is the sum of the reflected light (see Table 2.3).

Spectrophotometry takes advantage of the property of colored solutions to absorb light of specific wavelengths. To measure the concentration of a blue solution, light at about 590 nm is passed through it; the amount of yellow light absorbed varies directly as the concentration of the blue substances in the solution.

Beer's Law

When light of an appropriate wavelength strikes a cuvet that contains a colored sample, some of the light is absorbed by the solution; the rest is transmitted through the sample to the detector. The proportion of the light that reaches the detector is known as the per cent transmittance (%T) and is represented by the equation

$$\frac{I_t}{I_o} \times 100 = \%T$$

in which I_o is the intensity of light striking the sample, and I_t is the intensity of the light transmitted through the sample. In actual practice, the light absorbance of a Blank is substituted for I_o. The Blank may be water or the entire reagent mixture except for the sample.

As the concentration of the colored solution in the cuvet is increased, I_t and consequently %T, is decreased. The relationship between concentration and %T is not linear, as shown in Figure 2.3, but if the *logarithm* of the %T is plotted against the concentration, a straight line is obtained. The term *absorbance* (A) is used to represent the logarithm of $\frac{1}{\%T}$ or $(-\log\%T)$; absorbance increases linearly with concentration (Fig. 2.4).

The relationship of absorbance to concentration is expressed in the equation known as the Beer-Lambert Law (often referred to as Beer's Law):

$$\text{Equation 6: } A = a\,b\,c$$

in which A is absorbance, c is the concentration of the colored compound, a is the absorptivity coefficient (a constant) of the colored compound, and b is the length of the light path through the solution. Since a and b are constants in an assay, A is directly proportional to c.

In spectrophotometric procedures, the absorbance of an unknown concentration of a particular constituent is compared with that of a known concentration (a standard), which is reacted in the same way to produce a colored solution. The following relation holds:

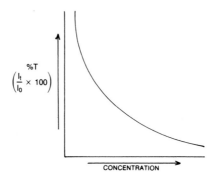

Fig. 2.3. Percent transmittance (%T) as a function of concentration.

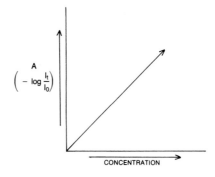

Fig. 2.4. Absorbance (A) as a function of concentration.

$$\text{Equation 7: } A_u/A_s = C_u/C_s$$

in which A_u and A_s are the respective absorbances of unknown and standard, and C_u and C_s are their respective concentrations. To solve the equation for C_u, the equation is rearranged to

$$\text{Equation 8: } C_u = A_u/A_s \times C_s.$$

Equation 8 is routinely used for calculations in spectrophotometric assays.

No assay gives a linear response between absorbance and concentration for all concentrations, going from very small to large, because sooner or later, some reactant becomes limiting and less color is formed. The range of linearity may be wide for some constituents and narrow for others. It is essential for any procedure to determine whether spectrophotometric response is linear for the usual concentrations to be measured and to determine the limits of linearity.

UV-Visible Spectrophotometry

For the last three decades, the simple spectrophotometer or colorimeter has been the workhorse of the hospital chemistry laboratory. *Colorimeter* has been the traditional name for an instrument that isolates specific wavelengths of light with interchangeable filters for the visible portions of the spectrum. In contrast to this, spectrophotometers have a continuously adjustable mono-chromator (prism or grating) and often can measure the intensity of light from the UV range through the visible. This capability is necessary today for the analysis of many enzymes and drugs.

The components of many spectrophotometers are basically the same. They consist of a power supply that provides current at the proper voltage and the components shown in Figure 2.5: (1) a lamp as a light source; (2) a mono-chromator to isolate the desired wavelength; (3) a sample holder or cuvet; (4) a photodetector which produces a current in response to the light impinging upon it; and (5) a meter or other readout device. These are discussed in detail below.

POWER SUPPLY. The power supply may be a simple transformer to convert line voltage to a constant low voltage required for the lamp, or it may provide current (both AC and DC) of different voltages for several components, such as the detector and readout devices, as well as the lamps.

LIGHT SOURCES. The light source is usually a tungsten lamp for wavelengths in the visible range (350 to 700 nm) and a deuterium or hydrogen lamp for

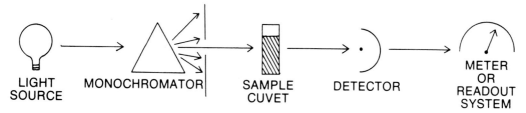

LIGHT SOURCE   MONOCHROMATOR   SAMPLE CUVET   DETECTOR   METER OR READOUT SYSTEM

Fig. 2.5. Basic components of a spectrophotometer.

ultraviolet light (below 350 nm). The lamps are positioned in the instrument in such a way that an intense beam of light is directed through the monochromator and the sample. A *selector switch* is used to shift a mirror to reflect the light from either the deuterium or the tungsten lamp into the monochromator as needed for the UV or visible light measurement. Although the lamps are prefocused and their positioning may be predetermined, intensities between lamps vary, and a spectrophotometer should be recalibrated whenever the lamp is changed. It is good practice to check the calibration daily at one wavelength with a didymium filter, and to check the spectrophotometers monthly for linearity at 3 wavelengths, using a commercial set of calibration standards. Wavelength-scanning instruments are checked against a holmium oxide filter for accuracy of both wavelength and absorbance.

MONOCHROMATORS. The next component in the system is the monochromator. Early colorimeters used glass filters that transmitted a wide segment of the spectrum (50 nm or more). Newer instruments employ interference filters that consist of a thin layer of a magnesium fluoride crystal with a semitransparent coating of silver on each side. The crystal transmits only light for which an exact multiple of the wavelength is equal to the thickness of the crystal. All other wavelengths are blocked. Interference filters have a bandpass of 5 to 8 nm.

The term *bandpass* defines the width of the segment of the spectrum that will be isolated by a monochromator; it is the range of wavelengths between the points at which the transmittance is equal to one-half the peak transmittance and is illustrated in Figure 2.6. In this example, the peak of light intensity occurs at 550 nm, the nominal wavelength setting of the monochromator. The bandpass is 20 nm because the intensity of light at 540 and 560 nm is one half that of the 550 peak. The inexpensive spectrophotometers commonly used in the laboratory often have a bandpass of about 20 nm, but the more sophisticated instruments may have a bandpass of 0.5 nm or less.

The monochromator consists of an entrance slit to exclude unwanted or "stray" light and a prism or diffraction grating preceded by a series of light

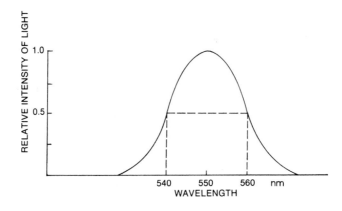

Fig. 2.6. Bandpass of a spectrophotometer. The curved line represents the intensity of the light at various wavelengths transmitted by a monochromator when it is set at 550 nm. The wavelengths at which the light intensity is one half as great as at peak height (550 nm) are 540 nm and 560 nm, respectively. These two wavelengths encompass the bandpass of the instrument (20 nm).

focusing lenses. An exit slit allows only a narrow fraction of the spectrum to reach the sample cuvet.

When polychromatic light enters a prism, it is refracted at the air-glass interface. The shorter wavelengths are refracted more than the longer ones, with the result that light is dispersed into a continuous spectrum. This dispersion permits the isolation of a particular wavelength band. Glass prisms and lenses are suitable for work in the visible range, but it is essential to use quartz or fused silica for the UV range.

Diffraction gratings consist of a series of parallel grooves cut into a surface all at exactly the same angle; there may be 10,000 to 50,000 grooves per inch. Light striking the grooves is diffracted and dispersed according to wavelength; a continuous spectrum is produced. If a grating is made from polished metal, the diffracted light is reflected and may be used for ultraviolet, visible, or infrared measurements. Diffraction gratings with the grooves cut into glass or quartz transmit the light through the grating.

Good quality gratings are difficult to cut and therefore expensive. A carefully made grating can serve, though, as a template for many replicas and lower the cost. A coating of epoxy is layered over the master grating and assumes the form of the original surface. The replicate surface is coated with a silvered layer to produce a reflectance grating. These replicas are relatively simple to make so that even inexpensive instruments may contain diffraction gratings of good quality.

An *exit slit* at the end of the monochromator chamber controls the width of the light beam and hence, the bandpass of the spectrophotometer. As the grating or prism is turned in response to the wavelength selector, different wavelengths of the spectrum strike the open slit and pass through it. The narrower the slit, the smaller the bandpass. It would seem that all instruments could have the ideal "narrow bandpass" just by making the exit slit very small. In actual practice, there is a practical limit. As the slit is narrowed, the total amount of light energy passing through the sample may become too small for accurate measurement.

Some spectrophotometers have a nonmovable slit; others have slits that can be manually adjusted. As the slit width is increased, more light reaches the detector, and the bandpass is also increased. Since many compounds have broad absorption peaks, this width poses no problem for routine tests.

In some instruments, the slit width varies automatically as the wavelength is changed. This serves to compensate for variations in the intensity of the different wavelengths of light produced by the lamp, and also for variations in the response of the phototube to different wavelengths. In such instruments, the sensitivity remains relatively constant over the whole wavelength range, although the bandpass varies.

THE SAMPLE CUVET. For accurate and precise readings, the cuvet must be clean, there must be no fingerprints, scratches, or any spills on the optical surfaces, and there should be no bubbles adhering to the inner surface of the filled cuvet. The solution in the cuvets must be thoroughly premixed, and when possible, be at room temperature so that no moisture condenses on the exterior optical surface.

Square or rectangular cuvets with flat optical surfaces are most desirable. These generally have an inside dimension (pathlength) of 1 cm and a capacity of 3 to 4 ml. For smaller sample volumes, cuvets are available with thicker sidewalls, yielding a long, narrow, rectangular cuvet. The pathlength is still 1 cm, but the required volume of solution is reduced to 1 ml or less.

Glass cuvets are used for readings in the visible light range. For measurements in the UV range, the cuvet must be made of quartz or fused silica.

Many of the less expensive spectrophotometers use round cuvets and have adaptors to accommodate cuvets of differing diameters. The round cuvets can be matched to about 1% tolerance in light transmission, which is adequate for routine testing.

A convenient accessory for the spectrophotometer is a flow-through or flush-out cuvet. (Each sample can be flushed out of the cuvet after it has been read.) Care must be taken to insure that the cuvet is clean: frequent checking of blanks for zero absorbance, thorough rinsing after use, and immediate clean up after over-filling. The cuvet should be checked frequently for visible signs of contamination or cloudiness.

For some tests requiring the periodic reading of absorbance, such as kinetic enzyme assays, cuvets are retained in the sample compartment for a prolonged time period, during which the sample must be maintained at a constant temperature. A constant temperature is provided in either of two ways: (1) the sample compartment is surrounded by a water jacket through which water is pumped from a constant temperature water circulator; or (2) the compartment is warmed by thermostatically-controlled heating coils.

DETECTORS. A wide range of photodetectors varying in sensitivity, amplification, and cost are available. All of them contain a light-sensitive surface that releases electrons in numbers proportional to the intensity of the light impinging upon it.

A *photocell* (barrier layer cell, selenide cell) is the simplest of the detectors. It consists of three layers sealed in a protective casing: (1) the bottom support layer consisting of a conductive metal such as iron, (2) a photosensitive layer of selenium or cadmium on top of the metal support, and (3) a transparent conductive layer covering the light-sensitive material. Light passing through the transparent layer to the selenium causes the release of electrons from the latter. The emitted electrons move to the clear upper layer, initiating a current that is measured as it flows through a circuit to the metal support layer of the cell. The current thus generated is a direct function of the light intensity striking the photocell. Photocells are simple and sturdy and seldom need replacement. They are generally not sensitive, and their output is not readily amplified, so they cannot be used for measuring low light levels or small changes in intensity. They are sensitive to temperature change (such as the heat of the lamp) and are slow to respond to changes in light intensity. The current produced by a photocell at any one wavelength is quite linear with light intensity, although the response to light in the shorter and middle wavelengths of the visible light range is much greater than that for the longer ones. The instrument and photocell should be warmed up for several minutes before use.

More sensitive than the photocell is the *phototube*, which consists of a curved cathode of metal coated with a photosensitive material. When light strikes its surface, the cathode emits electrons that are attracted to a positively charged anode, causing a current flow that is proportional to the light intensity. The response, or intensity of current, can be increased by increasing the voltage across the terminals. The phototube may be a vacuum tube, or gas-filled. In the latter, electrons emitted from the cathode strike gas molecules that release more electrons and thus amplify the current response to the light radiation. Although the increased sensitivity is counterbalanced by a shorter tube life, the advantages outweigh this limitation.

When low levels of light or quick bursts of light are to be measured, a *photomultiplier tube* is required. The photomultiplier tube is sensitive and fast in its response. It consists of a photosensitive cathode and an anode with several intermediate faces called dynodes. Each dynode is slightly more positive than the preceding one, and it attracts the electrons from the preceding dynode because of this increasing charge. The principle is simple: light striking the cathode causes emission of electrons which are attracted to the first dynode. Each impinging electron causes the release of secondary electrons; these electrons are attracted by the more positive charge on the second dynode, and so on. The process is repeated at several successive dynodes, with each step increasing the electron amplification; the total amplification factor may exceed 10^6. Photomultiplier tubes are used in spectrophotometers with a narrow bandpass (low light level) and in instruments that must record fast changes in light emission or absorbance—scanning spectrophotometers, for example.

Phototransistors and *photodiodes* are the newest entries among the light detectors. These are constructed of two types of semiconductors joined together so that there is a resistance to current flow between them. As light strikes the junction, the resistance is overcome, and current flows across the junction. Detectors of this type are small, durable, and capable of high amplification. They are relatively inexpensive and will prove to have wide application in spectrophotometry. One adaptation is the "diode array detector," as used in a Hewlett Packard spectrophotometer. It consists of several hundred diodes, each of which is associated with a capacitor. As current flows through a diode in response to light energy, the accompanying capacitor is discharged. The capacitors are recharged several times each second; the current required for the recharging is directly proportional to the quantity of light striking the diode detectors.

READOUT DEVICES. The magnitude of the current generated by a detector may be measured by any one of several types of readout devices: a galvanometer, ammeter with a meter needle, a recorder, or a digital readout. The signal may be transmitted to a computer or printout device.

The information from the readout may be presented as per cent transmittance, absorbance units, or as concentration of the constituent.

Double Beam Spectrophotometers

Double beam spectrophotometers operate like single beam spectrophotometers except that they are designed to compensate for possible variations in intensity of the light source. This is accomplished by "splitting" the light beam from the lamp and directing one portion to a reference cuvet and the other to the sample cuvet; any change in light intensity affects both cuvets simultaneously and thus, is cancelled out.

There are two types of dual beam systems. In the dual beam in space type, illustrated in Figure 2.7A, the light beam is split so that half of the monochromatic light is directed through the reference cuvet while the other half is

A.

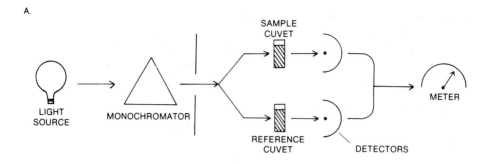

B.

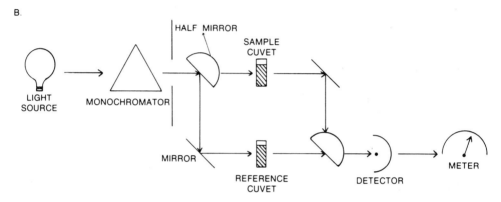

Fig. 2.7. Two configurations for double beam instruments. A, double beam in space. B, double beam in time.

transmitted directly through the sample cuvet. Separate detectors monitor the respective light intensities, and the ratio of sample light intensity to that of the reference cuvet is measured. The two detectors must be matched.

The configuration of the double beam in time is illustrated in Figure 2.7B. As it rotates, a semicircular mirror directs the monochromatic light alternately through the sample and reference cuvets. Another semicircular mirror directs the alternate beams of light from the sample and reference cuvets to a single detector. Again, the ratio of the intensities of sample to reference cuvets is measured (Fig. 2.8).

Double beam spectrophotometers are suited ideally for making a spectral scan because the instrument automatically corrects for the change in light transmission through the reference cuvet as the wavelength is changed. If a single beam instrument were used, it would be necessary at each change in wavelength to adjust the instrument to zero absorbance for a reagent blank before reading the absorbance of the sample.

The principles and components described pertain directly to absorption spectrophotometers, although they also apply to the other types of spectrophotometric instruments such as the flame photometer and the atomic absorption spectrophotometer.

Flame Photometry (Flame Emission Spectrophotometry)

The perfection of flame photometry as a clinical chemistry technique revolutionized the determination of the serum Na and K concentrations: cumber-

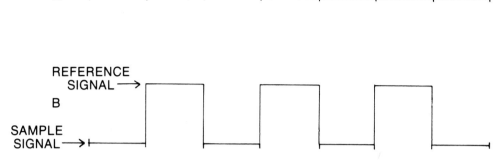

Fig. 2.8. Current flowing from a detector in a system that is "double beam in time." A, alternating signals produced by the detector when the reference solution and sample solution are the same, as when setting 100% T. B, alternating signals from the detector when the sample solution absorbs more light than the reference solution.

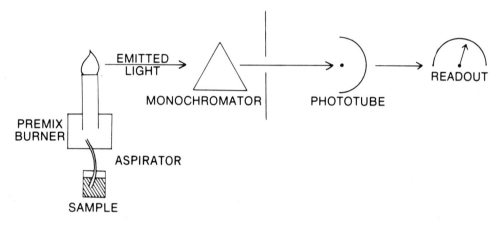

Fig. 2.9. Components of a flame photometer (atomic emission spectrophotometer).

some chemical analyses taking 2 or 3 days to perform were converted to simple procedures requiring only a minute or two.

COMPONENTS. The components of a flame photometer, illustrated schematically in Figure 2.9, consist of aspirator, premix chamber, flame, monochromator, detector, and readout. The last three components are similar to those of an ordinary spectrophotometer, but the flame serves as a light source and sample compartment; there are also entrance and exit slits in the monochromator to narrow the beam of light.

The monochromatic light is obtained by means of an interference filter. The detector is a phototube, the signal is amplified, and results are usually displayed by direct readout.

PRINCIPLE. The aspiration into a flame of a salt solution initiates the process shown in Figure 2.10. After evaporation of the water, the salt molecules are dissociated by the heat to an atomic vapor. A small percentage of the atoms are transformed to an excited state by the absorption of discrete packets of energy that displace orbital electrons to higher energy levels. The excitation is temporary, however; the atoms immediately return to the ground state and in the process, release the absorbed packets of energy in the form of light. The emitted light is of wavelengths specific for each element and can be quantitated under carefully controlled conditions.

An analysis is begun by diluting the sample (serum, urine, or other fluid) with a dilute solution of a nonionic detergent (wetting agent) containing a specified concentration of a lithium salt. The detergent reduces the viscosity of the solution and improves aspiration of the sample. The lithium serves as an internal reference, which is described below.

The diluted specimen is aspirated into a premix chamber by passing a stream of air at high velocity (compressed air) over the open end of a capillary tube that dips into the specimen solution. The sample is aspirated into the airstream by the partial vacuum produced at the capillary jet, broken into fine droplets, and carried as a mist to the premix chamber where the fuel (natural gas or propane) is mixed with the air. Larger droplets settle out and are drained off while the fine mist (about 2 to 5% of the aspirated specimen) is drawn into the flame with the fuel mixture.

The heat of the flame evaporates the water and vaporizes the salts. For sodium chloride, the sodium is converted to its elemental state, Na atoms; in a small percentage of the atoms, a valence electron absorbs some of the heat energy and temporarily moves into a higher orbit (the excited state). Light is emitted as the excited atoms return to the ground state (Fig. 2.10). If the flame is too hot, some of the sodium becomes ionized as electrons are lost.

The light emitted by the excited atoms is measured at a wavelength specific for that element. The characteristic emission spectral line for each element chosen for measurement is distinct from that of other elements that may be present in the solution. The special lines chosen for the Na and K analyses are 589 and 766 nm, respectively.

The concentrations of Na and K are usually measured simultaneously in current instruments. Lithium is incorporated as an internal standard in the diluting fluid in order to compensate for variations in sample feed, gas pressure, or fuel-air ratio which change the intensity of the emitted light, the background

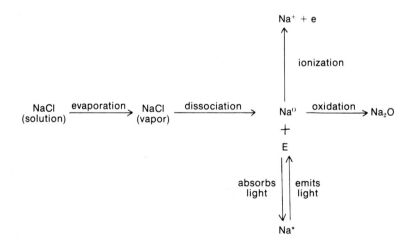

Fig. 2.10. Illustration of the principle involved in flame photometry. See text for the explanation of the production of atoms in the excited state and their emission of light as they return to the ground state. Na^0 represents a ground state sodium atom; Na* an excited atom; E, energy; and e, an electron.

light, or both.* A separate filter and phototube are required for each of the three elements as the photometer measures changes in the light intensity ratios of Na/Li and K/Li.

There is no way of calculating from theoretical considerations the light intensity that a certain concentration of Na or K should emit when aspirated into a flame. Standard solutions of the salts to be analyzed, in concentrations close to the unknown solutions, must be used in the same run to calibrate the instrument. The flame photometers are adjusted to read zero while aspirating the diluent and to give the expected values when aspirating the standards; then the unknowns are analyzed.

Atomic Absorption Spectrophotometry

Atomic absorption spectrophotometry (AAS) is the absorption of light by free metallic atoms. It is similar to flame photometry in that it utilizes the heat of a flame to dissociate molecules to free atoms, primarily at their lowest energy level (ground state), but differs drastically in what is measured. Flame photometry measures the intensity of *emitted* light when an activated atom returns to the ground state, whereas atomic absorption spectrophotometry measures the *absorption* of light of a unique wavelength by atoms in the ground state. The unique wavelength absorbed corresponds to the particular line spectrum for that element. With the proper light source, a particular cation can be analyzed in a mixture of many cations. Since at least 99.998% of the atoms are in the ground state at flame temperatures, atomic absorption spectrophotometry is far more sensitive than flame photometry.

PRINCIPLE. Monochromatic light for a particular element is produced by means of a hollow cathode lamp utilizing that element as the cathode. The monochromatic light is beamed through a long flame into which is aspirated the solution to be analyzed. The heat energy dissociates the molecules and converts the components to atoms; although some atoms are activated, most atoms remain in the ground state at the temperatures commonly used. The ground state atoms of the same element as in the hollow cathode cup absorb their own resonance lines; the amount of light absorbed varies directly with their concentration in the flame. The transmitted light that is not absorbed reaches the monochromator, which passes only the wavelengths close to the resonance lines of the particular element to be assayed. The transmitted light strikes a detector, and the decrease in transmitted light is measured.

COMPONENTS OF AN AAS SYSTEM. The components of an atomic absorption spectrophotometer parallel those of a good quality visible-UV light spectrophotometer (Fig. 2.11). The monochromator, photomultiplier tube, and readout devices are identical.

The light source, the hollow cathode lamp, is unique to the atomic absorption spectrophotometer (Fig. 2.12). The lamp is filled with an inert gas like argon

---

*In addition to serving as an internal standard, the lithium salt serves as an ionic buffer in the solution. When the lithium is not present, potassium ion excitation is affected by the number of sodium ions present. A solution of potassium in the presence of sodium will emit more light at 766 nm than a potassium solution of the same concentration with no sodium present; there is a transfer of energy from the excited sodium atoms to potassium atoms. The lithium atoms in the internal standard are present in a much greater concentration than either Na or K, so their interaction is minimized. The K atoms' environment is only slightly altered by the number of Na atoms present.

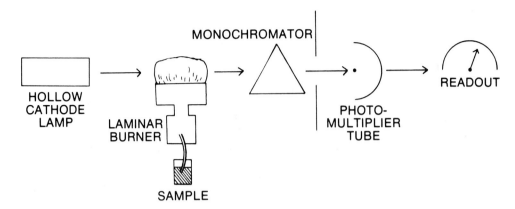

Fig. 2.11. Components of an atomic absorption spectrophotometer.

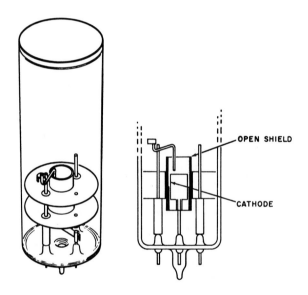

Fig. 2.12. Hollow cathode lamp. (Courtesy of Perkin-Elmer Corporation, Norwalk, CT 06856.)

or neon at a low pressure. When the lamp is fired, gas atoms between the electrodes are ionized and strike the cathode with high velocity, causing metal atoms to be "sputtered" from the cathode surface. Further collisions with gas ions produce excited metal atoms, which then emit light at the characteristic wavelengths (resonance lines) for that element. The emitted resonance lines are much narrower in wavelength than could be achieved with the finest monochromator.

The second unique component in the system is the burner, through which the sample is introduced (Fig. 2.13). The burner is comprised of three parts: nebulizer, premix chamber, and burner head. A sample in solution is aspirated and nebulized (reduced to a fine spray) by a stream of oxidant (air or oxygen) flowing across a sample capillary tube as in flame photometry. The mist is mixed with oxidant and fuel (commonly acetylene) in the premix chamber, where larger droplets are trapped and drained off. The mixture of gases and

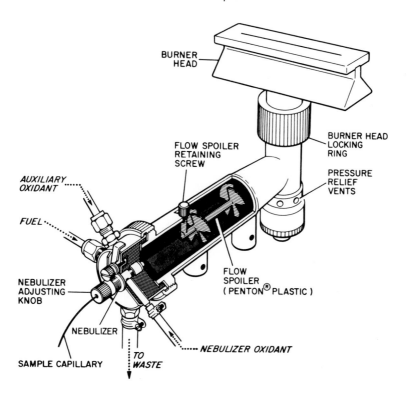

BURNER
HEAD

FLOW SPOILER
RETAINING
SCREW

BURNER HEAD
LOCKING
RING

AUXILIARY
OXIDANT

PRESSURE
RELIEF
VENTS

FUEL

FLOW
SPOILER
(PENTON® PLASTIC)

NEBULIZER
ADJUSTING
KNOB

NEBULIZER

SAMPLE CAPILLARY

TO
WASTE

NEBULIZER OXIDANT

Fig. 2.13. Premix burner with laminar burner head for atomic absorption spectrophotometry. (Courtesy of Perkin-Elmer Corporation, Norwalk, CT 06856.)

sample is directed into the burner head and the flame. The burner, specially designed for atomic absorption spectrophotometry, has a long, flat-topped head positioned directly below and parallel with the beam of light from the lamp. The gases flow through a 10-cm long slot in the top of the burner head so that a long, thin curtain of flame is produced. The light from the hollow cathode lamp passes through the full 10-cm length of the flame, greatly enhancing the absorption of light by the ground state atoms in the flame. The narrowness of the slot also concentrates the atoms and results in greater efficiency of light absorption. The aspirator is more efficient than in flame photometry; 10 to 20% of the aspirated sample is nebulized and enters the flame.

The function of the monochromator is to block light of wavelengths different from the desired resonance line from reaching the detector. The monochromatic light strikes the light-sensitive cathode of a photomultiplier tube, and the resulting current is converted to absorbance or concentration units. Like the flame photometer, the instrument must be calibrated with known standards.

One problem with atomic absorption spectrophotometers is the emission of light in the flame by excited atoms of the element analyzed as they return to the ground state. Since the resonance lines emitted are characteristic for that element, they cannot be filtered out. Hence, the light from the lamp is pulsed either by a chopper that cuts off the light signal at brief intervals or by a modulator that periodically interrupts the electric signal. The amplifier recognizes only the pulsed signal originating from the source lamp and rejects the steady light emitted by the flame. This effectively separates the essential light signal from the extraneous one.

SPECIAL APPLICATIONS. Atomic absorption spectrophotometry is frequently used for detecting small amounts of an element when concentrations are too small to be accurately measured by standard chemical means. Consequently, several means of enhancing the signal have been introduced. For each assay procedure, the fuel-air ratio, gas flow rate, sample aspiration rate, and burner height are optimized so that the greatest possible number of free atoms is present in the optical path. The diluent with which the sample is mixed is also an important factor. For example, calcium as it occurs in serum may be ionized, bound to protein, or complexed with citrate, phosphate, or other anions. Some of the complexes are only slightly dissociated by the heat of the flame; therefore, the Ca atoms in these complexes are not measured proportionally with other calcium forms. To eliminate this problem, the diluent for calcium determinations contains lanthanum chloride. Lanthanum, which is closely related to calcium in the periodic table, displaces calcium from the complexes in the solution and converts the calcium to a more readily dissociated and, therefore, measurable form.

It is necessary to have a specific lamp for every element to be assayed by AAS or to have a multielement lamp in which several different metals are incorporated to the same extent in the cathode. It is common to have a single lamp for both Ca and Mg.

FLAMELESS TECHNIQUES. Increased analytical sensitivity in atomic absorption spectrophotometry has been achieved by incorporating flameless techniques into the system. A graphite furnace (carbon rod) replaces the burner head and is positioned just below the light beam in the sample compartment. The sample is placed in a well in the rod, and an electric current is applied in three stages: small at first to evaporate the liquid, increased to ash the sample, followed by a large surge of current to instantaneously vaporize the entire sample and release the whole cloud of atoms in one burst. The absorption of light from the hollow cathode lamp occurs for one brief instant, too quickly for digital displays or meter needles to respond. A recorder is utilized to record the signal, a sharp peak whose height or area is proportional to the concentration of atoms in the sample. A series of standards is always included in the run.

In flameless systems, the surge of current used to vaporize the sample may also release a cloud of smoke or nonspecific vapor that blocks some light from reaching the detector. A background correction capability is advisable. For this correction, a deuterium continuum light is also directed through the flame, and its light is monitored at a wavelength other than that of the assay. Since extraneous vapor blocks light of all wavelengths, the instrument compensates for the amount of light that is deflected at both wavelengths and measures changes that occur only at the wavelength of the resonance line of interest.*

The flameless AAS technique has tremendously increased the sensitivity of measurement of quite a few cations in serum or blood. It is now possible to measure the concentration of blood lead with just microliter quantities. This

---

*Some instruments correct for background absorption at the wavelength of interest by passing light from the hollow cathode and deuterium lamps through the same monochromator. The lamps are pulsed at different frequencies, a procedure that permits electronic isolation of the two signals.

is of major importance when investigating a pediatric population for possible lead poisoning, a problem that occurs frequently in city slum areas, due to crumbling paint containing lead pigments.

Fluorometry

PRINCIPLES. Some molecules fluoresce, or emit light, after exposure to light of a certain wavelength. Light is emitted within a brief time ($<10^{-8}$ seconds), and is of lower energy (longer wavelength) than the light absorbed. The intensity of the fluorescence varies directly as the concentration of the solute; the sensitivity of the fluorometric assay may be 10^3 times that of absorption spectrophotometry. Since light of a particular wavelength is required for excitation of a given molecule and the emitted, fluorescent light is also of a wavelength characteristic for that molecule, a fluorescent analysis is more specific than the usual spectrophotometric method.

The process of fluorescence is illustrated in Figure 2.14. A molecule at the ground state energy level is excited by light absorption to a higher excited energy level (E2). Vibrational energy losses (collisions, heat loss) drop the molecule to a lower, yet still excited, energy level (E1); no light emission accompanies this drop. As the molecule quickly returns to the more stable ground state level, light is emitted. If this process is completed within 10^{-8} seconds, it is known as fluorescence.

Some molecules undergo a slower transition involving changes in electron spin before emitting light. The time required for these changes is at least 10^{-4} seconds; the emitted light is called phosphorescence. At present, phosphorescence measurements are not widely used in clinical chemistry laboratories.

The number of molecules that can fluoresce is limited; they are usually cyclic molecules with conjugated double bonds $(-C = C - C = C-)$. Some molecules that cannot fluoresce by themselves may be chemically converted to derivatives that are fluorescent. This conversion is usually accomplished by adding side chains that increase the freedom or mobility of the double bond electrons. As an example, addition of an NH_2 group enhances fluorescence, but adding an NO_2 group depresses the ability to fluoresce by attracting the electrons and reducing their mobility.

COMPONENTS. An instrument for measuring fluorescence differs from the other spectrophotometers discussed earlier by having two monochromators,

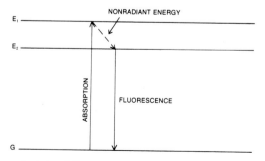

Fig. 2.14. Illustration of the principle of fluorescence. Light of the requisite frequency striking a molecule at the ground state energy level (G) immediately raises it to an excited energy level (E$_2$). Some energy is lost by collision with other molecules or as heat, dropping the molecule to a lower energy level (E$_1$). The molecule falls back to the ground state level, releasing energy as light (fluorescence).

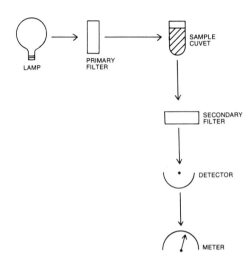

Fig. 2.15. Basic components of a filter fluorometer.

one to regulate the wavelength of light striking the sample and one to isolate the wavelength emitted from the sample. The basic components of a fluorometer are shown in Figure 2.15. The exciter lamp is usually either a mercury arc discharge lamp or xenon arc tube so that light in the UV range is provided. If no excitation at wavelengths shorter than 350 nm is required, a tungsten lamp is adequate. The monochromators may both be diffraction gratings (a fluorescence spectrometer) or filters in a simple filter fluorometer. The former has the advantage of providing narrow bandpass isolation at any wavelength of the spectrum for both exciting and emitted light, but the latter is a far less expensive instrument that can meet most of the needs of the clinical chemistry laboratory. The primary filter (or grating) allows the passage of light of the proper wavelength for absorption by the molecule; the secondary filter transmits light of the specific wavelengths emitted by the sample. Since the light is emitted equally in all directions, the detector is placed at right angles to the beam of light from the exciter lamp to the sample; this position prevents transmitted light originating in the lamp from reaching the detector. As the whole sample is diffused with light, the cuvet glows like a lamp, minimizing the effects of any surface scratches or imperfections. Glass cuvets are usable for exciting wavelengths greater than 350 nm in length; otherwise, quartz or fused silica is required. A phototube or photomultiplier tube is used as the detector, depending upon the sensitivity required.

FACTORS AFFECTING FLUORESCENCE. The intensity of fluorescence is theoretically linear with the concentration of the fluorescing molecule, but the following factors also affect the fluorescence:

1. *pH*: Changes in pH may induce changes in the ionic state of a molecule, with accompanying changes in fluorescing properties.
2. *Temperature*: Increased temperature enhances molecular motion, increasing molecular collisions and thus decreasing fluorescence.
3. *Length of time of light exposure*: Because molecules are excited more quickly than they fluoresce, there is an increase in the number of excited

molecules present with time, and consequently an increase in fluorescence.

4. *Concentration:* Fluorescence increases with concentration only in dilute solution. In more concentrated solutions, the light emitted may be absorbed by other molecules of the same compound as it passes through the solution in the cuvet. If a large number of absorbing molecules are present, a significant decrease in fluorescence (quenching) results.

Quenching of fluorescence may also rise from the action of other molecular species present in solution; this may involve absorption of either the exciting radiation or the fluorescent emission. Some common solvents, such as acetone and benzene, absorb light in the ultraviolet region where excitation occurs. Dichromate, used for cleaning glassware, absorbs light at 275 and 350 nm and may easily interfere with the excitation or emitted light of fluorescing compounds. Extraction and elution solvents and glassware cleaning solutions must therefore be chosen with care.

CLINICAL APPLICATION. Fluorometry is used in the clinical chemistry laboratory for certain classes of compounds, particularly when great sensitivity is required. Some of these classes are drugs (p. 387), hormones (p. 281), vitamins, amino acids, and porphyrins. The analysis of serum triglycerides, a routine test, is frequently performed by fluorometry (p. 341).

Chromatography

Chromatography is the separation of components in a solution by differences in migration rate as the solution mixture is passed through a porous medium. The separation may make use of one or more of the following physicochemical forces, depending upon the particular chromatographic system: differences in adsorption to the porous medium (the sorbent*), differences in the relative solubilities between a liquid (stationary phase) coating the inert medium and the liquid (mobile phase) percolating through the porous column, differences in ion exchange with the sorbent, and differences in molecular size (molecular sieving) as the solution percolates through a gel of very small pore size. Various types of chromatography are described below.

Chromatography is highly efficient and is widely used in the clinical chemistry laboratory for the identification in serum or urine of drugs (p. 386), sugars (p. 320), and amino acids. The process may be used for the purification of materials, for identification of compounds, and for quantitation.

Thin Layer Chromatography

Thin layer chromatography (TLC) is a simple technique that has been used in laboratories for separation and identification of urine sugars (p. 320), amino acids, drugs (p. 397), and other groups of compounds. In TLC the stationary phase consists of a thin layer of a finely divided substance applied to a sheet of glass or plastic backing. Sorbents commonly used, and commercially available as finished plates, include alumina, silica gel, and cellulose. Samples are applied to the plate as small streaks or spots near one edge of the plate. The

---

*Sorbent is the term applied to the stationary phase which may be liquid or solid. Solid sorbents are usually chosen for *adsorption, ion exchange,* or *molecular sieving* properties; a liquid sorbent is usually chosen for the *differential solubilities* of substances between it and the mobile phase.

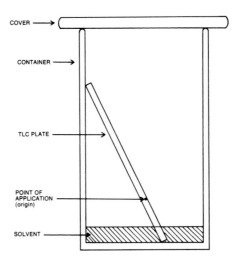

COVER →

CONTAINER →

TLC PLATE →

POINT OF
APPLICATION
(origin)

SOLVENT

Fig. 2.16. Apparatus for thin layer chromatography.

application area must be kept small; if necessary, repeated applications may be superimposed upon each other, with drying between applications. The plates are then placed in a closed chamber containing an appropriate solvent mixture (Fig. 2.16). Since there is no liquid stationary phase on the TLC plate, the solvent mixture consists of two liquids of different polarity (and therefore differing rates of migration); NH_4OH or acetic acid is frequently added to the mixture to modify the pH and change the relative solubilities of some components. The TLC plate is positioned so that the applied samples are along the lower edge, a little above the level of the solvent. As the solvent (the mobile phase) moves up the plate, the samples move upward too, separated by differing attractions for the sorbent and the components of the solvent mixture. When the solvent front has almost reached the top, the plate is removed from the chamber. The solvent front is marked and the plate is dried. The spots may be colored and therefore visible, or more likely they will need to be visualized by viewing under UV light or by spraying with color-producing reagents. The spots may be identified by their color and by their relative distance of migration from the point of application, or R_f *value*, which is a characteristic for each substance when chromatographed in a certain solvent system. The R_f value is calculated by the following formula:

$$R_f = \frac{\text{distance traveled by compound from origin}}{\text{distance traveled by solvent from origin}}$$

Known compounds should be applied to the same plate as the unknowns for comparison and confirmation of identification. A typical TLC plate is shown in Figure 2.17.

If the identity of the substance is known, a series of standards may be applied to the plate in order to estimate the concentration of the unknown specimen by comparing the size and density of the spot with the standards. A more quantitative result may be obtained by scraping the spots off the plates, eluting the colored compound, and reading the absorbance in a spectrophotometer.

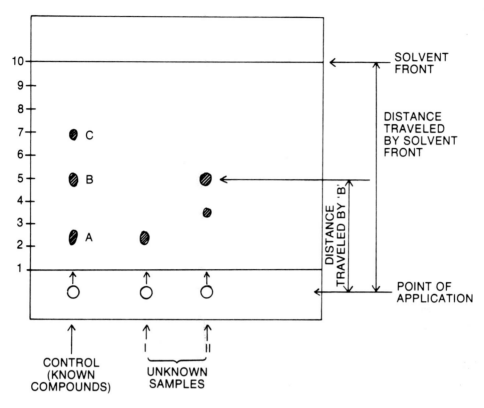

Fig. 2.17. A TLC plate after development in the tank and visualization of the spots. Sample I contains compound A; Sample II contains compound B and an unidentified compound. R_f for compound B = 0.5 in this solvent system.

Liquid Chromatography

The original work on chromatography utilized an inert material (sorbent) packed in a column. Plant pigments dissolved in heptane were placed on the column and strongly adsorbed to the sorbent. A more polar solvent (a mixture of heptane and ethanol, the mobile phase) was passed through the column, and the pigment bands began to separate upon the basis of their relative solubilities in the ethanol mixture; the more soluble pigments traveled further down the column. This separation of components by differential solubility is also known as *partition* chromatography. The final separation product is the *chromatogram*. The chromatogram may be extruded from the column and the different bands may be cut from it, or the materials may be eluted with an appropriate solvent and the liquid collected in a series of tubes. The resolution obtained in liquid chromatography depends upon the pH and ionic strength of the mobile phase and the relative solubilities of the constituents in the two phases.

In liquid chromatography, the mobile phase may be passed through a vertical glass column by gravity alone or assisted by a low pressure pump (column chromatography); it may also be forced through a narrow bore stainless steel column by a high pressure pump (high performance liquid chromatography, HPLC). The principles used in separating components are identical in column chromatography and HPLC, so these will be discussed together after a brief description of the apparatus used for each technique.

COLUMN CHROMATOGRAPHY. As the name implies, the system usually consists of a vertical glass column filled with a packing material that has been introduced carefully so that it is evenly distributed and has no channels coursing through any part of it. The column has been equilibrated with solvent before the sample is gently layered on top. Eluting solution is introduced above the sample, flows through the column by gravity, and carries along the components of the sample according to relative solubilities. The reservoir of eluting solvent is raised above the top of the column to create a "head" of pressure for faster flow. The eluting fluid may be unchanged throughout the separation, or a gradient flow may be utilized by progressively increasing the concentration of one component of the eluting solution during the separation.

A low pressure pump may be introduced into the system beyond the column to increase the rate of eluent flow.

HIGH PERFORMANCE LIQUID CHROMATOGRAPHY (HPLC). In HPLC, a high pressure pump is used to force the solvent and sample through a relatively short (often about 30 cm) and narrow bore column (2–4 mm). Pressures of about 1000 to 3000 psi are frequently used in routine separations.

A diagram showing the major components of a typical HPLC system is given in Figure 2.18. The elution solvent must be filtered and degassed before use and is constantly stirred as it is forced into the system by a reciprocating pump at a constant flow rate (for example, 2 ml/min); an auxiliary pump reservoir maintains the smooth, constant flow while the main pump is refilling.

The column contains the packing appropriate for the particular separation. The packing is very fine (5–10 μm in diameter) to greatly increase surface area per unit volume. These columns may be packed by the user but are usually purchased as prepacked units because of the expertise required for packing.

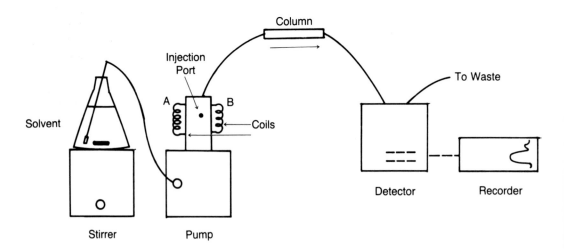

Fig. 2.18. The major components of a typical high performance liquid chromatography system.

The most common detector is an ultraviolet photometer; some models are factory-set to read absorbance at 254 nm. Other detectors include visible photometers, refractometers, solute transport detectors (flame ionization detector, see gas chromatography section, p. 49) and fluorometers. Several other types of detectors are designed to measure such properties as conductivity and radioactivity.

Normally, the solution is pumped through the longer coil A shown in Figure 2.18. When a sample is to be injected, the solvent flow is bypassed through the shorter coil B. Then the sample (a small volume, frequently under 10 μl) is injected into coil A. When the flow is returned to the regular route through A, the sample is swept by the solvent into the column.

The main advantages of HPLC are good resolution, which may be obtained by the proper selection of packing, solvent, and flow rate; and fast separation requiring minutes rather than hours for each sample.

An internal standard (similar in composition to the analyte) is generally added to each sample to compensate for variations in the size of sample injected, solvent evaporation, and other possible irregularities. The internal standard peak is used for calculation of sample concentrations. The peak height ratio for each standard is plotted against the concentration of the standard to construct the standard curve. Peak height ratios of the analytes to internal standard are used to determine their concentration from the curve (see Fig. 2.19).

$$\text{Peak height ratio} = \frac{\text{Height of analyte peak}}{\text{Height of internal standard peak}}$$

Internal standards are used in the same manner as for assays using gas chromatography.

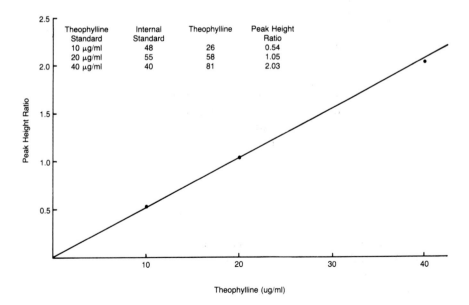

Fig. 2.19. A standard curve for an HPLC assay for theophylline. Peak height ratio is plotted against concentration.

Gel Filtration (Size Exclusion Chromatography)

Gel filtration (molecular sieve chromatography) separates molecules according to their size, although the shape of the molecule affects the filtration to some extent. The gels are in the form of beads containing a network of openings or pores through which small molecules may pass. The beads are available with different pore sizes; all molecules larger than the pores are excluded from the beads. Molecules that are much smaller than the openings enter the pores readily. The intermediate sizes may also flow into the pores but less frequently than the smaller ones. The molecules too large to enter the pores flow with the solvent between the beads and are eluted quickly. The small molecules take a long, devious route by moving in and out of many beads. Intermediate sizes enter fewer pores; their route is intermediate in length, and the time required for their passage through the column is a function of their molecular size. Three types of gel are used for molecular filtration: cross-linked dextran (Sephadex, Pharmacia), agarose (Sepharose, Pharmacia), and polyacrylamide (Bio-Gel P, Bio-Rad).* Each comes in a variety of pore sizes for separating compounds within different ranges of molecular weight.

The gels listed above cannot withstand the high pressures involved in HPLC; the most common size exclusion packing for use under pressure is a silica-based product (Waters), which is more resistant to pressure. Pressures above 800 psi tend to compact the gels, so they are seldom used in size exclusion chromatography.

Ion Exchange Chromatography

Molecules can be separated by their ionic charge in a process known as ion exchange chromatography. The sorbent or stationary phase consists of polymers with covalently bound ions. In cation exchange resins, these tightly bound ions are negatively charged and are associated with positive ions that are loosely attached by electrostatic charges. The positively charged substances to be separated from a mixture are first adsorbed to the sorbent, displacing the cations present in the resin. The mixture is buffered at a pH that facilitates binding and the column is washed with the same buffer to remove the nonbinding fractions. A slight change in the pH or ionic strength of the eluting buffer will cause the separation from the column of the adsorbed cationic substance, owing to competition from the buffer ions for the anionic sites. The buffers must be carefully chosen for good resolution. An anion exchanger operates in exactly the same way, except that its covalently bound ions are positively charged to attract the anions from the solution.

Partition Chromatography

In partition chromatography, the solute components are distributed between the mobile and stationary phases according to their relative solubility in each. Molecules that are readily soluble in the mobile phase are eluted quickly, while those with a greater affinity for the stationary phase are retained on the column for a longer time.

Historically, the early separations used a polar stationary phase with a less

---

*Complete names and addresses of sources of materials and equipment are given in Appendix 10.

polar mobile phase, and this pattern has become known as "normal phase" chromatography. For the separation of drugs and other nonpolar substances, a "reverse phase" chromatography has proved more useful. The "reverse phase" utilizes a nonpolar stationary phase with a relatively polar mobile phase.

Gas Chromatography

Gas chromatography (GC) or gas liquid chromatography (GLC), is designed to separate and quantitate volatile materials in a heated column, consisting of an inert support material, such as diatomaceous earth, coated with the stationary phase. The stationary phase must be a liquid at the elevated temperatures (up to 400° C) of the column, nonvolatile, and nonreactive with the samples or solvents moving through the column. The mobile phase is an inert carrier gas (for example, argon or nitrogen) that sweeps the volatile compounds through the column to a detector. The more volatile molecules, those with the lowest boiling points, tend to be swept most rapidly along by the carrier gas. Those that are soluble in the liquid phase are retained for periods in the liquid, and their forward movement is slowed. Therefore, both volatility and solubility differences serve to affect the rate of flow of sample molecules through the GC column.

A diagram of the main components of a gas chromatograph is shown in Figure 2.20. The carrier gas, which must be pure and dry, flows through the system to the detector. The sample enters the system through the injection port, a heated chamber at the beginning of the column. The temperature at the injection chamber is high enough to flash volatilize the sample so it can be swept into the column by the carrier gas. The column may be one to several meters in length, and about 0.5 centimeters in diameter, although both dimensions may vary widely. A separation may be run isothermally, with a constant oven temperature maintained, or the temperature may be steadily increased to speed the forward movement of the less volatile components. After separation on the column, the sample components enter the detector. There are several types of detectors, but the most commonly used are the thermal conductivity and the flame ionization detectors. The former consists of a metal block with two separate filaments similar to an incandescent lamp filament that is heated by an

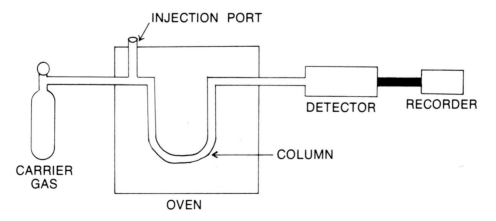

Fig. 2.20. The basic components of a gas chromatograph.

electric current. As the carrier gas flows over the reference filament, the gas conducts heat away from the filament, maintaining it at a cooler temperature. When a sample constituent eluted from the column is mixed with the carrier gas, the conductivity of the gas is reduced, there is less cooling effect on the sample filament, and its temperature rises. The difference in temperature changes the resistance and causes a current flow that is amplified and recorded as a "peak" by the recorder; the peak is proportional in size to the quantity of the eluted compound.

The flame ionization detector contains a small burner in which hydrogen gas mixed with the gas from the GC column is burned. As a carbon-containing compound from the sample is burned, negative ions are formed that are attracted to a positively charged wire. The current produced is measured and recorded.

Each of the detectors senses sample constituents as they emerge from the column. The signals are recorded as peaks: the distance from the solvent peak (eluted first) to sample peak designates the *retention time* for the compound, the time required for that constituent to pass through the column. For a given column with its packing, retention time for a substance is a constant, as long as all conditions (temperature, flow rate) are unchanged. The area of a peak is proportional to the amount of the substance eluted, and thus to its concentration in the original sample. Thus, the position of a peak identifies a substance; the area of the peak is used for quantitation. Since most peaks are symmetrical, peak height is often used for convenience.

Sample injections of one microliter size are not very reproducible, especially when they are in a volatile solvent. To compensate for this variability in sample size, an internal standard similar to the constituent to be quantitated is added to each sample-solvent mixture at a fixed concentration. If the sample size is slightly increased, all peaks from its constituents, including the internal standard, will be proportionally increased. Standard solutions containing known amounts of the compound being assayed are also chromatographed with an internal standard. A standard curve is drawn, using the peak-height ratio of sample to internal standard as the ordinate and concentration as the abscissa. The concentration of the unknown samples may be read directly from the curve.

Electrophoresis

A charged particle placed in an electrical field, migrates towards the anode or cathode, depending upon the net charge carried by the particle. The rate of migration in a porous medium varies with its net charge and the strength of the electrical field. Electrophoresis is the process of separating the charged constituents of a solution by means of an electric current.

The electrophoresis apparatus is designed so that the circuit between the two poles is bridged by the support medium holding the sample, and the current flow is partially carried by the components of the sample.

As shown in Figure 2.21, an electrophoresis chamber consists of two compartments separated from each other by a dividing wall; one side contains the anode, and the other the cathode (platinum wires). Each side is filled to the same level with a buffer (barbital buffer, pH 8.6, is often used for serum protein separation). A "bridge" across the top of the dividing wall holds a membrane or other support material so that each end is in contact with the buffer in one of the compartments. The only connection between the two compartments is

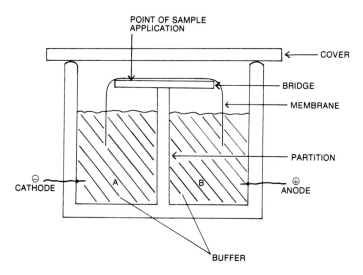

Fig. 2.21. Diagram of an electrophoresis chamber. Compartments A and B are completely separated from each other.

via the membrane. Prior to its use in the cell, the membrane is immersed in buffer, blotted, and placed in the chamber, and the sample is applied. When a voltage is applied to the cell, the current is carried across the porous membrane by the buffer ions. At a pH of 8.6, all of the serum proteins carry a net negative charge and tend to migrate towards the anode. Albumin carries the largest charge and therefore moves the fastest; the gamma globulins have the smallest net charge and move the least distance.

The actual distance of migration and the resolution of the separate bands are affected by many variables in the system. If the support medium has large pores (cellulose acetate, agar, paper) the separation is determined by the charge on the molecule; if a fine gel (acrylamide, starch) is used, the separation is also affected by the size of the molecules; the gel acts as a molecular sieve. Serum proteins separate into 5 bands on cellulose acetate and over 20 bands in acrylamide.

Selection of the proper buffer is important; it must provide the desired pH and must not react with the sample. As the buffer concentration is increased, the migration rate is decreased, but resolution of the components is improved. The reduction in sample migration distance is due to competition from the increased number of buffer ions available to carry the current. As the *ionic strength* of the buffer (the number of available ions) is increased, more current can be carried; the increased current results in elevated temperatures, which may denature some proteins and affect the separation.

The power supplies used for electrophoresis can be adjusted to provide either constant current or constant voltage. If the voltage is kept constant during the separation, the current will gradually increase, accompanied by an increase in temperature. For short-term electrophoresis (one-half hour or less), constant voltage is often used; for longer periods of time constant current is preferable because the temperature does not increase appreciably during the run. Increased temperature, as mentioned above, may result in poorer resolution of the separated components.

The support medium, buffer, and electrical field affect the final separation of components during electrophoresis. Other influences, not as readily apparent, also modify the results of electrophoretic migration. *Electroendosmosis* is a process taking place concurrently with sample migration (Fig. 2.22). The support medium (cellulose acetate, for example) develops a negative charge itself when exposed to an alkaline buffer. Associated with the negatively charged groups on the membrane are positive ions from the buffer. As these positive buffer ions move toward the cathode, they exert a counterflow to the protein molecules that are migrating toward the anode. All of the protein molecules are thus slowed in their progress, and the most weakly charged proteins, the gamma globulins, may be moved with the buffer, ending their migration at a point closer to the cathode than the point of application. The net effect of protein migration and electroendosmosis is similar to that of rowboats headed upstream against a river current: the strongest oarsmen make progress against the current if their forward motion is greater than the river flow; the weakest oarsmen are moved downstream in spite of their rowing in an upstream direction; between the two extremes are the boatmen who make little progress or can only hold their own against the current. The force of electroendosmosis does not prevent separation of the charged molecules, but it does affect their final position on the membrane.

Another force affecting the migration of the protein molecules is *wick flow*. There is a constant evaporation of water from the membrane at all points above the liquid level in the cell; as the water evaporates, buffer flows upward from the cell through both ends of the membrane to replace the lost moisture. This results in a buffer flow towards the center of the membrane from the ends, moving faster near the ends and more slowly as it approaches the center. Wick flow thus adds distance to the migrating molecules approaching the center, and retards those moving outward from the center. The net effect is a compression of the final pattern, with a reduction in the separation between the com-

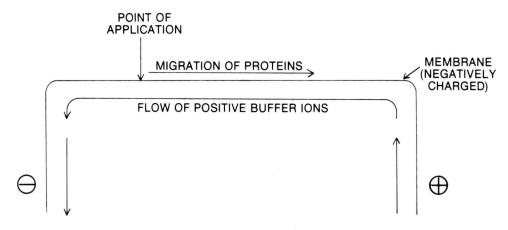

Fig. 2.22. Illustration of electroendosmosis. Buffer movement is counter to the current flow and migration of the serum proteins. See text for explanation.

ponents. A good electrophoresis procedure, then, is a satisfactory compromise among all of the competing forces.

Densitometry

A densitometer is a specialized colorimeter designed to scan and quantitate electrophoresis patterns. There are instruments for reading various gels as well as for cellulose acetate membranes. The latter will be used as an illustration of the principle.

After the serum proteins have been separated by electrophoresis, the proteins are precipitated and visualized by immersion in a fixative dye such as Ponceau-S. After removing excess dye and dehydrating the membrane, the membrane is cleared (from an opaque white) by carefully timed immersion in a dehydrating agent and a mixture of organic solvents. Heating the membrane to evaporate the solvents leaves a thin, transparent film with the protein bands stained by the red dye (see p. 162).

The cleared membrane is then scanned in a densitometer that contains a light beam passing through a filter to a photodetector. As the clear portion of the membrane passes through the light beam, the instrument records a baseline level. As the bands pass through the light beam, less light reaches the detector, and the instrument records the absorbance of light of the protein bands as a peak on the graph. Serum protein separations on cellulose acetate membranes generally have five bands that appear as five peaks on the densitometer graph (p. 151). The densitometer also integrates the area under each peak so that the relative concentration of each fraction (% of total protein) is recorded; if the concentration of the total protein is known, the actual concentration of each protein fraction (g/dl) can be calculated. Many densitometers are now equipped with microprocessors to do the calculations.

Clinical Applications

The electrophoresis procedure is used in the clinical laboratory for separation of serum proteins, lipoproteins, isoenzymes, hemoglobins, and other classes of macromolecules. The separation is similar in each case, although buffer, pH, and voltage may differ. The method of visualizing the bands is determined by the constituent to be measured (an oil soluble dye for fats, a suitable substrate for enzyme activity).

Electrodes

A major advance in clinical chemistry was the development of the pH electrode and, subsequently, other electrodes that make possible the selective measurement of particular ions when they are immersed in solution mixtures. The instruments utilizing these electrodes measure the potential difference that builds up at an interface when two different concentrations of the same ion are in contact with each other. Electrodes are the detectors that are sensitive to this potential difference.

A brief discussion of electrodes is presented in the next section. For more detail, the reader can refer to textbooks on laboratory instrumentation.[6-8]

When a metal is in contact with a solution of its ions, there is a tendency for the metal atoms to give up electrons and to enter the solution; there is also a tendency for the metal ions in solution to take up electrons and deposit as atoms on the metal surface. Charges build up near the interface, depending upon which reaction proceeds at the greater rate. When the first reaction proceeds faster than the second one, the net reaction will be a layer of electrons forming a negative charge on the metal surface and a layer of positive ions in the solution along the metal surface; an electrical potential exists across the interface owing to the opposing charges on either side. When another metal-solution interface with a different potential is introduced, and a circuit including both of them is completed, electrons flow through the system and produce an electric current. If an opposing voltage is adjusted so that no current flows, the total potential (voltage, electromotive force, EMF) of the system is equal to the imposed voltage and may be measured. This is the basic, greatly simplified principle of electrode measurements.

The interfaces through the system may be solid-solution as in the preceding instance, or the interface may exist between two solutions containing the same ion in different concentrations. The solutions are separated by a membrane (interface) that is permeable to the common ion. The potential across the membrane is dependent upon the relative rates of ion diffusion in each direction through the membrane.

Each of these solution interfaces is called a *half-cell,* and there must be two half-cells present in a circuit for a potential difference to be measurable.

Reference Electrodes

Some half-cell potentials are extremely stable and easy to reproduce. Since the potential of these half-cells can be established accurately, they serve as reference potentials against which unknown voltages are measured. The two most widely used reference electrodes (reference half-cells) are the silver-silver chloride electrode and the calomel (Hg_2Cl_2) electrode. The calomel electrode contains mercury in contact with mercurous chloride (calomel), which in turn is in contact with a KCl solution of known concentration. A small ceramic plug or fiber saturated with the KCl at the tip of the glass envelope serves as a membrane and carries the current into the surrounding solution to complete the circuit.

The silver-silver chloride electrode, consisting of a silver wire coated with a deposit of silver chloride, is immersed in a known KCl solution. As with the calomel electrode, the circuit is completed by a KCl junction with the surrounding solution.

When a reference electrode and the appropriate measuring electrode are immersed in a test solution and the circuit is completed, the voltage difference between them can be measured. The potentiometer is calibrated against a solution of known ion concentration, and the unknown solution is then introduced. The measured voltage change is related to the change in ion concentration; for H^+, a change of 1 pH unit causes a voltage change of 59.15 millivolts at 25° C.

Frequently, the reference electrode is made quite small to enable it to be included in the glass cylinder with the measuring electrode. This "combination" electrode is convenient and may be used with much smaller volumes of solution than with two separate electrodes.

pH Electrodes

The first ion-selective electrode to be widely used was the pH electrode designed to measure hydrogen ion concentration. This was made possible by the development of a special pH-sensitive glass. When a thin membrane of this glass separates two solutions of differing hydrogen ion concentrations, a hydrogen ion exchange takes place in the outer hydrated layers of the glass, causing a potential to develop across the glass membrane, which varies with the difference in hydrogen ion concentration between the two solutions. The pH electrode is filled with 0.1 N HCl; the HCl is in contact with a silver chloride layer on a silver wire completing the half-cell. The H^+ ion concentration of the solution surrounding the electrode determines the potential at the glass membrane surface. If a calomel half-cell is also immersed in the solution and the two are connected through a pH meter, the instrument can measure the potential difference between the electrodes and convert this to pH units.

Blood pH meters (Fig. 2.23) are based on exactly the same principle except that they are adapted to measure the hydrogen ion concentration of small samples. Since the measurement must be made anaerobically, the blood in the syringe is drawn into a fine capillary tube. The tube itself is constructed of the pH sensitive glass and is surrounded by the standard HCl solution. The system is usually maintained at 37° C to determine the pH of the blood as it existed in the patient's body. The temperature control is very important; pH is temperature dependent and decreases about 0.015 units for each degree rise in temperature.

The pCO_2 Electrode

Blood gas instruments in the laboratory are designed to measure the partial pressures of carbon dioxide (pCO_2) and of oxygen (pO_2) as well as blood pH.

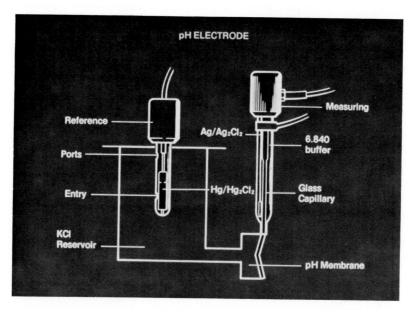

Fig. 2.23. Diagram of pH electrode from a blood gas instrument. (Courtesy of Instrumentation Laboratory, Inc., Lexington, MA 02173.)

Specialized electrodes designed for each gas determination are placed within the instrument so that one small blood sample suffices for both measurements. Some instruments measure all three parameters (pH, pCO_2, pO_2) from the same small aspirated sample; some have the pH electrode housed separately.

The pCO_2 electrode consists of a pH electrode with a CO_2-permeable membrane covering the glass membrane surface. Between the two is a thin layer of dilute bicarbonate buffer. The aspirated blood sample is in contact with the CO_2-permeable membrane, and as CO_2 diffuses from the blood into the buffer, the pH of the buffer is lowered. The change of pH is proportional to the concentration of dissolved CO_2 in the blood. The glass electrode responds to the buffer pH change, and the meter is calibrated to read the pCO_2 in mm of mercury. This type of pCO_2 electrode is known as the Severinghaus electrode (Fig. 2.24).

In a patient's blood the three values, pH, pCO_2 and bicarbonate are all interrelated according to the Henderson-Hasselbalch equation (Equation 4). If any two of the values are known, the third can be calculated. Since blood gas instruments designed today measure the pH and pCO_2, the bicarbonate value may be calculated. Often a "total CO_2," which includes dissolved CO_2, carbonic acid, and bicarbonate is ordered. The bicarbonate value can be calculated from the data, so the values of all three parameters are known as measured values.

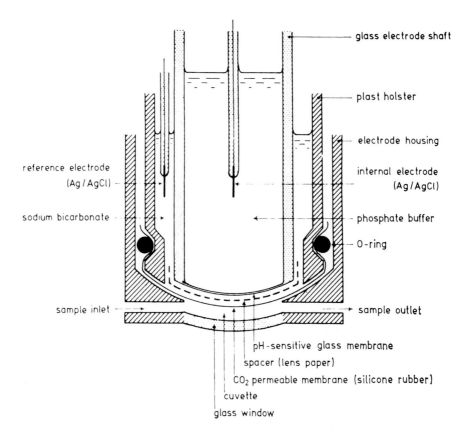

Fig. 2.24. Diagram of pCO_2 electrode. (From The Acid-Base Status of the Blood, 4th ed., by O. Siggaard-Andersen. Copenhagen, Munksgaard, 1974, p. 172.)

A quick substitution of the value into the Henderson-Hasselbalch equation then can serve as an extra cross check on the laboratory results.

The pO_2 Electrode

The Clark electrode for measuring the partial pressure of oxygen in the blood is based on a different principle from that of pH measurement. The latter measures a voltage difference when no current is flowing; the pO_2 electrode measures the current that flows when a constant voltage is applied to the system. The current is the stream of electrons that flow as the oxygen molecules are reduced at the cathode:

$$\frac{1}{2} O_2 + H_2O + 2e^- \rightarrow 2OH^-$$

The source of the electrons is the silver-silver chloride anode where the silver molecules are oxidized:

$$Ag \rightarrow Ag^+ + e^-$$

The amount of current that flows through the system is a direct measure of the number of electrons released to the oxygen and is consequently a measure of the number of oxygen molecules available for reduction. The current is directly linear with O_2 concentration as long as the constant voltage is maintained. The blood gas instrument measures the current flow produced by the loss or gain of electrons when the system is subjected to a polarizing current.

The electrode may be constructed as shown in Figure 2.25. A platinum wire forms the cathode; the anode is a silver wire in AgCl. The contact between the poles is an electrolyte solution that is separated from the test sample (blood) by a membrane permeable to O_2 molecules. Dissolved O_2 diffuses from the blood through the membrane and is reduced at the cathode. The rate-limiting factor in the system is the diffusion of oxygen molecules through the membrane. The diffusion rate depends directly upon the pO_2 of the sample, so the change in current flow offers a direct measurement of the pO_2.

Blood Gas Instruments

Blood gas instruments must be monitored constantly and calibrated frequently. All three electrodes (pH, pO_2, and pCO_2) are calibrated by setting with two standard concentrations. Two buffers in the physiologic range are used for pH calibration, and two gases (high and low concentrations of O_2 and CO_2) are used for the gas electrodes. The gases are bubbled through water in the instrument to saturate them with water vapor; gases dissolved in the blood would be comparably saturated. Corrections must be made for water vapor pressure and for the barometric pressure, which must be checked regularly throughout the day.

Since the pressure of gases and pH is dependent upon temperature, the temperature of the bath surrounding the electrodes must be carefully monitored and closely controlled. Usually the bath is maintained at $37° \pm 0.1°$ C.

Both the pO_2 and pCO_2 electrodes require regular maintenance to keep the membranes intact, taut, and clean. Obstruction to diffusion, such as protein buildup on the membrane, slows down the response and may give low results.

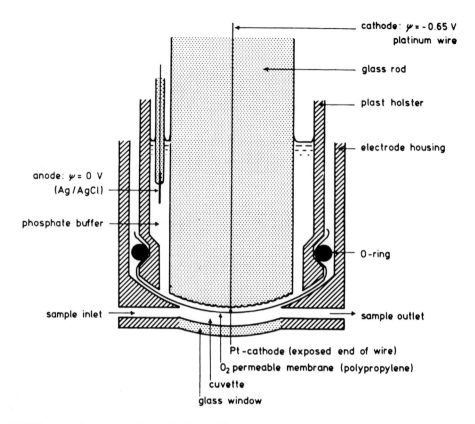

cathode: $\psi = -0.65$ V
platinum wire

glass rod

plast holster

electrode housing

anode: $\psi = 0$ V
(Ag/AgCl)

phosphate buffer

O-ring

sample inlet

sample outlet

Pt-cathode (exposed end of wire)

O$_2$ permeable membrane (polypropylene)

cuvette

glass window

Fig. 2.25. Diagram of an oxygen electrode. (From The Acid-Base Status of the Blood, 4th ed., by O. Siggaard-Andersen. Copenhagen, Munksgaard, 1974, p. 178.)

Ion-Selective Electrodes

Electrodes have been developed for the potentiometric measurement of many ions in addition to the hydrogen ion by the introduction of ion-selective membranes that are sensitive only to certain ions. They operate in a fashion similar to the glass electrode in its measurement of hydrogen ion concentrations. When an ion-specific membrane separates two solutions that differ in concentration of that ion, a potential is developed across the membrane; the size of the potential is dependent upon the difference in the ion concentrations. The membranes vary widely in type, including selective glass membranes, membranes consisting of a crystal, an immobilized precipitate, or a liquid layer.

The activity of any ion can be determined potentiometrically if an electrode can be developed that responds selectively to the ion of interest. A sodium-sensitive glass has been introduced that is insensitive to hydrogen ions and shows a selectivity for sodium over potassium of about 300 to 1. The potential is developed as the sodium ions undergo an ion exchange in the hydrated layer of the glass membrane. A potassium electrode incorporating a valinomycin membrane shows a selectivity for potassium over sodium of 1000 to 1. This specificity for potassium is developed because the ion exchange cavities in the valinomycin membrane are nearly equal in size to potassium ions. In these and similar systems, the potential developed is a function of the concentration of the ion being measured.

The development of an electrode selective for calcium ions has made possible the direct measurement of ionized calcium, the physiologically active form. Although total calcium measurements in the laboratory still far outnumber assays for ionized calcium, the latter determinations are becoming more important in the clinical picture.

Ion-selective electrodes have also been developed for many other ions, including lead, copper, ammonia, and chloride. Chloride electrodes have long been used in the laboratory for the measurement of chloride in sweat.

The use of selective electrodes in the clinical chemistry laboratory has grown by the skillful combination of enzymatic action with selective electrodes. Such an instrument (Beckman) has been devised for the measurement of serum glucose, urate, and cholesterol by utilizing specific oxidases to oxidize the substrate (glucose, urate, and cholesterol, respectively). Each reaction requires oxygen. The instrument has a pO_2 electrode to measure the rate of O_2 utilization which is proportional to the specific substrate concentration. In a similar manner, another instrument (Beckman; Fisher) measures serum urea concentration by a combination of an enzyme reaction and a conductivity meter. The enzyme urease splits off NH_3 from urea; the NH_3 diffuses through the membrane and is converted to NH_4^+ in the buffer. The change in conductivity of the solution is measured.

Radioactivity

An atom is identified by its atomic number, which is equal to the number of protons in its nucleus. Isotopes are atoms that have the same number of protons but different numbers of neutrons, resulting in a different atomic weight. Only certain ratios of neutrons and protons form stable nuclei; an excess of either of these particles results in an unstable nucleus that spontaneously changes to a more stable configuration. This nuclear alteration in the neutron to proton ratio is called *radioactive decay,* and its measurement forms the basis for the many radioassays performed in the clinical laboratories. Changes in the numbers of protons or neutrons—or both—in the nucleus of an unstable parent atom (a radionuclide) continues until a stable daughter atom is formed. Frequently, emission of additional energy as electromagnetic radiation is necessary before stability is achieved.

Radioactive decay of a radionuclide is accompanied by at least one of the three possible types of nuclear emissions: alpha (α) particles, beta (β) particles, and gamma (γ) rays. *Alpha particles* are the largest, consisting of two protons and two neutrons (equivalent to a helium nucleus). Their large size limits their penetration into matter, including skin, so danger from external exposure to alpha particles is slight; internal exposure is very serious, since their considerable energy would be dissipated in a small area. Alpha particles are not measured in the clinical laboratories. *Beta particles* consist of negatrons or positrons (negatively or positively charged particles, equivalent in mass to electrons) that are emitted from the nucleus of the atom. Emission of a negatron will transform a neutron to a proton, and emission of a positron will transform a proton to a neutron. Beta particles, because they are smaller and have greater velocity than alpha particles, can penetrate matter for short distances. This property is utilized in their measurment. Beta particles are an internal radiation hazard; external exposure at high levels may also present a hazard. *Gamma rays* have no measurable mass or charge; they are units of energy emitted to stabilize an atomic nucleus without changing atomic numbers or mass, although they may accompany such changes as beta emission or electron capture. Electron capture is the transfer of an inner orbital electron to a proton in the nucleus, converting the proton to a neutron. As stated above, these nuclear disruptions may be accompanied by emission of high energy electromagnetic radiation, the gamma rays. Gamma rays, like x-rays, can readily penetrate matter; because of this property, they can be used in extremely small amounts for *in vivo* scans of internal organs as well as *in vitro* laboratory assays. External exposure to high energy gamma rays can be extremely dangerous.

Half-Life of Radionuclides

The radionuclides used most commonly in the clinical chemistry laboratory for beta particle emission are carbon-14 (^{14}C) and tritium (^{3}H); the usual source of gamma radiation is iodine-125 (^{125}I). The rate of the radioactive decay of a radionuclide is measured in terms of its half-life ($t_{1/2}$). The half-life is a measure of the time it will take for half of its radioactive atoms to decay to a stable state. The half-life of ^{125}I is 60 days; thus one half of the radioactivity (measured as decays per second, dps) in a vial of ^{125}I will remain after 60 days and one fourth after 120 days. For assay purposes, one tries to select a radionuclide with a reasonable shelf-life so that its activity will remain sufficiently high for practical measurement. ^{14}C has a half-life of 5000 years; ^{3}H, 12 years; and ^{125}I, 60 days. The use of ^{131}I in radioassays has been virtually eliminated due to its short half-life of 8 days.

Measurement of Radioactivity

Radioactivity, as counts per minute (cpm), is measured in the laboratory by scintillation counters. The relationship between dps and cpm is a function of the efficiency of the counting instrument; not all of the decays are registered because of geometric configuration; the radiation is emitted in all directions and only the radiation hitting the sensor is counted.

Beta radiation is measured in *liquid scintillation counters*. For counting, the sample is dissolved or suspended in a scintillation fluid in a vial. The fluid contains chemicals called fluors, which emit flashes of light when struck by beta particles. The beta particles must be closely surrounded by these "scintillants," since their own travel path within the vial is so short. The scintillation vial is positioned between two photomultiplier tubes that count and register the flashes of light from the fluors as cpm.

Gamma radiation is detected by a scintillation crystal, a sodium iodide crystal activated with thallium. The sample tube is lowered into a well within the crystal; a gamma ray striking the crystal results in a flash of light that is registered by an adjacent photomultiplier tube. Gamma rays travel from the sample tube into the crystal for detection.

Both types of scintillation counters contain discriminators to isolate the voltage range typical of the energy emitted by the radionuclide of interest. These serve a purpose analogous to the monochromator in a spectrophotometer.

Radioactive elements are not indigenous to the biological samples assayed in the laboratory. "Tracers," molecules into which radioactive atoms have been inserted, are added to the sample for assay. These molecules compete with, and react like, the analogous nonradioactive molecules that are present in the sample. Measurement of the radioactive tracer provides information about the quantity of the nonradioactive species present. A detailed description of the radioimmunoassay incorporating this principle is found on page 253.

Automation

Hundreds of thousands of tests are performed each year in the chemistry laboratory of a large hospital. This monumental task is made possible by the use of automated instruments that perform all the functions involved in many test procedures, with far greater speed and better precision than is possible by manual operation.

There are three major approaches to clinical chemistry automation: continuous flow, discrete analysis, and centrifugal analysis. In *continuous flow*, reagents are pumped continuously through the system, samples are introduced sequentially at timed intervals and follow each other through the same network of tubing, coils, and colorimeter. For *discrete analysis*, the assay for each sample is contained in a separate, discrete tube or other reaction vessel, and there is minimal carryover or mixing between individual assays. *Centrifugal analysis* is really a discrete system incorporated into a centrifuge rotor. Transfer of solutions is carried out by the use of centrifugal force.

The first widely used automated instrument, the Auto Analyzer, introduced by Technicon, was based upon the continuous flow principle. The heart of the system was a peristaltic pump that continuously pumped reagents through the tubing system by the progressive squeezing and release of plastic tubing. Samples were sucked into the system at 1-minute intervals by a sampling probe. A series of air bubbles inserted at regular intervals "scrubbed" the tubing to minimize the carryover from sample to sample. In this system, six separate modules perform specific tasks, comparable to the manual steps in a procedure.
1. The *sampler* aspirates sample or standard, followed by a wash solution in a sequence timed to minimize carryover.
2. The *proportioning pump*, using tubings of various sizes, controls the relative volumes of sample and reagents introduced into the system. The manifold, seated astride the pump, consists of a series of coils and tubings that perform the mixing and time the interactions of the reagents as they are forced through the system by the peristaltic action of the pump.
3. The *dialyzer* separates large molecules, such as proteins, from the small molecules by means of a semipermeable membrane; the two streams are separated and the dialysate is taken for analysis.
4. A *heating bath* can supply a controlled optimal temperature (from 37° to 95° C) for the reaction, if needed.
5. A double beam *colorimeter* with interference filters measures the changes in absorbance as samples move through the flowcell.
6. A *recorder* converts the electric signal from the colorimeter to a graphic display.

Subsequent Technicon instruments have split the diluted sample stream so that as many as 20 tests may be done on each sample. The dialyzers and baths have been made much smaller and have been incorporated into manifolds that are stacked for simultaneous analysis of all the test constituents. The multitest continuous flow systems run all of the tests on every sample. Newer versions of the instrument can be programmed to report only the result of selected tests to the physician.

Numerous discrete analytical systems of varying capacities and intricacies are available to the clinical chemistry laboratory. In discrete systems, each sample is assayed in a separate container, so there is minimal carryover from sample to sample. In most instruments, the sampler probe picks up the needed volume of sample from a specimen cup, dilutes it, and aliquots it into the tubes needed for the requested assays. Required reagents are added to each tube, and the reaction mixtures are incubated at elevated or ambient temperatures until results are measured. Most measurements are made spectrophotometrically; some multitest instruments use fiber optics to transmit light from the source lamp to the cuvet, and from the cuvet to the detector. This arrangement eliminates the need for several detectors or for shifting different cuvets into the light path. Measurements may also be made by other devices, such as a fluorometer, flame photometer, or special electrodes. The data for all tests on each patient are usually collected, printed out, and/or fed into a central computer for reporting.

Many discrete systems operate by mimicking the manual procedures; however, some innovative discrete systems have been introduced. The DuPont Automatic Clinical Analyzer (ACA) system utilizes a transparent plastic pouch (test pack) as the discrete container for each test. The pouches are prepackaged with reagents for each assay and are coded with instrument instructions specific for the test procedure. To initiate an assay, a serum sample cup is placed in a rack together with patient identification and assay information, which is photographed to accompany the results. A test pack for each procedure is placed on the rack directly behind its specimen. The required volumes of serum are aliquoted into each pack before the pack is moved along the track where the assay is to take place. Within each pack there are small compartments carrying the reagents; the seals to these compartments are broken automatically to initiate the addition of the reagent at the proper time. Reagents are added and mixed, reaction mixtures are incubated, all at the direction of the coded information. When the reaction time has been completed, a cuvet with flat optical surfaces is molded from the pack itself. Absorbance of the reaction fluid is read in a colorimeter.

Since all of the reagents are prepacked within the packs that must be purchased from the manufacturer, there is no opportunity for the laboratory to substitute its own reagents or to change any methodology. The quality control on the ACA packs has been good, however, and laboratories have found the results dependable and the precision acceptable.

Another novel approach has been introduced by Eastman Kodak Company in its Ektachem System. Small cardboard slide mounts holding multilayered dry films form the discrete container for this system. A 10 µl drop of serum is automatically applied to the top, or "spreading," layer, which causes the sample to spread evenly across the slide before penetrating to the lower layers. The subsequent layers vary, each selected to suit the chemistry required for the test. A barrier layer impervious to large molecules may be included if it is desirable to remove proteins from the reaction mixture. In all cases, a reagent layer containing the reagents, often enzymes, forms the medium for the assay of the test analyte. As needed, there may be a buffer layer or a semipermeable membrane to allow only the desired products to penetrate to the next layer. After an incubation period, the final colored reaction product descends to a transparent support layer through which the reflectance density is read.

Ektachem has also set up another system with disposable electrodes imbedded in slides for the measurement of electrolytes. In the Ektachem instrument, the slides are stored in cartridges in dispensers from which they are selected as specific tests are programmed into the system.

Discrete analyzers are by far the most numerous of the automated systems, but there is a growing contingent of centrifugal analyzers. In these systems, which also utilize discrete chambers and cuvets for each sample, the reagent and sample are introduced into separate wells of chambers that radiate like spokes from the hub of the rotor. As the centrifuge begins to spin, all the samples are mixed simultaneously with reagent by centrifugal force. The reaction mixture flows outward to the cuvets that form the periphery of the rotor. At the completion of the reaction time, photometric readings are taken through the cuvets while the rotor is still spinning. Multiple readings from successive rotations are averaged for the final result. In the centrifugal systems, all tests in one run are assayed in parallel: the specimens are all mixed at once with reagent as the rotor spin is initiated, and all samples are read during the same spins of the rotor.

All of the automated systems perform, in different ways, assays that utilize chemistries that may be done individually, manually, or with special detecting systems. They have the advantages of speed and precision. The accuracy depends upon the chemistries involved and adherence to the special requirements of the chemical assay. It is never safe to assume that the instrument number is an accurate one. Instrument methodologies, like manual ones, require careful evaluation before they are accepted as routine laboratory procedures.

The approaches to automation are many and varied; selection of an appropriate system depends upon the needs of a particular laboratory. Cost is a primary consideration—the price of the instrument, as well as its cost of operation. A high capital cost may well be offset by the large volume of tests generated, if the workload is heavy. The operational costs (cost per test) is an important factor, for the instrument may utilize many disposables such as cuvets, sample cups, tips, rotors, pouches, columns, and reagents. Reagent costs vary widely and they are always expensive when bought already prepared. Some instruments pump reagent through the system continuously, whether or not a sample is being run. In discrete systems, reagent volumes per test can vary from 50 μl to 3 ml. For some instruments, you may make your own reagents, but, for others, you must use reagents supplied by the manufacturer. Large laboratories may find real savings in making their own reagents; smaller ones may find the technologists' time more expensive.

It is important to select an instrument that requires only a small volume of serum per test. This is essential for hospitals with a neonatal center and pediatric population and important for patients having extensive work-ups. It is aggravating for everyone to have insufficient serum for a series of tests.

Other considerations in selecting an automated instrument may be (1) the number of tests that can be run per batch; (2) the number of assays that may be run on each sample; (3) the time required per test: (4) the space (electrical, plumbing connections) required by the instrument; and (5) the ease of including stat samples as they come in.

In summary, then, an instrument must be selected with the goals of the laboratory in mind. Does the laboratory perform large volumes of single tests, so that an instrument capable of running many samples per batch is important? Is a standard profile of tests run on every admission or on large numbers of people? Are batches small, or is it important to select individual profiles of tests for each patient? Will there be other instruments for stat tests, or will it be desirable to interrupt the automated cycle to insert individual tests? Is the instrument rugged and dependable or does it break down frequently? Does the company provide good service in a hurry? Does the instrument come in components that can be replaced individually on short notice or does the whole instrument have to be shipped to the manufacturer for major repair? Does the company provide instruction classes for your laboratory when purchasing a new instrument? Are there good instruction and trouble-shooting manuals? There are instruments appropriate for many situations and for many laboratories. The task is to select the right tool for the job and then to verify that it is performing good chemistry assays with precision and accuracy.

Laboratory Computers

The proliferation of laboratory tests and the large increase in the number of test requests created a problem in data handling for most laboratories. Accordingly, the clinical laboratory has become more dependent upon computers for data processing, and the trend has accelerated in recent years as computer technology makes great advances. Technologists in most laboratories have to utilize a computer to some extent, so a basic understanding of computer operation should be helpful.

Components

A computer system is composed of three parts:
1. *Hardware:* This refers to the physical components and includes the central processing unit, memory, storage media, input and output devices. The central processing unit (the "brain") contains the circuitry to translate and execute the logic and arithmetic functions. Central memory is used to store the data and programs that are being executed. Slower forms of memory, such as magnetic tapes and disks, serve as repositories of long-term data and provide larger areas for temporary program manipulation. Input and output devices are the means of communication between a computer and the outside world. Line printers, character printers, and cathode ray tube terminals (CRT) are classified as input/output devices and are designed for communication between the computer and the user. The CRT's keyboard conforms with the standard typewriter key configuration. The dialogue between the user and the computer is displayed on the CRT screen. Printers are used to generate "hard copy" reports, worklists, labels, or other reports. The peripheral devices are connected to the computer through standard interfaces and print the requested information on the proper forms.
2. *Software:* Software embraces the set of instructions contained in a computer program for performing the specific functions assigned to it. These

programs are generally written in one of the higher level computer languages, such as FORTRAN, MUMPS, or BASIC, and control the operation of the computer circuits. The software applications package is the collection of interdependent programs designed to integrate the various computer tasks assigned to it. In addition, there must be an operating system and various utility and maintenance programs. The operating system is a set of programs for monitoring the overall performance of the machine and for providing certain essential functions that include management of data transfer from various parts of the computer and communication with peripheral devices.

3. *Personnel:* The successful operation of a computer depends upon the individuals who use and operate it. Medical technologists, who enter and retrieve data, and specialized computer personnel are the ones who are usually involved with a computer system. In the larger installations, a programming staff may be needed to modify and enhance the software package; initial data entry concerning patients and test requests may be entered into the computer by clerks instead of medical technologists. On-site programmers are not necessary when the system is "turn-key" (runs as purchased from the software vendor). Computer operators are required to perform functions on the system such as loading tape and making backup images of the system. A specially trained medical technologist may serve as a laboratory data manager and supervise the systems data files, which include test information tables, codes for laboratory tests and security codes. A data manager also helps to integrate the transmission of data from the laboratory to the physician and to trace and eliminate problems when they occur.

Tasks

Computers in a clinical laboratory are utilized for many purposes, which are primarily determined by computer capacity. The extent of use varies among the different laboratories from simple test reporting to highly sophisticated systems for retrieving test results according to patient age, sex, disease, or other parameters or for performing complicated statistical analyses. The following describes some typical tasks performed by computers in many laboratories, but it is far from complete.

1. *Specimen management:* Upon receipt of a test request, the test(s) are entered into the computer with appropriate patient information, which usually includes patient name, identification number, age, sex, location in hospital, and physician. This information is used to consolidate and facilitate result reporting, to assign appropriate reference values according to age and sex, and to facilitate the transmission of results to the correct location. After completion of the above step, the computer assigns an accession number to the specimen and prints sample collection labels and aliquot labels and generates work lists for individual work stations. The work lists may assign instrument cup numbers for patient specimens, standards and controls. When the assay is completed, results may be entered into the computer by the technologist or they may be fed "on-line" by an automated instrument communicating directly to the computer. The computer may be programmed to check for values outside

defined reference and critical ranges and, if a discrepancy is found, to flag the result. The computer can also order additional confirmatory tests if so programmed.

2. *Information processing:* A large variety of informational reports can be produced from the input of test data. *Stat reports* can be printed directly in patient care units (intensive care units, emergency room). *Interim ward reports* may be printed that contain all new results that have been generated since the last report, grouped by ward and by patient. *Daily cumulative reports* that list results and pending tests in cumulative format for the individual patient may be printed. Upon *discharge,* reports are generated containing the completed set of data in cumulative form.

3. *Laboratory management:* A good computer system helps the laboratory to become aware of tests that have not been performed or of results that have not been reported. A *pending list* identifies specimens with no reported results; *overdue reports* call attention to results delayed too long; and *quality control reports* may be accompanied by graphic displays and statistical evaluation, which help to maintain a good quality control system.

4. *Administrative functions:* Billing information for laboratory tests is usually generated and may be used to produce bills locally or be transmitted to a large computer in the hospital business office that prepares the bills. With proper programs, reports may be generated that compile the numbers of tests that are performed daily, monthly, or yearly, workload statistics, turnaround times for results to reach wards, or other such statistics.

The computer system is but one link in a chain of events that originates with a request for a laboratory test and ends with a test result in the hands of a physician. Quality control and high performance are just as essential for personnel working with the computer as they are for the technologists performing the analysis. The good teamwork of technologists and computer personnel is essential for achieving the desired aim—the delivery of accurate and precise test results to the medical team responsible for patient care.

QUALITY CONTROL

Every analysis in the laboratory generates a number. The validity of that number cannot be taken for granted, however, without some body of supporting evidence. The most convincing evidence is the establishment of a rigorous, comprehensive quality control program that is faithfully followed. This is the only way to guard against the possible deterioration of an analytic system, to become alerted to imprecision of results when they occur, and to have confidence in test results when everything is in control. As professionals, it is just as important for us to be assured that our analyses are correct as it is for the physician to have confidence in the laboratory result.

A comprehensive quality control (QC) program consists of all the means used by a laboratory to assure the reliability of every assay performed on specimens arriving in the laboratory. It includes much more than the analysis of a control serum with every run, and it involves the laboratory director as well as the technologists and all other personnel who are an integral part of the specimen collecting and analytical system.

It is the responsibility of the director or supervisor of the clinical chemistry laboratory to:

1. Select the most accurate and precise analytical methods that are feasible for performing the tests in a time period that is most helpful to the physicians.
2. Adequately train and supervise the activities of laboratory personnel.
3. Make available printed procedures for each method, with explicit directions, an explanation of the chemical principles, a listing of reference values (normal range), and a listing of common conditions in which the test results may be high and those in which they are low. This information packet should also include data on the linearity, precision, and sensitivity of the method, as well as a list of substances that may interfere in the assay.
4. Select good instruments and institute a regular maintenance program.
5. Institute a good quality control program, make available a supply of control serum, regularly inspect the control charts, and instill in the entire staff an awareness of the importance of a good quality control system.
6. Conduct continuing education sessions with the technologists to sustain their professional interests and keep them abreast of changing technology.

It is incumbent upon the technologists to:

1. Follow assay directions explicitly.
2. Use the proper control serum for each run, chart the results, and take appropriate action when the control serum result is beyond the established limits.
3. Always employ sound, analytical techniques.
4. Be conscientious in instrument maintenance.
5. Notify the supervisor immediately when analytical problems develop and when unusual or life-threatening results are obtained on a patient.

There may be some gaps in the blood collecting or delivery system, however, that are not under laboratory supervision and may generate error. If blood is drawn from the wrong patient, samples are switched in processing, or blood is drawn in a chemically contaminated syringe, it is obvious that the laboratory findings will not reflect the real condition of the patient. The laboratory has to cooperate with other members of the health service team in an attempt to eliminate this type of error and also to make them conscious of the problem of quality control in laboratory testing.

Precision and Accuracy

The first step in the establishment of a quality control program is to ascertain the limits of uncertainty for each test. Every measurement, every analysis carries with it a degree of uncertainty, a variability in the answer as the test is performed repeatedly. It is essential to determine the *precision* of each test, which reflects the *reproducibility* of the test (the agreement of results among themselves when the specimen is assayed many times). The less the variation, the greater is the precision. Precision must not be confused with *accuracy*, which is the deviation from the *true* result. An analytical method may be precise but inaccurate because of a bias in the test method; an example was the Folin-Wu glucose method, which determined certain nonglucose-reducing substances as glucose.

The degree of precision of a measurement is determined from statistical

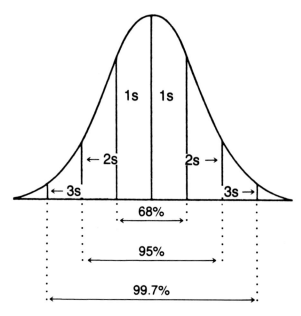

Fig. 2.26. Normal frequency curve. ± 1 s encompasses 68% of the values; ± 2 s 95%, and ± 3 s 99.7%.

considerations of the distribution of random error; it is best expressed in terms of the *standard deviation*. A normal frequency curve (bell-shaped, Gaussian curve) is obtained by plotting the values from multiple analyses of a sample against the frequency of occurrence, as shown in Figure 2.26. From statistical considerations, the standard deviation, s, is derived from the following formula:

$$\text{Equation 9: s} = \sqrt{\frac{\Sigma(\overline{x} - x)^2}{N - 1}}$$

where s = 1 standard deviation, Σ = sum of, $\overline{x}$ = mean (average value), x = any single observed value, and N = total number of observed values. With a normal distribution, 68% of the values are encompassed by ± 1 s, 95% by ± 2 s, and 99.7% by ± 3 s.

The procedure for calculating the s of the example shown in Table 2.4 follows:

1. Calculate the mean of all values.
2. Find the difference of each individual value from the mean (Column 2). Use the absolute value; ignore negative or positive differences.
3. Square the differences (Column 3).
4. Add the entries in Column 3 to obtain the sum of the squares of the differences.
5. Find the standard deviation (s) by using Equation 9.

For the sake of brevity, the example given in the table includes only ten values. The precision data for a test should be acquired from at least 20 separate assays.

In the example given in Table 2.4, 68% of the future assays are expected to fall within ± 1.6 units of the mean value of 20, and 95% of the values within 20 ± 3.2, when the method is functioning well. Thus one assay in 20 may

TABLE 2.4
Calculation of Standard Deviation (s) and Coefficient of Variation (CV)

ASSAY VALUES	$\bar{x} - x$	$(\bar{x} - x)^2$	
18	2	4	$N = 10$
20	0	0	$\bar{x} = 20$
21	1	1	$\Sigma (\bar{x} - x)^2 = 24$
17	3	9	
22	2	4	$s = \sqrt{\dfrac{\Sigma(\bar{x} - x)^2}{N - 1}}$
19	1	1	
20	0	0	
20	0	0	
21	1	1	$s = \sqrt{\dfrac{24}{9}} = 1.6$
22	2	4	
			$CV = \dfrac{s}{\bar{x}} \times 100 = \dfrac{1.6}{20} \times 100 = 8\%$

produce a value further than 3.2 units from the mean, even when reagents, standards, and instruments are all acceptable, and there is *no* technologist error.

The standard deviation is greater when a method is less precise; it is generally greater when the mean of the assay values is a larger number. A larger deviation from the mean (and therefore a larger s) should be expected when the mean value is 200 than when it is 20 or 2. The *coefficient of variation (CV)* expresses the standard deviation as a percentage of the mean value and is a more reliable means for comparing the precision at different concentration levels:

$$CV = \frac{s}{\bar{x}} \times 100 \text{ and is expressed as } \%.$$

The precision of a method varies inversely with the CV; the lower the CV, the greater is the precision.

Control Serum and Quality Control

The most practical way of monitoring the analysis of serum constituents is to have a generous supply of serum on hand that is analyzed with every run or batch of tests; this serves as the control serum and is a better way of testing all phases of an analytical method than by the use of an aqueous solution of the single constituent being measured. Control serum in lyophilized form is available from commercial sources and comes in two forms: assayed (constituents analyzed by a group of reference laboratories, with the range of values on the label) or unassayed. The latter is cheaper, but the laboratory itself bears the responsibility for determining the concentration of the various constituents. For those who purchase control serum, it is advisable to obtain at least a year's supply of the same lot number and avoid the extra labor and cost of performing multiple analyses to establish new mean values and their standard deviations at more frequent intervals.

It is also possible to prepare one's own control serum pool by collecting and pooling unused serum samples or by utilizing outdated blood bank plasma.

This type of control involves investment of laboratory time, requires ample freezer space to store the pool aliquots, presents a greater danger from infectious hepatitis as patients' specimens are collected, and has a shorter usable period than lyophilized material. It does not pay to prepare your own pool unless a minimum of 3 months' supply is prepared at one time. For some special purposes, such as for control of enzyme assays, it may be advisable to prepare your own by preparing extracts from appropriate human tissues and diluting with control serum to the activity levels desired.

Always treat control serum, whether from a commercial source or your own pool, as a potential source of infectious hepatitis, although most commercial control sera are negative for the Australian antigen (hepatitis B antigen).

Control sera should be included with every run of each assay; with long runs they should be interspersed at regular intervals among the patient samples. Inclusion of control sera at both abnormal and normal concentrations will provide information about the performance of the assay over a range of values. The selection of a control serum with values conforming to medical decision levels can provide reassurance to both the laboratory and the physician; if a drug is considered toxic above a certain serum concentration, the laboratory should strive to verify its results by inclusion of a control serum containing the critical drug concentration.

Before being used for quality control, a control serum should be analyzed at least 20 times, preferably on separate days, and the mean values and s should be calculated for each constituent. When these are established, the control serum may be used to monitor the daily performance of the various procedures. Every assay should include an aliquot of the control serum in the batch. The method is considered to be in control if the value obtained on the control serum is within ± 2 s of its mean value. If the value for the control serum is greater than ± 3 s from the mean, there is an obvious error in the method, and no patient results should be reported until the batch is repeated with a new control serum aliquot that is in control. One should look for obvious sources of error before repeating the batch, such as incorrect calculations, wrong reagents, a new reagent made up incorrectly, measurement at an incorrect wavelength, or the use of wrong pipets. The value for the control serum should be expected to fall between ± 2 and ± 3 s 5% of the time; this is considered to be not in control, but the trend in the control chart should be looked at because this may be the statistical one in 20 chances for the value to be out of control.

Westgard[9] has described a protocol for reducing the number of false rejections of an analytical run, while maintaining a high probability of error detection. This method requires the use of two control sera of different concentrations in every run. When one control value exceeds the limits set at $\bar{x}$ ± 2 s, the method is suspected to be out of control, so the following test checks are applied to determine whether to reject the run:

1. If both serum controls exceed the 2 s limit, either in the same direction or in opposite directions (one high and the other low), the run is out of control and should be rejected.
2. If the same control material exceeds the 2 s limit in two successive runs, the method is out of control and the run should be rejected.
3. If four consecutive control values are greater than one standard deviation from the mean, all in the same direction, the method is out of control.

Reject the run. This criterion can be based on one control or on the last four measurements of both controls.

4. If ten consecutive control values fall either above or below the mean, the method is out of control. Reject the run.

In summary, one control value exceeding the $\bar{x} \pm 2$ s limit is considered out of control if any one of the above test checks is also out of control. Initiate troubleshooting procedures to determine the cause of the problem. If *all* of the four additional checks show the run to be in control, the patient values can be reported. In the event that a batch of test results should be held up because of a quality control problem, grossly abnormal results on patients' specimens coming from the emergency service should be telephoned to the attending physician, with the statement that the results are provisional, pending a repetition of the test. The provisional result may confirm a diagnosis and permit the start of treatment.

Quality Control Charts

A quality control chart is established for each constituent in the control serum; the concentration is plotted on the ordinate, with a black line drawn across the chart at the mean value, blue lines at ± 1 s, orange lines at ± 2 s and red lines at ± 3 s. The days of the month are plotted on the abscissa. The chart is hung in a convenient location or kept in a notebook at the workbench, and each value obtained on the control serum is recorded on the chart every time an analysis is made. It is important to keep a permanent record of each value that is out of control and the corrective action that was taken (Fig. 2.27A).

When control values are plotted regularly, it is sometimes possible to see that a method is getting out of control even while the values are still within 2 s of the mean (Fig. 2.27B and C). Figure 2.27B shows a shift in the values as more than 6 consecutive results have suddenly fallen below the mean. Such a shift is an early warning indication that something has gone wrong, possibly an improperly made or contaminated reagent or standard or an incorrectly calibrated or functioning instrument. It calls for corrective action. In a similar fashion, Figure 2.27C illustrates a trend in control values that may indicate a steadily deteriorating component of the system. Thus, the control charts become a source of help in detecting problems and a source of pride when all goes well. They have to be kept up daily and inspected rigorously.

The mean and s of a control chart should be updated monthly as more analytical results come in; they become more meaningful when they rest upon a larger number of analyses.

Additional Quality Control Procedures

In addition to analyzing the control serum with every batch, several other devices should be employed to improve the quality control program.

1. The use of "blind" controls. A control serum pool disguised as a patient's sample is employed occasionally because of the human tendency to be more careful with a known control than with the patient's serum. The recycling or splitting of a patient's serum sample serves the same purpose.
2. Participation in one or more proficiency testing systems. Federal, state, and private agencies provide proficiency testing services by periodically sending out sera for the analysis of particular constituents. After a period

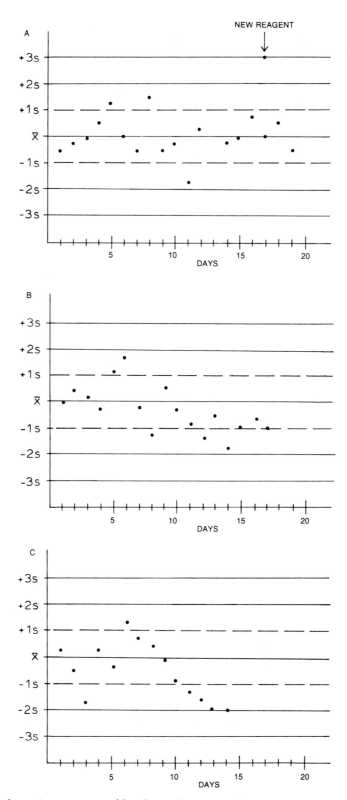

Fig. 2.27. Control charts: A, one unacceptable value, with notation of the correction made; B, a shift in values; C, a downward trend in the control serum values.

of time, a summary of the laboratory test results compared to those of both reference and participating laboratories is provided. Comparison of various methodologies and suggestions for improvement of the analytical procedures may be included with the reports. These testing services help to provide assurance in the work when all goes well and to call attention to problem areas. Some of the federal and private agencies engaging in proficiency testing are the Center for Disease Control, the College of American Pathologists, the Institute for Clinical Science, and the American Association for Clinical Chemistry (Therapeutic Drug Monitoring, and Endocrinology & Metabolism Programs). Some states have their own proficiency testing service for laboratories in their states.

Another phase of quality control requires a regular maintenance program for the various laboratory instruments to insure that they are in top working condition.[10] This includes regular calibration of spectrophotometer wavelengths, continuous recording of refrigerator and freezer temperatures to insure that requisite cold temperatures are maintained, testing of water purity with a resistance meter, checking water bath temperatures regularly, and calibration of micropipets. These seemingly small details may greatly affect performance.

The maintenance of a good quality control program is costly both in time and money. Control serum is not cheap, and a great deal is used in a year. The time invested by technologists in carrying out tasks that bring in no revenue (analyzing control serum, repeat testing of samples in a run not in control, and calibrating glassware) is considerable. It is estimated that a good quality control system adds 20 to 25% to a laboratory's operating cost.

Evaluation of New Methodology

Every laboratory seeks to improve its methodology as technology advances, new techniques are discovered, and new methods introduced. New methods cannot be accepted on faith, however; they must be rigorously tested and shown to have some advantages over existing methods. If it is a completely new test, it should be shown to be accurate, precise, and reasonably free of interferences from substances likely to be found in the serum of hospital patients, and it must not be too costly to carry out.

In considering a new test, precision studies similar to those described for control serum should be carried out. The measurement of within-run precision is the first step; if this is as satisfactory as the current method in use, then the day-to-day precision over a 20-day period should be tested and found acceptable before proceeding further.

The new method is then compared with the existing method by analyzing a series of patients' serum samples (not fewer than 40) by each method, covering a wide range of concentrations. The results (each mean of duplicate analysis) are plotted on a graph, with the existing or reference method results along the abscissa and the new method results along the ordinate; thus, a single point on the graph represents the results obtained by each method when a perpendicular line is drawn from the point to each axis. Figure 2.28 illustrates the ideal situation in which the points are evenly distributed about a line with a slope of 1; this represents a perfect correlation between the two methods.

Two similar graphs are shown in Figure 2.29. The slope of the line in Figure 2.29A is 1, but the line does not pass through the origin. The results by the

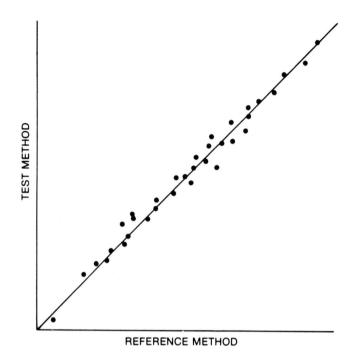

Fig. 2.28. An ideal plot comparing patient test results by a new methodology (test method) and an approved reference method. Slope of the line, which passes through the origin, is 1, showing 1.00 correlation between the two methods. Units for both coordinates are identical.

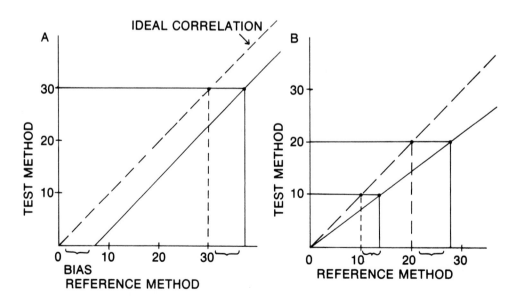

Fig. 2.29. Two plots representing patient sample comparisons with poor correlation. A, Case of constant error, with the new method giving results that are lower by a consistent difference; B, case of proportional error where the discrepancy increases as the concentration is increased. In each graph, the ideal correlation lines are broken; actual lines are solid.

new method are consistently lower than the reference method by a constant amount; there is a bias, and if the existing or reference method previously has been found to be accurate, the bias is in the new method.

Figure 2.29B illustrates a graph with poor agreement between the two methods. The line passes through the origin, but the slope is less than 1. Again, the results of the new test are lower than those of the reference method, but the error is proportional; the difference becomes larger at higher concentrations.

The best way to assess the accuracy of a new method is to compare it with a recognized reference method in a manner similar to that just described.

Another approach to accuracy is to assess the percentage of recovery of pure substance added to a serum pool that has been analyzed repeatedly.[12] An example is shown in Table 2.5 in which the glucose concentration of aliquots of a serum pool is increased by 50, 100, and 150 mg/dl, respectively; the original glucose concentration of the pool was 45 mg/dl. After analysis of the three supplemented aliquots in triplicate, as well as of the original pool, the mean concentration of glucose in the serum pool is subtracted from the glucose concentration found in the supplemented samples to give the total actual recovery in g/dl of glucose.

$$\text{The \% recovery} = \frac{\text{actual recovery}}{\text{added glucose}} \times 100$$

A recovery of 95% or better is generally considered acceptable.

A brief summary of the major steps in evaluating a new method has been presented here. In practice, much data need to be collected and analyzed, including information on interferences, limits of error at critical medical decision levels, and comparisons with the "state of the art." An excellent series of articles by J.O. Westgard presents a practical plan of action for the selection and evaluation of new methods.[11]

SAFETY

People working in clinical laboratories are exposed to various potential hazards. These dangers can be reduced by eliminating hazards where possible, establishing clean, safe work habits, taking proper precautions, and maintaining awareness of safety practices.

TABLE 2.5
Recovery of Serum Glucose

SAMPLE	ASSAYED VALUE (mg/dl)	GLUCOSE RECOVERED (mg/dl)	GLUCOSE ADDED (mg/dl)	% RECOVERED
Serum Pool	45	—	—	—
Pool + 50 mg/dl glucose	92	47	50	94
Pool + 100 mg/dl glucose	145	100	100	100
Pool + 150 mg/dl glucose	199	154	150	103

The mean glucose value of a serum pool was 45 mg/dl. In three separate experiments 50, 100, and 150 mg/dl, respectively, were added to it, assays were made, and percentages recovered were calculated.

The federal government passed an Occupational Safety and Health Act (OSHA) in 1971 that was designed to make places of work, including clinical laboratories, safer places in which to labor. Many states have passed similar legislation to supplement the federal laws. Although these laws make it easier to force recalcitrant or dilatory employers to take the necessary steps to make the work place a safer environment, no clinical laboratory will be completely safe unless the technologists become informed of the dangers and their remedies; they must establish good habits and take proper precautions at all times. Forrey, Delaney, and Ruff[14] have written a good article on laboratory safety.

The main potential physical dangers in a clinical chemistry laboratory are (1) fire, (2) infection, (3) contact with corrosive chemicals, (4) exposure to toxic fumes, (5) cuts from glassware with broken lips, (6) exposure to carcinogenic compounds, and (7) possible exposure to low-level radioactivity. These potential hazards are discussed more fully in the following sections.

Fire

The danger of fire is always present as long as there are volatile, flammable solvents in the laboratory; a static spark or the spark from a thermostat can set off an explosion if the proper mixture of ethyl ether and air is present. The danger of fire (or explosion) with flammable solvents varies inversely with the boiling point; the more volatile the solvent is, the greater is its vapor pressure at any given temperature and hence, the more quickly the vapor concentration reaches a combustion level. Of the solvents commonly present in a clinical chemistry laboratory, the following are listed in ascending order of their boiling points: ethyl ether (34° C), petroleum ether (40 to 60° C fraction), acetone (56° C), methanol (65° C), ethyl acetate (77° C), ethanol (78.5° C), isopropanol (96° C), toluene (111° C), and xylene (139° C).

A fire in a chemistry laboratory can wreak tremendous damage because of the quantities of flammable liquids that are frequently used and the presence of gas cylinders under high pressure. A raging fire in the laboratory could cause these cylinders to explode like huge bombs. The following precautions are advised:

1. When possible, substitute solvents with higher boiling points for the more dangerous ones with lower boiling points. Avoid the use of ethyl ether if at all possible.
2. Always work in a fume hood with the blower on when using low boiling (below 100° C), flammable solvents.
3. Keep the laboratory supply of flammable solvents to a minimum and store them in a flameproof cabinet.
4. Do not permit smoking in the laboratory.
5. Utilize fiber glass heating mantles when possible to heat flasks containing volatile solvents. Work in the fume hood and do *not* use gas flames. Hot plates can also be dangerous if the solution should bump and land on the hot surface; the solution usually ignites.
6. When it is necessary to employ a flame, use a mechanical igniter instead of matches to light the flame.
7. Place CO_2 fire extinguishers at strategic locations in the laboratory and have drills to make certain that all personnel know where they are located

and how to use them. Seconds are precious in extinguishing an incipient fire or preventing its spread.

8. All connections of flammable gases to instruments such as the flame photometer must be leak proof.

9. All gas cylinders, whether full or empty, must be chained securely so that there is no chance for them to be knocked over.

10. Never store flammable solvents in a refrigerator or deep freeze in which the thermostat is inside the compartment; a spark may explode the vapor mixture with tremendous force. Laboratory refrigerators and freezers should be converted so that thermostats are outside the cold compartment.

11. The laboratory should cooperate with fire inspectors to reduce all dangers to a minimum, to educate the staff to be constantly alert and careful, and to hold periodic fire drills.

Infection

The hospital is always filled with sick people, some of whom have contagious diseases. With the usual precautions of cleanliness (frequent hand washings with soap), no eating or smoking in the laboratory, and the avoidance of mouth-pipetting, there is little danger of chemistry personnel contracting a patient's disease, with the exception of infectious hepatitis. A few extra precautions have to be taken against this disease because occasionally it has spread to laboratory personnel. Infectious hepatitis may be contracted by entrance of the virus through breaks in the skin (cuts, accidental needle punctures), or through the entrance of contaminated material into the gastrointestinal tract; the latter is probably more common. The following precautions should be taken to reduce the chances of infection of any type:

1. When processing all blood sera, serum should not be poured from one tube to another because a drop or two of the serum may roll down the outside of the tube and contaminate all who handle it. Pasteur pipets or other devices should be used for transferring serum.

2. Warning labels (red or yellow tape) should be pasted on the tubes of samples from patients known to have infectious hepatitis, and those who handle the tubes should wear disposable plastic gloves.

3. The person first opening a vacuum tube (Vacutainer) containing blood should point the stopper away from himself and remove it carefully to avoid inhaling an aerosol.

4. No raw blood material should be poured down the sink. All blood tubes (with clots) and the excess serum when no longer needed for possible rerun should be autoclaved before washing or discarding. All specimens, including urines, from known hepatitis cases should be autoclaved before disposing of them.

5. A spill of patient serum or urine on the laboratory desk should be wiped up with a hypochlorite solution (bleaching solution) or other disinfectants to inactivate the hepatitis virus if it should be present.

6. Commercial control sera should be treated with the same precautions as patients' sera. Some control sera are positive for the Australian antigen (virus particle of one type of infectious hepatitis).

7. Frequent hand washing with an antiseptic soap after handling blood specimens is a good habit to cultivate. Never eat or even smoke before washing

the hands. By following these precautions and establishing good hygienic habits, the danger of getting infectious hepatitis in the work environment is reduced to a low level.

Corrosive Chemicals

Strong acids and alkalies are the most common corrosive chemicals to which clinical technologists are exposed. Most of the danger occurs through the splashing of reagents during their preparation. Injury to the eyes is the greatest danger because it takes time to wash away the chemical completely, and the cornea is easily injured; the chemical is even more difficult to remove quickly if the person is wearing contact lenses. The wearing of protective goggles is a necessity whenever preparing a caustic or irritating solution of any type. Care must be taken in the pouring of reagents in order to minimize splashing.

Toxic Fumes

In the clinical chemistry laboratory, it is occasionally necessary to prepare extracts with solvents whose vapors are toxic. This is particularly true of the toxicology section, which frequently extracts with chlorinated hydrocarbons (CH_2Cl_2, $CHCl_3$, CCl_4), chemicals that cause liver damage after a certain amount of exposure. Other solvents may depress bone marrow activity (production of leukocytes, the white blood cells), and a few are listed as possible carcinogens. The precautions are simple; keep the exposure to a minimum.
1. Always work in a fume hood with good ventilation whenever pouring or using organic solvents.
2. Avoid contamination of the skin with the solvents because they are slowly absorbed. If some should get on the skin, wash it off with soap and water.
3. See that the laboratory is well-ventilated. Exposure to a low concentration of vapor for a long period of time is also dangerous.

Broken Glassware

Beakers and flasks with broken lips are a hazard in the laboratory, particularly to the personnel who wash them. The remedy is simple: remove from circulation and destroy the partially broken glassware. If it is urgent to use a beaker or flask with a broken lip because no replacement is available immediately, the sharp edge of the broken section may be dulled by stroking it with a file or striking it with a wire screen; this is a temporary measure until a replacement is obtained.

Carcinogens

The danger of contracting cancer by exposure to carcinogenic chemicals in a clinical chemistry laboratory is low because most of the known carcinogens are not used there. The original 1974 OSHA list of 14 carcinogens appears in Table 2.6 (a larger, tentative list appeared in 1978). They are used widely in industry, and production workers may be exposed to dangerous concentrations.

Few of these compounds are in general use in a clinical chemistry laboratory. Benzidine has been employed in the measurement of *plasma* hemoglobin concentration, a test which is seldom requested. It is now possible to substitute a noncarcinogenic compound, an all ortho-substituted tetramethyl-benzidine (Aldrich), for benzidine.[13] Benzidine derivatives, such as o-tolidine and dian-

TABLE 2.6
Regulated Chemical Carcinogens (OSHA)*

2-Acetylaminofluorene	Methyl chloromethyl ether
4-Aminodiphenyl	4,4'-Methylene bis (2-chloroaniline)
Benzidine	Alpha-naphthylamine (1-naphthylamine)
3,3'Dichlorobenzidine (and its salts)	Beta-naphthylamine (2-naphthylamine)
bis-Chloromethyl ether	4-Nitrobiphenyl
4-Dimethylaminoazobenzene†	N-nitrosodimethylamine
Ethyleneimine	Beta-propiolactone

*Federal Register, Part III, Vol. 39, No. 20, Jan. 29, 1974.
†Töpfer's reagent.

isidine, which are weakly carcinogenic, have been used as chromogens in enzymatic reactions in which H_2O_2 is produced (with glucose oxidase and uricase). It is possible, however, to substitute other redox indicators that are not carcinogenic in these reactions (see glucose oxidase method, p. 314).

Naphthylamines (both alpha and beta) have been used as standards in some enzyme tests that employed derivatives of naphthylamines as substrates; the naphthylamine that was split off was converted into an azo dye whose absorbance was measured. It is far better to discontinue the use of these methods and substitute one that carries no exposure to a potential carcinogen.

Dimethylaminoazobenzene is the indicator dye used in Töpfer's reagent for estimation of gastric acidity (see p. 369). This test is not used very often because the test for free acid in gastric juice can be made with the indicator, methyl orange, or with a pH meter and thus eliminate the need for using a carcinogen.

In May, 1976, the Department of Health, Education, and Welfare (HEW), drew up a supplementary list of chemical carcinogens; HEW safety standards for laboratory operations involving these chemicals are mandatory. The only ones that may be of concern to clinical chemistry laboratories are hydrazine, chloroform, carbon tetrachloride, o-tolidine, and o-dianisidine. The latter two compounds have been discussed with benzidine. Hydrazine does not have to be utilized in a laboratory test because an adequate substitute or a better method can be found. Chloroform ($CHCl_3$) and carbon tetrachloride (CCl_4) were discussed in the section concerning exposure to toxic fumes. Both solvents produce liver damage if ingested or if the vapors are inhaled for a period of time. If adequate precautions are taken by working in a fume hood, there is little danger in working with these solvents.

If a reagent containing a chemical on the list of carcinogens has to be prepared, take care not to spill the powder or breathe its dust when weighing or transferring the chemical. Wash hands with soap and water immediately if any material should get on them. The carcinogens do not jump up and bite you; they have to be absorbed and the exposure usually has to be above a certain minimal concentration for a period of time before a cancerous growth is induced.

Radioactivity

Federal regulations permit the use of radioactive isotopes only in facilities that are under the direction of a person authorized to do so. Carefully specified procedures for storage, work space, monitoring programs, waste disposal, and operations have to be followed. The details may appear onerous but they are designed for the protection of personnel and containment of the radioactivity. A course in radiation safety should be a prerequisite for all persons working with radionuclides.

For most clinical chemistry laboratory test procedures employing radionuclides, for example, in radioimmunoassay (p. 253), only tracer doses are used. This means that the level of radioactivity is quite low and the exposure danger is small if proper precautions are taken in storing, handling, and disposing of the radioactive material. Gamma rays are penetrating, so gamma emitters should be stored behind a lead shield. Beta rays have little penetrating power, so they are safe if stored in their containers in an out-of-the way corner of the laboratory or in the refrigerator.

Disposable gloves should be worn when handling the concentrated material and care taken to avoid contamination or spills of any kind. Periodic monitoring of personnel, clothes, and work areas for contamination should be standard practice. Careful disposal of the waste is a necessity. There is little danger when good, careful work habits are practiced and exposure is kept to a minimum.

The health hazards in a clinical chemistry laboratory are far less than those in the chemical industry. If you observe reasonable precautions in the laboratory and are equally alert when driving an automobile, the chances are good that you will eventually collect and enjoy your pension as a senior citizen and later die of old age.

REFERENCES

1. Quantities and Units in Clinical Chemistry. International Union of Pure and Allied Chemistry and International Federation of Clinical Chemistry. Information Bulletin Number 20. Oxford, IUPAC, 1972.
2. Young, D.S., and Mears, T.W.: Measurement and standard reference materials in clinical chemistry. Clin. Chem. 14, 929, 1968.
3. Meinke, W.W.: Standard reference materials for clinical measurements. Anal. Chem. 43, 28A, 1971.
4. Catalog of NBS Standard Reference materials. NBS Special Publication 260, U.S. Department of Commerce/National Bureau of Standards.
5. Christian, G.D., and Feldman, F.J.: Atomic Absorption Spectrophotometry: Applications in Agriculture, Biology and Medicine. New York, Krieger, 1979.
6. Geddes, L.A., and Baker, L.E.: Principles of Applied Biomedical Instrumentation. New York, John Wiley & Sons, Inc., 1968.
7. Christian, G.D.: Analytical Chemistry. Waltham, MA, Xerox College Publishing, 1971.
8. Lee, L.W.: Elementary Principles of Laboratory Instruments, 4th Ed. St. Louis, C.V. Mosby, 1978.
9. Westgard, J.O., Barry, P.L., Hunt, M.R., and Groth, T.: A multi-rule Shewhart chart for quality control in clinical chemistry. Clin. Chem. 27, 493, 1981.
10. Dharan, M.: Preventive maintenance of instruments and equipment. In Selected Methods of Clinical Chemistry, Vol. 9, Washington, D.C., American Association for Clinical Chemistry, 1982.
11. Westgard, J.O., et al.: Concepts and practices in the evaluation of laboratory methods. Am. J. Med. Technol. 44, 290, 420, 552, 727, 1978.
12. Grannis, G.F., and Caragher, T.E.: Quality control programs in clinical chemistry. Crit. Rev. Clin. Lab. Sci. 7(4), 327, 1977.
13. Garner, R.C., Walpole, A.L., and Rose, F.L.: Testing of some benzidine analogues for microsomal activation to bacterial mutagens. Cancer Letters 1, 39, 1975.
14. Forrey, A.W., Delaney, C.J., and Ruff, W.L.: Safety. In Standard Methods of Clinical Chemistry, Vol. 9, edited by W.R. Faulkner and S. Meites, Washington, D.C., American Association for Clinical Chemistry, 1982, p. 43.

Chapter 3

WATER BALANCE, OSMOLALITY, ELECTROLYTES, pH, AND BLOOD GASES

The following definitions will aid in understanding the material:

1. *Electrolyte:* In medical usage, the term *electrolyte* is applied to any of the four ions in plasma (Na^+, K^+, Cl^-, and HCO_3^-) that exert the greatest influence upon water balance and acid-base relationships. We shall follow medical custom in this chapter; other ions that are important in clinical chemistry (Ca^{2+}, Mg^{2+}, Li^+, phosphate, Fe^{3+}, Cu^{2+}, and Zn^{2+}) are discussed in Chapter 12, "Mineral Metabolism," pages 347 to 363.
2. *Internal environment:* This refers to the concentration of ions and other constituents in body fluids; it also includes H^+ and therefore, the pH.
3. *Homeostasis:* The maintenance of a steady state in the body, and a relatively constant concentration of ions, pH, and osmotic pressure in the various body fluids.
4. *Osmolality:* Osmolality is a measure of the number of dissolved particles (ions and undissociated molecules, such as glucose or proteins) per unit weight of water.
5. *Osmotic pressure:* When two compartments of different osmolalities are separated by a semipermeable membrane that allows only small molecules to pass through it, the osmotic pressure represents the hydrostatic pressure that would have to be applied to the compartment of higher osmolality to prevent water passage into it from the lower one.
6. *Active transport:* Active transport is the transfer of ions or molecules across a cell membrane against a concentration gradient. It involves energy-linked, enzymatic reaction(s) because energy in the form of ATP is usually required.

BODY WATER

The protoplasm of multicellular organisms contains a high percentage of water. Water is found both inside and outside cells and can be divided into two main spaces or compartments: intracellular water that is within cells and the extracellular water that is outside of cells. The extracellular water can be further divided into the fluid contained in blood vessels (plasma) and the interstitial fluid between cells. Body water, with its dissolved salts and other substances, constitutes the internal environment. The approximate distribution of body water in an adult is shown in Table 3.1.

TABLE 3.1
Distribution of Body Water in the Adult*

COMPARTMENT	PERCENT OF BODY WEIGHT	VOLUME (L)	PERCENT OF TOTAL BODY WATER
Extracellular			
Plasma	5.0	3.5	8
Interstitial	15.0	10.5	25
Intracellular	40.0	28	67
Total Body Water	60.0	42	100

*For the average 70 kg person. Body water as percent of body weight varies inversely with the fat content; an obese person has the same *amount* of water but at a lower *percentage of body weight.*

TABLE 3.2
Daily Water Loss

SITE	VOL/DAY ml
Skin	500
Expired Air	350
Urine	1500
Feces	150
TOTAL	2500

Water Compartments

The content of body water is about 60% of total body weight. Approximately 67% of the water is contained within cells (intracellular water), while 33% is outside of cells (extracellular water). The extracellular water is divided further: about 5% of the body weight (3.5 L) is confined within the vascular system (arteries, veins and capillaries) as blood plasma; the remainder is distributed between the cells as interstitial fluid and amounts to 15% of body weight (10.5 L).

In many illnesses, the water and electrolyte balance is severely disturbed because of heavy fluid and electrolyte losses caused by vomiting, diarrhea, fistulas, or excessive urination. Although fluid or electrolyte imbalance is seldom listed as a cause of death, it certainly contributes to the seriousness of an illness and could hasten death in the critically ill. It is important to understand how water and electrolytes are distributed and the consequences of sudden imbalances.

Water Balance

An adult drinks and takes in with his food about 2500 ml of water daily. The same amount is lost daily in urine, feces, sweat, and expired air (Table 3.2). Individuals remain in water balance by regulating the intake (fluids) to compensate for daily losses. An exquisitely sensitive thirst mechanism normally regulates the desire for fluids so that the balance is maintained within relatively close limits. Problems occur when there is an inability to swallow.

Examples of losses are by vomiting, diarrhea, excessive urination, excessive sweating, or through fistulas.

ELECTROLYTE DISTRIBUTION

The concentration of electrolytes in extracellular and intracellular water are shown in Table 3.3. There are striking differences in composition between the two fluids. Na^+ is the principal cation of the extracellular fluid and comprises over 90% of the total cations while it has a low concentration in intracellular water and constitutes only 8% of the total cations. K^+, by contrast, is the principal cation within cells and has a low concentration in extracellular fluid. Similar differences exist with the anions. Cl^- and HCO_3^- predominate in the fluid bathing the cells, while phosphate is the principal anion within cells. These differences in ionic concentration are maintained within the cell by the expenditure of energy in the form of active transport mechanisms. For example, the sodium pump removes much of the Na^+ that diffuses into cells and replaces it with K^+. Energy is required to move these ions against concentration gradients.

The osmotic pressure of extracellular fluids is therefore determined by the concentration of Na^+ since, with its associated anions, it accounts for over 90% of the particle concentration (osmolality). Na^+ concentration likewise determines the extracellular fluid volume, because water flows from or into other compartments to restore osmotic homeostasis if disturbed. Water that shifts in and out of cells to equalize osmolality affects cell size and if extensive, can affect cell function. K^+ similarly determines intracellular osmolality to a large extent and cells shrink (lose water) when large K^+ losses are incurred (by prolonged vomiting, diarrhea, or hypersecretion of aldosterone).

Plasma Volume

The blood plasma is a subdivision of the extracellular fluid compartment (p. 82); as such, its volume is determined primarily, but not entirely, by the concentration of Na^+. The capillary walls are freely permeable to ions and water, and the Na^+ concentration in plasma is essentially the same as in the rest of the extracellular water. The hydrostatic blood pressure tends to force plasma

TABLE 3.3
Concentration of Cations and Anions in Extracellular and Intracellular Water

| CATION | CONCENTRATION IN WATER | | | | ANION | CONCENTRATION IN WATER | | | |
| | EXTRACELLULAR | | INTRACELLULAR | | | EXTRACELLULAR | | INTRACELLULAR | |
	meq/L	%*	meq/L	%*		meq/L	%*	meq/L	%*
Na^+	154	92	15	8	HCO_3^-	29	17	10	5
K^+	5	3	150	77	Cl^-	111	67	1	1
Ca^{2+}	5.4	2	27	14	HPO_4^-	2	1	100	51
Mg^{2+}	2.6	2	2	1	SO_4^-	1	1	20	10
					Organic Acid	7	4		
					Protein	17	10	63	33
TOTAL	167	100	194	100	TOTAL	167	100	194	100

*% of total ions

water out of the capillary bed, but it is counterbalanced by an osmotic pressure generated by plasma protein molecules that cannot diffuse out of the capillaries. The osmotic pressure produced by plasma proteins (sometimes called oncotic pressure) attracts water to the capillary bed and is sufficient to counteract the dispersing effect of the blood pressure. Albumin is the plasma protein that contributes the most to the oncotic pressure because of its high plasma concentration and relatively low molecular weight (most mol/L of proteins in plasma). The osmotic pressure contribution of the plasma proteins amounts to only 0.5% of the total plasma osmotic pressure, however; the body content of Na^+ is the main determinant of extracellular fluid volume and hence, plasma volume. The concentration of plasma albumin has to fall by about 50% before the change in plasma osmotic pressure becomes manifested as edema (passage of sufficient plasma water into the interstitial spaces to become noticeable as swelling).

Correction of Disturbed Osmolality

The body responds to disturbances in osmolality by instituting changes in water intake and excretion and not by changing body electrolytes. A dehydration (high osmolality) promotes the feeling of thirst, so the subject drinks fluids if they are available; it stimulates the posterior pituitary secretion of ADH (p. 112), which promotes renal tubular water reabsorption (less water excreted as urine). An overhydration (low osmolality) inhibits the thirst mechanism, so that no fluids are taken in, as well as the secretion of ADH, which results in the excretion of a large volume of dilute urine. The following is a summary of osmolality changes and water shifts:

1. Movement of water from one compartment to another is rapid and follows osmotic pressure gradients, from low to high. No energy is required.
2. The concentration of ions in intracellular water is markedly different from that in the extracellular fluid. Movement of ions against a concentration gradient is made possible by active transport. Some of the ions within cells are bound by protein and cannot move freely.
3. Cell volume varies directly with its water content; large changes can handicap cell function or destroy cells.
4. The plasma Na^+ concentration is the principal determinant of extracellular fluid osmolality and hence body water osmolality; a steady state exists and the osmolality of all compartments equalize by water shifts.
5. The body content of Na^+ determines the extracellular fluid volume, including the plasma volume.
6. The osmotic pressure of the plasma proteins, even though it is small, counterbalances the blood pressure in the capillaries and helps to maintain the plasma volume.
7. The body responds to disturbances in osmolality by regulating water intake and excretion.

SODIUM

The body of the average-sized adult contains about 80 g of sodium, 35 g of which are present in the extracellular fluids. The amount of sodium in the body is relatively constant, despite variation in intake. Although the average person ingests about 3 g daily of sodium as the chloride, sulfate, or other salt, he also excretes this amount daily. Since the sodium in plasma is in equilibrium with that in the interstitial fluid, the determination of serum sodium concentration is representative of its extracellular fluid concentration.

When an ingested sodium salt is absorbed, there is a temporary increase in extracellular fluid volume, as the absorbed sodium ion equilibrates between plasma and interstitial fluid. There is a small, temporary exchange of sodium for potassium inside the cell. In Table 3.3, it can be seen that the concentration of sodium in the fluid outside the cells is about ten times that of sodium inside the cell. The cell, however, is permeable to sodium ion, and this differential concentration is maintained by a "sodium pump."

Plasma sodium is filtered in the renal glomerulus and approximately 70% is reabsorbed in the proximal tubule; most of the remainder is absorbed in the distal tubule under the influence of aldosterone, a hormone secreted by the adrenal cortex (p. 114). Aldosterone accelerates the exchange of Na^+ for K^+ across all cell walls including those of the distal tubule. This promotes the retention of Na^+ and the excretion of K^+. The reverse situation occurs if there is a deficiency of aldosterone. Some gonadal steroids may cause a temporary retention of salt and water; this sometimes occurs premenstrually.

REFERENCE VALUES. The concentration of sodium in normal serum is 136 to 145 meq/L (136–145 mmol/L).

INCREASED CONCENTRATION (HYPERNATREMIA). Elevated levels of serum sodium are found in (1) severe dehydration owing to inadequate intake of water, irrespective of cause, or to excessive water loss; (2) hyperadrenalism (Cushing's syndrome), in which excessive reabsorption of sodium in renal tubules occurs as a result of overproduction of adrenal steroids; (3) comatose diabetics following treatment with insulin as some Na^+ in cells is replaced by K^+; (4) hypothalamic injury interfering with thirst mechanisms; (5) nasogastric feeding of patients with solutions containing a high concentration of protein, without sufficient fluid intake; and (6) diabetes insipidus (deficiency of antidiuretic hormone) without sufficient intake of water to cover the fluid loss.

DECREASED VALUES (HYPONATREMIA). Most low serum sodium values are found in the following situations: (1) a large loss of gastrointestinal secretions occurring with diarrhea, intestinal fistulas, or in severe gastrointestinal disturbances of any sort; (2) the acidosis of diabetes mellitus before the coma stage, when large amounts of Na^+ and K^+ are excreted into the urine as salts of the keto-acids, with replacement of water because of thirst; (3) renal disease with malfunction of the tubular ion-exchange system of Na^+ for H^+ and K^+ (salt-losing nephritis); (4) Addison's disease, with depressed secretion of aldosterone and corticosteroids; and (5) diabetes insipidus (posterior pituitary deficiency) with compensatory intake of water.

DETERMINATION OF NA. Since Na^+ and K^+ are usually determined simultaneously with an emission flame photometer, the description of the method will appear at the end of the section on potassium.

POTASSIUM

Potassium is the cation having the highest concentration within cells and is approximately 30 times higher than in the extracellular fluids. Approximately 2 to 3 g of potassium are ingested and excreted daily in the form of salts. Potassium salts in the diet are absorbed rapidly from the intestinal lumen but have little effect upon the plasma concentration; the rise is slight and transitory.

After tissue needs are met, the remainder is excreted by the kidney. The excretion process consists of glomerular filtration, absorption in the proximal tubule, and finally, excretion primarily by exchange for sodium ion in the distal tubules (p. 110). The kidney does not have the ability to reduce the potassium excretion to nearly zero, as it has for sodium.

The close control of the concentration of potassium in extracellular fluids is essential because elevated concentrations of K^+ ($>$ 7.0 mmol/L) may seriously inhibit muscle irritability, including the heart, to the point of paralysis or cessation of heartbeat. Low serum potassium values are also dangerous because they increase muscle irritability and can cause cessation of the heartbeat in systole (contraction); low serum potassium concentration can be rectified by intravenous injection of appropriate solutions. It is essential for the laboratory to notify immediately the attending physician whenever a seriously high or low potassium value occurs so that appropriate action can be taken in time. The above changes in cardiac muscle irritability caused by either high or low potassium concentration may be reflected in altered electrocardiographic patterns.

REFERENCE VALUES. The normal serum concentration of potassium varies from about 3.8 to 5.4 meq/L (3.8–5.4 mmol/L). It may be a little higher in the newborn, but it soon adjusts to adult values.

INCREASED CONCENTRATION (HYPERKALEMIA). Since the concentration of potassium within cells is so great, its concentration in plasma rises when it leaves the cells at a greater rate than the kidney can excrete it. This overload occurs in conditions of anoxia and acidosis. It also occurs when there is a decreased output of urine, with normal intake of potassium. Conditions of shock or circulatory failure usually produce hyperkalemia. Adrenal cortical insufficiency, particularly a decreased production of aldosterone, is accompanied by an elevation of serum potassium. Elevated serum potassium values commonly accompany chronic renal insufficiency because a tubular malfunction interferes with the exchange of sodium for hydrogen or potassium ion and promotes potassium ion retention.

DECREASED CONCENTRATION (HYPOKALEMIA). A decreased concentration of potassium in serum occurs as a result of either a low intake over a period of time or an increased loss of potassium through vomiting, diarrhea, gastrointestinal fistulas or long-term therapy with diuretics. The fluids of the gastrointestinal tract contain relatively high concentrations of potassium, and their removal or loss can produce serious deficits. Increased secretion of adrenal steroids, primarily aldosterone, results in excessive potassium loss through the kidneys and a low serum potassium concentration. Certain carcinomas that secrete ACTH (adrenocorticotropic hormone) cause a lowering of serum K^+ concentration through stimulation of the adrenal cortex to produce excessive amounts of steroids (p. 282). Some diuretics promote K^+ excretion.

Serum Sodium and Potassium

By Flame Photometry

The sodium concentration in plasma or serum is commonly determined by means of a flame emission spectrophotometer or by ion-selective electrodes. Most of the flame photometers measure the sodium and potassium concentrations simultaneously, using lithium or cesium as an internal standard. The directions for carrying out the tests vary somewhat with the make of the instrument, so the manufacturer's directions should be followed. The principles are the same, however, for all makes and types of flame photometers.

PRINCIPLE. A dilute solution of plasma or serum is aspirated into a hot flame. Some of the atoms of the alkali metal group (sodium, potassium, lithium) are temporarily activated in the hot flame as the electrons move into a higher orbit. Upon return to the ground state, they emit the light characteristic for the particular element. With a sodium filter (590 nm), the intensity of the emitted light is proportional to the concentration of the sodium ion. Most of the instruments have three separate phototubes with interference filters for Na^+, K^+, and Li^+, respectively, positioned around the emitted light beam; the Na^+ and K^+ concentrations are measured simultaneously, using the intensity of the lithium light as an internal standard.

Serum is usually diluted 1:100 or 1:200 with a dilute solution of nonionic detergent, depending upon the particular instrument used. When lithium is employed as an internal standard, a specified concentration of $LiNO_3$ or LiCl is incorporated into the diluting solution. The detergent facilitates the aspiration of sample and the formation of an aerosol prior to introducing the dilute sample into the flame, and the internal standard serves to minimize the effects of fluctuating gas pressure upon emission light intensity (p. 36).

The main areas in which differences may be found among commercially available flame photometers are the following:

1. Use of lithium or cesium as an internal standard.
2. Manual or automatic dilution of serum.
3. Type of fuel mixture (natural gas or propane with air or oxygen).
4. Manual or automated introduction of sample.

PRECAUTIONS. No matter which instrument is used, the following precautions must be observed:

1. All glassware used in preparing solutions and standards must be scrupulously clean and rinsed with deionized water. Bottles of polyethylene or borosilicate glass should be used for storage of solutions. Soft glass containers must not be used because sodium ion can be leached out of the glass.
2. Disposable plastic cups are recommended for sample handling in order to reduce the possibility of contamination by washing solution.
3. The aspirator line and burner should be flushed and cleaned periodically.
4. Proper pressures of gas and air (or oxygen) must be maintained.
5. As with all quantitative analyses, the pipetting of sample and diluent must be accurate and reproducible.
6. Standards and control sera must be run frequently in order to check on the various aspects of the system.

By Ion-Selective Electrodes (p. 58)

Various automated instruments are available commercially that measure all or some of the serum electrolytes by means of ion-selective electrodes.

PRINCIPLE. Serum is diluted in a high ionic strength buffer solution. The Na electrode has a glass membrane that selectively exchanges Na^+ 300 times as rapidly as K^+ and that is insensitive to H^+. By contrast, the K electrode is coated with a valinomycin membrane that exchanges K^+ about 1000 times as fast as Na^+ and that is insensitive to H^+. As Na^+ and K^+ enter their respective electrodes, a potential (voltage) change occurs that is directly proportional to the selective ion concentration. The electrodes are calibrated with solutions of known concentrations.

The serum Na^+ concentrations by ion-selective electrode usually run about 1 to 2 meq/L lower than those obtained by flame photometry; this discrepancy also gives a lower value for the anion gap, a figure calculated by subtracting the sum of $[Cl^-]$ and $[HCO_3^-]$ from $[Na^+]$.

The serum K^+ concentration by ion-selective electrode correlates very closely with the value determined by flame photometry.

Sodium and Potassium in Body Fluids

The concentration of Na^+ and K^+ in other body fluids is determined by flame photometry or ion-selective electrodes in a manner similar to that in serum; the dilution may have to be modified according to the concentration of these ions. Usually, the Na^+ concentration in cerebrospinal fluid (CSF), exudates, transudates, and in juices collected from various types of fistulas (pancreatic, duodenal, bile) is within the range of the instrument if treated as serum. The same applies to K^+, except that fluid from the ileum may have to be diluted 2 or 3 times more than serum to be in range because values for K^+ in this fluid may vary from 6 to 29 meq/L.

In urine, however, the amount excreted depends more directly upon the intake, particularly for Na^+.

On an average diet, the 24-hour excretion of K^+ usually varies between 30 and 90 mmol, but it certainly could go higher. For Na^+ the 24-hour excretion on a usual diet may vary from 40 to 220 mmol, but it could fall to low levels for the patient with a severely restricted salt intake. With some instruments, the urinalysis for Na^+ and K^+ can be performed exactly as for serum except that different standards are used, ones that are close to the urine concentration.

CHLORIDE

Chloride is the extracellular anion in the highest concentration in serum and plays an important role in the maintenance of electrolyte balance, hydration, and osmotic pressure. Since Cl^- cannot accept H^+ at physiologic pH (HCl is a strong acid), it cannot act as a buffer. Its concentration does vary inversely with HCO_3^- at times because electrochemical neutrality must be maintained always. In metabolic acidosis, the $[Cl^-]$ rises as the $[HCO_3^-]$ decreases, whereas the reverse is true in metabolic alkalosis. With the exception of the red cell, chloride ion does not enter cells and is confined to the extracellular space.

About 2.5 g of chloride are ingested daily in the normal diet, as a salt of Na, K, Ca, or Mg. Chloride is readily absorbed in the intestine and is removed from the body by excretion in the urine and in sweat. Excessive amounts of both Na and Cl may be lost during periods of intense perspiration, requiring a supplementary intake of NaCl to prevent a deficit from occurring.

Serum chloride concentration is usually measured when an electrolyte determination is requested, but it provides the least clinical information of the four constituents. The best estimate of the plasma osmolality is represented by the $[Na^+]$, which usually constitutes about 90% of the total cations. The main benefit of the Cl^- determination is in making an estimate of the residual fraction or ion gap that remains when the sum of the Cl^- and CO_2 concentrations is subtracted from the $[Na^+]$. Under normal circumstances, the ion gap is about 8 to 12 meq/L, but can rise to much higher values when other anions, such as lactate or those from keto-acids, increase. If the ion gap is low in a whole series of determinations on a group of patients, the chances are great that there is some analytical error in one of the constituents. A large ion gap, however, is to be expected in conditions in which organic acids, HPO_4^{--} and SO_4^{--} increase, such as in severe renal disease, in severe diabetic acidosis, or in lactic acidosis. It is important to measure the $[Cl^-]$ in hypokalemic alkalosis because the condition cannot be corrected without the addition of both potassium and chloride ions. When the plasma $[K^+]$ is low, K^+ diffuses from tissue cells, including the renal tubular cells, and is replaced by H^+. This makes more H^+ and less K^+ available for excretion into the urine. An increase of $[H^+]$ in the urine leads to an effective increase in the reabsorption of HCO_3^-; the loss of H^+ and the increase of HCO_3^- cause the alkalosis, which cannot be reversed until K^+ become available for entry into cells and Cl^- is increased to displace HCO_3^-.

REFERENCE VALUES. The normal Cl^- concentration is 98 to 108 meq/L (98–108 mmol/L).

INCREASED CONCENTRATION. High concentrations of Cl^- are usually found in dehydration, certain types of renal tubular acidosis, and in patients who lose CO_2 by overbreathing (respiratory alkalosis) following stimulation of the respiratory center by some drugs, hysteria, anxiety, or fever.

DECREASED CONCENTRATION. Decreased concentrations of serum Cl^- occur in metabolic acidosis of various types. In uncontrolled diabetes, there is an overproduction of keto-acids whose anions replace Cl^-; in renal disease, phosphate ion retention accompanies impaired glomerular filtration, with a concomitant decrease in plasma Cl^- concentration. A deficit of body Cl^- and a decreased serum $[Cl^-]$ accompany prolonged vomiting, whether caused by pyloric stenosis or high intestinal obstruction. Gastric secretions contain a high concentration of H^+ and Cl^-. Low values are also found in salt-losing nephritis and in metabolic alkalosis, in which the $[HCO_3^-]$ is increased and $[Cl^-]$ falls reciprocally. Low values are usually encountered during a crisis in Addison's disease (adrenal cortical deficiency); both Na^+ and Cl^- are low.

Serum Chloride

Serum chloride may be determined by one of several methods: the Cotlove coulometric titration (semiautomated); the Schales and Schales mercuric titration (manual); the continuous flow colorimetric method employing mercuric thiocyanate, or the ion-selective electrode.

Coulometric Method

REFERENCE. Cotlove, E.: Chloride. *In* Standard Methods of Clinical Chemistry, Vol. 3, edited by D. Seligson. New York, Academic Press, 1961, p. 81.

PRINCIPLE. The coulometric method of Cotlove is semiautomated and employs an instrument that generates silver ions at a constant rate from a silver wire anode that is immersed in a solution containing the sample to be measured (serum, urine, cerebrospinal fluid). The silver ions combine with the chloride ions in the sample to form the very insoluble AgCl salt. When Cl$^-$ is completely precipitated, the first excess of Ag$^+$ greatly increases the conductivity of the solution, and the subsequent current surge triggers a relay circuit to shut off the current and stop an electric timer. This provides an accurate measure of the time of the current flow or, when properly calibrated, can give a direct readout of the chloride concentration. The amount of silver ion generated is proportional to the time of current flow when the current strength is kept constant; the [Cl$^-$] is equal to the [Ag$^+$] generated when corrected for a blank determination.

PROCEDURE. Because the procedure differs slightly, according to the make of the instrument (American Instrument, Buchler, Corning, London Co.) the manufacturer's directions should be followed. In general, 0.10 ml of sample is placed in a vial together with 4 ml of an acid-gelatin mixture. The silver anode is immersed in the solution, and the current flow and timer are started while the solution bathing the electrode is mixed. The current and timer are stopped automatically at the end point. An adjustment (blank delay) is made for the short titration time of a chloride-free (blank) solution. The instrument is calibrated by running a standard solution containing 100 meq/L of Cl$^-$. For instruments that have a direct readout, the readout is adjusted to 100 meq/L with the standard after a blank correction. Some of the older instruments may merely indicate the elapsed time of the titration in seconds. The Cl$^-$ concentration of an unknown is then equal to $t_u/t_s \times 100$, where t_u is the titration time of the unknown in seconds and t_s is the titration time of the 100 meq/L standard.

Titration Method

REFERENCE. Schales, O.: Chloride. *In* Standard Methods of Clinical Chemistry, Vol. 1, edited by M. Reiner. New York, Academic Press, 1953, p. 37.

PRINCIPLE. Chloride ions combine with mercuric ions to form soluble, undissociated mercuric chloride. The proteins in serum are precipitated with tungstic acid (Folin-Wu precipitant), and an aliquot of the filtrate is titrated with an acidic solution of mercuric nitrate, using s-diphenylcarbazone as an indicator. The indicator turns to a violet-blue color with the first excess of Hg^{2+}.

REAGENTS

1. Hg(NO$_3$)$_2$ Solution, approximately 8.5 mmol/L. Dissolve 2.9 to 3.0 g of reagent grade Hg(NO$_3$)$_2 \cdot$H$_2$O in about 200 ml of water in a 1000-ml volumetric flask. Add 20 ml of 2 mol/L HNO$_3$ and make up to volume with water.
2. Chloride Standard, 10.0 mmol/L. Dry reagent grade NaCl at 120°C. Dissolve 584.5 mg in water and make up to 1000 ml volume.
3. Standardization of Hg(NO$_3$)$_2$. Titrate 2-ml portions of the chloride standard with Hg(NO$_3$)$_2$, as described for serum.

The concentration of the Hg^{2+} solution is calculated as follows:

$$[Hg^{2+}] = \frac{\text{ml standard Cl solution}}{\text{ml Hg solution}} \times 10.0 = \frac{20.0}{\text{ml Hg}^{2+} \text{ solution}} \text{ in meq L}$$

If the titration of the standard Cl solution required 1.20 ml of the Hg solution, the concentration of the latter equals 20/1.2 = 16.7 meq/L or

8.35 mmol/L. The Hg solution is stable when refrigerated, but it is good practice to check the standardization each time the test is run.

4. H_2SO_4, 41.6 mmol/L: Transfer 2.35 ml concentrated H_2SO_4 to approximately 700 ml of water in a 1000-ml volumetric flask, with stirring. Cool and make up to volume with water. The concentration is 1/12 normal or 41.6 mmol/L.

5. Sodium Tungstate, 100 g/L: Dissolve 100 g of $Na_2WO_4 \cdot 2H_2O$ in water and make up to 1000 ml volume.

6. Indicator Solution, 1 mg/ml in ethanol. Dissolve 100 mg s-diphenylcarbazone in 100 ml 95% ethanol. Store in refrigerator in a brown bottle. Stable for about 1 month.

Note: The protein precipitation reagents (#4 and #5) are those of Folin and Wu as modified by Haden,[1] who combined the 7 volumes of water used for dilution with the sulfuric acid solution and reduced its concentration from 0.333 mol/L to 0.0416 mol/L. Eight volumes of Haden reagent are taken per volme of serum; this provides the same amount of acid and water as in the Folin-Wu method but eliminates one pipetting.

PROCEDURE

1. Transfer 0.5 ml serum to a test tube. Precipitate the proteins by adding 4 ml of sulfuric acid reagent and 0.5 ml of sodium tungstate solution. Mix well, let stand 5 minutes and centrifuge.

2. Transfer 2.0 ml of supernatant fluid to a suitable titration vessel, add 2 drops of diphenylcarbazone indicator, and titrate to a violet end point with the mercuric nitrate solution. Use a microburet with a fine glass or plastic tip.

   Serum may be titrated without prior deproteinization, but the end point is not as sharp.

CALCULATION. The calculation of the serum chloride concentration must take into account the dilution of the sample from deproteinization. Using the $Hg(NO_3)_2$ from the example above:

$$\text{Serum } [Cl^-] \text{ in meq/L} = \frac{\text{ml of Hg solution} \times [Hg^{2+}]}{\text{ml serum taken}} = \frac{\text{ml Hg} \times 16.7}{0.5 \times \frac{2}{5}}$$

$$= \frac{\text{ml Hg} \times 16.7}{0.2} = \frac{83.5 \times \text{ml Hg to titrate}}{\text{serum sample}}$$

If a serum sample required 1.25 ml of Hg solution, the chloride concentration is:

$$83.5 \times 1.25 = 104 \text{ meq/L}$$
$$= 104 \text{ mmol/L}$$

Colorimetric Method with $Hg(SCN)_2$

PRINCIPLE. Mercuric thiocyanate $[Hg(SCN)_2]$ is a poorly soluble, relatively unionized salt that does dissociate more than mercuric chloride, however.

Serum Cl⁻ is determined by adding serum to a mixture of $Fe(NO_3)_3$ and $Hg(SCN)_2$. The Hg^{2+} dissociates from the $Hg(SCN)_2$ to form the less ionized $HgCl_2$, and liberates an equivalent amount of SCN⁻ in the process. Fe^{3+} combines with 3 SCN⁻ to form a colored complex whose absorbance is measured at 480 nm. The [Cl⁻] is proportional to the color intensity and is obtained by comparing the absorbance with that produced by appropriate standards.

The colorimetric method can be used by continuous flow analyzers[2] or by discrete analyzers. Details differ according to the instrumentation used.

Cerebrospinal Fluid (CSF) Chloride

The chloride concentration in normal CSF is higher than in serum because the protein concentration in cerebrospinal fluid is low; hence, there are practically no proteinate anions. The normal concentration of Cl⁻ in CSF ranges from 120 to 132 meq/L. The [Cl⁻] in cerebrospinal fluid falls to approximately that of serum in cases of bacterial meningitis when the protein concentration in cerebrospinal fluid becomes greatly elevated. The CSF Cl⁻ concentration is determined as in serum. Protein precipitation may be omitted in the Hg titration method.

Urine Chloride

The amount of urinary Cl⁻ varies greatly with the intake. An adult eating an average diet may excrete from 110 to 250 meq/d. Patients on low salt diets, however, excrete very little Cl⁻.

Procedure

The test should be performed only upon an accurately timed collection. Test the pH of the urine and adjust to approximately pH 3 with dilute nitric acid. Then measure the chloride concentration as for serum. For a sample containing a high concentration of chloride, dilute as necessary to bring into the range at which an accurate analysis can be carried out.

Sweat Chloride

Significant amounts of Na⁺ and Cl⁻ appear in sweat even though the concentration is much lower than in serum. There is a congenital disease, cystic fibrosis, in which the concentration of these ions is significantly elevated over normal. A sweat Cl⁻ test is usually requested to screen for this disease when suspected. The test is performed primarily on a pediatric population.

It is not easy to collect an adequate sweat sample from a small child, and an attempt to induce total body sweating is dangerous. An instrument is available, however, to introduce a sweat-inducing drug, pilocarpine, into a limited area of skin by means of an electric current flowing between two electrodes attached to a limb (child) or back (infant). This technique, called iontophoresis, moves the pilocarpine from a pad under the positive electrode into the skin toward the negative electrode. Local sweating will be induced in the skin area where the pilocarpine penetrates. After 5 minutes of iontophoresis, the current is turned off, and the electrodes are removed. The pilocarpine area is quickly sponged off and treated as follows, depending upon the method of measuring the sweat chloride. These are:

Ion-Selective Electrode Method[3]

An instrument (Orion) is commercially available that induces local sweating by iontophoresis of pilocarpine into a small area of skin and then measures the sweat Cl^- with an electrode that is selective for chloride ion. If this instrument is available, then carry out the following steps:

1. Introduce pilocarpine into the skin (forearm, leg, or back) by means of iontophoresis (electric migration).
2. Place a small plastic cap over the pilocarpine area; tape down so that the sweat can accumulate and not evaporate.
3. Standardize the electrode with known Cl standards of 20 and 100 mmol/L. Then test a series containing 10, 40, and 60 mmol/L to see that it is reading properly.
4. After 10 minutes, remove the cap from the skin and quickly place the electrode on the moist skin and take a reading.

Chloridometer (See Coulometric Method for Serum Cl)

1. After removing the electrode after the iontophoresis, sponge the skin and wipe dry.
2. With forceps, place an accurately weighed (to 0.1 mg) 25-mm filter paper disc on the pilocarpine area, cover with the plastic cap, and tape down. Prior to this, the filter paper is weighed in a closed vial or weighing bottle.
3. After 15 minutes, quickly remove the plastic cap, and with forceps place the damp filter paper in the weighing bottle, close the cover, and weigh. Calculate the weight of sweat. With the chloridometer, the chloride concentration in as little as 10 mg of sweat can be determined, but the results are more satisfactory if 50 mg or more can be obtained.
4. Two ml of the acid-gelatin solution are added to the weighing bottle containing the filter paper which is immersed in it. After 15 minutes with agitation, the filter paper is squeezed and removed, and the chloride concentration is determined as for serum on the chloridometer.

Hg Titration (At least 50 mg of sweat are necessary)

1. Carry through steps 1, 2, and 3 as in the chloridometer method. Extract the chloride in the filter paper with 2 ml of deionized water acidified with a drop of dilute HNO_3.
2. Titrate with the $Hg(NO_3)_2$, using diphenylcarbazone as an indicator.
3. Modify the calculation in accordance with the amount of sweat taken.

NORMAL VALUES. Normal sweat chloride concentration varies from about 5 to 40 meq/L (5 to 40 mmol/L).

ELEVATED VALUES. The sweat Cl concentration of most of the patients with cystic fibrosis is above 60 mmol/L and may reach as high as 100 to 140 mmol/L. Those with a concentration in the range of 35 to 50 or 60 mmol/L should be rechecked upon several occasions to make certain of the result; cystic fibrosis is too serious a diagnosis to make without adequate laboratory verification.

ACID-BASE HOMEOSTASIS AND BLOOD BUFFER SYSTEMS

A review of some definitions would be of benefit before proceeding with acid-base homeostasis:

1. *Buffer Base (BB):* The sum of the concentrations of *buffer anions* present in whole blood, principally of bicarbonate, phosphate, hemoglobin, and plasma protein. Normal arterial BB = 46 to 50 meq/L.
2. *Base Excess (BE):* The deviation of buffer base from the normal. Normal arterial BE = 0 to ± 2.5 meq/L.
3. *Oxygen Tension (pO$_2$):* The partial pressure of oxygen in blood.
4. *Oxygen Content:* The volume occupied by the oxygen bound to hemoglobin plus that dissolved in 100 ml of whole blood, when completely liberated. It is expressed as ml/100 ml blood.
5. *Oxygen Saturation (Capacity):* The actual O$_2$ content of blood expressed as a percentage of the O$_2$ content of the same blood when fully oxygenated. The latter is obtained experimentally by exposing the blood in a rotating, thin film to air and then measuring its O$_2$ content.

The pH of pure water that has been boiled to eliminate any CO$_2$ absorbed from the atmosphere is 7.0. The pH of the extracellular fluids of the body, however, is 7.4 ± 0.05, which is slightly on the alkaline side of neutrality. Acidosis and alkalosis are terms indicating relative acidity and alkalinity, respectively, compared to the normal blood range. Acidosis refers to a blood pH < 7.35 while alkalosis denotes a pH > 7.45.

Blood Buffer Systems

When fats, carbohydrates, and proteins are catabolized for energy purposes, the carbon atoms in the molecules are converted to CO$_2$ if the oxidation is complete. Although CO$_2$ forms a weak acid, H$_2$CO$_3$, when dissolved in water, the reaction is reversed in the lung alveoli, and the CO$_2$ is rapidly eliminated by the lungs during respiration. This reversal is fortunate for us because it entails the elimination of approximately 20,000 mmoles of CO$_2$ per day. Incomplete oxidation of metabolites, however, causes the formation of nonvolatile acids (those that cannot be exhaled); this imposes the necessity for the kidney to eliminate about 50 mmoles of acid daily when all systems are functioning normally. The amount of acid may be greatly increased in certain diseases. The kidney is the organ involved in the excretion of nonvolatile acids, and the lung is responsible for the elimination of H$_2$CO$_3$ as the latter decomposes into CO$_2$ and H$_2$O.

When fats, proteins, and carbohydrates are catabolized, the hydrogen atoms of the carbon chains are converted to H$^+$. The H$^+$ is transported with an electron

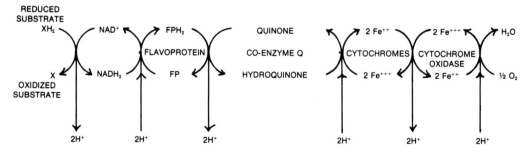

Fig. 3.1. Transport of electrons and H$^+$ in the respiratory chain in mitochondria. Reduced substrates such as lactate and malate are oxidized to pyruvate and oxaloacetate, respectively.

along a chain of coenzymes in mitochondria (the respiratory chain) until finally the hydrogen ions with electrons are transferred to atmospheric oxygen to form water (Fig. 3.1). This vital process is known as electron transport. Under normal circumstances, the supply of oxygen is adequate and there is no appreciable increase in H^+. In strenuous exercise, however, there may be a temporary shortage of O_2, with local buildup in the concentration of H^+.

In conditions of oxygen deficiency (anoxia), acidosis develops, owing to the body's inability to transfer all of the H^+ to O_2 to form water. Anoxia may be caused by poor gaseous exchange in the lungs, an obstruction in the airways, or by a poor delivery system (low blood pressure, heart failure, severe anemia). Hence, the development of a system for reducing the concentration of H^+ was a biologic necessity in order to protect tissue cells and vital functions. Several buffer systems preserve the internal environment by reducing the concentration of H^+. This reduction is accomplished by the reaction of buffer salts with H^+ to form a weak (relatively undissociated) acid. The principal body buffer systems are the following:

Bicarbonate/Carbonic Acid Buffer System

$H_2CO_3 \longleftrightarrow H^+ + HCO_3^-$ is the most important buffer system in plasma because of (1) the high concentration of HCO_3^- and (2) the readiness with which H_2CO_3 may be increased through diminished lung activity and decreased by blowing off CO_2 through increased pulmonary ventilation. The equilibria are the following:

$$\text{Equation 1: } CO_2 + H_2O \xrightarrow{\text{(a)}} H_2CO_3 \xrightarrow{\text{(b)}} H^+ + HCO_3^-$$

Reaction (a) proceeds very slowly to the right in plasma, but carbonic anhydrase, an enzyme in erythrocytes and renal tubular cells, greatly speeds up the formation of H_2CO_3. The equilibrium point of reaction (b) is far to the right, so the net effect in renal and red cells is the formation of $H^+ + HCO_3^-$ from CO_2 and water as both reactions proceed to the right. The increased $[HCO_3^-]$ in red blood cells is followed by the diffusion of HCO_3^- into plasma, accompanied by the passage of Cl^- into the erythrocyte to maintain ionic balance. This is the chloride shift, a process reversed in the lungs when the reactions in Equation 1 proceed to the left as CO_2 is exhaled. CO_2 is always being produced through the catabolism of fat, carbohydrate, or protein. Its rate of elimination by the lungs depends upon the rate and/or depth of respiration.

Hemoglobin Buffer System

Hemoglobin, the protein present in high concentration in red blood cells (erythrocytes), binds oxygen in the lungs and releases it in tissues. The oxygenated form of hemoglobin is a much stronger acid than the deoxygenated form. Consequently, when the oxygenated hemoglobin ($KHbO_2$) reaches the tissues where CO_2, and therefore H_2CO_3 are generated, the following reaction takes place:

$$KHbO_2 + H_2CO_3 \rightarrow HHb + K^+HCO_3^- + O_2$$

The weak acid, deoxygenated hemoglobin (HHb), is formed as O_2 is released and made available to the tissues. As described above, the HCO_3^- diffuses into plasma and is replaced in the red blood cell by Cl^-. The reaction is reversed in the lungs. This buffer system accounts for about 30% of the buffering capacity of whole blood, and the bicarbonate system accounts for about 65%; in plasma, however, the bicarbonate system supplies about 95% of the buffering capacity.

Phosphate Buffer System

The phosphate buffer system is a minor component of the total blood buffer system, but it does play an important role in the elimination of H^+ in the urine.

$$HPO_4^{--} + H^+ \longleftrightarrow H_2PO_4^-$$

In the plasma at pH 7.4, 80% of the phosphate is in the form of HPO_4^{--}, but in an acid urine, the bulk of it exists as $H_2PO_4^-$ as the above reaction shifts to the right.

Plasma proteins can also accept H^+ to a minor extent and, therefore, serve as a buffer, but the total buffering capacity from this source is negligible compared to the contributions of bicarbonate and hemoglobin.

Effects of CO_2, HCO_3^-, and H_2CO_3 upon pH

As related before, CO_2 is the ever-present product of oxidative metabolism. It is also the source of formation of plasma HCO_3^- as H_2CO_3 dissociates into H^+ and HCO_3^-. The H^+ may react with a buffer salt to form an undissociated acid, as happens in erythrocytes and body fluids, or be excreted by the renal tubular cells into the urine either in exchange for Na^+, or as an undissociated keto-acid. The H^+ in the lumen usually converts HPO_4^{--} to the less dissociated form, $H_2PO_4^-$; in acidosis when NH_3 is formed and excreted by tubular cells, H^+ combines with NH_3 to form NH_4^+. Under normal circumstances, there are no keto-acids available for excretion, and little NH_3 is formed by the kidney, but both processes assume great importance when acidosis and ketosis are present.

The bicarbonate buffer system is the dominant buffer system in plasma. The pH of the plasma depends upon the ratio $[HCO_3^-]/[H_2CO_3]$ as indicated by the Henderson-Hasselbalch equation (p. 24):

$$pH = 6.1 + \log \frac{[HCO_3^-]}{[H_2CO_3]}$$

Since the normal pH of plasma is 7.4, the log of $[HCO_3^-]/[H_2CO_3]$ must be 1.3. The antilog of 1.3 is 20, which means that the pH of plasma is 7.4 when the ratio of $[HCO_3^-]/[H_2CO_3]$ is 20:1. In normal adults, the concentration of H_2CO_3 is usually 1.35 mmol/L; the concentration of HCO_3^- is 20 times as great, or 27 mmol/L.

Deviations from the normal blood pH occur when the ratio of $[HCO_3^-]/[H_2CO_3]$ departs from the 20:1 ratio. In metabolic acidosis, the concentration of HCO_3^- falls more than 20 times as much as the H_2CO_3 concentration, thereby lowering the blood pH. In metabolic alkalosis, the $[HCO_3^-]$ increase greatly exceeds that of $[H_2CO_3]$, with a consequent rise in pH. Thus, metabolic acidosis is characterized by a decrease in $[HCO_3^-]$, and metabolic alkalosis by an increase of $[HCO_3^-]$.

The concentration of H_2CO_3 responds rapidly to the rate and depth of respiration, which is another way of saying that it is responsive to the rate at which CO_2 is eliminated by the lungs. Changes in pH attributed primarily to increases in $[H_2CO_3]$ are referred to as respiratory acidosis and as respiratory alkalosis when $[H_2CO_3]$ is decreased. Respiratory acidosis may occur if there is an impediment to gaseous exchange in the lungs, as in pneumonia (fluid in the lungs), blockage of air passages, or in central nervous system depression of respiration which could happen as a result of overdosage with opiate drugs or other depressants. In these instances, the concentration of H_2CO_3 in blood rises proportionately more than that of HCO_3^-, so the ratio $[HCO_3^-]/[H_2CO_3]$, falls even though the $[HCO_3^-]$ may be higher than normal. The reverse effect upon blood pH occurs in situations of overbreathing, whether caused by anxiety, drugs (as in aspirin overdosage), or by central nervous system stimulation. Even though the concentration of HCO_3^- may be lower than normal, the concentration of H_2CO_3 falls even more, relative to the $[HCO_3^-]$, causing an increase in the ratio of $[HCO_3^-]/[H_2CO_3]$ and hence in the pH.

Primary changes in $[HCO_3^-]$ arising from alterations in nonvolatile acids or bases cause a disturbance in the $[HCO_3^-]/[H_2CO_3]$ ratio, thereby producing a *metabolic* (nonrespiratory) acidosis or alkalosis.

Compensation of Acid-Base Disturbances

Disturbances in blood pH are usually compensated to a greater or lesser extent by appropriate responses of the respiratory and renal systems, insofar as they are able. For example, an accumulation of nonvolatile acids, as in renal failure or diabetic ketosis, results in a metabolic acidosis manifested by a decreased $[HCO_3^-]$; to minimize the resultant fall in pH, pulmonary ventilation is increased, which reduces the $[H_2CO_3]$, increases the $[HCO_3^-]/[H_2CO_3]$ ratio, and raises the pH. The kidneys also try to compensate by increasing the rate of formation of NH_3 from glutamine and excreting NH_4^+ salts in the urine. Conversely, metabolic alkalosis may be compensated in whole or in part by renal excretion of HCO_3^- and depression of the respiration rate, both of which decrease the $[HCO_3^-]/[H_2CO_3]$ ratio and lower the pH. The compensatory changes produced by alterations in respiratory rate are rapid in contrast to the relatively slow modifications induced by renal changes.

CO_2 is transported in the blood by both erythrocytes and plasma. It exists in the form of HCO_3^-, H_2CO_3, and carbamino-bound CO_2. This latter fraction consists of CO_2 reacting with the free amino group of a protein thusly:

$$CO_2 + R\text{-}NH_2 = R\text{-}NHCOOH$$

The amount of CO_2 bound as a carbamino complex in serum proteins is small, but the amount carried as a carbamino group in hemoglobin may amount to 10 to 30% of the total. The carbamino CO_2 cannot be measured as a separate entity in a routine clinical chemistry laboratory but usually appears lumped together with the concentration of HCO_3^-.

The following relations exist:

$$\text{Equation 2: Total } CO_2 \text{ or } CO_2 \text{ content} = [HCO_3^-] + [H_2CO_3]$$

where H_2CO_3 represents the sum of the concentrations of undissociated H_2CO_3 and CO_2, which is physically dissolved in the plasma. Since both of these concentrations are proportional to the partial pressure of CO_2, designated as pCO_2, then $[H_2CO_3] = k \times pCO_2$. The value for k has been determined and appears in Equation 3.

Equation 3: $[H_2CO_3] = 0.0301 \times pCO_2$
Equation 4: When Equations 2 and 3 are combined, the following is obtained:
$[HCO_3^-] = CO_2$ content $- [H_2CO_3] = CO_2$ content $- 0.0301 \times pCO_2$

Substitution of the above in the Henderson-Hasselbalch equation (p. 96) gives:

$$\text{Equation 5: pH} = 6.1 + \log \frac{CO_2 \text{ content} - 0.0301 \times pCO_2}{0.0301 \times pCO_2}$$

All three components, pH, $[HCO_3^-]$, and pCO_2 can be measured in the laboratory. If only two are measured, the third can be obtained by calculation.

Blood pH

REFERENCE. Gambino, S.R.: pH and pCO_2. Standard Methods of Clinical Chemistry, Vol. 5, edited by S. Meites. New York, Academic Press, 1965, p. 169.

PRINCIPLE. The pH is usually measured directly upon heparinized blood collected anaerobically. Modern instruments have a probe or capillary tip that leads directly to a micro glass electrode enclosed in a water jacket maintained at 37° C. Directions for operating the instrument vary with the make, but they all operate upon the same principle.

An electric potential is generated when a thin glass membrane separates two solutions of different H^+ concentration. When blood is one of the solutions, the difference in potential between that of the glass electrode and a reference electrode is measured, amplified, and converted into a direct pH reading on the meter. The instruments in common use register the pH directly by means of a digital readout (p. 55).

REAGENTS AND MATERIALS. The pH meter is usually calibrated with each of two buffers whose pH can be prepared with exactitude and which lies within the physiologic range (6.80 to 7.80). These may be obtained commercially or can be prepared in the laboratory.

1. Phosphate Buffer, pH 7.386 at 37° C and 0.1 ionic strength. Dissolve 1.179 g of KH_2PO_4 and 4.302 g of Na_2HPO_4 in distilled water that contains no CO_2 or ammonia and dilute to 1000 ml at 25° C.
2. Standard Phosphate Buffer, pH 6.840 at 37° C and 0.1 ionic strength. Dissolve 3.388 g KH_2PO_4 and 3.533 g of Na_2HPO_4 in water containing no CO_2 or ammonia and dilute to 1000 ml at 25° C.
3. Distilled water containing no CO_2, NH_3, bacteria, or mold.
4. NaCl 8.5 g/L for rinsing purposes.
5. Electrode Wash Solution, 8.5 g/L NaCl. Add 1 ml or 1 g of concentrated neutral detergent per 100 ml saline solution. (Dreft or Lux liquid detergent is satisfactory.)

COLLECTION OF BLOOD. Arterial blood is desired for measurement whenever pO_2 is included in the group. Venous blood may be taken for just pH and pCO_2

if blood is drawn without stasis (no tourniquet) and without the patient clench-ing the fist. "Arterialized" venous blood may be obtained by heating the hand and forearm in water at 45° C for 5 minutes and then drawing the blood from the dilated veins on the back of the hand. Capillary blood is arterialized by warming the ear, finger, or heel at 45° C before taking the sample. When drawing blood with a needle and syringe, the dead space in the needle should be filled with sterile anticoagulant to prevent the formation of an air bubble. To keep the blood anaerobic, the tip of the needle should be occluded by sticking it into a cork or rubber stopper immediately after drawing the sample. The needle and syringe with the blood sample should be placed in a tray of ice and sent to the laboratory for analysis where it should be kept in the ice bath until analyzed. The sample should be well mixed before measurement of the pH.

PROCEDURE

1. Warm up the pH meter and water bath for at least 15 minutes before analyzing a sample.
2. Check for proper grounding and for cleanliness of all surfaces.
3. Introduce the 6.840 buffer into the electrode and calibrate the meter with this buffer.
4. Remove the buffer by washing out with saline solution.
5. Replace with the 7.386 buffer and set the meter reading to 7.386 ± 0.010. Recheck first buffer (should read 6.840 ± 0.005).
6. Remove the buffer by washing with saline.
7. Introduce the blood sample without air bubbles and record the meter reading.
8. Remove the blood by washing with saline and prepare for the next blood sample.
9. Recheck the meter reading with 7.386 buffer after completing a blood pH determination. The electrode should be rinsed with distilled water before introducing the standard buffer. If the reading of the buffer is more than ± 0.01 units from the stated value, the calibration and the blood pH measurement should be repeated.
10. After removal of the buffer solution with saline, the electrodes should be washed with the neutral detergent.
11. Rinse with distilled water and leave the electrode filled with either saline or distilled water.

Notes: In all systems, the accuracy of the measurement depends upon the following:

1. Accurate standardization of the pH meter with two different standard solutions at pH levels within those compatible with life. Two solutions that are widely used at present are those at pH 6.84 and 7.38.
2. Collection of blood anaerobically and without stasis so that there is no loss of CO_2 from the blood and no buildup of CO_2 in the vein as the result of stagnant flow.
3. Adequate cleaning of the probe and glass electrode in order to avoid the buildup of a protein film or the growth of bacteria.
4. Accurate and constant maintenance of the temperature of the glass elec-trode, since the pH is temperature dependent; the temperature of the water jacket should be maintained at 37 ± 0.05° C. It is also important to correct the blood pH to that of the patient's body temperature, since significant

deviations occur in those with high fever or in those with a low body temperature. The pH of the blood varies inversely with the temperature so that for patients with a fever, the correction is *negative* while for those in hypothermia (as in open heart surgery) the correction is *positive*. The correction factor for whole blood is $0.0146 \times (37.0\text{-}t°)$ and $0.012 (37.0\text{-}t°)$ for plasma, where $t°$ is the body temperature of the patient.

BLOOD pH REFERENCE VALUES. The normal range of blood pH is 7.35 to 7.45, with a mean in arterial blood of 7.40. Venous blood may be about 0.03 pH units less than arterial.

INCREASED pH. The blood pH is increased when there is a primary increase in $[HCO_3^-]$ as in metabolic alkalosis (usually following prolonged ingestion of antiacids, overdosage with some steroid hormones, or in patients with tumors producing ACTH). It is also present following a primary decrease in $[H_2CO_3]$ or pCO_2 caused by hyperventilation. Some degree of compensation may take place in an attempt to restore a normal ratio of $[HCO_3^-]/[H_2CO_3]$.

DECREASED pH. A decrease in blood pH always occurs when the ratio of $[HCO_3^-]/[H_2CO_3]$ falls below 20:1. This takes place in conditions of metabolic acidosis typified by a fall in $[HCO_3^-]$ (commonly found in chronic renal disease and in uncontrolled diabetes mellitus). The primary defect can also be respiratory, with the greatest change being an increase in $[H_2CO_3]$ or pCO_2; hypoventilation and pulmonary disease are the principal causes.

CO$_2$ Content (Total CO$_2$)

The CO_2 content consists of the sum of the concentrations of dissolved CO_2, undissociated H_2CO_3, and carbamino-bound CO_2; the carbamino fraction is low in serum but appreciable in whole blood because of the presence of hemoglobin. Plasma or serum is usually taken for the determination of CO_2 content because extraneous factors such as erythrocyte count or degree of oxygen saturation do not affect it; the determination of CO_2 content in whole blood is influenced greatly by these latter two factors. In general, the CO_2 content of serum or plasma is obtained by automated, continuous flow colorimetric methods or by ion-selective electrode.

In former years, the CO_2 content was measured manometrically (pressure of a constant volume of CO_2 liberated from a blood sample) in a Natelson microgasometer,[4,5] but electronic blood gas instruments have replaced this cumbersome technique.

Automated, Continuous Flow Colorimetric Method[2]

The CO_2 appearing in serum as HCO_3^-, H_2CO_3, or carbamino-bound is released by the addition of acid. The gaseous CO_2 is dialyzed through a silicone-rubber gas-dialysis membrane into a buffer solution of cresol red at pH 9.2. The CO_2 diffuses through the membrane and lowers the pH of the buffered cresol red solution. The decrease in color intensity is proportional to the CO_2 content and is measured in a spectrophotometer at 430 nm. Other continuous flow methods have used phenolphthalein as the indicator.

Note: The serum cups must be covered to avoid loss of CO_2 while waiting for the analysis to begin. The loss of CO_2 from an uncovered sample cup can be as high as 10% in a 3-hour period.

By Ion-Selective Electrode

A Severinghaus electrode is capable of measuring the serum total CO_2 content in a blood gas instrument or in one dedicated to electrolyte measurements. A measured volume of serum is transferred to a reaction cup containing an electrode coated with a silicone rubber membrane. Serum CO_2 diffuses through the membrane and lowers the pH of a bicarbonate solution between the membrane and the pH electrode. The rate of pH change is proportional to the CO_2 content in the sample. The rate signal is compared to an identical electrode used as a reference and the difference is converted to serum CO_2 concentration. Many different instruments are available commercially.

REFERENCE VALUES. The CO_2 content of serum varies from about 24 to 30 mmol/L for healthy adults. The concentration in premature newborns varies from 18 to 26 mmol/L; for infants it ranges from 20 to 26 mmol/L.

INCREASED CONCENTRATION. The CO_2 content of serum is increased in metabolic alkalosis, in compensated respiratory acidosis, and frequently in the alkalosis accompanying a large potassium deficiency (Table 3.4).

DECREASED CONCENTRATION. The CO_2 content of serum is decreased in metabolic acidosis and in compensated respiratory alkalosis (Table 3.4).

Blood Gases: pH, pCO_2, and pO_2

We have seen in the previous section how intimately involved the plasma CO_2 content is with acid-base balance and the great role that the lung plays in the elimination of CO_2. To recapitulate briefly, the average person produces

TABLE 3.4
Changes in Acid-Base Parameters of Whole Arterial Blood in Acid-Base Imbalances

PATHOLOGICAL STATE	pH		pCO_2		$[HCO_3^-]$		$[H_2CO_3]$	
	UN-COMP.	COMP.*	UN-COMP.	COMP.	UN-COMP.	COMP.	UN-COMP.	COMP.
Metabolic Alkalosis Vomiting; potassium depletion; primary hyperaldosteronism; excessive alkali ingestion	↑	→	→	↑	[↑]	[↑]	→	↑
Respiratory Alkalosis Hyperventilation	↑	[→]	[↓]	↓	→	↓	[↓]	↓
Metabolic Acidosis Uncontrolled diabetes mellitus; severe renal disease; renal tubular failure: severe diarrhea; acute MI*; starvation; NH_4Cl ingestion	↓	→	→	↓	[↓]	[↓]	→	↓
Respiratory Acidosis Pneumonia; emphysema; congestive heart failure; lung lesions; drug overdosage (CNS* depressants); anesthesia	↓	→	[↑]	[↑]	→	↑	[↑]	[↑]

Increase denoted by ↑, decrease by ↓ and no change by →. Arrows enclosed in rectangle show the primary change.
*MI = myocardial infarction; CNS = central nervous system; COMP. = fully compensated

and exhales daily around 20,000 mmoles of CO_2 as fat, carbohydrate, and protein are oxidized for energy purposes.

Pulmonary ventilation is adjusted by reflex action to the body needs for elimination of H_2CO_3 as CO_2 gas; the respiratory control centers are located in the brain (medulla) and in chemoreceptors in the aortic and carotid bodies. The respiratory center and chemoreceptors are sensitive to changes in $[H^+]$ and arterial pCO_2; the latter receptor is also sensitive to changes in arterial pO_2 and alters pulmonary ventilation according to the need as reflected by the O_2 tension. An oxygen defiency (anoxia or hypoxia) results in a localized or even general acidosis because of the inability of the electron transport system to transfer all of the accumulated hydrogen ions plus electrons to molecular oxygen. Approximately 25,000 mmoles of O_2 are required daily. Thus, a partial deficiency in the tissue oxygen supply leads to acidosis, and a more severe deprivation may result in loss of consciousness or death.

Three physiologic conditions must be fulfilled in order to supply the tissues with adequate oxygen:

1. *Introduction into the lung alveoli of an adequate, oxygen-containing mixture.* The altitude is an important factor in oxygen intake, since the pO_2 of the atmosphere decreases with the elevation. In order to bring atmospheric air into the alveoli, the diaphragm and rib muscles of the subject must function properly, and the airway must be free from obstructions.

2. *Properly functioning lung alveoli.* There must be proper gas exchange between the alveolar blood capillaries and the gas contained in the alveoli. Pathologic changes such as fluid accumulation in the lungs or thickened alveolar membranes greatly interfere with the transfer of oxygen into the blood capillaries and the removal of CO_2 from the blood.

3. *A properly functioning loading and distribution system for the gases.* CO_2 is quite soluble in water, but it does require the presence of the enzyme, *carbonic anhydrase,* in the red blood cells to facilitate the conversion to H_2CO_3. O_2 is rather poorly soluble in plasma and is transported almost entirely by binding to hemoglobin in the erythrocyte. A severe anemia greatly restricts oxygen transport because of the lowered red cell count and the hemoglobin deficiency. Even with a normal erythrocyte count, the cardiovascular system must function properly to deliver sufficient oxygen-laden blood to the tissues. The blood pressure and blood flow must be adequate to fulfill the body needs; problems arise when there is coronary insufficiency (heart failure) or inadequate pulmonary circulation.

Thus, there are many conditions in which a knowledge of the state of the gas exchange of O_2 and CO_2 is highly desirable. These include all of the respiratory disorders and situations in which the patient is in a respirator or an oxygen tent, has a tracheotomy, is recuperating from cardiac surgery, or is in a nursery for prematures where the incidence of respiratory problems is high. Knowledge of the pCO_2 when measured with the blood pH is useful in the treatment of acidosis.

In a gas mixture, such as air, each constituent gas exerts its partial pressure independently of the others; the total pressure of the mixture is the sum of the component partial pressures. In an average day at sea level, the air pressure is close to 760 mm Hg. The bulk of this is contributed by nitrogen, an inert gas;

the partial pressure of oxygen is about 20% of the total or 150 mm Hg. Water vapor contributes about 2% of the total pressure, whereas that due to atmospheric CO_2 is only a trace. The gases in the alveolar sacs are not completely exchanged or blown out with each respiration so there is a certain amount of intermixture of newly inspired air with that in the dead space left over from the previous exhalation. This reduces the pO_2 of the alveolar air to approximately 100 mm Hg. Since the saturation of hemoglobin with oxygen is nearly complete in normal circumstances, arterial pO_2 is usually around 95 mm of Hg.

REFERENCE. Fleischer, W.R., and Gambino, S.R.: Blood pH, pO_2 and Oxygen Saturation. Chicago, American Society of Clinical Pathologists, Commission on Continuing Education, 1972.

PRINCIPLE. Measurements are usually made upon arterial blood drawn anaerobically into a syringe that is capped. Capillary blood that has been "arterialized" by intensively warming the area for 5 minutes prior to sampling is a good substitute. A hyperemic ear (for children and adults) or a warmed heel (infants) may be taken as the site for blood collection when arterial blood cannot be obtained readily.

Measurements are made simultaneously of blood pH, pCO_2, and pO_2 on a single blood gas instrument (Corning, Instrumentation Laboratory, Radiometer).

pH. The pH is measured by a micro glass electrode as described on page 55. All electrodes are enclosed and kept at a constant temperature (37° C) by circulating warm water.

pCO_2. The partial pressure of blood CO_2 is measured by a modified pH electrode. The electrode is separated from the blood sample by a thin Teflon or silicone membrane that is permeable to CO_2 gas but not to ions. The CO_2 diffuses into a layer of bicarbonate solution on the other side of the membrane and thereby changes the pH. The glass electrode measures the change in pH; when calibrated with CO_2 gas of known composition, the instrument records the measurement directly as pCO_2.

pO_2. The partial pressure of blood O_2 is measured amperometrically by a Clark electrode. A membrane (Teflon, polyethylene) that is permeable to O_2, but not to most of the blood constituents, separates the electrode from the blood. The electrode consists of a platinum cathode and an Ag/AgCl anode. A negative electric charge is impressed upon the Pt electrode, and the oxygen which diffuses there is rapidly broken down, causing a change in ionic current. The change in current flow between anode and cathode is measured after suitable amplification; conversion of change in current to pO_2 is made by calibrating the electrode with known concentrations of pO_2 (p. 57).

[HCO_3^-] AND CO_2 CONTENT. After measurement of the pH and pCO_2, it is also possible to obtain the [HCO_3^-] and CO_2 content by nomogram (Fig. 3.2) or as a direct readout from a calculation programmed in the instrument. Since the methodology varies somewhat with the particular instrument selected, one should follow the manufacturer's directions. The following precautions should be observed:

1. Heparinized arterial blood should be used when possible.
2. It should be drawn and stored anaerobically and without air bubbles.
3. The sample should be placed in an ice bath to minimize metabolic changes and transported to the laboratory immediately.

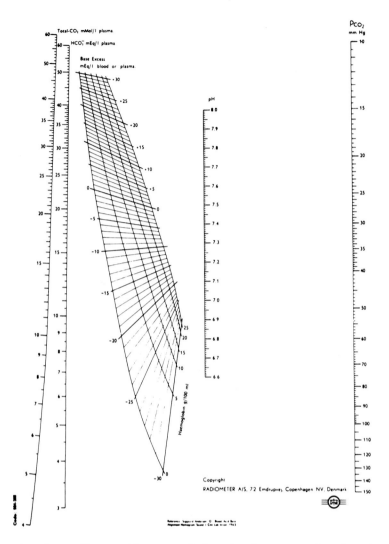

Fig. 3.2. Siggaard-Andersen Alignment Nomogram. By knowing the concentration of any two constituents of the following three—pCO$_2$, pH, HCO$_3^-$ (or CO$_2$ content)—one can obtain the third by laying a straight edge on the two values and reading the third from the proper column. (Courtesy of Radiometer A/S, Copenhagen, Denmark.)

4. The sample should be run as soon as possible.
5. The blood gas instruments must be in good repair, cleaned well between samples, and the membranes surrounding the electrode must be kept moist.
6. The instrument must be checked periodically with known gas mixtures and with tonometered blood with known pCO$_2$ and pO$_2$.

REFERENCE VALUES. Analytic values for acid-base constituents are particularly susceptible to the physiologic activity of the host (whether completely at bed rest or ambulatory), to the specimen collection technique, to the length of time elapsed before the specimen is analyzed, and to instrumental variation (temperature control, calibration, instrument bias) so that it is not surprising that the "normal range" varies even among the first-rate laboratories. The age and sex of the subjects affect some of the parameters. The reference values in Table 3.5 were taken from Fleischer and Gambino because the work was done carefully and is representative of the usually accepted normal ranges.

The direction of the changes in values of the acid-base parameters in various states of imbalance are summarized in Table 3.4. The parameters treated are pH, pO_2, HCO_3^-, and H_2CO_3 concentrations, and the four major states of acid-base imbalance listed are metabolic and respiratory alkalosis and metabolic and respiratory acidosis. Two columns appear under each constituent; in the first appear the changes that would be found in the completely uncompensated pathologic state; the second of the two columns lists the changes when compensation is complete. With many patients, there is only partial compensation and the resulting changes will be intermediate between those listed in the uncompensated and fully compensated columns.

To summarize, the common clinical situations in which the pCO_2 is elevated and the pO_2 decreased are the various respiratory disorders, such as pneumonia, hyaline membrane disease of premature newborns, paralysis of the respiratory muscles, intake of drugs that depress the respiratory center, emphysema, congestive heart failure.

The pCO_2 is depressed in hyperventilation (conscious overbreathing, anxiety states, emotional states), and in various types of fully or partially compensated metabolic acidosis (compensated by overbreathing).

The pO_2 is elevated only when the patients are breathing a gas mixture containing a high percentage of oxygen (> 20% by volume or > 150 mm Hg).

Blood Oxygen Content and Oxygen Saturation

Although it is not part of the "blood gas package," oxygen content and oxygen saturation values are sometimes requested. Most of these requests come from the heart catheterization laboratory, where the information is used for diagnostic purposes. In heart catheterization, a small catheter is inserted into a large blood vessel in the arm and threaded all of the way into various chambers of the heart or other particular positions in that area. The exact location of the

TABLE 3.5
Normal Values (38°C) of Acid-Base and Blood Gas Constituents

CONSTITUENT	ARTERIAL or CAPILLARY WHOLE BLOOD	VENOUS PLASMA (separated at 30°C)	
pH	7.37–7.44	7.35–7.45	
pCO_2			
Men	34–45 mm Hg	36–50 mm Hg	
Women	31–42 mm Hg	34–48 mm Hg	
CO_2 Content (Plasma)			
Men	24–30 mmol/L	26–31 mmol/L	
Women	21–30 mmol/L	24–29 mmol/L	
pO_2	85–95 mm Hg	30–50 mm Hg	Whole
O_2 Saturation	94–97%	50–70%	Blood
O_2 Content	20 ml/dl	15 ml/dl	

From Fleischer, W.R., and Gambino, S.R.: Blood pH, pO_2, and Oxygen Saturation. Chicago, American Society of Clinical Pathologists, Commission on Continuing Education, 1972, p. 30.

catheter tip is ascertained by means of fluoroscopy. Heparinized blood samples are withdrawn anaerobically from specific locations and brought to the laboratory for analysis of their oxygen contents. This may be done while the patient is breathing room air or oxygen, depending upon the circumstances and the desired information.

The oxygen content and saturation of whole blood may be measured by one of several different techniques which vary considerably in the sophistication of the instrumentation and in the length of time required to perform the measurement. These may be done gasometrically,[4] spectrophotometrically, or polarographically. The classic manometric method devised by Van Slyke was the most widely used gasometric method, but in recent years gasometric methods have been supplanted by spectrophotometric and polarographic methods because of greater ease in performance without a sacrifice of accuracy.

By Polarography

The Lex-O_2-Con (Lexington Instruments) has been devised to replace the manometric determination of O_2 content or saturation. This instrument sweeps the oxygen contained in a measured amount of blood to a cell where the O_2 is electrolytically (polarographically) reduced to hydroxyl ion, setting up a flow of electrons. The current change is measured, and after proper calibration, the instrument provides a direct readout of O_2 content. The percentage of O_2 saturation is obtained by calculation after measuring the O_2 content of the blood sample before and after saturating completely with oxygen by exposing a rotating film of blood to the atmosphere.

REFERENCE VALUES. The oxygen content of arterial blood in normal persons varies from 15 to 23 ml/dl blood. Arterial blood is from 94 to 97% saturated with oxygen (Table 3.5).

The % oxygen saturation is also related to the pO_2 and pH in a complex fashion. If the body temperature, pH, and pO_2 are known, the % O_2 saturation can be calculated from nomograms (see reference in next paragraph); modern blood gas instruments calculate the % O_2 saturation from the data they have.

Transmission Spectrophotometry

REFERENCE. Fleischer, W.R., and Gambino, S.R.: Blood pH, pO_2 and Oxygen Saturation. Chicago, American Society of Clinical Pathologists, Commission on Continuing Education, 1972, p. 61.

PRINCIPLE. Oxyhemoglobin and deoxyhemoglobin (reduced hemoglobin) have different light transmission spectra. Upon plotting absorbance against wavelength, the absorbance curves of oxyhemoglobin and deoxyhemoglobin cross at several points, called isobestic points, where the absorbance for oxyhemoglobin and deoxyhemoglobin is identical. The percentage of oxyhemoglobin, which is the same as percentage of oxygen saturation, is obtained by measuring the absorbance of a hemolyzed blood sample at two different wavelengths, one of which is at an isobestic point and one where the difference in absorbance between the oxygenated and deoxygenated states is large. In practice, the choice of 805 nm for the isobestic point and 650 nm for the second wavelength is widely accepted, although other pairs of wavelengths have been employed.

REAGENTS AND MATERIALS

1. A Narrow Bandpass Spectrophotometer, with square cuvets.

2. Silica Precision Cuvets, 10 mm light path, with 9 mm spacers to reduce the effective light path to 1 mm.
3. Triton X-100-Borate Solution. Prepare a solution containing 0.1 mol/L of borate by dissolving 3.8 g of $Na_2B_4O_7 \cdot 10H_2O$ in 100 ml water. Add 3 ml of Triton X-100 to 67 ml of borate solution. Store in a plastic bottle at room temperature.

PROCEDURE

1. Draw up 0.05 ml of Triton-borate solution into a heparinized plastic syringe. Then draw up arterial blood in the same syringe to the 1 ml mark and mix well.
2. Attach a small plastic tube to the tip of the #21 needle on the syringe, and fill a cuvet containing a 9-mm spacer with the hemolyzed blood, starting at the bottom of the cuvet in order to avoid bubbles.
3. Read the absorbance at 650 and 805 nm against a water blank.
4. Calculate the 650/805 absorbance ratio and obtain the per cent oxygen saturation from the calibration curve.

PREPARATION OF CALIBRATION CURVE. Blood is prepared with zero per cent oxygen saturation by blowing nitrogen over a thin film of blood. Blood that is 100% saturated is prepared by exposing a thin film of blood to atmospheric air. The absorbances of both bloods are measured at 650 and 805 nm, and the absorbance ratio 650/805 is calculated. The absorbance ratio is plotted as ordinate against the percentage of O_2 saturation as abscissa and a straight line is drawn between the two points (Fig. 3.3). An instrument is available commercially, the Instrumentation Laboratory Co-oximeter, which measures the absorbance of hemolyzed blood at three different wavelengths. It is programmed to solve three simultaneous equations and to provide as a readout the percentage

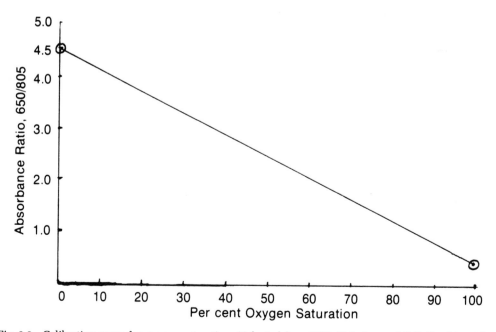

Fig. 3.3. Calibration curve for oxygen saturation. (Adapted from W.R. Fleischer and S.R. Gambino: Blood, pH, pO₂ and Oxygen Saturation. Chicago, American Society of Clinical Pathologists, 1972, p. 62.)

of O_2 saturation and the percentage of carboxyhemoglobin (hemoglobin binding carbon monoxide).

REFERENCES

1. Haden, R.L.: A modification of the Folin-Wu method for making protein-free blood filtrates. J. Biol. Chem. *56*, 469, 1923.
2. Kenny, M.A., and Cheng, M.H.: Rapid, automated simultaneous determination of serum CO_2 and chloride with the "Auto Analyzer I." Clin. Chem. *18*, 352, 1972.
3. Szabo, L., Kenny, M.A., and Lee, W.: Direct measurement of chloride in sweat with an ion-selective electrode. Clin. Chem. *19*, 727, 1973.
4. Natelson, S.: Techniques of Clinical Chemistry, 3rd ed., Springfield, IL, Charles C Thomas, 1971, pp. 220 and 527.
5. Natelson, S.: Carbon dioxide content by microgasometer. *In* Selected Methods of Clinical Chemistry, Vol. 9, edited by W.R. Faulkner and S. Meites. Washington, D.C., American Association for Clinical Chemistry, 1982, p. 131.

Chapter 4

THE KIDNEY AND TESTS OF RENAL FUNCTION

When life on earth was confined to the sea, the primitive creatures had few problems of water or electrolyte balance because the internal environment (tissue fluids and cells) had ready access to the external environment, the sea water, for equilibration; they were bathed in an aqueous solution of a relatively constant composition of the various ions. As sea water flowed through the gill system of fishes, oxygen was extracted from it and wastes were excreted or diffused into it; the circulating blood and tissue fluids were maintained at the same concentration of salts and pH as the bathing fluid by simple exchange mechanisms.

When the early creatures began to leave the sea, survival depended upon their ability to utilize the oxygen of the atmosphere and to preserve the internal environment of the cells. In the course of evolution, the higher forms of life developed complex mechanisms for conserving body water, for maintaining a relatively constant pH in body fluids despite the continual production of acids (H^+) as a result of metabolic activity, and for maintaining a relatively constant concentration of certain electrolytes despite large fluctuations in intake. The formation of primitive lungs and primitive kidneys, which through the millennia evolved into the complex structures present today in mammals, made possible the adaptation of sea creatures to a terrestrial life in which the external environment is far different from that bathing the cells.

MAINTENANCE OF THE INTERNAL ENVIRONMENT

Humans have elaborate mechanisms and control systems for preserving the internal environment of the cells. Even though the system is quite efficient in ordinary circumstances, various diseases or certain pathologic processes upset the delicate balance by affecting organs or processes playing an important role in this steady state condition (homeostasis). When the internal environment is seriously disturbed, it becomes a matter of grave concern to personnel caring for patients. Impairment of kidney function is particularly serious because of its importance in water and electrolyte balance.

Role of the Kidneys

The kidney is a highly specialized organ that has evolved to perform two main functions: eliminate soluble waste products of metabolism and preserve the internal environment of the cells (maintain water balance, pH, ionic equilibrium, and fluid osmotic pressure). Although the kidney does perform other functions [synthesis of erythropoietin, a hormone stimulating red blood cell

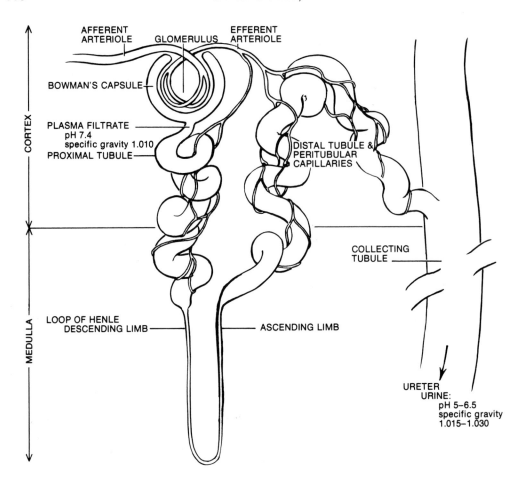

Fig. 4.1. Representation of a nephron and its blood supply.

production, and 1-hydroxylation of 25-hydroxy-vitamin D_3 to form the biologically active vitamin D-hormone (p. 348)], this chapter will deal only with the kidney's role as an excretory organ and its function in maintaining a constant internal environment.

There are two kidneys, one on each side of the body. The working unit of the kidney is the nephron, which is shown schematically in Figure 4.1; each kidney contains approximately one million of these working units. A nephron is composed of a glomerulus, a proximal convoluted tubule, a long loop of Henle (thin-walled descending loop and thick-walled ascending loop), distal convoluted tubule, and collecting tubules that merge to form collecting ducts. The urine flows from the collecting ducts into the renal pelvis and via the ureter (one per kidney) to the bladder.

Formation of Urine

A large volume of arterial blood flows through the glomerulus, which may be considered to be a very specialized blood filter. The red and white blood cells, platelets, and molecules of molecular weight >50,000 Daltons (most proteins and constituents tightly bound to proteins) are retained in the capillaries, while ions and small molecules like water and glucose pass through the capillary membrane into the surrounding capsule. About 20% of the plasma is filtered in one single pass of the blood flowing through the kidney. This is the first step in the formation of urine. The conservation and selection processes begin immediately as the filtrate commences its journey through the nephron, and these will be described in detail. The internal environment of the cells is preserved by complex renal and hormonal mechanisms that (1) efficiently eliminate accumulated acids and excess solutes, (2) preserve the water balance by regulation of the outflow, (3) excrete various waste products, and (4) selectively excrete or retain certain substances and ions.

The relatively large blood flow through the kidneys (approximately 20% of the blood pumped out by each heartbeat) is necessary to produce a urine that eliminates waste products. As the blood flows through each working nephron, illustrated schematically in Figure 4.1, the filtration process begins in the glomerulus. The arteriole bringing the blood to the glomerulus branches out to form a tuft of capillaries that then recombine to exit as an arteriole with a smaller diameter. The leaving (efferent) arteriole then branches out to provide an abundant blood supply to the tubules.

The change of lumen diameter between the entering (afferent) arteriole and the leaving (efferent) arteriole causes a rise in blood pressure in the capillary tuft, which aids in the filtration process. For filtration to take place, the effective filtration pressure must exceed the osmotic forces drawing water into the protein-rich plasma (oncotic pressure); a back pressure created by an obstruction to urine flow from the tubules decreases the net effective filtration pressure and hinders glomerular filtration.

The filtration process is passive and requires no input of energy for it to occur. The capillary walls of the glomerular tuft are thin and well adapted to their function as a filter; water and small molecules pass easily through the pores to form an ultrafiltrate of plasma. The ultrafiltrate is large in volume (about 200 L/day for the average adult) and its constituents are of the same concentration as in plasma. Ninety-nine percent of the ultrafiltrate water and a large percentage of its constituents must be reabsorbed before the urine leaves the collecting tubules on its way to the bladder.

The second step in the formation of urine is the passage of the ultrafiltrate down the proximal convoluted tubule where the various selective processes of absorption begin. The filtrate starts out with the same specific gravity as a protein-free filtrate of plasma, 1.010, and the same pH, 7.4. About 70% of water, Na^+, and Cl^- and all except negligible amounts of glucose, amino acids, and K^+ are reabsorbed in the proximal tubules; some substances, such as urea, phosphate, and Ca^{++}, are incompletely reabsorbed. H^+ is exchanged for Na^+ throughout the tubule, while K^+ is exchanged for Na^+ only in the distal tubule; the K^+-Na^+ exchange is regulated by the hormone aldosterone. The filtrate is reduced to about one third of its original volume in the proximal tubule, but because water and salts have been absorbed together, there is no change in the osmotic pressure of the filtrate. The physiologic need, however, is to conserve water and to eliminate salts and H^+ by excreting a hyperosmolar urine that is much more acidic than plasma. This is accomplished by some unique anatomic features, an interplay of physical forces, and a finely regulated hormonal control. Water is always reabsorbed passively when the osmotic pressure outside of a semipermeable membrane is higher than inside; it follows an osmotic gradient toward restoration of equilibrium. The osmotic gradient is usually produced by active Na^+ transport (the sodium pump), an energy-requiring process.

The anatomic features that promote the independent reabsorption of water and Na^+ from the distal and collecting tubules are: (1) a thickened wall in the ascending limb of the loop of Henle that is impermeable to water but not to ions; (2) the extension of the loop of Henle and large sections of the distal and collecting tubules from the kidney cortex into the medulla; (3) the passage of the blood capillaries into the interstitial tissue, surrounding in countercurrent fashion the loops of Henle, the distal tubules, and collecting ducts; and (4) the impermeability to water of the walls of the distal and collecting tubules unless acted upon by the antidiuretic hormone, ADH.

The interstitial fluid in the kidney increases in osmotic pressure as one goes from the cortex to the medulla; this osmotic gradient is caused by an increasing concentration of Na^+ and Cl^- as the Na^+ is pumped out of the ascending loop of Henle by an active process. The osmotic gradient in the interstitial fluid surrounding the descending loop of Henle is the physical force which greatly accelerates the reabsorption of water from the lumen of the descending loop. As the filtrate moves up the ascending loop, the reabsorption of water stops because the lumen wall is impermeable to water. The reabsorption of Na^+ is considerable because of pump action; Cl^- moves passively with it. The net result is the production of a *hyposmolar* urine (greater loss of Na^+ and Cl^- than water) by the time the distal tubule is reached. The process of creating a high osmolality in the kidney medulla while reducing the osmolality of the urine leaving the loop of Henle is known as countercurrent multiplication. The water and ions that pass into the interstitial fluid are reclaimed by absorption into the blood capillaries surrounding the loops of Henle and the tubules.

The walls of the distal and collecting tubules are impermeable to water unless acted upon by ADH, the antidiuretic hormone. Under the influence of this hormone, the final reabsorption of water takes place, with the production of hyperosmolar urine. The hyperosmolar interstitial fluid surrounding the collecting tubules accelerates the water uptake. ADH secretion is stimulated by a high plasma osmotic pressure and inhibited by decreased osmotic pressure so that the net result is conservation or excretion of body water according to osmotic needs (Fig. 4.2). This process essentially controls the volume of urine finally excreted.

REGULATION OF ADH OUTPUT. ADH is a hormone (vasopressin) produced by the hypothalamus but stored in the posterior pituitary gland. Receptor cells in the hypothalamus are sensitive to changes in plasma osmotic pressure and transmit nerve impulses to the posterior pituitary. An elevated osmotic pressure stimulates the posterior pituitary gland to secrete ADH and thus promotes water retention by increasing renal reabsorption of water (decrease in urine volume); a decrease in osmotic pressure inhibits the secretion of ADH and promotes water loss by decreasing the renal reabsorption of water (increase in urinary volume). An increase in osmotic pressure also stimulates the feeling of thirst, so more water is usually drunk if it is available.

REABSORPTION AND EXCRETION PROCESSES. Many different mechanisms are involved in the various absorption processes; some are passive in the sense that

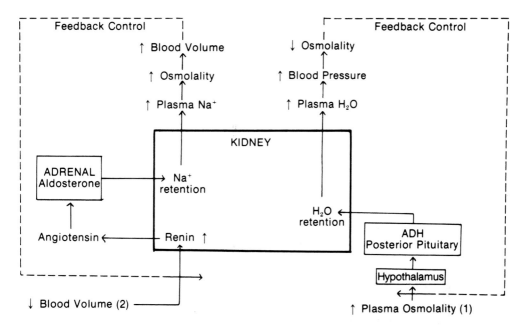

Fig. 4.2. ADH and aldosterone in the control of the reabsorption of water and Na⁺, respectively, by the kidney. ADH secretion by the posterior pituitary is stimulated by increased plasma osmolality; aldosterone secretion by the adrenal cortex is triggered by a decreased blood volume.

they can be explained by concentration gradients, or by osmotic pressure differences in the case of water, whereas others require active transport mechanisms involving the expenditure of energy. The active processes are usually coupled with some enzyme action to transfer substances from the lumen into cells against a concentration gradient. Water in the filtrate is absorbed passively in the proximal tubule, but glucose and amino acids are almost completely reabsorbed by active transport. About 70% of the sodium is reabsorbed in the proximal tubule by an active process, primarily in conjunction with chloride. In the distal tubule, however, a specialized ion exchange mechanism participates in the conservation of sodium and in the excretion of potassium and hydrogen ions. Bicarbonate is converted to carbon dioxide and water by the hydrogen ion in the filtrate, and the CO_2 diffuses into the tubular cells where it is converted by the enzyme, carbonic anhydrase, back to bicarbonate ion. The effective reabsorption of HCO_3^- occurs throughout the entire length of the tubule. Creatinine, on the other hand, is not absorbed at all by the renal tubule; in fact, small amounts may be secreted by the tubular cells into the urine. Potassium ion is almost completely reabsorbed in the proximal tubule, but a portion is exchanged for Na⁺ in the distal tubule where the potassium ion is secreted into the lumen of the tubule. Phosphate ion is reabsorbed by an active process in the proximal tubule, but the reabsorption is never complete, and fine control is achieved by secretion of the parathyroid hormone, which inhibits its reabsorption. Urea, on the other hand, diffuses passively into the blood as it passes down the tubule. Under ordinary circumstances, about 60% of the urea in the filtrate is excreted in the urine, but this varies with the glomerular filtration rate and the rate of filtrate flow in the tubules.

TABLE 4.1

Average Filtration, Reabsorption and Excretion of
Certain Normal Constituents of Plasma*

(1) CONSTITUENT	(2) FILTERED/24 hr (g)	(3) EXCRETED/24 hr† (g)	(4) REABSORBED/24 hr	
			(g)	%
Water	180,000	1,800	178,200	99.0
Chloride	630	5.3	625	99.2
Sodium	540	3.3	537	99.4
Bicarbonate	300	0.3	300	100
Glucose	140	0	140	100
Amino acids	72	1	71	98.6
Urea	53	32	24	39.6
Potassium	28	4	24	85.7
Uric acid	8.5	0.8	7.7	90.6
Phosphate	6.5	1	5.5	84.1
Creatinine	1.4	1.4	0.0	0
Total protein‡	?	0.06	?	?

*Modified after R. W. Berliner and G. Giebisch in Best and Taylor's Physiological Basis of Medical
Practice, 10th ed., Baltimore, Williams & Wilkins, 1979, p. 5–35.
†Typical normal values, but greatly dependent upon dietary intake.
‡Many different proteins appear in the urine in trace amounts. Most of them have a molecular weight
lower than 70,000, but a few are as high as 160,000.

In Table 4.1 it can be seen that the kidneys perform a tremendous amount of work in forming normal urine. Approximately 180 L (180,000 g) of plasma are filtered in a 24-hour period, 178 to 179 L of which are reabsorbed. Stated another way, about 99.2% of the 180 L of water in the ultrafiltrate return to body fluids. The amount of solutes in the filtrate varies from 540 g of sodium and 630 g of chloride to 0.06 g of a mixture of proteins (primarily albumin and glycoproteins). Most of the sodium and chloride ions and amino acids are reabsorbed; glucose is reabsorbed virtually completely in normal circumstances, whereas creatinine is not reabsorbed at all. There is a *threshold* or limit, however, to the reabsorption capacity for glucose. When its concentration in plasma and consequently, in the glomerular filtrate exceeds 180 mg/dl, some of the glucose escapes absorption and appears in the urine. The high plasma glucose levels in uncontrolled diabetes mellitus explain the customary finding of appreciable amounts of glucose in the urine of these patients.

Table 4.1 also shows in column (3) the amounts of the different substances excreted by a person eating an average diet, and in column (4) the amounts reabsorbed or reclaimed for use. The degree of reabsorption of the water and salts appearing in the glomerular filtrate is subject to fine control mechanisms responsive to body needs.

The control mechanism for the regulation of sodium reabsorption rests with the hormone, aldosterone, a steroid produced by the adrenal cortex. This hormone responds to changes in blood volume and accelerates the reabsorption of sodium in the distal tubule by facilitating its exchange for potassium ions. Thus, with excess production of the hormone, sodium is retained and potassium is excreted. With a deficiency of this hormone, the reverse takes place; there is increased excretion of sodium, with less retention in the body, and decreased excretion of potassium. The feedback control of sodium excretion by means of aldosterone secretion makes possible a variation in the excretion of Na^+ from zero to large amounts, according to the needs of the body. The necessity for the conservation of Na^+ becomes acute in those areas of the world where sodium chloride is scarce or in situations in which the loss of NaCl is great because of profuse sweating, with inadequate intake to compensate for the loss. Thus, there was evolutionary pressure to develop sodium-conserving mechanisms as well as mechanisms for excreting surplus Na^+ when the intake was high.

REGULATION OF ALDOSTERONE OUTPUT. Receptor cells in the kidneys are sensitive to changes in renal blood flow. When the renal blood volume is reduced (either by reduction in total blood volume or by reduction in blood pressure), the receptor cells release renin, a protease. Renin acts upon plasma angiotensinogen (an α_2-globulin) and splits off angiotensin I (a decapeptide-10 amino acids). Angiotensin I loses two amino acids from its chain by action of a lung peptidase and becomes angiotensin II, a powerful vasoconstrictor. Angiotensin II not only produces an immediate rise in blood pressure but also stimulates the adrenal cortex to secrete the hormone aldosterone. Aldosterone promotes the retention of Na^+, which increases the fluid volume by its osmotic effect.

Thus, the regulation of fluid volume and osmotic pressure is coordinated by the interplay of the two hormones, aldosterone and ADH, which control the degrees of reabsorption of Na^+ and water, respectively, in the distal and col-

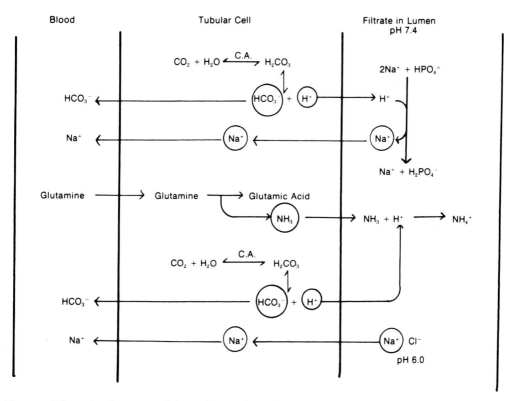

Fig. 4.3. Schematic illustration of the exchange of Na^+ for H^+ in the renal tubule, and of the formation of ammonia. C.A. represents the enzyme, carbonic anhydrase.

lecting tubules. The thirst mechanism is also an important factor in regulation, but sick patients may not have ready access to a supply of fluids, and in some cases may not be able to retain ingested fluid.

The control of potassium excretion, however, is not so highly refined because there was no evolutionary pressure to develop a conservation mechanism; most natural foodstuffs contain potassium, making it virtually impossible for a mammal eating such a diet to suffer from a K^+ deficiency. A K^+ deficiency has become possible only in the last century for man, with the production of highly refined foods devoid of potassium and with the introduction of intravenous saline and glucose therapy, innovations for which the kidney has no built-in protective mechanism. The kidney is able to excrete large amounts of K^+ when the intake is high, but it cannot reduce the K^+ excretion below a certain level even if the intake falls to zero. Aldosterone, the hormone that stimulates the reabsorption of Na^+ in the kidney, has the opposite effect on K^+ by promoting its excretion, an exchange of K^+ for Na^+ across the tubular wall.

Bicarbonate ion cannot be absorbed directly from the tubular lumen because the wall is impermeable to it, but it is absorbed indirectly. As the glomerular filtrate becomes acid, HCO_3^- is converted to CO_2 and H_2O. The CO_2 diffuses into the tubular cell where it is recombined with water and reaches an equilibrium with H^+ and HCO_3^-. The HCO_3^- is absorbed into the blood capillaries, a process that achieves the same result as direct reabsorption of HCO_3^-. The equilibria may be summarized as follows:

1. In filtrate:

$$\uparrow \text{ diffusion into tubular cell}$$
$$HCO_3^- + H^+ \longleftrightarrow H_2CO_3 \longleftrightarrow H_2O + CO_2$$

2. In tubular cell (see Fig. 4.3):

$$\uparrow \text{ diffusion into blood}$$
$$CO_2 + H_2O \xrightarrow{\text{carbonic anhydrase}} H_2CO_3 \longleftrightarrow HCO_3^- + H^+$$
$$\downarrow$$
$$\text{diffusion into tubular lumen}$$

Under conditions of metabolic alkalosis, however, the urine is alkaline, and reaction 1 does not take place. Appreciable amounts of HCO_3^- are then excreted into the urine, a compensatory mechanism that helps to lower the plasma concentration of HCO_3^- as well as the pH.

EXCRETION OF ACIDS. The principal renal mechanisms for excreting excess hydrogen ions or protons, summarized in Figure 4.3, are the following:
1. Exchange of Na^+ for H^+.
2. Trapping the H^+ by the following reactions:
    a. $HPO_4^{2-} + H^+ \rightarrow H_2PO_4^-$.
    In the plasma at pH 7.4, 80% of the phosphate is in the form of HPO_4^{2-}. Protons (H^+) are secreted into the renal tubular lumen where the above reaction takes place. At pH 6.0, practically 100% of the phosphate will be in the form of $H_2PO_4^-$.

    b. $NH_3 + H^+ \rightarrow NH_4^+$.

In the tubular cell, glutamine is converted into glutamic acid and ammonia (NH_3) by the enzyme glutaminase. The NH_3 diffuses readily into the lumen where it reacts with H^+ to form NH_4^+, which is excreted. An equivalent amount of Na^+ is returned to the tubular cell in exchange for the H^+, with the net effect of excreting NH_4^+ instead of Na^+. This becomes a significant mechanism for disposing of excess H^+ in chronic acidosis, since the capacity for producing NH_3 from glutamine is greatly increased as synthesis of the glutaminase enzyme is stimulated by the acidosis.

    3. Excretion of undissociated acids such as ketoacids in uncontrolled diabetes mellitus or prolonged fasting.

In summary, urine is formed by a continuous process that starts with the ultrafiltration of plasma followed by reabsorption of water and other constituents to a greater or lesser extent. Reabsorption may take place by many mechanisms: passive diffusion, special transport mechanisms, energy-linked processes, and ion exchange. Fine regulation of the final product to meet body needs is achieved by the secretion or withholding of essential hormones that act upon tubular cells. The final composition of the urine is attained by the secretion of some constituents from the blood plasma into the distal tubular lumen. The net effect of this complicated process is to excrete waste products of metabolism and to preserve the volume, ionic concentration, balance, and pH of the body fluids. The normal urine contains more dissolved substances than the plasma ultrafiltrate (higher specific gravity) and is much more acid.

LABORATORY TESTS OF RENAL FUNCTION

Laboratory tests play an important role in the diagnosis and assessment of renal disease because the clinical signs and symptoms of renal disease may be quite vague or absent. Some of the renal function tests reveal primarily disturbances in glomerular filtration, and others reflect dysfunction of the tubules, but damage is seldom confined solely to a particular portion of the nephron. The anatomic portions of the nephron are closely related and have a common blood supply so that damage to one portion of the nephron gradually involves the nephron as a whole. Thus, with chronicity, both glomerular and tubular portions of the nephron become involved irrespective of the site of the original lesion. Not all of the nephrons are affected at the same time and to the same extent, however, and a reserve capacity is provided for kidney function. The pathologic condition of the kidney must be considerable before the tests of renal function become abnormal.

The formation of a normal urine at a normal rate requires the presence of properly functioning kidneys receiving an adequate blood supply at a sufficiently high blood pressure, with no obstruction to urine outflow. Thus, a malfunction in the formation and/or elimination of urine may be attributable to prerenal, renal, or postrenal causes. The prerenal factors are those affecting blood volume, blood flow, or blood pressure, such as hemorrhage, shock, dehydration, intestinal obstruction, prolonged diarrhea, and cardiac failure. When these defects are corrected, kidney function usually returns to normal. The renal factors are those within the kidneys themselves that may affect the glomerular filtration rate, the various tubular activities, or the renal blood vessels.

TABLE 4.2

Abnormal Laboratory Findings in Glomerular and Tubular Dysfunction

TYPE OF DYSFUNCTION	SERUM		URINE
Glomerular [a] ↓ GFR	↑ Creatinine* ↑ Phosphate ↑ Potassium ↓ Bicarbonate ↓ Creatinine clearance*	↑ Urea* ↑ Urate ↓ Calcium ↓ pH	Urinalysis* ↓ Volume ↑ Specific gravity ↑ Osmolality ↑ Urea concentration, ↓ amount ↓ Sodium concentration
Tubular [b]	↓ Potassium ↓ Bicarbonate When glomeruli also involved: ↑ Creatinine*	↓ Phosphate ↓ pH ↑ Urea*	Urinalysis* ↑ Volume ↑ pH ↓ Specific gravity and osmolality Casts may be present ↑ Sodium concentration ↓ Urine concentration test*

↑ indicates increased concentration, ↓ decreased
*Principal tests of renal function
   a. Findings do not apply to the nephrotic syndrome because the lesion is in the glomerular membrane; serum proteins are lost but the GFR (glomerular filtration rate) is normal.
   b. Acute tubular damage may trigger an output of renin locally that reduces the glomerular blood flow and GFR.

The postrenal factors decreasing renal function are those obstructing the flow of urine, such as renal calculi (stones), carcinomas or tumors that may compress the ureters, urethra, or the bladder opening, or an enlarged prostate gland that partially occludes the urethra. Chemistry tests alone, however, are incapable of differentiating between the three prime causes of renal dysfunction, but a careful history and physical examination of the patient, in combination with a few other tests, lead to an elucidation of the problem.

Some of the common renal diseases that produce a diffuse involvement of the kidneys are glomerulonephritis, the nephrotic syndrome, pyelonephritis, and arteriolar nephrosclerosis.

Glomerulonephritis, a diffuse, inflammatory disease, affects first the glomeruli but rapidly produces degeneration of the tubules. In its *acute* form, the disease is manifested suddenly by the appearance of hematuria and proteinuria, with varying degrees of hypertension (high blood pressure), renal insufficiency, and edema. The causative agent for the disease is usually a prior infection with a Group A, β-hemolytic streptococcus. It is believed tha an autoimmune re-action to the antibody produced against the streptococcus is responsible for the damage to the glomerular capillaries. Most patients recover completely from the acute phase, but a few do not. These latter undergo progressive loss of renal function and finally die in renal failure.

There is also a *chronic* form of glomerulonephritis that varies greatly in its degree of severity. In some patients, the disease is progressive, with continuing loss of renal function; in others, there are remissions and relapses that may go on for many years.

The *nephrotic syndrome* is characterized by a heavy proteinuria, hypoal-buminemia, edema, and hyperlipidemia; lesions in the glomerular membrane allow proteins to escape. Laboratory tests reveal little or no impairment of renal function. The syndrome is frequently associated with systemic lupus erythe-matosus, proliferative glomerulonephritis, amyloidosis, or syphilis, but frequently, the causative factor is not known. Many cases respond to treatment with adrenocortical steroids.

Pyelonephritis is an inflammatory renal disease caused by infectious organisms that have ascended the urinary tract and invaded kidney tissues. Chronic or repeated infections may lead to replacement of renal cells by scar tissue, with some loss in renal function.

Arteriolar nephrosclerosis is characterized by a thickening of the inner lining of the renal arterioles, resulting in a decreased lumen and an increased blood pressure. As the blood vessels become necrotic, there is progressive loss in both glomerular and tubular function. Scar tissue replaces the damaged cells, and the kidney becomes contracted. Proteinuria is common. In some patients, the high blood pressure cannot be controlled (malignant hypertension), and the impairment in renal function becomes progressive and rapid; the outcome is fatal. In the benign form, the blood pressure can be reduced to reasonable values by appropriate drugs, so the loss in kidney function is relatively mild and slow in its progression.

Since the kidney is the primary organ involved in the excretion of certain nitrogenous wastes (creatinine, urea, uric acid), in the regulation of water and electrolyte balance, and in the excretion of fixed acids, renal dysfunction is characterized by changes or abnormalities in one or more of these parameters. Some of these abnormalities (electrolyte or acid-base imbalance) are nonspecific, since they may also occur in quite a few other disease states. Clinical chemistry tests for the diagnosis of renal dysfunction are confined, therefore, to those few which are more indicative of renal disease: serum creatinine, serum urea nitrogen, creatinine clearance, and urinalysis (Table 4.2). Some other tests are useful for confirmation or for intelligent management of the patient and may be ordered for this purpose. Urinalysis will be described first because it yields useful information and is a routine procedure for all patients.

URINALYSIS[14]

Examination of the urine as an aid to diagnosis of a number of diseases has been carried out for centuries by medical practitioners. Some of the current methods of examination are still traditional, such as noting the appearance and odor of the specimen and making a microscopic examination of the urinary sediment. The main advances have been in providing dipsticks or strips for the semiquantitation of a group of constituents and in measuring urine osmolality as an indication of total solute concentration. The visual and microscopic examination may yield useful clinical information and must not be neglected because it is not "quantitative."

A routine urinalysis usually consists of an examination of a morning specimen (upon arising) for color, odor, specific gravity, or osmolality; the performance of some qualitative or semiquantitative tests for pH, protein, glucose or reducing sugars, ketones, blood and perhaps bilirubin, urobilinogen, and nitrite; and a microscopic examination of the urinary sediment. Tests that are especially useful in evaluating renal function or renal disease will be described in detail; those that are employed for the diagnosis of other diseases will be treated more fully in the appropriate chapters.

Macroscopic and Physical Examination

Volume

Knowledge of the daily urinary output may be of value in the study of renal disease, but this test requires a timed specimen and a good collection. The daily output of urine depends largely upon the fluid intake and many other factors such as degree of exertion, temperature, salt (NaCl) intake, and hormonal control, but a good figure to remember is that the average excretion of a normal adult is approximately 1 ml/min or about 1400 ± 800 ml/24 hr. A decreased urinary output is called *oliguria;* an increased output is referred to as *polyuria.* Oliguria may be caused by *prerenal* (low blood pressure, shock, hemorrhage, fluid deprivation), *renal* (acute tubular necrosis, certain poisons, renal vascular disease, precipitation of certain compounds in nephrons), or *postrenal* (calculi, tumors compressing ureters, prostatic hypertrophy) factors. Polyuria may be caused by the excretion of a large amount of solutes, with obligatory excretion of water (after excessive salt intake, in diabetes mellitus with glycosuria), by a deficiency or depression of the antidiuretic hormone (ADH), or by the excessive ingestion of fluids or diuretic substances.

Color

It is important to call attention to an abnormally colored urine even though this does not occur frequently. Fresh blood or hemoglobin may impart a reddish color, whereas old blood makes the urine look smoky; both are indications of bleeding in the genitourinary tract. Bile pigments produce a green, brown, or deep yellow color that signify liver or biliary tract disease (p. 230). A dark brown urine may be caused by homogentisic acid excreted in a rare genetic disease, alkaptonuria. Some drugs or dyes may also contribute color to the urine.

Odor

Fresh urine has a characteristic odor that may be affected by foods, such as asparagus. In diabetic acidosis, there may be a fruity odor caused by the ketoacids and acetone (p. 307). In maple syrup disease, a rare genetic defect, the urine has the odor of caramelized sugar or maple syrup. When urine specimens are old, or when there is a *Proteus* infection, there is usually a strong odor of ammonia. A putrid odor usually means that the urine has undergone bacterial decomposition because it stood around too long without refrigeration.

Specific Gravity

The specific gravity of the urine varies directly with the grams of solutes excreted per liter. It provides information concerning the ability of the kidney to concentrate the glomerular filtrate. The physiologic range of specific gravity varies from 1.003 to 1.032, but the usual range for a 24-hour specimen varies from 1.015 to 1.025. The most concentrated specimen is obtained upon arising in the morning. In renal tubular disease, the concentrating ability of the kidney is among the first functions to be lost.

The specific gravity of urine may be determined directly with a urinometer or indirectly through measurement of its refractive index.

BY URINOMETER (HYDROMETER). The urinometer is a hydrometer designed to fit into and float in a narrow cylinder filled with urine. The urinometer has a slender neck, with a specific gravity scale wrapped around it that usually covers the range from 1.000 to 1.040. The urinometer should be calibrated by testing it with a solution of known specific gravity.

Procedure
1. Fill the cylinder about three fourths full with specimen and place on a level surface.
2. Insert the urinometer into the cylinder and spin slightly so that it will float freely.
3. Read the specific gravity directly from the scale on the stem at the lowest point of the meniscus of the urine surface.
4. *Temperature correction:* If the urine is not at the urinometer calibration temperature, add 0.001 to the specific gravity for every 3° C that the urine temperature is above this value and subtract 0.001 for every 3° that it is below this temperature.

By REFRACTOMETRY. The refractive index of a solution also varies with the amount of dissolved substances and hence, is related to the specific gravity. Because the urine must be clear for this measurement, it is usually carried out upon a centrifuged specimen. There are commercial refractometers available with a scale that gives a direct readout in specific gravity.

Procedure
1. Place a small drop of clear urine on the glass surface of the refractometer and close the lid.
2. Look through the meter directly toward a light source.
3. Record the specific gravity at the point where the line separating the light area from the dark crosses the specific gravity scale.

Osmolality

Osmolality is a measure of the moles of dissolved particles (undissociated molecules as well as ions) contained in a kg of solvent; it reflects the total concentration of solutes (p. 81).

When substances are dissolved in a solvent, they affect some of the properties of the solvent and cause some physical changes that can be measured. These are a lowering of the freezing point, a decrease in the vapor pressure, and an increase in the boiling point of the pure solvent. There are commercial instruments available that make use of the first two properties for the measurement of the osmolality of body fluids. The most commonly used ones make use of the freezing point depression whereby one mole of each ionic species and each nonionized solute per kg of water lowers the freezing point by 1.86° C.

PROCEDURE
1. Follow the manufacturer's directions for the particular instrument available to you.
2. Calibrate the instrument with known standards.
3. Centrifuge the specimen well to eliminate suspended matter.
4. Measure the freezing point lowering or the vapor pressure lowering, as the case may be.
5. Record the osmolality.

REFERENCE VALUES

Serum:	278 to 305 mosm/kg
Urine—random specimen:	40 to 1350 mosm/kg
On normal fluid intake, 24-hour specimen:	500 to 800 mosm/kg
During maximal urine concentration:	850 to 1350 mosm/kg

Note: The osmolality is a more accurate reflection of the concentration of dissolved substances than is the specific gravity because in various diseases the urine may contain relatively large amounts of glucose or protein. These substances have a much higher molecular weight than the salts commonly found in urine and hence, affect the specific gravity much more than they affect the osmolality. The receptors in the body respond to osmolality or changes in solute concentration.

Qualitative or Semiquantitative Tests

A routine urinalysis also includes tests for the measurement of pH and for the detection of protein, glucose or reducing sugars, ketone bodies, and frequently for bilirubin and blood. There are commercial dipsticks (Ames; Bio-Dynamics) to test for a single urinary constituent or for two or more (up to eight) different constituents on a single test strip. The test strips are of plastic but at spaced intervals contain a small paper or cellulose mat impregnated with chemicals and/or indicators that turn various shades of color, depending upon the concentration of a particular constituent. The eight constituents that may be detected in this manner are pH, protein, glucose, ketones, blood, bilirubin, urobilinogen, and nitrite. Of these, only those most commonly used for routine testing will be discussed.

pH

The urine has a physiologic pH range of 4.6 to 8.0, with a mean around 6.0. Starvation and ketosis increase the acidity of the urine. Acid-producing salts are sometimes administered for the treatment of urinary tract infections. The urine is seldom alkaline but becomes alkaline in alkalosis, after the ingestion of alkali over a period of time for the treatment of ulcers, or from bacteria in the urine that generate ammonia. The pH is usually measured by means of a paper strip impregnated with an indicator or by using a commercial dipstick that contains a mat of cellulose or paper impregnated with an indicator. The color produced on the spot or test strip is compared visually with that of a color chart calibrated for pH.

Protein

A small amount of protein (50 to 150 mg/24 hr) appears daily in the normal urine. Some of this protein comes from a small amount of albumin that is filtered in the glomerulus but not reabsorbed in the tubules; the rest of it is due to glycoproteins coming from the linings of the genitourinary tract. Normally, the protein concentration in urine is below 10 mg/dl and is not detectable by the usual urinalysis methods.

Proteinuria (the presence of detectable amounts of protein in the urine) usually is an indication of injury to the glomerular membrane, which permits the filtration or escape of protein molecules. It has to be differentiated, however, from a transient proteinuria that may take place during the course of a high fever or from a harmless condition, *orthostatic proteinuria* , which occurs only when a patient is active and on his feet. Since the latter condition does not occur when the subject is recumbent, the first urine specimen upon arising in the morning should have no detectable protein. Protein in urine may be measured by dipstick, sulfosalicylic acid, or heat coagulation tests.

DIPSTICK. The basis for the protein test is the "protein error" of indicators, a term applied to the change in ionization and hence pH when an indicator dye is adsorbed to protein; this shift in pH causes a change in the color of the indicator. The paper spot in the dipstick is impregnated with citrate buffer (pH 3.0) containing bromphenol blue indicator, which is yellow at pH 3.0 and blue at pH 4.2. At pH 3.0, the indicator is mostly unionized. If protein is present in the urine into which it is dipped, the ionized fraction will bind to the protein, causing more dye to ionize until equilibrium is reached; hence, the impregnated strip will have less yellow and more blue color as the protein concentration increases, which will be seen visually as a change from yellow to green (a mixture of yellow dye plus blue dye appears green). The color is compared with that of a color chart, which provides a crude estimation of the protein content from 30 mg/dl to about 1000 mg/dl.

Note: False positives may be obtained with a buffered, alkaline urine. There are no false positives with x-ray contrast media, sulfonamides, or other drugs or medication that may cause turbidity with other tests for protein (sulfosalicylic acid test or heat precipitation at pH 5).

SULFOSALICYLIC ACID. In a chemical test for urine protein, transfer approximately 3 ml of centrifuged urine to a test tube, hold at an angle, and let 3 drops of 250 g/L sulfosalicylic acid run down the side of the tube. The acid forms a layer underneath the urine; do not mix.

Examine the urine-acid interface for turbidity after about 1 min. A barely perceptible turbidity is reported as a "trace" or ±, and heavier amounts are graded from 1 + to 4 + . A protein concentration of 5 mg/dl usually registers as a trace, but false positive results may be encountered with the urine from patients who had been injected recently with x-ray contrast media or who were receiving sulfonamides, tolbutamide, or other medications.

HEAT COAGULATION TEST. Clarify urine by centrifuging, and transfer about 7.5 ml to a 16 mm borosilicate test tube. Add 3 drops of 5 M acetate buffer, pH 4.0 (24 ml glacial acetic acid plus 7.4 g of anhydrous sodium acetate plus water to 100 ml), and mix. If a precipitate forms at this point, it is mucin (a mucoprotein) and should be removed by centrifuging. Heat the upper half of the urine column to boiling and then examine for turbidity, comparing the upper half with the lower half. A barely visible turbidity is graded as ± or trace; larger amounts are rated up to 4 + . A distinct turbidity is 1 + (10 to 30 mg/dl), a moderate turbidity is 2 + (40 to 100 mg/dl), a heavy turbidity is 3 + (200 to 500 mg/dl), and a flocculent precipitate is 4 + (> 500 mg/dl).

Glucose

Although glucose may appear in the urine as a result of renal disease (renal glycosuria), this is not a common occurrence. The usual rationale for including a test for glucose in a routine urinalysis is to detect unsuspected diabetes or to check the efficacy of insulin therapy in diabetics. This test will be treated in detail on page 318. Only the dipstick test will be described in this section.

The dipstick paper or spot is impregnated with two enzymes, glucose oxidase and peroxidase, and with two chemicals, one of which is an organic peroxide and the other is o-tolidine, a colorless hydrogen donor that turns blue when oxidized (loses 2 H atoms). The glucose oxidase oxidizes the glucose and produces H_2O_2 in the process. In a coupled reaction, the H_2O_2 is decomposed by peroxidase, with the simultaneous production of H_2O and the oxidation (dehydrogenation) of o-tolidine to the blue chromogen.

Ketone Bodies

Although ketonuria may accompany situations of carbohydrate deprivation (fasting, or high-fat, high-protein diets), it derives its clinical importance primarily from its occurrence in uncontrolled diabetes. Accordingly, it will be treated in greater detail on page 307.

The dipsticks contain a strip impregnated with sodium nitroprusside and an alkaline buffer. In the presence of acetoacetate or acetone, a lavender color is produced that is compared with that of a color chart.

Blood

The presence of small amounts of occult blood (blood cells or hemoglobin that do not visibly color the urine) may be detected by appropriate dipsticks. The reaction is based upon the enzymatic action of hemoglobin in decomposing peroxides, which in the presence of a hydrogen donor, o-tolidine, produces a blue color. This end reaction is identical to the color obtained in the glucose oxidase reaction. Hemoglobin in the urine indicates hemolysis in the blood stream or the lysis of red blood cells in the urinary tract. It is a common accompaniment of various renal disorders or conditions affecting the urinary tract.

The appropriate dipstick (Ames; Bio-Dynamics) is impregnated with a buffered organic peroxide and o-tolidine. A blue color appears within 30 s if hemoglobin is present and may be graded by the intensity of the color.

Bilirubin, Urobilinogen, and Nitrite

Tests for bilirubin, urobilinogen, and nitrite appear on some dipsticks, but not all laboratories test for these substances routinely. Tests for bilirubin and urobilinogen have their greatest application in liver disease and are discussed on page 230. Nitrite in the urine suggests the possible presence in the urinary tract of bacteria capable of reducing nitrate to nitrite.

Microscopic Examination of the Urine Sediment

Microscopic examination of urinary sediment is most important because it yields information that may be helpful in making a diagnosis. For best results, obtain a concentrated specimen (upon arising) that has been clean-voided. The specimen should be examined within an hour of voiding because cells deteriorate upon standing, a process that may be delayed by refrigeration or by the addition of formalin (0.2 ml/100 ml urine). About 12 ml of a well-mixed urine sample is centrifuged for 5 min at 1500 rpm, and all but 0.2 to 0.3 ml of the urine is poured off. Suspend the sediment in the residual fluid by flicking the bottom of the tube on a test tube rack and take a drop for microscopic examination.

It is beyond the scope of this book to give instructions regarding the use of the microscope and the identification of the various formed elements. Suffice it to say that abnormal cells, oval fat bodies, casts, and crystals can be identified if they are present.[11,12]

NORMAL FINDINGS

1. Squamous and epithelial cells are present in all urine, especially in those from females; they have no pathologic significance.
2. An occasional or rare red or white cell has no pathologic significance.
3. Some hyaline casts may be found in normal urines, particularly after stress, exercise, or fever, in the absence of renal disease.
4. Bacteria may be present as an external contamination; clean-voided specimens that are examined when they are fresh help to eliminate possible confusion.

ABNORMAL FORMED ELEMENTS

1. *Cells*
   a. *Red blood cells:* More than an occasional red blood cell is abnormal and requires further investigation. Red blood cells may originate from any location in the urinary tract. In the female, this may also be a contribution from menstruation.
   b. *White blood cells:* A few white cells are not abnormal, but large numbers in a freshly voided specimen indicate the presence of an infection somewhere in the genitourinary tract.
   c. *Yeasts:* These are recognized by their budding and oval shape. They are common contaminants but can cause infections.*
2. *Oval fat bodies:* These are usually present to some degree in all types of diseases affecting the renal parenchyma but are most numerous in the nephrotic syndrome. They are thought to be degenerated tubular epithelial cells that have become filled with fat droplets.
3. *Casts:* Casts are formed by precipitation of mucoprotein in the lumen of renal tubules and collecting ducts; later, they are passed into the urine. Cellular elements frequently are entrapped in the cast.
   a. *Red blood cell casts:* These casts show presence of red cells in the protein matrix. They are reddish-brown or orange in color from the hemoglobin leaking out of broken-down red blood cells. The presence of red blood cell casts is always a pathologic condition because it denotes glomerular inflammation and bleeding and is associated with glomerulonephritis, systemic lupus erythematosus with kidney involvement, or other glomerular diseases.
   b. *White blood cell casts:* These are casts containing imbedded leukocytes and signify an infection (pyelonephritis).
   c. *Hyaline casts:* Simple hyaline casts contain protein alone and are glass-clear because they have almost the same refractive index as urine. They may be present in the urine of normal individuals, particularly after strenuous exercise, but they are also found in the urine in many disease states, usually when there is a proteinuria. Hyaline casts may contain some cellular debris or fat.

---

*In diabetics with glycosuria, in debilitated patients, or in those who are treated vigorously with antibiotics.

d. *Granular casts:* The presence of epithelial cellular debris in a cast makes it appear granular.

e. *Fatty casts:* These appear like oval fat bodies but are cylindrical in shape. This is an abnormal finding indicative of renal parenchymal disease.

f. *Waxy casts:* These are cellular casts that have degenerated and look "waxy" or like ground glass. They may be present in any one of a number of kidney diseases.

g. *Broad casts:* These are short and wide casts that have formed in the broad collecting tubules and are found only in renal failure.

4. *Crystals:* A variety of crystals may be found in normal urine and have no particular significance. These are calcium phosphate, triple phosphate, calcium oxalate, amorphous phosphates, sodium or ammonium urate, and sometimes calcium carbonate. Large amounts of urate or uric acid crystals should be noted because they may indicate excessive breakdown of tissue cells (nucleoproteins) or be an accompaniment of gout. Unusual crystals such as cystine or sulfa drug crystals must be noted.

TESTS OF NONPROTEIN NITROGEN CONSTITUENTS (NPN)

The concentrations of several constituents of blood plasma are elevated when there is impaired formation or elimination of urine, irrespective of the cause. These compounds are relatively small nitrogen-containing molecules that are collectively called the "nonprotein nitrogen constituents" of plasma or serum. The nitrogen compounds whose serum concentrations yield useful information in kidney disease are creatinine and urea, primarily, and, to a much less extent, uric acid. The latter component is of secondary value in the diagnosis of renal disease.

In addition to creatinine and urea, other NPN constituents in serum, as well as serum phosphate, tend to be elevated when there is failure to form or excrete urine properly. The measurement of NPN today is obsolete. It was a useful test in the early days of clinical chemistry because it was technically simple to measure the amount of nitrogen contained in serum constituents after precipitation of the proteins.

The NPN test has been superseded by the analysis of urea-N (comprising about 45% of the total NPN) and creatinine. The assay for uric acid, a minor component of NPN, is discussed on page 138.

Creatinine

Creatinine is a waste product formed in muscle from a high energy storage compound, *creatine phosphate (phosphocreatine)*. Adenosine triphosphate (ATP) is the immediate source of energy for muscular contraction as it is hydrolyzed to ADP. ATP cannot be stored in sufficient quantity to meet the energy demand of intense muscular activity, however; but creatine phosphate can be stored in muscle at approximately four times the concentration of ATP and is utilized for this purpose. When needed for energy, creatine phosphate and ADP are converted by enzymatic action to creatine and ATP (Fig. 4.4). A side reaction occurs, however, and a small portion of the creatine phosphate loses its phosphate as phosphate ion, with closure of the ring to form creatinine, as illustrated

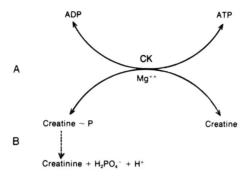

Fig. 4.4. Schema for the formation of creatinine in muscle as a side reaction from the spontaneous breakdown of creatine phosphate. Reaction A illustrates the reversible storage of high energy phosphate as creatine phosphate; Reaction B, the side reaction. CK = creatine kinase; $\sim$ is high energy bond.

in Figure 4.4. This reaction is not reversible, and the creatinine is excreted in the urine as a waste product.

The amount of creatinine excreted daily is a function of the muscle mass and is not affected by diet, age, sex, or exercise. It amounts to approximately 2% of the body stores of creatine phosphate and is roughly 1 to 2 g per day for an adult. Women excrete less creatinine than men because of their smaller muscle mass.

A small amount of preformed creatinine is ingested as a constituent of meat, but this has little effect upon the concentration of creatinine in serum. Elevated concentrations occur only when renal function is impaired.

Creatinine appears in the glomerular filtrate and is not reabsorbed by the tubule; hence, any condition that reduces the glomerular filtration rate will result in a lessened excretion from the body, with a consequent rise in the concentration of creatinine in plasma. Since the excretion rate of creatinine is relatively constant ($\pm 15\%$ for an individual per day) and since its production rate is not influenced by protein catabolism or other external factors, the concentration of creatinine in the serum is a good measure of renal glomerular function.

The serum creatinine concentration is elevated whenever there is a significant reduction in the glomerular filtration rate or when urine elimination is obstructed. The kidney reserve is such, however, that about 50% of kidney function must be lost before a rise in the serum concentration of creatinine can be detected. The concentration of serum creatinine is a better indicator of renal function than either that of urea nitrogen or uric acid because it is not affected by diet, exercise, or hormones, factors that influence the levels of the two latter constituents. A small percentage of the creatinine appearing in the urine may be derived from tubular secretion. This is negligible at normal serum levels of creatinine but becomes larger as the concentration in serum rises.

Creatinine metabolism and methodologies are reviewed by Narayanan and Appleton.[13]

Serum Creatinine

REFERENCE. Clinical Chemistry: Principles and Technics, edited by R. Henry, D.C. Cannon, and J.W. Winkelman. Hagerstown, MD, Harper & Row, 1974, p. 543.

PRINCIPLE. After removing the serum proteins, creatinine reacts with alkaline picrate solution to give a reddish color. This is known as the Jaffé reaction, which is not specific for creatinine; about 80 to 90% of the color produced by this reaction in plasma or serum, however, is due to creatinine. The noncreatinine chromogens that react with alkaline picrate include ketones, glucose, and ascorbic acid. Some methods measure "true" creatinine by utilizing an enzymatic method specific for creatinine,[1] by eliminating the effect of the noncreatinine chromogens by adsorbing creatinine on Lloyd's reagent[2] or on a cationic exchange resin,[3] and by manipulation of reaction conditions.[4,5] Continuous flow methods for serum creatinine utilize dialysis for separating creatinine from the proteins and measure the color produced by the Jaffé reaction.

REAGENTS

1. Picric Acid, 0.036 mol/L. Dissolve 9.16 g reagent grade picric acid (containing 10 to 12% added water as a safety feature*) in warm water, cool, and make up to 1 L volume.
2. Tungstic Acid in Polyvinyl Alcohol. Place 1 g polyvinyl alcohol (Elvanol 70–05, DuPont) in 100 ml water and heat to dissolve but do not boil. Transfer to a 1-L volumetric flask containing 11.1 g $Na_2WO_4 \cdot 2H_2O$ in 300 ml water. Then add 2.1 ml concentration H_2SO_4 in 300 ml water. Mix and dilute with water to 1 L volume. This solution is stable at room temperature for 2 years and does not require refrigeration.
3. NaOH, 1.4 mol/L. Dissolve 54 g NaOH in water and dilute to 1 L volume. Store in a polyethylene bottle.
4. Creatinine Stock Standard, 0.111 mg/ml. Dissolve 111.0 mg of creatinine in 1 L of 0.1 mol/L HCl.
5. Working Standard A. Dilute stock standard 1:50 with water. This will be equivalent to a serum creatinine of 2.0 mg/dl if 3.0 ml of working standard A is treated as a serum filtrate from 0.5 ml serum as described in the procedure below.
6. Working Standard B. Dilute stock standard 1:20 with water which makes its concentration equivalent to 5.0 mg/dl if treated as described below.

PROCEDURE

1. Precipitate the serum proteins by adding 0.5 ml serum to 4.0 ml tungstic acid solution contained in a 16 × 100 mm test tube. Shake vigorously and centrifuge for 10 minutes.
2. Set up in test tubes the following: blank—3.0 ml H_2O; standard—3.0 ml standard; unknown—3.0 ml protein-free centrifugate.
3. Add 1.0 ml picric acid to each and mix well.
4. Add 0.5 ml of 1.4 mol/L NaOH to the first tube, mix and set a timer for 15 minutes. Add NaOH to the remaining tubes at 30 second intervals.
5. Read absorbances of standards and unknowns against blank at 515 nm exactly 15 minutes after adding the NaOH. Read tubes at 30-second intervals.

$$\text{mg creatinine/dl} = \frac{A_u}{A_s} \times C$$

---

*Anhydrous picric acid is explosive. Picric acid crystals must never be desiccated or heated.

where C = concentration of standard; A_u and A_s are absorbances of unknown serum and standard, respectively.

$$\text{For Standard A: } C = \frac{0.111}{50} \times 3 \times \frac{4.5}{3.0} \times \frac{100}{0.5}$$

$$= 2 \text{ mg/dl when treated as a filtrate.}$$

$$\text{For Standard B: } C = \frac{0.111}{20} \times 3 \times \frac{4.5}{3.0} \times \frac{100}{0.5}$$

$$= 5 \text{ mg/dl when treated as a filtrate.}$$

$$\text{For SI units: } \mu\text{mol/L} = \text{mg/dl} \times 10 \times \frac{1000}{113} = \text{mg/dl} \times 88.4$$

REFERENCE VALUES

Adult males: 0.7 to 1.4 mg/dl or 62 to 125 μmol/L
Females: 0.6 to 1.3 mg/dl or 53 to 115 μmol/L
Children: 0.4 to 1.2 mg/dl or 36 to 107 μmol/L

The values for "true" creatinine are about 0.1 to 0.2 mg/dl lower than those listed.

INCREASED CONCENTRATION. The concentration of serum creatinine rises when there is impaired formation or excretion of urine, irrespective of whether the causes are prerenal, renal, or postrenal in origin.

Values that are 2 standard deviations above the upper limit of normal suggest possible renal damage, and the test should be repeated; levels of 1.5 mg/dl and

TABLE 4.3

Typical Correlation of Serum Creatinine Concentrations
with the Creatinine Clearance and Patient Status

SERUM CREATININE (mg/100 ml)	CREATININE CLEARANCE (ml/min)	PATIENT'S STATUS
0.6–1.3	100 ± 20	Normal
		Some patients with proteinuria
1.4–2.4	61–99	Capable of performing usual types of activity
2.5–4.9	24–60	Difficulty in performing strenuous physical activity
5.0–7.9	12–23	Unable to perform all of daily physical activities except on part-time basis
		Acidosis, sodium loss, serum Ca ↓
8.0–12	7–12	Severe limitation of physical activity
		Serum K ↑, Ca ↓, Na ↓
>12	6 or less	May be disoriented or in coma

Modification from American Heart Association, Council on Kidney in Cardiovascular Disease.

NOTE: When there is retention of creatinine in serum, there is always retention (elevation) of serum phosphate, urate, and urea. The degree of rise in urea concentration depends to a large extent upon the amount of protein in the diet.

greater indicate impairment of renal function. Minor changes in concentration may be of significance, and the serum level usually parallels the severity of the disease. Table 4.3 shows the relation between the serum creatinine concentration, the creatinine clearance value and the patient's status. The prerenal factors causing an increased serum creatinine level are congestive heart failure, shock, salt and water depletion associated with vomiting, diarrhea, or gastrointestinal fistulas, uncontrolled diabetes mellitus, excessive use of diuretics, diabetes insipidus, and excessive sweating with deficient salt intake. Renal factors may involve damage to glomeruli, tubules, renal blood vessels, or interstitial tissue. Postrenal factors may be prostatic hypertrophy, neoplasms compressing the ureters, calculi blocking the ureters, or congenital abnormalities that compress or block the ureters.

The serum creatinine concentration is monitored closely following a renal transplant because a rising concentration, even though small, may be an indication of transplant rejection.

DECREASED CONCENTRATION. Low serum creatinine concentrations have no clinical significance.

Urine Creatinine

Since the concentration of creatinine in urine (it approximates 1 mg/ml) is much higher than in serum, a dilution of urine is in order. For urines of normal protein content, dilute 1:200 with water and treat the same as a serum filtrate—that is, take 3.0 ml for analysis. If there is proteinuria, precipitate the proteins as follows: Add 0.5 ml urine + 0.5 ml water to 4.0 ml tungstic acid solution; mix and centrifuge. Dilute the filtrate 1:20 with water and take 3.0 ml for analysis as for serum. Thus both types of urine, those with and those without proteinuria, are diluted 1:200. Develop color in standards A and B as for serum. There is very little noncreatinine chromogen in urine.

CALCULATION

$$\text{For Standard A: } C = \frac{A_u}{A_s} \times \frac{0.111}{50} \times 3 \times \frac{100}{3/200} \text{ mg/dl}$$

$$= \frac{A_u}{A_s} \times \frac{0.111}{50} \times 3 \times \frac{100}{0.015} = \frac{A_u}{A_s} \times 44.4 \text{ mg/dl}$$

REFERENCE VALUES. The urinary excretion of creatinine depends upon the muscle mass of the individual, but the following are rough reference values:

Men:	2.0 ± 0.5 g/24 hr or 23 ± 2 mg/kg/24 hr	
Women:	1.6 ± 0.6 g/24 hr or 20 ± 2 mg/kg/24 hr	
Children:	Newborn	7 to 12 mg/kg/24 hr
	1.5 to 22 mo	5 to 15 mg/kg/24 hr
	2.5 to 3.5 yr	9 to 15 mg/kg/24 hr
	4 to 10 yr	15 to 25 mg/kg/24 hr

The factor for converting mg/kg/day to mmol/kg/day = 0.00884.

Creatinine Clearance

The most sensitive chemical method of assessing renal function is the creatinine clearance test. This clearance test provides an estimate of the amount of plasma that must have flowed through the kidney glomeruli per minute with complete removal of its content of creatinine to account for the creatinine per minute actually appearing in the urine. The test requires the complete collection of the urine formed in an accurately measured period of time in addition to the analyses of the creatinine concentrations in the urine and in the serum. The creatinine clearance is calculated as

$$\frac{U}{S} \times V$$

where U is the urine concentration of creatinine, S is the serum creatinine concentration, and V is the volume of urine excreted per minute. U and S must be measured in the same concentration units, although it does not matter whether the measurement is in mg/dl or in SI units. The dimension of the clearance thus becomes expressed as ml/min, since the dimensions of U/S cancel each other out. The creatinine clearance is practically the same as the glomerular filtration rate; the tubules excrete a small amount of creatinine, a factor that may become significant when serum creatinine concentrations are elevated. The creatinine clearance value is closer to the glomerular filtration rate when "true" creatinine is measured because noncreatinine chromogens increase the serum concentration but not of urine creatinine.

The importance of obtaining the total excretion of urine in an accurately timed period cannot be overstressed because any error committed in the collection and measurement of the volume excreted per minute is carried over into the calculation of the clearance. The following points should be emphasized. Those who administer the test (nurses, ward clerks, technologists) must thoroughly understand the procedure and communicate in a simple way to the patient so that the patient understands and cooperates. The test may be carried out over *any accurately timed period* if the urine flow is adequate, even though the usual requests are for 2-, 4-, or 24-hour periods. The elapsed time should be recorded to the minute. It is necessary to start timing the test when the bladder is completely empty.

PROCEDURE

1. Promote a good urine flow by having the patient drink about 500 ml of water 10 or 15 minutes before the test starts.
2. *When the patient is able to void,* have him/her completely empty the bladder, *note down the time exactly,* and *discard this urine.*
3. When the patient seems to have a full bladder and can void (in approximately 2 hours), have him/her completely empty the bladder into a container, note the time exactly, and send this urine to the laboratory, along with a blood sample, for measurement of the clearance. If the patient's height and weight are provided, the laboratory can correct the clearance for standard body surface area from nomograms in Figure 4.5. Body surface area may also be calculated from the formula:

$$\log A = 0.425 \log W + 0.725 \log H - 2.144$$

where A = body surface in m², W = weight in kg, H = height in cm.

Clinical Chemistry

NOMOGRAMS FOR CALCULATING BODY SURFACE AREA *

Nomogram for Calculating the Body Surface Area of Children¹

Nomogram for Calculating the Body Surface Area of Adults¹

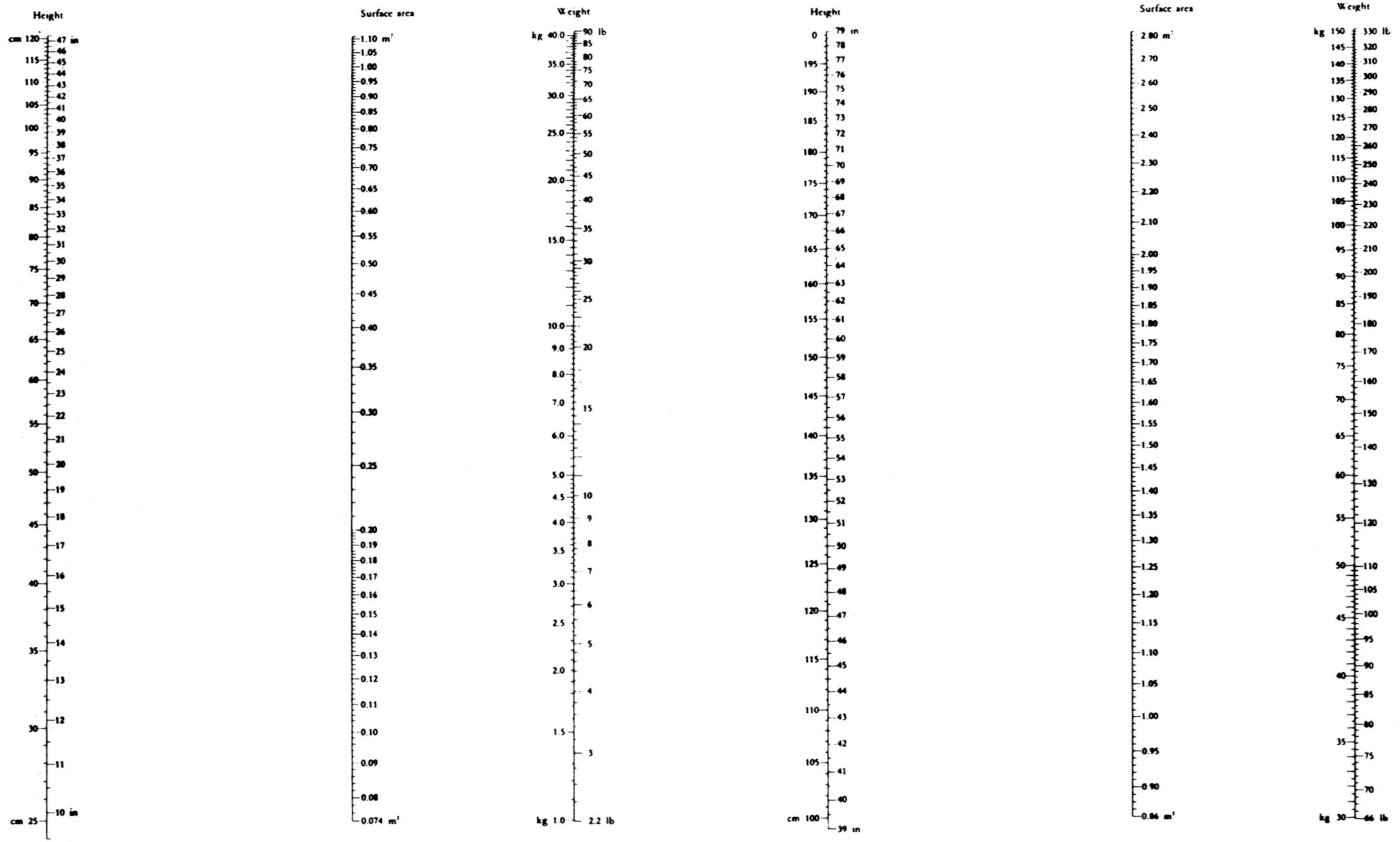

1) From the formula of Du Bois and Du Bois, Arch. intern. Med. 17, 863, 1916. $S = W^{0.425} \times H^{0.725} \times 71.84$, or log $S = 0.425$ log $W + 0.725$ log $H + 1.8564$, where $S =$ body surface area in square centimeters, $W =$ weight in kilogram, $H =$ height in centimeters.

*From *Documenta Geigy Scientific Tables*, 6th ed., pp. 632–633. By permission of J. R. Geigy S.A.

Fig. 4.5. Nomogram for the determination of body surface area. (Reproduced from Documenta Geigy Scientific Tables, 7th ed. Courtesy of Ciba-Geigy Limited, Basel, Switzerland.)

4. *Calculation of Creatinine Clearance*

$$\text{Uncorrected: Clearance} = \frac{U}{S} \times V$$

$$\text{Corrected for surface area: Clearance} = \frac{U}{S} \times V \times \frac{1.73}{A}$$

Where U is the urine creatinine concentration, S is the serum creatinine concentration, V is the urine flow in ml/min, A is body surface area in m², and 1.73 is the standard body surface area of the average adult.

Notes:

1. It is important that the patient understands what is required and cooperates by completely voiding when necessary and records the time to the minute. Thus, an order for a 2-hour collection should be recorded as the total elapsed time in minutes because the time between the two voidings is rarely 2 hours exactly.
2. A 24-hour collection is *not* more accurate than a 2-hour collection and is far more inconvenient. Moreover, there is a good chance of losing a urine specimen occurring with a bowel movement or losing a urine specimen at night if the patient has not been cautioned or may not be too alert.
3. The total specimen should be sent to the laboratory for measurement. Measurements made on the ward or clinic may not be reliable. If aliquots are to be removed for other tests, it should be done by laboratory personnel *after mixing and measuring the complete collection.*
4. It is important to correct for body surface area in small individuals (infants, children, and small adults) and also in large adults.

REFERENCE VALUES. The values obtained for the creatinine clearance depend to some extent upon the methodology because of a larger percentage of non-creatinine chromogen in plasma or serum than in urine. For nonspecific creatinine methods, the adult normal creatinine clearance for males is 105 ± 20 ml/min and 97 ± 20 ml/min for females. When a specific creatinine method is employed, these values become 120 ± 25 ml/min for males and 112 ± 20 ml/min for females.

In children above the age of 1 year, the creatinine clearance is about the same as for adults *when corrected for body surface area.* The creatinine clearance of prematures and newborns is about 40 to 65 ml/min/1.73 m² while that of 6-month-old infants is about 75 ml/min/1.73 m².

INCREASED CREATININE CLEARANCE VALUES. An increased value has no medical significance. If the value should be greatly increased, suspect some error in the collection and/or timing of the clearance. Perhaps the bladder was not completely empty when the timing was started.

DECREASED CREATININE CLEARANCE VALUES. When the clearance test is carefully executed, a decreased creatinine clearance is a very sensitive indicator of a decreased glomerular filtration rate. The reduced glomerular filtration rate may be caused by acute or chronic damage to the glomerulus or to any of its components. Reduced blood flow to the glomeruli may also produce a decreased creatinine clearance. Acute tubular damage may result in a decreased creatinine

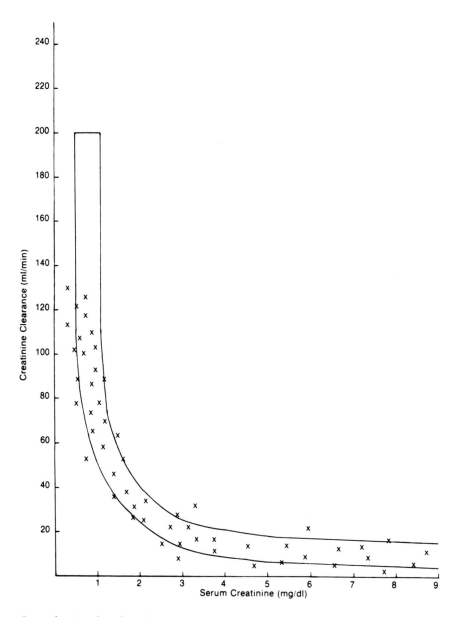

Fig. 4.6. Curve showing the relation between serum creatinine concentration and the creatinine clearance.

clearance as blood flow to the glomeruli is drastically reduced in response to osmolar changes. Table 4.3 lists typical values of the creatinine clearance and serum creatinine concentrations in patients with varying degrees of renal damage; in Figure 4.6, creatinine clearance values are plotted against serum creatinine.

Note: If a low creatinine clearance is obtained in a patient with a normal serum creatinine concentration, suspect either an error in the specimen collection (incomplete urine voiding, incorrect timing, loss of some specimen, incorrect volume measurement) or in the creatinine analyses. The former is more likely.

Urea Nitrogen

When proteins are ingested, they are hydrolyzed to amino acids that can be used for anabolic or catabolic purposes. Protein cannot be stored in the body to any appreciable extent, so when the intake is in excess of body requirements for the synthesis of structural and functional components, the surplus amino acids are largely catabolized for energy purposes as illustrated schematically in Figure 4.7.

The α-amino group of all amino acids that are broken down in the mammalian body, whether derived from the diet or endogenous sources, end up in the compound, urea, whose structure is shown in Figure 4.8. It is a waste product that is soluble in water and excreted solely by the kidney. Urea is the characteristic nitrogenous end product of protein catabolism in mammals, but this is not true for many other forms of life; birds and reptiles excrete uric acid as the end product, and for some bony fish, the chief product is ammonia.

For historic reasons, it has been customary to express the serum urea concentration in terms of its nitrogen because in the early days of clinical chemistry, nitrogenous substances were analyzed by the Kjeldahl method, which determined the nitrogen content. The molecular weight of urea is 60, and each g mole contains 28 g of nitrogen. The distinction between urea-N and urea becomes meaningless in the SI, which expresses the urea concentration as

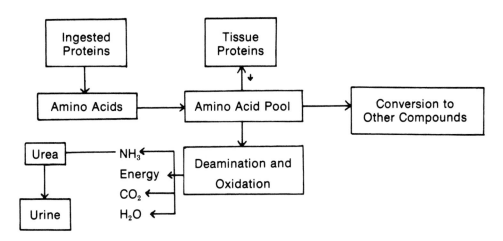

Fig. 4.7. Illustration of the fate of ingested protein. Hydrolysis to amino acids occurs in the intestine. The absorbed amino acids are carried to the liver and other tissues where they may be utilized for the synthesis of new proteins, converted to other compounds, or utilized for energy. When catabolized, the amino group ends up in urea, a waste product excreted by the kidney.

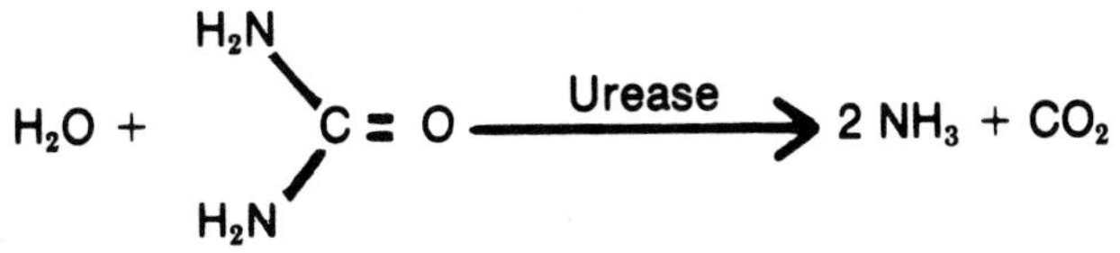

$$H_2O + \begin{array}{c} H_2N \\ \\ \\ H_2N \end{array} \hspace{-6pt} C = O \xrightarrow{\text{Urease}} 2\,NH_3 + CO_2$$

Fig. 4.8. The reaction that occurs when urea is split by urease.

mmol/L. Thus, a serum concentration of 28 mg/dl of urea-N is equivalent to 60 mg/dl of urea or 10 mmol/L of urea or urea-N in SI units.

Urea appears in the renal glomerular filtrate in the same concentration as in plasma, but unlike creatinine, some urea is absorbed as the urine-to-be traverses the renal tubule. Under conditions of normal flow and normal renal function, about 40% of the filtered urea is reabsorbed; when the flow rate is decreased, the actual and relative amount reabsorbed increases. As with creatinine, the serum concentration of urea (or urea-N) rises with impaired renal function.

The serum concentration of urea-N is influenced by factors not connected with renal function or urine excretion. It is affected strongly by the degree of protein catabolism, whether produced by a high protein diet or by hypersecretion or injection of adrenal steroids that results in the mobilization of protein for energy purposes. In the case of diet, a change to a high protein diet can double the serum urea-N concentration, and a low protein intake can reduce it by half. In like manner, the injection or ingestion of steroids produces a rise in serum urea-N as do stressful situations (e.g., breaking an arm) that cause the adrenal gland to secrete additional cortisol. For these reasons, the measurement of serum creatinine is a better indicator of kidney status than is that of urea-N, although in many cases they go up and down together. The various prerenal, renal, and postrenal factors that affect the concentration of serum creatinine also influence the level of serum urea-N. In the case of the latter constituent, one always has to assess the possible influence of dietary or hormonal factors.

Serum Urea-N

Two fundamentally different chemical approaches are used for determining serum urea or urea-N. The first employs the enzyme urease, which splits off ammonia from the urea molecule in a highly specific reaction. The ammonia may then be measured by reacting it with phenol and alkaline sodium hypochlorite, with sodium nitroprusside as a catalyst, to produce a very stable, intensely blue solution[6]; this is called the Berthelot reaction. The ammonia may also be measured spectrophotometrically by an NADH-dependent reaction with α-ketoglutarate, catalyzed by the enzyme glutamic dehydrogenase.[7] The method employing the Berthelot reaction is well suited to manual determination, whereas the reaction coupled with the glutamic dehydrogenase enzyme may be performed manually or by automation.

The second fundamental approach to the measurement of the urea or urea-N is that of reacting a protein-free solution of urea with diacetyl monoxime at 100° C in the presence of strong acid and an oxidizing agent.[8] This method is popular for automated instruments, but is less suitable as a manual method, because the color is not stable and the timing is critical.

REFERENCES. Chaney, A.L., and Marbach, E.P.: Modified reagents for determination of urea and ammonia. Clin. Chem. *8*, 130, 1962.
Kaplan, A., and Teng, L.L.: Serum urea (urease-Berthelot). In Selected Methods of Clinical Chemistry, Vol. 9, edited by W.R. Faulkner and S. Meites. Washington, D.C., American Association for Clinical Chemistry, 1982, p. 357.

PRINCIPLE. Urease hydrolyzes serum urea and only urea into NH_3 and CO_2. The liberated NH_3 is converted into an indophenol dye (blue) by treatment with phenol and sodium hypochlorite in alkaline solution containing a catalyst (sodium nitroferricyanide). The intensity of the blue color is measured spectrophotometrically. It is not necessary to precipitate serum proteins.

REAGENTS

1. Phenol-nitroferricyanide Solution (0.5 mol/L phenol and 0.8 mmol/L sodium nitroferricyanide). To approximately 500 ml water in a liter volumetric flask, add 50 g reagent grade phenol and 0.25 g sodium nitroferricyanide, $Na_2Fe(CN)_5NO\cdot2H_2O$. Dilute to the mark with water. The reagent is stable for at least 2 months when stored at 4 to 8° C.

2. Alkaline Hypochlorite (0.5 mol/L NaOH, 30 mmol/L NaOCl). In approximately 600 ml water in a liter volumetric flask, dissolve 25 g NaOH pellets. When cool, add 43 ml of a commercial hypochlorite solution containing 52 g/L NaOCl. Dilute to volume with water. The reagent is stable for at least 3 months when protected from the light and stored at 4 to 8° C.

3. *Phosphate buffer-EDTA solution* (20 mmol/L phosphate and 10 mmol/L EDTA), pH 6.9 at 25° C. To approximately 900 ml of water, add 3.36 g of the disodium salt of (ethylenedinitrilo) tetracetic acid (EDTA) and 3.48 g of K_2HPO_4. Stir until dissolved and then adjust the pH to 6.90–6.95 with HCl. Make up to 1 L volume with water and store at 4 to 8° C.

4. Stock Urease Solution, approximately 40 Sigma U/ml (456 IU/ml). Dissolve sufficient high purity urease (Type IX, Sigma) to provide 2000 Sigma units (approximately 300 mg) in 25 ml water and add 25 ml glycerol. The solution is stable for at least 6 months at 4 to 8° C.

5. Dilute Urease Solution, 0.4 Sigma U/ml. Dilute 1 ml stock urease to 100 ml with phosphate buffer-EDTA solution. The enzyme is stable for at least 3 weeks when stored at 4 to 8° C.

6. Stock Urea-N Standard, 500 mg/100 ml. Dissolve in water 1.0717 g Bureau of Standards urea that has been stored in a desiccator, add 0.05 g sodium azide as a preservative, and make up to 100 ml volume. Store at 4 to 8° C.

7. Working Urea-N Standards:
   a. 15 mg/dl. Dilute 3.0 ml of stock standard to 100 ml with 0.5 g/L sodium azide solution.
   b. 50 mg/dl. Dilute 10.0 ml of stock standard to 100 ml with 0.5 g/L sodium azide solution.

PROCEDURE

1. Label tubes for standard, blank, control, and unknowns.

2. Pipet 1.0 ml of urease working solution into each tube.

3. Transfer 10 μl water and 10 μl working standard to the blank and standard tubes, respectively, and 10 μl appropriate sera to the control and unknown tubes. Mix and incubate all tubes for 5 min at 55° C (or for 15 min at 37° C).

4. Add 1.0 ml phenol-nitroferricyanide to each tube, mix, and then add 1.0 ml alkaline hypochlorite solution to each. Mix and incubate at 50 to 55° C for 5 minutes (or 20 minutes at 37°).

5. Add 7 ml water to all tubes, mix, and record the absorbance at 630 nm using the blank as a reference. If an intense color is obtained with the serum of a uremic patient, the mixture may be diluted with up to 3 volumes of water and read against a blank diluted in a similar fashion. The final

concentration obtained by multiplying by the appropriate dilution factor obeys Beer's Law.

6. Calculation:

$$\frac{A_u}{A_s} \times C = mg\ urea\text{-}N\ per\ dl$$

where A_u and A_s are absorbances of unknown and standard, respectively, and C = concentration of standard.

For SI units, mmol/L = mg/dl × 0.357.

REFERENCE VALUES: 8 to 18 mg/dl with normal protein intake*
2.9 to 6.4 mmol/L

INCREASED CONCENTRATION. High protein diet, administration of cortisol-like steroids, stressful situations, and prerenal, renal, and postrenal factors as described for creatinine increase the concentration of urea-N.

DECREASED CONCENTRATION. The urea concentration is low in late pregnancy when the fetus is growing rapidly and utilizing maternal amino acids, in starvation, and in patients whose diet is grossly deficient in proteins.

Uric Acid

Uric acid is a purine compound that circulates in plasma as sodium urate and is excreted by the kidney. It is derived from the breakdown of nucleic acids that are ingested or come from the destruction of tissue cells; it is also synthesized in the body from simple compounds. The formation of uric acid

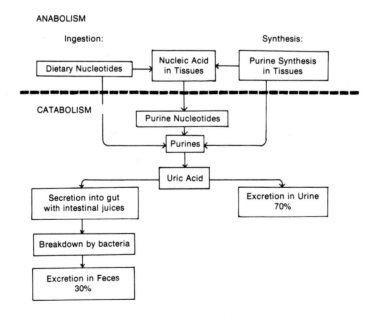

Fig. 4.9. Schematic representation of uric acid formation and excretion.

---

*The normal curve is skewed to the right and many apparently healthy individuals may have a serum urea-N concentration as high as 27 mg/dl.

is illustrated in Figure 4.9. Urate is the end product of purine metabolism in man and the higher apes but is oxidized to allantoin by most other mammals. Urate appears in the glomerular filtrate and is partially reabsorbed in the tubules. Urate is of low solubility in plasma, and uric acid is less; there is a danger of precipitation of uric acid crystals where there is a local rise in $[H^+]$ in tissues. Urate deposition in the kidney may ultimately lead to renal failure.

The measurement of serum uric acid is *not* used as a primary test for the evaluation of kidney function because creatinine and urea serve this purpose much better, but when multichannel analyses are carried out, the concentration of serum urate, in general, should reflect the same changes that occur with creatinine and urea-N. Thus, it may serve as a confirmatory check on the analyses of these constituents; the serum concentration of urate is usually elevated when there is improper formation or excretion of urine, irrespective of the cause.

The main value of the serum uric acid test is in the diagnosis of gout or for following the treatment of patients with this disease. It also is used sometimes as an indicator of a large-scale breakdown of nucleic acid such as occurs in toxemia of pregnancy, after massive irradiation for tumors, or following the administration of cytotoxic agents in the treatment of malignancies.

Gout, an ancient and painful disease, is characterized by the precipitation of uric acid crystals in tissues and joints, particularly in joints of the big toe. Although the deposition of crystals in the joints is responsible for the pain and debilitation of gout, the deposition of urate may damage the kidneys and presents the greatest danger to the patient.

Serum Uric Acid

REFERENCE. Jung, D.H., and Parekh, A.C.: An improved reagent system for the measurement of serum uric acid. Clin. Chem. 16, 247, 1970.

PRINCIPLE. The classic chemical method for the determination of serum urate is based upon the reduction of phosphotungstic acid by urate to a blue phosphotungstate complex, which is measured spectrophotometrically. The reaction is not specific for urate because other substances present in serum (reducing agents) also produce a small amount of phosphotungstate blue under the conditions of the test; the substances that do this are referred to as "nonurate chromogens." "True urate" values may be obtained by utilizing an enzymatic method for the determination of serum urate.[9] An added enzyme, uricase, catalyzes the oxidation of urate to allantoin, H_2O_2, and CO_2. The serum urate may be determined by measuring the absorbance at 292 nm before and after treatment with uricase, since urate absorbs light at this wavelength and allantoin, CO_2, and H_2O_2 do not,[9] or by measuring the amount of oxygen consumed in the uricase reaction.[10]

REAGENTS
1. Trisodium Phosphate, 10 g/L. Dissolve 1.0 g $Na_3PO_4 \cdot 12H_2O$ in water and make up to 100 ml volume.
2. Phosphotungstic Acid, 50 g/L. Transfer 50 g of molybdenum-free sodium tungstate ($Na_2WO_4 \cdot 2H_2O$) in 350 ml water, add 20 ml of 85% H_3PO_4, and reflux for 2 hours. Add 1 drop of liquid bromine, cool to room temperature, and make up to 1 L volume with water.

3. Carbonate-Urea-Triethanolamine (CUTE). Dissolve 100 g Na_2CO_3, 200 g urea, and 50 g triethanolamine in water and dilute to 1 L volume.
4. Uric Acid Stock Standard, 1.0 mg/ml. Dissolve 100 mg Li_2CO_3 in 100 ml water in a 200-ml volumetric flask while warming in a 50° water bath. Add 200.0 mg uric acid and stir until dissolved. Add 4 ml of formaldehyde solution (37 to 40 g/100 ml) and 1 ml of glacial acetic acid slowly, while shaking, and make up to 200 ml volume with water.
5. Working Uric Acid Standard, 5.0 mg/100 ml. Dilute 5.0 ml stock standard uric acid to 100 ml volume with water.

PROCEDURE

1. In 3 ml centrifuge tubes labeled "Standard," "Blank," "Control," and "Unknown," transfer 0.2 ml working uric acid standard, water, control serum, and unknown sera, respectively.
2. Add 0.2 ml trisodium phosphate reagent to each; mix and let stand for 5 minutes. This alkaline treatment destroys any ascorbic acid that may be in the sample, a nonurate chromogen.
3. Add 0.6 ml phosphotungstic acid, mix well, and centrifuge for 5 minutes to remove proteins.
4. Transfer 0.5 ml of the supernatant fluid to a 12 × 75-mm colorimeter tube or to a square cuvet of 10-mm light path.
5. Add 0.2 ml of the phosphotungstic acid and 1.5 ml of CUTE reagent. Mix by inversion.
6. In 20 to 50 minutes read the absorbance of sample against the blank at 680 to 700 nm.
7. *Calculation*

$$\frac{A_u}{A_s} \times 5.0 = \text{mg/dl uric acid}$$

For SI units:

$$\text{mmol/L} = \text{mg/dl} \times 10 \times \frac{1}{168} = \text{mg/dl} \times 0.0595$$

PRECAUTIONS

1. The pH for precipitation of the proteins should be below 3 in order to avoid loss of uric acid.
2. The precipitated proteins *should be removed by centrifugation* rather than by filtration in order to avoid loss of uric acid by adsorption to the filter paper.
3. The pH for color development must be quite alkaline (at least, pH 9), but there is danger of turbidity caused by precipitation of phosphotungstate if the solution is too alkaline. The method recommended above utilizes triethanolamine as a stabilizing agent at a reaction pH of 10.5.

REFERENCE VALUES

Males: 3.5 to 7.5 mg/dl or 0.210 to 0.445 mmol/L
Females: 2.5 to 6.5 mg/dl or 0.150 to 0.390 mmol/L
Children: 2.0 to 5.5 mg/dl or 0.120 to 0.330 mmol/L

INCREASED CONCENTRATION. Serum urate concentration is elevated in most patients with gout, in renal disease, after increased breakdown of nucleic acid or nucleoproteins (in leukemia, polycythemia, toxemia of pregnancy, resolving pneumonia and after irradiation of x-ray sensitive carcinomas). Some individuals have an elevated concentration of serum urate despite the absence of disease (idiopathic hyperuricemia).

DECREASED CONCENTRATION. The concentration of serum urate may be decreased after the administration of ACTH or cortisol-like steroids, certain drugs that decrease the reabsorption of urate by renal tubules (aspirin, probenecid, penicillamine) or by drugs (allopurinol) that block a step in the formation of uric acid.

Urine Uric Acid

The concentration of uric acid in the urine is influenced by the purine content of the diet. The average person on an average diet usually excretes from 250 to 750 mg/24 hr. The method for its determination in urine is identical to that in serum except that (1) the urine must be diluted because its urate concentration is much higher than in serum, and (2) there is usually no protein precipitate when the phosphotungstate is added.

PROCEDURE

1. Mix the urine well to suspend any sediment, take an aliquot and warm to 55 to 60° for 10 minutes in order to dissolve any precipitated uric acid, and centrifuge.
2. Dilute the urine 1:10 with water and treat the diluted urine in the same way as described for serum, beginning with Step 1. In Step 3, omit the centrifugation if there is no precipitate.
3. Calculate as for serum uric acid except multiply by 10 because of the dilution.

$$\frac{A_u}{A_s} \times 5 \times 10 = \frac{A_u}{A_s} \times 50 = \text{mg/dl uric acid}$$

To convert to mmol/L, multiply mg/dl × 0.0595.

URINE CONCENTRATION TEST

The ability of the kidney to concentrate urine is a test of tubular function that can be carried out readily with only minor inconvenience to the patient. A urine concentration test is widely used to test tubular function.

REFERENCE. Jacobson, M.H., et al.: Urine osmolality. A definitive test of renal function. Arch. Int. Med. *110*, 83, 1962.

The original concentration tests requiring water deprivation for 24 hours have been replaced by a 14-hour period that gives similar results. The test should *not* be performed on a dehydrated patient.

PROCEDURE

1. The patient eats an early supper and is allowed no food or water after 6 P.M. on the night preceding the test. Discard any urine voided during the night.

2. On the test day, the first specimen is voided at 7 A.M., the bladder emptied completely and the specimen discarded.
3. A second specimen is collected at 8 A.M., 14 hours after the commencement of the test and the osmolality or specific gravity is measured. If the osmolality exceeds 850 mosm/kg or the specific gravity is >1.022, the patient has adequate renal concentrating power and the test is over.
4. If the values are less than 850 mosm/kg or 1.022 for osmolality and specific gravity, respectively, a urine sample is collected at 9 A.M. and assayed.
5. Interpretation: The renal concentrating ability is impaired if neither urine sample reached an osmolality of 850 mosm/kg or a specific gravity of 1.022. Normal values may be as high as 1350 mosm/kg or a specific gravity of 1.032. If the concentrating ability should fall to zero, the specific gravity would be 1.010 and the osmolality close to 300 mosm/kg.

URINARY TRACT CALCULI (STONES)[14]

Calcium and/or magnesium salts frequently form insoluble calculi (stones) in the kidney. About 20% of the people with renal stones are found to have an elevated serum calcium concentration due to hyperparathyroidism. The host is unaware of a stone until it blocks a tubule or a ureter and then the dilated vessel produces intense pain. Prolonged deposition causes irreversible renal damage. Stones are formed by the concentric deposition of poorly soluble compounds around some nuclei. The nuclei may be blood clots, fibrin, bacteria, or sloughed epithelial cells. Many times, the precipitation of the relatively insoluble compounds may be initiated or aggravated by an infection, dehydration, excessive intake or production of the compound, urinary obstruction, and other factors. Once formed, the calculi tend to grow by accretion unless they happen to be dislodged and are sufficiently small to travel down the urinary

TABLE 4.4
Frequency of Occurrence of the Various Chemical Constituents in Calculi*

CHEMICAL GROUP	% OCCURRENCE
Calcium	97
Phosphate	88
Oxalate	65
Urate	15
Carbonate	12
Magnesium	25
Ammonium	20
Cystine	2
Iron	Rare
Sulfate	Rare
Xanthine	0.5
Cholesterol	Rare
Indigo	Rare
Sulfonamides	Rare

*Based upon a study of calculi analyzed by U.S. Naval Medical School, Bethesda, Maryland laboratory through the year of 1956.

tract to be excreted. The larger calculi may remain in the kidney or become stuck in a ureter from which they must be removed by surgery. Whether a stone is passed in the urine or is removed by surgery, information concerning the type of calculus is of value; changes in diet may be recommended.

The analysis of calculi is not difficult if there is sufficient material with which to work. Specimens obtained by surgery usually present no problem, but stones that are passed in the urine are quite small; there is so little material to analyze that the analyst should look first for the most likely constituents before running out of specimen (see Table 4.4).

PRINCIPLE. The calculus is weighed and size, shape, color, surface appearance, and consistency are noted. The calculus is pulverized in a mortar if it is larger than 25 mg and crushed in a test tube with a glass rod if it is less than 25 mg. The powder is dissolved, if possible, in 1 mol/L HCl, and chemical spot tests are made for different constituents on both the acid solution and the residue or powder.

REAGENTS AND MATERIALS

1. Sodium Carbonate, 2 mol/L. Dissolve 20 g Na_2CO_3 in 100 ml water.
2. Phosphotungstic Acid. Same reagent as for the determination of uric acid (see p. 139).
3. Ammonium Molybdate in HNO_3. Dissolve 5 g ammonium molybdate in 123 ml water and add 12 ml concentrated HNO_3 with stirring.
4. NH_4OH, concentrated.
5. Sodium Cyanide, 50 g/L. Dissolve 5 g NaCN in 100 ml water and add 0.2 ml concentrated NH_4OH.
6. Sodium Nitroferricyanide. Dissolve 5 g $Na_2FE(CN)_5NO \cdot 2H_2O$ in 100 ml water. Discard when color fades.
7. HNO_3, concentrated.
8. HCl, 1 mol/L. Add 8 ml concentrated HCl to water and make up to 100 ml volume.
9. Ammonium Thiocyanate. Dissolve 3 g NH_4SCN in 100 ml water.
10. Sodium Oxalate. Saturate 100 ml water with $Na_2C_2O_4$ by adding 5 g of the salt, shaking well, and allowing it to settle.
11. NaOH, 5 mol/L. Dissolve 20 g of NaOH pellets in water and make up to 100 ml volume when cool.
12. Titan Yellow (Clayton Yellow, National Aniline). Dissolve 0.1 g dye in 100 ml water. Make alkaline by the addition of 3 drops 5 mol/L NaOH. Store in amber bottle and prepare freshly every 30 days.
13. Manganese Dioxide, MnO_2, powdered, reagent grade.
14. Phenol-Nitroferricyanide Reagent. (Same as for urea-N determination, solution 1, p. 137).
15. Alkaline Hypochlorite. (Same as for urea-N determination, solution 2, p. 137).
16. Sodium Nitrite, $NaNO_2$, 0.1 g/dl water. Prepare just before use.
17. N-(1-naphthyl) Ethylenediamine Dihydrochloride (Eastman). Dissolve 0.1 g in 100 ml water. Prepare on day of use.
18. Chloroform.
19. Acetic Anhydride.
20. Concentrated H_2SO_4.
21. p-Methylaminophenol Sulfate. (Elon, Eastman).

22. Spot plates, 12-hole, white spot plate; 3-hole, black spot plate.

PROCEDURE

1. If the stone weighs less than 25 mg, place it in a test tube and crush it with a glass rod. Leave the rod in the tube and add 2 ml 1 mol/L HCl; stir to dissolve. Observe and treat the supernatant fluid for the constituents listed below. If the stone weighs more than 25 mg, pulverize in a mortar and place about 5 mg powder in a test tube and mix with 2 ml HCl.

    a. *Carbonate.* Observe the supernatant fluid above for bubbles when the acid is added. Effervescence indicates the presence of carbonate.

    b. *Calcium.* Place 3 drops of supernatant fluid in a black spot plate. Add 4 drops sodium oxalate and 2 drops NH_4OH. A white precipitate is positive for calcium.

    c. *Magnesium.* Place 3 drops of supernatant fluid in a white spot plate. Add 3 drops of NaOH and 1 drop of Titan Yellow solution. A blood red precipitate is positive for Mg (disregard any other color).

    d. *Ammonia.* Transfer 3 drops of supernatant fluid to a white spot plate. Add 2 drops NaOH, 1 drop phenol-nitroferricyanide solution (#14) and 1 drop alkaline hypochlorite solution (#15). Mix and place in a 37° incubator for 3 to 5 minutes. A blue color is positive for the presence of an ammonium salt (the Berthelot reaction).

    e. *Oxalate.* Add a pinch of MnO_2. The "explosive release" of tiny gas bubbles is positive for oxalate.

2. a. *Phosphate.* Place a few mg of powder in a white spot plate; add 2 drops ammonium molybdate solution and 2 drops Elon solution (#21). Prepare a reagent blank by excluding the powder. A much deeper blue color for the sample over the blank is positive for phosphate. For very small stones, use 3 drops of supernatant fluid instead of the powder.

    b. *Urate.* To a bit of the stone powder or residue, add 1 drop Na_2CO_3 plus 2 drops phosphotungstate solution. A prompt deep blue color is positive for urate.

    c. *Cystine.* To a few mg of stone powder in a spot plate (or to residue from acid solution) add 1 drop NH_4OH, 1 drop NaCN, wait 5 minutes, and then add 2 drops of sodium nitroferricyanide. A beet red color indicates cystine.

As can be seen from Table 4.4, one or more of the above constituents are found in 99% of all renal calculi; calcium is present in 97%.

3. In the event that all of the preceding tests are negative, test for the following rare constituents:

    a. *Iron.* Pulverize the calculus. Add 3 drops HNO_3 plus 3 drops NH_4SCN. A red color is positive for iron.

    b. *Xanthine.* Place some of the powdered stone in an evaporating dish. Add 0.5 ml concentrated HNO_3 and heat to dryness in a fume hood. Add 3 drops concentrated NH_4OH to the residue. A yellow residue that turns orange when the NH_4OH is added is positive for xanthine.

    c. *Cholesterol.* Place some powder in a small test tube. Stir with 5 drops chloroform. Add 10 drops acetic anhydride and 1 drop concentrated sulfuric acid. A bluish-green color is positive for cholesterol.

d. *Sulfonamide.* To some powdered stone in a spot plate, add 2 drops HCl, 2 drops $NaNO_2$, 3 drops ammonium sulfamate, and 3 drops of coupling reagent (#17). A red color is positive for sulfonamides.

REFERENCES

1. Larsen, K.: Creatinine assay by a reaction-kinetic principle. Clin. Chim. Acta *41*, 209, 1972.
2. Owen, J.A., Iggo, B., Scandrett, F.J., and Stewart, C.P.: The determination of creatinine in plasma or serum and in urine; a critical examination. Biochem. J. *58*, 426, 1954.
3. Rockerbie, R.A., and Rasmussen, K.L.: Rapid determination of serum creatinine by an ion-exchange technique. Clin. Chim. Acta *15*, 475, 1967.
4. Lustgarten, J.A., and Wenk, R.E.: Simple, rapid kinetic method for serum creatinine measurement. Clin. Chem. *18*, 1419, 1972.
5. Raabo, E., and Walloe-Hansen, P.: A routine method for determining creatinine avoiding deproteinization. Scand. J. Clin. Lab. Inv. *29*, 297, 1972.
6. Chaney, A.L., and Marbach, E.P.: Modified reagents for determination of urea and ammonia. Clin. Chem. *8*, 130, 1962.
7. Rubin, M.: Fluorometry in clinical chemistry. Progress in Clinico-chemical Methods, Vol. 3, edited by O. Wieland. Basel, Switzerland, S. Karger, 1968, p. 6.
8. March, W.H., Fingerhut, B., and Miller, H.: Automated and manual direct methods for the determination of blood urea. Clin. Chem. *11*, 624, 1965.
9. Remp, D.G.: Uric acid (uricase). *In* Standard Methods of Clinical Chemistry, Vol. 6, edited by R.P. MacDonald. New York, Academic Press, 1970, p. 1.
10. Meites, S., Thompson, C., and Roach, R.W.: Two ultramicro-scale methods for plasma uric acid analysis with uricase. Clin. Chem. *20*, 790, 1974.
11. Spencer, E.S., and Pedersen, I.: Hand Atlas of the Urinary Sediment, 2nd ed., Baltimore, University Park Press, 1976.
12. Hamernyik, P.: Manual on Clinical Urinalysis Procedure, Seattle, Department of Laboratory Medicine, University of Washington, 1977.
13. Narayanan, S., and Appleton, H.D.: Creatinine: A review. Clin. Chem. *26*, 1119, 1980.
14. Freeman, J.A., and Beeler, M.F.: Laboratory Microscopy/Urinalysis, 2nd Ed., Philadelphia, Lea & Febiger, 1982.

Chapter 5

PROTEINS IN BODY FLUIDS

PROTEINS

The proteins are made up of *polypeptide chains of amino acids* linked together by *peptide bonds*. There are some 20 naturally occurring amino acids, of which 12 can be synthesized by humans and the remaining 8 must be provided in the diet of humans. The structure of a typical amino acid appears in Figure 5.1. The features common to all amino acids are a terminal carboxyl group (-COOH) and an adjacent amino group (-NH$_2$), although the rest of the structure may vary. Amino acids are linked together by an enzymatic reaction that condenses the -NH$_2$ group of one amino acid with the carboxyl group of another amino acid, as shown in Figure 5.2. A protein may consist of one or more intertwined polypeptide chains, some of which may contain 100 or more amino acids. Hemoglobin, for example, the principal protein in the red blood cell, is composed of two sets of two different polypeptide chains.

Protein synthesis takes place in cells by means of the information stored in coded form in the genes (DNA in the nucleus) of that cell. This is a complex mechanism involving both DNA and RNA. The code contained in the DNA provides directions for the sequential, step-by-step assembly of amino acids into a polypeptide chain, which is a component of a specific protein. These proteins are the primary constituents of the cellular machinery.

The building blocks of the proteins, the amino acids, are derived primarily from dietary protein which may be of vegetable (e.g., grains, beans) or animal

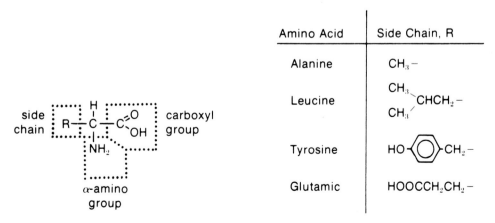

Fig. 5.1.   Illustration of the general structure of amino acids. All amino acids contain the terminal carboxyl and adjacent amino group. Some of the common side chains are shown.

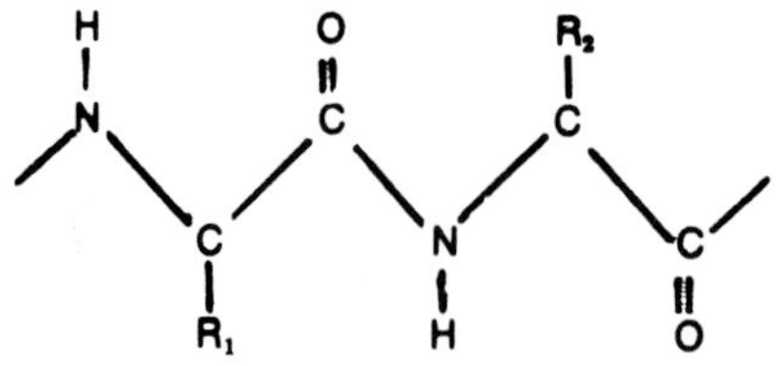

Fig. 5.2. Linkage of amino acids by means of peptide bonds to form a polypeptide chain. The R side chains may carry positive or negative charges, or none, depending upon the polarity of the group.

(meat, dairy products, eggs) origin. After ingestion, the hydrolysis of protein begins in the stomach where pepsin, the *protease* secreted there, acts in the highly acid gastric juice. As the food bolus leaves the stomach, pancreatic juice raises the pH of the partially digested mixture and contributes several proteases (trypsin, chymotrypsin) and various *peptidases* that together with peptidases contributed by the intestinal mucosa, complete the hydrolysis by breaking down the various protein residues into their component amino acids.

The amino acids are absorbed into the capillary blood of the intestines and finally enter the portal vein where they are carried first to the liver, which is the great metabolic factory of the body. Many of the plasma proteins are synthesized in the liver, but the amino acids that are not utilized or transformed there are carried by the circulatory system to the rest of the cells of the body where they become available for use as needed. Proteins coming from cells that are destroyed in the body are also hydrolyzed by proteases and contribute their amino acids to the available pool. Some types of cells (nerve cells, ova) are not replaced during the life of the individual, but other types have a relatively rapid turnover, i.e., they have limited life, become senescent, and die as replacements are formed. An example of this is the red blood cell (erythrocyte), which has a lifetime of approximately 120 days, or the white blood cell (leukocyte), which may live only 15 to 30 hours. The proteins contained within the senescent cells are digested by phagocytic cells, with the resulting amino acids becoming available to the general pool. Amino acids in the body pool may follow either an anabolic pathway and be utilized for the synthesis of protein or a catabolic pathway to yield energy.

Proteins are present in all body fluids, but the protein concentration normally is high (3 g/dl) only in blood plasma, lymph, and in some exudates. Protein concentration in the cerebrospinal fluid of normal subjects is less than 45 mg/dl, whereas the urine contains only a trace.

Plasma is the clear fluid that is obtained when blood is drawn into a tube containing an anticoagulant and is centrifuged. The formed elements (red and white blood cells) are heavier than the fluid medium and appear as a layer in the bottom half of the centrifuged tube. The plasma is a complex fluid that contains all of the dissolved substances in whole blood: inorganic compounds, small organic molecules, such as glucose and urea, and hundreds of different proteins, all dissolved in water. Water constitutes about 92% of the plasma, by weight. Serum is similar to plasma except that the blood has been allowed to clot before centrifuging. The only differences between serum and plasma are the following: (1) *plasma* contains the protein fibrinogen (0.2 to 0.4 g/dl), which is missing from serum because it has been converted into the insoluble fibrin clot; (2) *plasma* is the blood component existing in the body whereas *serum* is the fluid obtained *outside* the body when the blood is allowed to clot; and (3) *plasma* obtained outside the body always contains an anticoagulant, a chemical that prevents the blood from clotting (heparin, oxalate, citrate, EDTA*). Most of the anticoagulants function by removing Ca ion from the plasma, either bound to an undissociated complex (citrate, EDTA) or as an insoluble salt (oxalate); Ca^{2+} is essential for the clotting process.

SERUM PROTEINS

Analyses for proteins are usually performed on serum because this is the fluid medium generally used in the chemistry laboratory. Concentration of total protein in serum usually ranges from 6.0 to 8.2 g/dl and is about 0.3 g/dl higher for plasma. It is not feasible to express the protein concentration in mmol/L because serum is a mixture of so many proteins with differing molecular weights. The molecular weight may vary from 40,000 to 50,000 for some of the smaller globulins to about one million for the macroglobulins. Some of the lipoproteins may have molecular weights of 10 to 20 million.

The proteins are large molecules which cannot pass through the membranes lining the blood vessels. Since water, other small molecules, and most ions can diffuse in both directions through the capillary walls, the presence of the proteins in the vascular bed draws water into the system and creates an osmotic pressure, which is usually referred to as oncotic pressure.

The plasma proteins have many functions that may be summarized as follows:

1. They affect the distribution of extracellular fluid between the vascular bed and the interstitial fluid by means of the oncotic pressure they generate. This is a general property of proteins, but the most important protein in this respect is albumin because of its relatively small size (molecular weight 69,000) and high concentration, about 60% of the total plasma proteins.
2. They serve as carriers for various cations and some compounds that are relatively insoluble in water, such as bilirubin, fatty acids, steroid hormones and lipids.
3. They function as antibodies to provide a defense system for the body against foreign proteins, viruses, and bacteria. This role is reserved for the gamma globulins, as will be described later.
4. They form part of the endocrine system. Many of the hormones are proteins or polypeptides, which reach their target organ by circulating in plasma.
5. They protect against damage to the vascular system by forming a complex blood-clotting system.
6. They provide tissues with a source of nutrients for building materials, or calories. Some of the proteins can be catabolized for energy purposes or reutilized for structural or other purposes by contributing to the amino acid pool.
7. Some proteins function as enzymes. Most plasma enzymes are derived from intracellular sources and appear in plasma during the course of

---

*EDTA, ethylenediamine tetraacetic acid, a chelate that prevents blood clotting by forming an unionized complex with calcium.

ordinary breakdown and replacement of cells. Because this process is greatly increased during tissue necrosis, the plasma enzyme pattern may provide clues concerning specific tissue or organ pathology in some disease states.

Proteins may be classified into different groups according to function (such as immunoglobulins, clotting factors), composition (glycoproteins, mucoproteins, lipoproteins), relative migration distance in an electric field, or sedimentation rate when subjected to a centrifugal force. Although all of the above types of nomenclature are employed, the most widely used general classification system is to name the protein groups after the five main categories or bands that are obtained on cellulose acetate or paper when separation takes place in an electric field.

The technique for separating proteins by means of an electric current is called *electrophoresis* and is described on page 161. At pH 8.6, all of the serum proteins carry a negative charge to a greater or lesser extent. If a small serum sample is placed upon a wet (pH 8.6 buffer) support medium, such as cellulose acetate, paper, or agar gel suspended between two electrodes immersed in pH 8.6 buffer, and subjected to an electric charge of several hundred volts, the proteins will move toward the positive pole (anode). The migration distance varies directly with the charge carried by the protein but is modified to a slight extent by buffer flow.

The buffer flow (electro-osmotic or endosmotic flow) is caused when the support medium, such as paper, cellulose acetate, or agar gel, has a negative charge when the capillaries or pores become wet, leaving the liquid medium (buffer) with a small positive charge. When the current is turned on, there is a flow or movement of buffer toward the negative electrode (cathode), and this movement passively carries with it to a small extent the proteins that tend to migrate toward the anode because of their negative charge.

When serum proteins are fractionated on cellulose acetate, five bands are usually obtained at the end of the electrophoretic run when the support medium is stained with a dye that adsorbs to protein. These bands are known as albumin, alpha one (α_1), alpha two (α_2), beta (β), and gamma (γ) globulins, respectively, starting with the greatest migration toward the anode (Fig. 5.3, A,B,C,E). Because of endosmotic flow, the gamma band is usually displaced from the point of application slightly toward the cathode. The slowest of the beta fraction (fibrinogen, if plasma is used) usually is at the point of application, and all faster fractions move toward the anode.

Serum albumin consists of a single species of protein and is the most prominent band in an electropherogram; it constitutes approximately 60% of the total serum protein and migrates the farthest toward the anode. It is synthesized exclusively in the liver, functions as a regulator of blood oncotic pressure, as a carrier for many cations and water-insoluble substances (calcium, bilirubin, fatty acids), and as a pool of amino acids for caloric or synthetic purposes. It has a half-life in plasma of about 17 days, which means that its plasma concentration would fall about 3% per day if synthesis were completely halted. Hepatocellular damage usually results in a decrease in the serum albumin concentration, but the change is relatively slow. The concentration of serum albumin is decreased in the following situations:

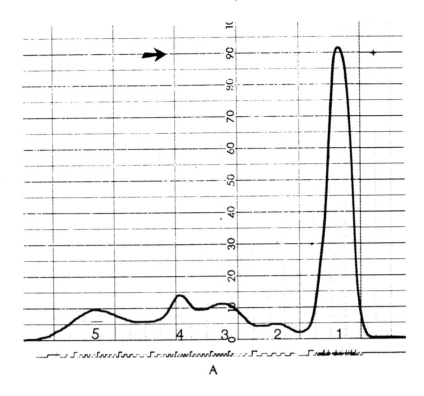

A

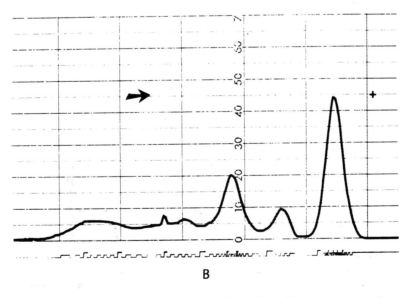

B

Fig. 5.3. Typical serum protein electrophoresis patterns obtained on cellulose acetate membranes. A, normal serum protein pattern (band 1 = albumin, 2, 3, 4, and 5 are α_1-, α_2-, β-, and γ-globulins, respectively); B, nephrotic syndrome.

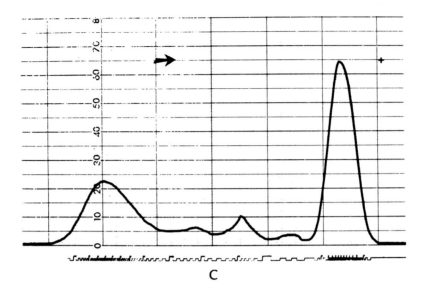

C

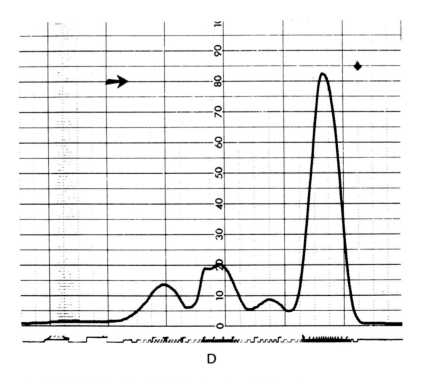

D

Fig. 5.3. *continued*   C, infectious hepatitis; D, hypogammaglobulinemia.

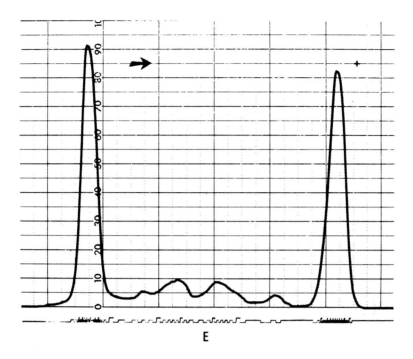

E

Fig. 5.3. *continued* E, multiple myeloma. The arrow indicates the direction of the electrophoresis and +
indicates the anodal end.

1. *Decreased synthesis,* whether caused by (a) damaged hepatic cells, (b)
   deficient protein intake, as in malnutrition or starvation, or (c) impaired
   digestion or absorption of protein products (sprue).
2. *Extensive protein loss,* whether (a) through the kidneys, as in the nephrotic
   syndrome, (b) through the skin, following extensive burns or severe skin
   lesions, as in exfoliative dermatitis, or (c) through the gastrointestinal tract,
   as in protein-losing enteropathies (protein-losing intestinal diseases).
3. *Shift to ascitic fluid,* which may happen in chronic liver disease with
   cirrhosis. Ascites, the accumulation of fluid in the peritoneal cavity, is
   frequently caused by portal hypertension (increased portal vein blood
   pressure in the liver) and a lowered plasma osmotic pressure (lowered
   albumin concentration), by obstruction to hepatic vein flow, or by a combination of factors occurring in hepatocellular disease.

The remaining four bands in serum electrophoresis are globulins. The globulins are less soluble than albumin in various salt solutions and formerly were
separated as a single group from albumin by precipitation with a concentrated
Na_2SO_4 or Na_2SO_3 solution. Much more clinical information is obtained by an
electrophoretic separation.

The second fastest band, the one adjacent to albumin but closer to the cathode,
is the α_1-globulin fraction. It is a mixture of many proteins, some of which have
been identified and characterized. Many in this group are combined with carbohydrate and are called glycoproteins; others exist in combination with lipids
and are named α_1-lipoproteins. Some of the prominent individual proteins in
the α_1-band are α_1-acid glycoprotein, thyroxine-binding globulin, and α_1-anti-

Clinical Chemistry

trypsin. In general, the α_1-globulin band increases as a nonspecific response to inflammation (acute phase reactants) arising from a variety of causes: infection, trauma, and neoplasms. The α_1-globulins are synthesized in the liver.

The third protein band visible on cellulose acetate electrophoresis is composed of the α_2-globulins. This fraction also contains some α_2-lipoproteins. Some of the well-known proteins in this fraction include the haptoglobins, which are genetically determined proteins that bind hemoglobin, and ceruloplasmin, a copper-binding protein that has oxidase activity. The mucoproteins in the α_2-fraction also are increased in inflammatory conditions.

The fourth band is the β-globulin fraction, which contains the β-lipoproteins, the iron-transporting protein (transferrin), fibrinogen, and other lesser known proteins. Any condition that increases the β-lipoproteins makes this band more prominent because the lipoproteins predominate in this fraction.

The slowest band is the γ-globulin fraction, which contains the immunoglobulins, or circulating antibodies, that are so essential for defense against foreign proteins of all sorts. These proteins are synthesized by cells of the reticuloendothelial system: plasma cells and lymphocytes.

There are five known classes of immunoglobulins: IgG, IgA, IgM, IgD, and IgE. These are related structurally in that each member of a group is composed of two light polypeptide chains and two heavy chains linked to each other by disulfide bonds. The same two light chains, designated as kappa (κ) and lambda (λ) chains are utilized in all of the five classes of immunoglobulins, but the heavy chains are unique and distinctive for each class. Both light chains are the same for any particular molecule, but each class of immunoglobulin contains a mixture of κ and λ light chains. The binding of an antigen to an immunoglobulin involves binding sites upon both the heavy and light chains.

Only IgG, IgA, and IgM are present to any appreciable extent in normal serum, and these are the only ones that are usually measured in the laboratory. IgD and IgE are normally present in such small amounts that they can be detected only by the most sensitive methods. Several cases of multiple myeloma have been reported, however, in which the abnormally elevated immunoglobulin was IgD or IgE.

Some of the properties, composition, and serum concentrations of the im-

TABLE 5.1
Classes of Immunoglobulins

CLASS	IgG	IgA	IgM	IgD	IgE
Heavy chains	gamma (γ)	alpha (α)	mu (μ)	delta (δ)	epsilon (ϵ)
Molecular weight	53,000	64,000	70,000	68,000	75,000
Light chains	kappa (κ) lambda (λ)	κ, λ	κ, λ	κ, λ	κ, λ
Molecular weight	22,500	22,500	22,500	22,500	22,500
Percent carbohydrate	2.9	7.5	11.8	11	10.7
Molecular weight	150,000	180,000–500,000	950,000†	180,000	196,000
Serum concentration* mg/dl	700–1680	140–420	50–190	3–40	0.01–0.14

*In normal individuals.
†Exists as a pentamer, with each of the 5 subunits composed of 2 heavy and 2 light chains.

munoglobulin classes are shown in Table 5.1. It should be noted that all of the immunoglobulins contain appreciable amounts of carbohydrate.

The IgG class is the one present in the highest concentration in serum and comprises about 75 to 80% of the γ-globulin fraction; most of the circulating antibodies belong to this class. IgG is the only immunoglobulin to cross the placenta and appear in the plasma of newborns. The enzyme systems for synthesizing immunoglobulins are not developed in a newborn infant and do not appear until several weeks to 3 months after birth. Synthesis of IgG in plasma cells starts about 3 or 4 months after birth and reaches the adult level in about 1 year. When there is a delay in synthesis, which sometimes occurs, the infant usually suffers from a series of respiratory infections.[1]

IgA is a class of immunoglobulins devoted primarily to defense of the mucosal linings of the gastrointestinal and respiratory tracts and the various orifices. IgA is synthesized in reticuloendothelial (R-E) cells below the mucosal layer and helps to prevent invasion of the mucosal barrier. About 15 to 20% of the circulating immunoglobulins are of the IgA class.

The IgM class is composed of the largest molecules; each molecule consists of a cluster of five subunits, each containing two heavy and two light chains. The IgM immunoglobulins are synthesized by lymphocytes. They are the first species of immunoglobulins to be synthesized by a newborn and begin to appear around the second week of life, reaching adult levels in 6 to 12 weeks. They comprise about 5 to 7% of the circulating immunoglobulins, and their increase is rapidly stimulated by the presence of foreign, particulate bodies, such as bacteria. Both IgG and IgM can bind complement. The change of concentration of the immunoglobulins with age is shown in Table 5.2.

The plasma concentration of the gamma globulins as a group usually rises in chronic infection, with an increase in IgG, IgA, and IgM. There are some diseases, however, in which there may be an increase in only one class of the immunoglobulins. These are called monoclonal diseases because a single class of immunoglobulin is synthesized in excess by a proliferation of lymphocytic cells. The cell proliferation is believed to arise from a single clone that results in a mass of lymphocytes all producing the same class of immunoglobulin. An abnormal immunoglobulin is called a *paraprotein*.

The monoclonal diseases are *multiple myeloma* (myelomatosis), a malignant

TABLE 5.2
Reference Values* for Serum Immunoglobulins in Various Age Groups (mg/dl)

AGE	IgG RANGE	IgA RANGE	IgM RANGE
Cord serum	775–1665	0.05–9	4–25
0.5–3 months	305–835	3.5–6.3	16–125
3–6	145–970	5–86	19–114
6–12	425–1120	15–90	45–215
1–2 years	360–1185	14–113	38–232
2–3	500–1245	24–131	51–197
3–6	575–1355	36–199	53–207
6–9	670–1505	30–271	52–220
9–12	645–1570	63–282	67–268
12–16	695–1520	85–242	47–247

*Modified from the table, "Immunoglobulins," in Pediatric Clinical Chemistry, edited by S. Meites, Washington, D.C., American Association for Clinical Chemistry, 1981, p. 284.

proliferation of plasma cells that results in an abnormally high concentration of serum immunoglobulins, usually IgG or IgA, *Waldenström's macroglobulinemia,* a malignant disease of the lymphoid elements, characterized by a high serum concentration of IgM, *cryoglobulinemia,* a condition in which plasma or serum proteins (IgM) precipitate when cooled below body temperature, and some cases of *lymphomatous diseases.*

When immunoglobulins are synthesized in the lymphocytic cells, the heavy-chain polypeptides are usually assembled on a different ribosome from the light chain. The process operates synchronously as two light chains are attached to two heavy chains to form the finished immunoglobulin. In the disease multiple myeloma, it frequently happens that light chains are produced in excess over the heavy chains and enter the bloodstream; since they are of relatively low molecular weight, they pass through the glomerular membrane and appear in the urine. These protein chains of low molecular weight are known as *Bence Jones proteins* and have peculiar solubility properties, which were first reported by Dr. Henry Bence Jones. The Bence Jones proteins precipitate when the urine is heated from 45 to 60° C and redissolve when the heating is continued above 80° C. A myelomatosis with Bence Jones protein in the urine is now called "light chain disease." The chain may be of either the κ or λ type.

The reverse can also occur, the production of heavy chains being in excess over the light chains, but this happens much less frequently. Since the heavy chains are of a sufficiently high molecular weight to be retained by the intact glomerular membrane, it requires a kidney lesion in addition to heavy chain overproduction before they are present in the urine. Heavy chain disease involving each of the three types (γ, α, and μ) has been described.[5]

Total Serum Proteins

The total serum protein concentration in an ambulatory adult varies from 6.0 to 8.2 g/dl. The values are about 0.5 g/dl higher in an ambulatory person than in one at bed rest because the erect position produces a pooling of fluid (extravasation of protein-free fluid from the blood vessels to interstitial fluid) in the lower extremities. The total protein concentration is lower in a newborn infant (see Table 5.3), but slowly reaches the adult level in about a year or so.

The total serum protein, as its name implies, represents the sum total of

TABLE 5.3*

Normal Values† of Serum Total Protein and the Protein Electrophoretic Fractions of Infants at Different Ages

	NEWBORN (1st wk)	3–4 mo	12 mo	Over 4 yr
Total protein	6.0 ± 1.6	5.8 ± 1.6	6.4 ± 0.8	6.8 ± 0.9
Albumin	4.2 ± 1.3	3.9 ± 1.1	4.5 ± 0.6	4.6 ± 0.8
Alpha-1-globulin	0.17 ± 0.08	0.23 ± 0.16	0.22 ± 0.12	0.22 ± 0.12
Alpha-2-globulin	0.38 ± 0.08	0.57 ± 0.26	0.54 ± 0.26	0.57 ± 0.22
Beta globulin	0.38 ± 0.22	0.57 ± 0.26	0.62 ± 0.24	0.62 ± 0.12
Gamma globulin	0.87 ± 0.52	0.43 ± 0.32	0.55 ± 0.20	0.81 ± 0.48

Cellulose acetate electrophoresis; Beckman Microzone

*From Normal Values for Pediatric Clinical Chemistry, an American Association for Clinical Chemistry special publication (1725 K St. NW, Washington, D.C. 20006), 1974.

†Expressed as g/dl (mean ± 2 SD).

numerous different proteins, many of which vary independently of each other. The total protein concentration must be measured when performing an electrophoresis in order to calculate the concentration of each protein fraction from its percentage (p. 164). Aside from this situation, the determination of total protein supplies limited information except in conditions relating to changes in plasma or fluid volume, such as shock, dehydration, possible overhydration, and hemorrhage. The need for fluids is revealed by an elevated serum protein concentration that shows hemoconcentration. It is also useful to measure the total serum protein level when determining the calcium concentration because the nondiffusible calcium fraction is bound to protein and varies directly as the serum protein (p. 348).

Biuret Method

REFERENCES. Doumas, B.T.: Standards for total serum protein assays—a collaborative study. Clin. Chem. *21*, 1159, 1975.
Peters, T., Jr., and Biamonte, G.T.: Protein (total protein) in serum, urine and cerebrospinal fluid; albumin in serum. *In* Selected Methods of Clinical Chemistry, Vol. 9, edited by W.R. Faulkner and S. Meites, Washington, D.C., American Association for Clinical Chemistry, 1982, p. 317.

PRINCIPLE. Cu^{2+} reacts in alkaline solution with the peptide linkages of proteins to form a violet-colored complex. The intensity of the color produced is proportional to the protein concentration. A structure containing at least two peptide linkages (-CONH) is required to form the complex; hence amino acids do not react. Alkaline copper reagent forms a similar complex with biuret, (NH_2-CO-NH-CO-NH_2), from which the reaction takes its name.

REAGENTS

1. *NaOH, 6.0 mol/L:* Dissolve 240 g NaOH in CO_2-free water (freshly distilled) and dilute to 1 L. Store in a tightly closed polyethylene bottle.
2. *Biuret reagent:* Dissolve 3.00 g (12 mmol) $CuSO_4 \cdot 5H_2O$ in 500 ml freshly distilled water. Add 9.0 g (32 mmol) KNa tartrate·$4H_2O$ and 5.0 g (30 mmol) KI. After solution is completed, add 100 ml of 6 mol/L NaOH and dilute to 1 L with water. Store in a tightly closed polyethylene bottle at room temperature. The reagent is stable for at least a year but the authors recommend that it not be used after 6 months.
3. *Biuret blank reagent:* Dissolve 9.00 g KNa tartrate·$4H_2O$ and 5.0 g KI in 800 ml water. Add 100 ml 6 mol/L NaOH and dilute to 1 L with water. Store in polyethylene bottle at room temperature. Reagent is stable for approximately 6 months but discard earlier if there is evidence of biological growth.
4. *Protein Standard:* Dry some bovine serum albumin (BSA, Cohn Fraction V) obtained from Metrix Clinical Diagnostics in a desiccator for several days. Prepare a solution containing 6.0 g/dl by dissolving 6.0 g albumin plus 0.05 g sodium azide in water. Adjust the pH to 6.5 to 6.8 and make up to 100 ml volume with water. Dilute appropriate volumes with 0.05% (w/v) sodium azide to make a series of standards, 2, 3, and 4 g/dl, respectively. Check the absorbance of this BSA standard with that of a standardized bovine albumin solution SRM 927, available from the National Bureau of Standards.

PROCEDURE

1. Pipet 2.5 ml of biuret reagent into a *Test* series of tubes and 2.5 ml of biuret blank into a *Blank* series.
2. Pipet 50 μl of specimen (protein standard, serum or control) into one tube of each series and mix thoroughly. Also prepare a *reagent blank* by pipetting 50 μl of water into 2.5 ml of biuret reagent.
3. Incubate all tubes for 30 min at room temperature or for 10 min at 37° C.
4. Set zero absorbance at 540 nm with biuret blank reagent. Read the absorbances of the *Blank* series, followed by the absorbances of the reagent blank and the *Test* series.
5. *Calculations:*
   a. Substract both reagent blanks and sample blanks from the corresponding absorbances (A) obtained for the *Test* series.
   $$A_{net} = A_{test} - (A_{reagent\ blank} + A_{sample\ blank})$$
   The absorbance of the reagent blank for 1 cm light path $= 0.100 \pm 0.005$.
   Specimen blanks usually read <0.010 if the sera are clear.
   b. Protein concentration, g/dl $= A_{u\ net}/A_{s\ net} \times C$, where $A_{u\ net}$ and $A_{s\ net}$ are net absorbances of specimen and standard, respectively, and C = concentration of standard, g/dl.

REFERENCE VALUES. The concentration of total serum proteins in normal adults ranges from 6.0 to 8.2 g/dl. The values in plasma are about 0.2 to 0.4 g/dl higher because of the presence of fibrinogen. The serum concentrations in infants for the first three or four months of life are about 1.0 g/dl lower than in adults (see Table 5.3).

INCREASED CONCENTRATION. The total protein concentration of serum is usually increased in patients with dehydration, monoclonal disease (multiple myeloma, macroglobulinemia, cryoglobulinemia) and in some chronic polyclonal diseases (liver cirrhosis, sarcoidosis, systemic lupus erythematosus, and chronic infections). An abnormal, monoclonal protein is called a *paraprotein*.

DECREASED CONCENTRATION. The serum total protein concentration is decreased in inadvertent overhydration, in conditions involving protein loss through the kidneys (nephrotic syndrome), from skin (severe burns), or gut (protein-losing enteropathies), or in failure of protein synthesis (starvation, protein malnutrition, severe nonviral liver cell damage).

Measurement of the Refractive Index

The refractive index of an aqueous solution increases directly with an increase in solute concentration. Even though changes in concentration of different substances may affect the refractive index unequally, the concentrations of electrolytes and small organic molecules in serum usually do not vary greatly; hence, a change in refractive index is indicative of a change in protein concentration. The use of this method provides a rapid and direct way of measuring the serum total proteins and requires only one drop of serum. Most of the time the values obtained by refractive index correlate well with those found by the biuret method, but discrepancies should be expected when abnormal serum proteins (paraproteins) are present because of their different properties.

REFERENCE. Barry, K.G., McLaurin, A.W., and Parnell, B.L.: A practical temperature-compensated hand refractometer (the TS Meter): its clinical use and application in estimation of total serum proteins. J. Lab. Clin. Med. 55, 803, 1960.

PRINCIPLE. The refractive index of serum varies directly with the total protein concentration. Icteric, lipemic, or grossly hemolyzed sera should not be used; the serum must be clear.

APPARATUS. A refractometer (TS Meter, Model 10400, American Optical).

PROCEDURE. Spread one drop of serum on the plate between the prism and cover. Hold the end up to a bright light and read the protein concentration from a scale where the demarcation line between the bright and dark areas crosses it. Wipe off the film of serum with lens paper, wash with a squirt of water from a wash bottle, and dry.

Serum Protein Fractions

As pointed out earlier, the determination of the total serum protein concentration alone provides little clinical information except as to the state of the patient's hydration. Some additional information is obtained if the total proteins are separated and determined as the albumin and globulin fractions. The most useful data are provided by an electrophoretic separation of the serum proteins into albumin and the four globulin fractions of α_1-, α_2-, β-, and γ-globulins, respectively. By the latter procedure, one can readily detect abnormalities and in what fractions; monoclonal disease becomes obvious and so does hypogammaglobulinemia, a condition in which the concentration of gamma globulins is low.

Estimation of the concentration of the albumin and globulin fractions may be obtained by measuring the total protein concentration by the biuret reaction and the albumin by a dye-binding technique (bromcresol green); the globulin value is obtained by difference. Conversely, the globulin fraction may be measured directly, based on the tryptophan content of the protein, and the albumin concentration can be obtained by difference.

Serum Albumin—Dye-binding Technique with Bromcresol Green

REFERENCE. Doumas, B.T., and Biggs, H.G.: Determination of serum albumin. *In* Standard Methods of Clinical Chemistry, Vol. 7, edited by G.R. Cooper, New York, Academic Press, 1972, p. 175.

PRINCIPLE. Bromcresol green, an anionic dye, binds tightly to albumin when added to serum, and the complex absorbs light much more intensely at pH 4.20 and 628 nm, than does the unbound dye. The increase in light absorption is directly proportional to the albumin concentration. Bromcresol green is the dye of choice because it is bound so tightly to albumin that it is not displaced by bilirubin and does not bind significantly to other serum proteins. The method can be performed manually or by automation.

REAGENTS

1. Succinic Acid, 42 mmol/L. Dissolve 5 g succinic acid in water and make up to 100 ml volume.
2. NaOH, 2.5 mol/L. Dissolve 10 g NaOH in water and make up to 100 ml volume.
3. Succinate Buffer, 0.10 mol/L, pH 4.15. Dissolve 11.8 g succinic acid and 500 mg sodium azide (NaN_3) in 800 ml water. Adjust the pH to 4.15 with NaOH, 2.5 mol/L, and make up to 1000 ml volume with water.
4. Bromcresol Green (BCG) Stock Solution, 0.60 mmol/L: Transfer 432 mg of the sodium salt of BCG to a 1000-ml volumetric flask and dilute to

volume with water. The same concentration of BCG could be prepared by using 419 mg of the free acid instead of the sodium salt, provided 10 ml of 0.1 mol/L of NaOH were added to make the BCG soluble. Store at 4 to 8° C.

5. Working BCG Solution. Mix 250 ml of stock BCG solution with 750 ml of succinate buffer. Add 4 ml of 30% Brij 35 solution (Alpkem, Fisher) and adjust the pH to 4.20 ± 0.05. If the pH is greater than 4.25, use the succinic acid solution, 42 mmol/L for adjustment. Store at 4 to 8° C.

6. Albumin Standard Solutions. Use standard albumin solution, 6.0 g/dl, prepared for total protein. Prepare a series of standard solutions containing 2.0, 3.0, 4.0, and 6.0 g/dl by appropriate dilution with a NaN_3 solution, 50 mg/dl (w/v).

PROCEDURE

1. Pipet 5.0 ml of working BCG solution into a number of test tubes for unknowns, controls, standards and reagent blank.
2. Add 25 μl of serum or working standards to the appropriate tubes, and add 25 μl water to the reagent blank.
3. Mix contents and allow tubes to stand for 10 minutes at room temperature.
4. Read the absorbance at 628 nm against the reagent blank.
5. Calculation:

$$\text{Serum albumin} = \frac{A_u}{A_s} \times C = \text{g/dl}$$

where A_u = absorbance of unknown, A_s = absorbance of standard, and C = concentration of standard.

$$\text{Globulin} = \text{T.P.} - \text{Alb.}$$

where T.P. = total proteins and Alb. = Albumin.

The Doumas and Briggs reference also describes the method for use with a continous flow type of automation. Pinnell and Northam[9] prefer to substitute bromcresol purple as the dye for the serum albumin determination because of closer agreement with more specific methods for measuring albumin concentration.

REFERENCE VALUES. The normal concentration of serum albumin varies from 3.5 to 5.2 g/dl. For about the first four months of life, the serum albumin concentration of most infants is about 10 to 12% lower than in adults (Table 5.3). The discrepancy may be a little larger for premature infants.

INCREASED CONCENTRATION. Conditions of hyperalbuminemia are rarely seen except in the presence of acute dehydration or shock. An increase in albumin concentration will only be temporary, as interstitial water is drawn into the vascular bed by increased osmotic forces.

DECREASED CONCENTRATION. The same conditions that produce a lowered total protein concentration also lower the serum albumin concentration, as has been discussed earlier in this chapter.

Globulins

The tryptophan content of human serum globulins is about 2 to 3%; that of albumin is only 0.2%. The globulin concentration of serum may be estimated from the determination of the tryptophan content. The measurement is easily made by reacting the protein solution with glyoxylic acid in an acid solution (the Hopkins-Cole reaction).

REFERENCE. Goldenberg, H., and Drewes, P.A.: Direct photometric determination of globulin in serum. Clin. Chem. *17*, 358, 1971.

PRINCIPLE. Tryptophan in a polypeptide chain reacts with glyoxylic acid in an acid solution to give a purple color. The concentration of globulin is proportional to the intensity of the color, which is measured at 540 nm.

REAGENTS

1. Globulin Reagent. dissolve 1.0 g $CuSO_4 \cdot 5H_2O$ in 90 ml of water contained in a 1000-ml volumetric flask. Add 400 ml glacial acetic acid and then 1.0 g of 98% glyoxylic acid monohydrate (MC/B), and mix. Without delay, carefully add 60 ml concentrated H_2SO_4, with stirring. Cool to room temperature and dilute to the mark with glacial acetic acid. Mix well. The reagent is stable for at least one year if stored at 4 to 8° C.

2. Standard Globulin Solution, 3.0 g/dl. Place some purified human gamma globulin (Cohn Fraction II, Sigma HG-II) in a desiccator for several days. Transfer 3.0 g human gamma globulin, 4.5 g human albumin (Cohn Fraction V, Calbiochem), and 50 mg NaN_3 to a 100-ml volumetric flask. Dissolve in water and make up to 100 ml volume.

PROCEDURE

1. Place 4 ml of globulin reagent in a series of tubes.
2. To each tube, add 20 μl of unknown serum, control serum, standard, or water, respectively.
3. Mix contents of all tubes and place in a boiling water bath or 100° C heating block for 5 minutes.
4. Cool tubes in tap water for 3 minutes, mix, and read against a reagent blank at 540 nm.
5. Calculation:

$$\text{Serum globulin} = \frac{A_u}{A_s} \times C = \text{g/dl}$$

where A_u = absorbance of unknown, A_s = absorbance of standard and C = concentration of standard.

Note: Albumin is added to the standard globulin solution in order to compensate for the color contributed by albumin in the unknown sera. Without this compensatory effect, the globulin determination would be about 7 to 10% too high.

REFERENCE VALUES. The normal concentration of total globulins varies from 2.3 to 3.5 g/dl. The actual concentrations of globulins and albumin are of physiologic and clinical importance, not their ratio, which may be misleading.

INCREASED CONCENTRATION. The serum globulins are usually elevated in the monoclonal diseases, some forms of liver disease, in sarcoidosis, collagen disease, and chronic infection.

DECREASED CONCENTRATION. Low values of the total globulins are usually associated with hypogammaglobulinemias. See the discussion of electrophoretic separation which follows.

Fractionation by Electrophoresis

As mentioned earlier in this chapter, serum proteins carry a negative charge at pH 8.6 and move in an electric field toward the positive electrode. The migration distance varies directly with the charge on the protein when there is no impediment to free movement. In practice, electrophoresis is carried out on a support medium wet with buffer that is connected by a bridge or wick with each electrode. The first support commonly used for electrophoresis in the clinical chemistry laboratory was paper, but this was superseded by cellulose acetate. Agar or agarose is also used in some instances, but cellulose acetate is more popular for protein electrophoresis because it is so convenient and technically simple to use.

Gels made from starch or acrylamide can also be used as a support medium for electrophoretic separation of proteins; these gels have very fine pores that serve as a molecular sieve. Since this causes larger proteins to move through the medium more slowly than smaller proteins, separation is a function of both electric charge and size of the protein. Electrophoresis of serum yields 20 to 30 different bands with these gels, which complicate the interpretation of the patterns. Acrylamide gels may be the support medium of choice for checking homogeneity of a protein solution, for quantitating certain isoenzymes, or for analyzing the genetic distribution of particular serum proteins, but they are not well suited for routine serum protein fractionation.

PRINCIPLE. Serum proteins are separated on a cellulose acetate membrane in an electric field at either constant voltage or constant current. The membrane is then placed in a solution containing trichloroacetic acid to denature the proteins and Ponceau S dye to stain the proteins. After washing and clearing the membrane, the membrane is run through the densitometer, and the light absorbance by the dyed protein bands is recorded. The absorbance of light is proportional to the relative protein concentration. The protein concentration of each band, in g/dl, is obtained by multiplying the total protein concentration by the per cent protein in each fraction.

It is not too difficult to build certain components of an electrophoresis system,[3] but there are several convenient and economical systems available commercially (Beckman, Helena Laboratories, Gelman) that permit the separation of eight microsamples (less than 0.5 μl each) simultaneously on a small membrane. The following directions are adequate for most systems.

REAGENTS
1. Barbital Buffer, pH 8.6, ionic strength 0.075. Dissolve 15.40 g sodium diethyl barbiturate and 2.76 g barbituric acid in water and make up to 1000 ml volume.
2. Fixative-Dye Solution. Dissolve 2 g Ponceau S dye (Allied Chemical) and 30 g trichloroacetic acid in water and make up to 1 L volume.
3. Rinse Solution, 50 g/L acetic acid in water. Add 100 ml glacial acetic acid to water, and make to 2 L volume with water.
4. Alcohol Rinse. Methanol, absolute, A.C.S. reagent grade.
5. Clearing Solution for making the membranes transparent. This varies with the different commercial brands of membranes and should be prepared shortly before use. The relative proportions of the solvents may be modified to suit individual preferences or to accommodate changes in membrane manufacture. In general, increasing the relative proportion of glacial acetic acid makes the membranes softer and speeds the clearing process.

    (a) For Beckman, Millipore, and Schleicher and Schuell membranes, mix 75 ml of methanol with 25 ml of glacial acetic. Adjust the ratio slightly, if necessary.

    (b) For Sepraphore III membranes, use 90 ml of methanol plus 10 ml of glacial acetic acid.

6. Parafilm (American Can) Squares. Approximately 3 × 3 cm.
7. Sample Applicator. Several of the companies making electrophoresis instruments have micro applicators that will pick up and deliver about 0.25 μl of serum to the membrane. Some applicators pick up and apply the eight samples simultaneously.
8. Cellulose Acetate Membranes, 57 × 145 mm, with punched holes (Beckman, Gelman, Schleicher and Schuell, Helena Laboratories, and Instrumentation Laboratories).

PROCEDURE

1. Fill the chamber with buffer of 0.075 ionic strength, and level the liquid.
2. *Soaking of membrane.* Place a small amount of buffer in a tray or shallow dish and float the membrane on the surface. The buffer should be drawn up evenly through the pores of the membrane without trapping air bubbles. A consistent gray color shows that the membrane is impregnated evenly (requires only a few seconds). Then immerse the membrane completely by agitating the tray.
3. *Application of serum sample.* Remove the membrane from the buffer bath and gently blot between two pieces of blotting paper. Immediately position the membrane in the chamber according to the manufacturer's directions. Check that the reference hole is aligned with the number-one selection groove, that the correct amount of buffer is in the chamber and is level, and that the ends of the membrane are immersed in the buffer. Close the lid and allow the membrane and cell to equilibrate. Place a drop of serum on a square of Parafilm and move the extended applicator tip across its top surface. Retract the tip, place the rider in the positioning groove, position the strip selection pin in the groove adjacent to the proper strip selection number, and depress the tip. Allow the tip to be in contact with the membrane for about 5 seconds, retract, clean the tip, and prepare to apply the succeeding samples. Close the lid of the chamber when the eight samples have been applied.
4. *Electrophoresis run.* The electrophoresis is performed at constant voltage. When 250 v is indicated on the meter, turn the indicator switch and read the current. The starting current should be between 3.5 and 5.8 ma. Return the switch to the voltage reading and run at 250 v for 20 minutes.
5. *Dyeing of protein bands.* Turn off the current, remove the membrane, float it on the fixative-dye solution until the protein bands are stained and then immerse it. Leave it in dye solution for about 7 minutes.
6. *Rinsing of membrane.* Remove the excess dye by gently agitating the membrane in three rinses of acetic acid solution (50 g/L).
7. *Clearing of membrane.* Blot excess liquid off the membrane and then dehydrate by immersing it in absolute methanol for about 1 minute. Then place the membrane in the clearing solution positioned on a glass plate, and remove after 60 seconds. Remove any wrinkles by gentle stroking of

the membrane with a squeegee or glass rod. Place in an oven at 70 to 80° C for 15 or 20 minutes.

8. *Mounting of membrane.* When cool, remove the membrane from the glass plate and mount in a plastic envelope.

9. *Quantitation of protein fractions.* Mount the transparent membrane in an appropriate holder and pass through a recording densitometer with an integrator. Use a 520 nm filter.

The percentage of protein in each band (from trough to trough) is obtained either as direct readout on some instruments or is calculated from the integration pips in each band and expressed as a percentage of the total number of integration pips for all bands. The absolute amount of protein in each fraction is obtained by multiplying the total protein concentration (from the biuret reaction) by the per cent of protein in each fraction.

For laboratories that do not use such a microsystem, cellulose acetate electrophoresis may be performed on 2.5 × 30 cm or 5 × 20 cm strips.[2,3] The running time of the electrophoresis for the longer strips is usually 45 to 60 minutes at 300 to 400 volts. The clearing process and densitometry are the same as described for the microsystem. For laboratories without access to a densitometer, the intensity of dye in each band may be determined by eliminating the clearing step, cutting the bands individually from the dyed strip, eluting the dye from each band with 2.0 ml of 0.10 mol/L NaOH, and then reading the absorbance at 520 nm. The absorbance of each band is read against a blank made by elution of an unstained portion of the membrane.

$$\% \text{ total protein in a band} = \frac{\text{absorbance of that band}}{\text{sum of absorbance of all bands}} \times 100$$

REFERENCE VALUES. The normal adult has a serum protein concentration that varies from 6.0 to 8.2 g/dl. Upon separation by electrophoresis, the following concentrations in g/dl and percentages are usually found:

FRACTION	CONCENTRATION	PERCENT OF TOTAL
Albumin	3.5–5.2	50–65
α_1-globulin	0.1–0.4	2–6
α_2-globulin	0.5–1.0	6–13
β-globulin	0.6–1.2	8–15
γ-globulin	0.6–1.6	10–20

The actual concentrations are more important clinically than the percentages because the concentrations give more information concerning an excess or deficit of a protein class. See Table 5.3 for changes in concentration of the various protein fractions with age. A normal electrophoretic pattern is illustrated in Figure 5.3A. Separated bands are shown in Figure 5.4.

ABNORMAL VALUES

1. *Albumin:* Changes in serum albumin concentration in various disease states have been described earlier in this chapter.

2. *α_1-globulin:* The α_1-globulin fraction is frequently increased when the serum albumin concentration falls, particularly in infections and inflammatory diseases. Its concentration is lowered in acute hepatitis and in

TABLE 5.4

Changes in Total Serum Protein and Electrophoretic Fractions in Various Diseases*

DISEASE	TOTAL PROTEIN	ALBUMIN	GLOBULIN			
			α_1	α_2	β	γ
Rheumatoid arthritis	→	↓	→↑	↑	→	↑
Lupus erythematosus	→↑	↓	→	↑	→	↑
Acute glomerular nephritis	→	→↓	→↑	↑	→	→↑
Nephrotic syndrome	↓	↓	→↓	↑	→↑	↓
Hepatitis						
Acute	→↓	↓	→↑	↓	→	↑
Chronic (cirrhosis)†	→↓	↓	→	↓	↑	↑
Biliary obstruction	→	→	→	↑	↑	→
Acute infection	→	→↓	↑	→↑	→	→
Chronic infection	→↓	↓	↑	↑	→	↑
Malnutrition	↓	↓	→	→	→	↓
Multiple myeloma‡	↑	↓	→	→↑	→↑	→↑
Hypogammaglobulinemia	↓	→	→	→	→	↓

*A decrease in concentration is denoted by ↓, an increase by ↑ and a normal value by →. These patterns, with a few exceptions, are not sufficiently specific to be diagnostic of a particular disease but are compatible with the ones listed.

†Area between β- and γ-bands is "bridged" because of increase in IgA and IgM.

‡The abnormal globulin may migrate with the velocity of an α_2-, β or γ- globulin but it appears most frequently in the γ band.

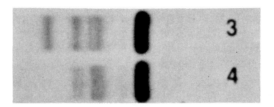

Fig. 5.4. Photograph of two membrane patterns obtained by serum protein electrophoresis. The upper band is from a patient with multiple myeloma while the lower one is from a patient with hypogammaglobulinemia. (Courtesy of Beckman Instruments, Inc., Clinical Instruments Division, Brea, CA 92621.)

familial α_1-antitrypsin deficiency, which is a cause of pulmonary emphysema.

3. α_2-*globulin:* The concentration of this fraction is greatly increased in the nephrotic syndrome owing to a large increase in α_2-macroglobin, a protein too large to pass through the lesions in the glomerular basement membrane which occur in this disease. The α_2-fraction is frequently increased in inflammatory conditions such as rheumatoid arthritis, lupus erythematosus, or after a myocardial infarct. In rare cases of multiple myeloma, the paraprotein (γG or γA) may migrate in the α_2-band. The serum concentration of α_2-globulins is decreased in acute hepatocellular disease.

4. β-*globulin:* The β-globulin fraction is usually elevated in hyperlipemias of various types. Sometimes the paraprotein of multiple myeloma may migrate with the β fraction; when it does, it appears as a sharp spike in the electrophoretic pattern.

5. γ-*globulin:* It has been mentioned earlier than the paraproteins of the monoclonal diseases are abnormal γ-globulins. Each one is unique and

has a different amino acid composition from the normal immunoglobulins. Hence, they may migrate further toward the anode than usual if amino acids with a greater negative charge should be substituted in the molecule. Most of the paraproteins migrate in the γ fraction, some as a slow γ and some as a fast γ; a small percentage of the paraproteins may migrate in the β-band, and only rarely will they be found in the α_2-fraction. In a monoclonal disease, such as multiple myeloma, the abnormal clones, whether of γG or γA, suppress the synthesis of the normal immunoglobulins. The monoclonal diseases exhibit a sharp spike in the electrophoretic pattern. The immunoglobulins may be elevated as a group in various polyclonal diseases and appear as a broad, dark band on the cellulose acetate membrane after staining, i.e., the densitometer tracing will appear as a broad hump. This is the pattern seen in viral hepatitis, sarcoidosis, rheumatoid arthritis, chronic infections, and some leukemias and lymphomas. Some typical patterns are shown in Figure 5.3 and a portion of a membrane photograph in Figure 5.4. Changes found in various diseases are summarized in Table 5.4.

Decreased serum γ-globulin concentration (hypogammaglobulinemia) occurs when the synthesis of these proteins by lymphocytes is impaired; the condition may be congenital (an inherited defect in the γ-globulin synthesizing system) or acquired. The acquired type may be caused by drugs to which the patient is sensitive; the γ-globulin concentration can be restored to normal by withdrawal of the offending drug. The body defense mechanism is impaired in severe hypogammaglobulinemia, but can be partially restored by the periodic intramuscular injection of γ-globulins.[1] The γ-globulin concentration may be decreased in the terminal stages of Hodgkin's disease, a malignant disease involving the lymph nodes; in the early stages, the γ-globulin concentration may be moderately elevated.

Specific Serum Proteins

The serum concentration of specific plasma proteins are sometimes requested as diagnostic aids. The list may include the different classes of immunoglobulins (IgG, IgA, IgM), polypeptide hormones of the pituitary, pancreas or parathyroid glands, cancer markers such as α-fetoprotein or carcinoembryonic antigen, transport proteins such as transferrin (iron) or haptoglobin (plasma hemoglobin) or a host of other proteins. Killingsworth presents an excellent review of plasma protein patterns in health and disease.[10] The analyses are most readily accomplished by some form of immunotechnique described on p. 253. For those plasma proteins of concentration > 1 mg/L, immunonephelometric[7] (p. 253) or fluorescent immunoassay[8] (p. 258) techniques may be most suitable to use. Radioimmunoassay (RIA, p. 253) is used most extensively for the polypeptide hormones but other immunotechniques are beginning to compete with it. Commercial kits are available for many of these tests. The topic is covered in more detail on pages 253 to 258.

CEREBROSPINAL FLUID PROTEIN

Cerebrospinal fluid (CSF) is a clear, colorless liquid that circulates in the brain over the cerebral hemispheres and downward over the spinal cord. It is a combination of an ultrafiltrate of plasma and a secretion from the choroid plexus. The cerebrospinal fluid finally enters the venous blood after absorption by some villi (p. 365).

Cerebrospinal fluid is usually obtained by lumbar puncture of the spinal column; only a limited amount can be removed without causing a headache. It is a very precious specimen because of the difficulty and inconvenience in obtaining a good specimen and because several laboratories (microbiology, hematology, and chemistry) usually have to work with the same specimen, which is limited in amount (5 to 10 ml). Every effort should be made to conserve material, and to save all surplus material in the refrigerator for at least a 2-week period in case further checking should be required.

The common tests on cerebrospinal fluid requested of the clinical chemistry laboratory are protein, glucose, and chloride. The two latter constituents are discussed in the carbohydrate and electrolyte chapters, respectively. The protein concentration of the cerebrospinal fluid is very low relative to that in plasma, ranging between 15 and 45 mg/dl, or about 0.4% of the plasma concentration. The protein concentration of cerebrospinal fluid rises in various types of infection of the meninges, the membranes covering the brain and spinal cord; such infections are known as *meningitis*. The protein concentration of cerebrospinal fluid also rises in inflammatory lesions of the brain, after trauma, or in conditions that produce an elevated pressure (tumor, brain abscess). In addition to the analysis for protein, other laboratory observations provide useful information, i.e., the presence of blood or other chromogens, white cells in large numbers, or bacteria.

The protein concentration of cerebrospinal fluid is too low to measure conveniently by the biuret method but can be determined by turbidimetric methods. A bloody sample gives falsely high results because of the high protein content of the contaminating plasma.

Total CSF Protein

REFERENCE. Meulemans, O.: Determination of total protein in spinal fluid with sulphosalicylic and trichloroacetic acids. Clin. Chim. Acta 5, 757, 1960.

PRINCIPLE. The sulfosalicylic acid-Na_2SO_4 reagent produces a fine suspension of protein when it is added to CSF. The protein concentration is proportional to the turbidity that is measured by the decrease in light transmittance.

REAGENTS
1. Sulfosalicylic Acid-Na_2SO_4 Reagent. Dissolve 70 g Na_2SO_4 and 30 g sulfosalicylic acid in 800 ml water and dilute to 1000 ml volume. Add a few drops of chloroform as a preservative.
2. NaCl, 9 g/L (saline solution). Dissolve 9.0 g NaCl in water and make up to 1000 ml volume.
3. Standard Solutions. Transfer to a 100-ml volumetric flask 2.0 ml of a control serum whose protein concentration has been accurately measured. Add 50 mg sodium azide (NaN_3), and make up to volume with 9 g/L NaCl solution. Prepare a 1:2 and 1:4 dilution of this standard with saline solution and preservative as above. The concentration of the standards will be 0.02, 0.01, and 0.005 times the control serum total protein concentration, respectively.

PROCEDURE

1. Produce a cell-free spinal fluid by centrifuging the CSF.
2. Pipet 0.5 ml portions of CSF, standards, control, and water (for a blank) into appropriately marked tubes.
3. Add 2.0 ml of sulfosalicylic acid-Na_2SO_4 reagent to each. Cap with Parafilm and mix by gentle inversion.
4. Measure the absorbance at 450 nm when read against the blank.*
5. Calculation:

$$\text{CSF protein in mg/dl} = \frac{A_u}{A_s} \times C$$

where A_u and A_s are the absorbances of unknown and standard, respectively, and C is the concentration of the standard closest to that of the unknown.

Note: If the protein concentration is too high to read accurately, dilute the sample with an equal volume of 9 g/L NaCl and repeat. If the protein concentration should exceed 1000 mg/dl, the protein should be measured by the same biuret method that is used for serum total protein.

REFERENCE VALUES. 15 to 45 mg/dl.

INCREASED CONCENTRATION. The CSF protein concentration is elevated in various types of meningitis, in neurosyphilis, in some cases of encephalitis, in some brain tumors and abscesses, and frequently after cerebral hemorrhage. Lesions that inflame the meninges or increase CSF pressure usually increase the CSF protein concentration.

DECREASED CONCENTRATION. This is of no clinical significance.

Fractionation of CSF Protein by Electrophoresis

In patients with multiple sclerosis, the γ-globulin fraction of the protein in cerebrospinal fluid is frequently elevated, even though the total protein concentration may be within normal limits. The elevation in γ-globulin, if present, is revealed by a protein electrophoresis of the cerebrospinal fluid.

When the total protein in cerebrospinal fluid is within normal limits, the fluid has to be concentrated at least 100-fold in order to carry out a successful electrophoresis on cellulose acetate. Concentration methods have been described by Kaplan and Johnstone[4]; ultrafiltration at reduced pressure gave the best results. With concentrated cerebrospinal fluid, electrophoresis is carried out exactly as for serum with the exception that two or more applications of the cerebrospinal fluid concentrate may have to be made on the membrane if the cerebrospinal fluid is not sufficiently concentrated. Normal γ-globluin in cerebrospinal fluid is less than 11% of the total protein, but in multiple sclerosis it usually exceeds 18%. A normal pattern is shown in Figure 5.5A.

URINE PROTEIN

A semiquantitative method of determining urine protein using dipsticks or strips is described on page 123. To evaluate some renal diseases, it is necessary

---

*If specimen is icteric, read against a sample blank prepared by diluting 0.5 ml CSF with 2 ml NaCl solution.

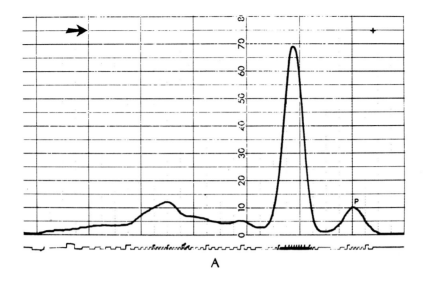

A

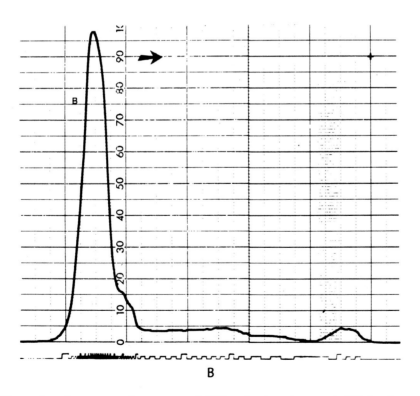

B

Fig. 5.5. Electrophoresis patterns of concentrated CSF and urine. A, normal CSF (total protein = 30 mg/dl. Concentrated 100 × and two applications made). Note the prealbumin band (P), which is typical of CSF electrophoresis. B, urine containing Bence Jones protein (total protein = 350 mg/dl. Concentrated 25 ×). The Bence Jones protein (B) predominates and migrates in this sample as a gamma globulin.

to know how much protein is lost in the urine in a 24-hour period, and a quantitative method must be used.

Total Protein

REFERENCE. Henry, R.J., Sobel, C., and Segalove, M.: Turbidimetric determination of proteins with sulfosalicylic and trichloroacetic acids. Proc. Soc. Exp. Biol. Med. 92, 748, 1956.

PRINCIPLE. Trichloroacetic acid precipitates the protein in urine as a fine suspension. The turbidity of the suspension is proportional to the amount of protein precipitated.

REAGENTS
1. Trichloroacetic Acid, 12.5 g/dl. Dissolve 12.5 g of trichloroacetic acid in water and make up to 100 ml volume. Prepare fresh reagent each month.
2. NaCl, 9 g/L.
3. Protein Standard. Accurately measure the protein concentration in a control serum pool. Dilute with 9 g/L NaCl to obtain a protein concentration of 25 mg/dl. To check linearity, run a standard curve with protein concentrations of 10, 25, 50, 100 and 200 mg/dl, respectively.

PROCEDURE
1. Urine must be clear (not turbid). Centrifuge if necessary to obtain a clear specimen.
2. Pipet 4 ml of standard to a test tube and 4 ml portions of urine to each of 2 tubes (unknown and urine blank).
3. To unknown and standard, add 1.0 ml of trichloroacetic acid solution and mix immediately by inversion. Add 1.0 ml of 9 g/L NaCl to the urine blank.
4. Between 5 and 10 minutes later, record the absorbance of standard versus water and unknown versus urine blank at 420 nm.
5. Calculation: Results are reported as grams per total volume in milliliters (T.V.) of urine in the sample.

$$\text{mg protein/100 ml} = \frac{A_u}{A_s} \times 25$$

$$\text{g protein/total volume} = \frac{A_u}{A_s} \times 25 \times \frac{T.V.}{100} \times \frac{1}{1000} = \frac{A_u}{A_s} \times \frac{T.V.}{4000}$$

REFERENCE VALUES. A normal urine may contain 0.05 to 0.1 g in a 24-hour period, even though this may not be detected by routine methods.

INCREASED EXCRETION. Protein may be lost in large quantities in the nephrotic syndrome in which the lesion is in the basement membrane of the glomerulus; lesser amounts of protein are excreted in other diseases producing renal lesions. Bence-Jones protein, the light chains of immunoglobulins, appears in the urine in many cases of myelomatosis. The amount excreted depends to a large extent upon the stage and severity of the disease. Some individuals have an orthostatic proteinuria, protein appearing in the urine when they stand erect or walk for any period of time. This is a benign condition.

Bence Jones Protein

As mentioned on page 156, Bence Jones protein is a protein with peculiar thermal solubility properties. It is found in the urine of about 50% of the patients with multiple myeloma. The protein is a polypeptide consisting of a light chain of either the κ or λ type.[5] The heat test for Bence Jones protein, which depends on a protein precipitating when the urine is heated to 40 to 60° C, but which dissolves when heating is continued beyond 80° C, is not reliable because there are too many false negatives. The heat test is also unsatisfactory because it is difficult to be certain of a precipitate when the urinary concentration of the Bence Jones protein is low, and secondly, many cases of myelomatosis are accompanied by glomerular damage, with leakage of plasma proteins into the urine; the plasma proteins are irreversibly denatured by heat and obscure any redissolving of Bence Jones protein.

The Bence Jones protein also has peculiar solubility properties in different concentrations of 1-propanol. Jirgensons and his associates[6] devised a method based upon the fact that the Bence Jones protein at pH 5 precipitates maximally in 20% (v/v) propanol, compared to 10%, 30%, and 40%. In 10% propanol, serum globulins produce minimal turbidity, which gradually increases in intensity as the alcohol concentration is increased to 40%; albumin does not precipitate until the propanol concentration exceeds 50%. This method is more satisfactory than the heat test, but some false negatives may be obtained when the urine concentration of Bence Jones protein is low.

The definitive test for Bence Jones protein is by immunodiffusion in agar gels containing antibodies to κ and λ chains, respectively. A precipitation line in the gel occurs when the antigen (Bence Jones protein in the urine sample) diffuses out of the sample well in sufficient quantity to react and precipitate with the specific antibody. This test not only reacts positively with Bence Jones protein, but identifies the type (p. 250).

Not all laboratories are set up for immunodiffusion, so the identification of a Bence Jones proteinuria by electrophoresis will be described.

PRINCIPLE. The urine is concentrated by vacuum ultrafiltration or by other suitable means[4] until the total protein concentration is between 2 and 6 g/dl. An aliquot is taken for electrophoresis on cellulose acetate exactly as for CSF protein. After the electrophoresis run, the membrane is stained and cleared as described before. The presence of a Bence Jones protein is demonstrated as a sharp band on the membrane. When a graph of the absorbance is made with a densitometer, the Bence Jones protein appears as a sharp spike somewhere in the globulin region (Fig. 5.5B).

PROCEDURE. Estimate the protein concentration of the urine by turbidity measurement. Concentrate the urine in the manner described for cerebrospinal fluid so that the final protein concentration will be between 2 and 6 g/dl. Perform the electrophoresis as described for cerebrospinal fluid and look for a sharp spike in the globulin area of the electrophoresis tracing.

REFERENCES

1. Gitlin, D., Gross, P.A.M., and Janeway, C.A.: The gamma globulins and their clinical significance. N. Engl. J. Med. *260*, 21:72; 121; 170, 1959.
2. Kaplan, A., and Savory, J.: Cellulose acetate electophoresis of proteins of serum, cerebrospinal fluid and urine. *In* Standard Methods of Clinical Chemistry, edited by R.P. MacDonald, Vol. 6. New York, Academic Press, 1970, p. 13.

3. Nerenberg, S.T.: Electrophoretic Screening Procedures. Philadelphia, Lea & Febiger, 1973, p. 14.
4. Kaplan, A., and Johnstone, M.: Concentration of cerebrospinal fluid proteins and their fractionation by cellulose acetate electrophoresis. Clin. Chem. *12*, 717, 1966.
5. Pruzanski, W., and Ogryzlo, M.A.: Abnormal proteinuria in malignant disease. *In* Advances in Clinical Chemistry, Vol. 13, edited by O. Bodansky and C.P. Stewart. New York, Academic Press, 1970, p. 335.
6. Jirgensons, B., Ikenaka, T., and Gorguraki, V.: Concerning chemistry and testing Bence Jones proteins. Clin. Chim. Acta 4, 876, 1959.
7. Savory, J.: Nephelometric immunoassay techniques. *In* Ligand Assay, edited by J. Langan and J.J. Clapp. New York, Masson Publishing, U.S.A., 1981, p. 257.
8. Ullman, E.F.: Recent advances in fluorescence immunoassay techniques. *In* Ligand Assay, edited by J. Langan and J.J. Clapp. New York, Masson Publishing U.S.A., 1981, p. 113.
9. Pinnell, A.E., and Northam, B.E.: New automated dye-binding method for serum albumin determination with bromcresol purple. Clin. Chem. *24*, 80, 1978.
10. Killingsworth, L.M.: Plasma protein patterns in health and disease, CRC Critical Rev. Clin. Lab. Sci. *10*, 1, 1979.

Chapter 6

ENZYMES

NATURE OF ENZYMES

Biological Catalysts

Most chemical reactions, particularly those for the oxidation or transformation of organic compounds, proceed imperceptibly or not at all at 37° C, because they require elevated temperatures for their action. Living cells cannot be subjected to high temperatures because of the delicate nature of protoplasm; most protein solutions coagulate at temperatures above 56° C. Biological systems, however, developed a number of catalysts that enabled the necessary metabolic reactions to occur at body temperature. The biological catalysts speed up reaction rates and are called *enzymes*.

Enzymes are present in all body cells and each one functions in a specific reaction. Enzymes catalyze all essential reactions (oxidation, reduction, hydrolysis, esterification, synthesis, molecular interconversions) that supply the energy and/or chemical changes that are necessary for vital activities (muscle contraction, nerve conduction, respiration, digestion, growth, reproduction, maintenance of body temperature). As catalysts, enzymes are not used up in the process and do not change the equilibrium point of the reaction; they merely accelerate the rate for reaching equilibrium.

All known enzymes are proteins and are synthesized in the body in the same manner as all other proteins; the synthesis of each particular enzyme is under the control of a specific gene. Many genetic defects are manifested by the absence or low production of a specific enzyme and can be demonstrated by appropriate laboratory tests.

Enzyme Action

The properties, mode of action, and kinetics of enzymes are treated more extensively in many textbooks of biochemistry. Supplementary reading in the books listed in references 1 through 3 will be helpful.

The first step in the action of an enzyme is the formation of an enzyme-substrate complex—that is, the binding of the substrate molecule to the large protein enzyme. Some enzymes require the presence of a coenzyme, a nonprotein molecule, before they can function. The majority of the coenzymes are small, heat-stable compounds that are usually derivatives of particular members of the vitamin B family, such as thiamine phosphate (B_1), pyridoxal (B_6), nucleotides of riboflavin (B_2), nucleotides of nicotinamide, and others. Enzyme assays requiring nicotinamide adenine dinucleotide or its phosphate as part of the system are commonly used in the clinical chemistry laboratory because the formation or disappearance of the reduced coenzyme is easily followed by measuring the absorbance change at 340 nm. The reduced coenzyme (NADH, NADPH) has a strong absorbance at 340 nm, while the oxidized form (NAD+, NADP+) has none. The great sensitivity of this measurement allows the use of small serum samples.

Enzymes have an active site, the actual place where particular bonds in the substrate are strained and ruptured. The active site has a specific, three-dimensional configuration, depending upon the enzyme's amino acid sequence, which determines the distribution of charges around the site. Part or all of the substrate fits into the active site in the same manner as a key matches a lock, a process that confers specificity upon the enzyme. Certain enzymes require the presence of a particular ion to induce the necessary configuration for proper binding to the substrate; such inorganic ions are called *activators*. Some common cations are calcium, magnesium, and zinc but other cations are also activators (copper, manganese, iron). Anions may be required by some enzymes for full activity; chloride ion is essential for amylase action.

After the enzymatic action (rupture of a particular bond in the substrate) has occurred, the products of the reaction are no longer bound and the enzyme is recycled as the products diffuse away. The free enzyme is able to bind another substrate molecule and repeat its action.

Inhibition

Since the catalytic action of an enzyme depends upon its specific configuration as a protein, many substances are able to inhibit this activity by changing the configuration of the enzyme (binding of heavy metal cation, drugs, or other charged *molecules* to the protein).

Some inhibitors closely resemble the binding portion of the substrate in shape and charge and therefore can bind to the enzyme. These molecules are competitive inhibitors because the substrate and inhibitor compete for the same binding site; the addition of more substrate results by mass action in the displacement of inhibitor molecules by substrate molecules, with a consequent increase in enzymatic activity. Competitive inhibitors of common metabolic reactions are also called *antimetabolites*. The noncompetitive types of inhibitors alter the configuration of the enzyme in such a way as to reduce or abolish access of the substrate to the active site and thus cause a decrease in enzyme activity. The addition of more substrate has no effect upon this type of enzyme inhibition because it cannot reverse the structural alteration that has occurred.

Factors Affecting the Velocity of Enzyme Reactions

The velocity of enzyme reactions is affected by a variety of factors, both general and specific: concentration of reactants, pH, temperature, ionic strength, presence of specific ions (activators or inhibitors), and presence of inhibitors. Any substance that alters the configuration of an enzyme affects its activity.

Enzyme Concentration

When the concentration of substrate is high and essentially constant during an enzyme reaction, the velocity of the reaction is directly proportional to the enzyme concentration. The reaction proceeds more rapidly when more enzyme molecules are present to bind the abundant substrate.

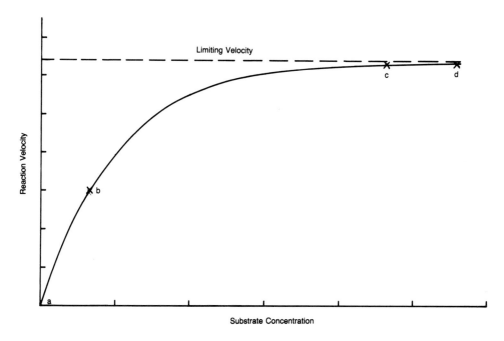

Fig. 6.1. Illustration of the relation between substrate concentration and enzyme reaction velocity.

Substrate Concentration

When the amount of enzyme in a reaction mixture is fixed and the reaction rate is plotted against increasing amounts of substrate, the type of curve illustrated in Figure 6.1 is obtained. The reaction rate varies directly with substrate concentration when its concentration is low, as illustrated by the linear segment *ab* of the curve. The direct proportionality between reaction rate and substrate concentration is known as a *first order reaction.* At higher substrate concentrations, there are fewer binding sites available for attachment, so the increase in reaction velocity for each added increment of substrate is progressively diminished (segment *bc,* Fig. 6.1). The reaction velocity becomes independent of the substrate concentration when the binding sites are saturated; the reaction proceeds at maximum velocity. At maximum velocity, the *rate of increase* of the velocity falls to zero and this reaction state is called a *zero order reaction.* The majority of enzyme tests employed in the clinical chemistry laboratory follow zero order kinetics—that is, a large excess of substrate is present so that the amount of enzyme activity is the only rate limiting factor in the assay. The term, *enzyme kinetics,* is used in connection with the study of the rates of enzyme reactions.

pH

Enzymes, being proteins, carry a net charge on the molecule that is sensitive to alterations in pH; the binding of substrate to enzyme may be greatly affected by the environmental pH. In general, the optimum pH range for an enzyme is relatively small and the curve produced by plotting enzyme activity versus pH is usually steep, with a rapid falling off after the optimum point. Also, there is a distinct limit to the range of permissible pH before protein denaturation

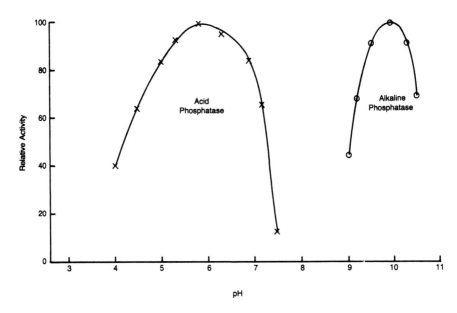

Fig. 6.2. Relative reaction velocities of alkaline and acid phosphatase, respectively, at various pH's.

of the enzyme affects activity. The effect of pH upon the relative activities of acid and alkaline phosphatase are illustrated in Figure 6.2. Note that the optimum pH range is narrower for alkaline phosphatase and that the fall off is steeper.

Temperature

The reaction velocities of most chemical reactions increase with temperature and approximately double for each 10° C rise. Again, the protein nature of enzymes places very strict limits upon the permissible temperature range because most proteins become denatured and insoluble within minutes of being subjected to temperatures of 60° C and above. Most of the body enzymes have a temperature optimum close to 37 to 38° C and have progressively less activity as the temperature rises above 42 to 45° C. All enzyme activity ceases when the protein is completely denatured. Low temperatures also decrease enzyme activity and they may be completely inactive at temperatures of 0° C and below. The inactivity at low temperatures is reversible, so many enzymes in tissues or extracts may be preserved for months by storing at −20 or −70° C. It is inconvenient for the laboratory to measure enzymes at a variety of temperatures. Usually the waterbaths and instruments are set to conduct enzyme measurements at 25°, 30°, or 37° C. Unfortunately, there is not universal agreement upon the exact temperature to use and this is one of the factors responsible for variation in results from laboratory to laboratory. It is essential to control the incubation temperature to an accuracy of ±0.1° C because of appreciable change of enzyme activity with temperature.

Electrolyte Environment

Many enzymes are sensitive to certain buffer ions present in the incubation mixture; hence, different buffers must be tested before deciding upon optimum conditions for measuring activities. Certain cations at a particular concentration range may be necessary for complete activation of the enzyme, and the total ionic strength of the incubation mixture may also be a factor affecting activity.

Inhibitors

For some enzymes, a product of the reaction may inhibit enzyme activity. Inorganic phosphate, a reaction product formed during the hydrolysis of organic phosphate esters by alkaline phosphatase, is a competitive inhibitor of this enzyme. Also, various drugs administered to patients under treatment may be competitive or noncompetitive inhibitors of certain of the serum enzymes.

In the ideal situation, assay of an enzyme's activity should be carried out under optimum conditions for substrate concentration, pH, buffer selection, coenzyme, activators if needed, temperature, ionic strength of the solution, and time of incubation. The real world of the clinical laboratories is far from the ideal situation, and many variations of methods are in use. Variations in reference values reflect differences in assay conditions. National and international clinical chemistry groups, however, are attempting to optimize and standardize enzyme methods.

Commercial kits for enzyme measurements should be tested before adoption in order to select those that provide conditions close to optimum and are stable and reliable.

Measurement of Enzyme Activity

The body contains hundreds of different enzymes, most of which are located and function primarily on or within cells—that is, attached to cell membranes, nuclei, various organelles such as mitochondria, ribosomes, lysosomes, and endoplasmic reticulum. Other enzymes may exist free in the cytoplasm.

Enzymes also are found in low concentration in body fluids: plasma, cerebrospinal fluid, urine, and exudates. These come primarily from the disintegration of cells during the normal process of breakdown and replacement (wear and tear), but certain enzymes appear in these fluids in much higher concentration following injury or death of tissue cells. Some of the intracellular enzymes from injured or necrotic cells diffuse into the plasma because all tissues are engulfed in a network of blood capillaries. In addition, the altered membrane permeability that may occur with inflammation may be sufficient to permit the diffusion of some cytoplasmic enzymes from cells into body fluids. Serum is the usual fluid taken for enzyme measurement for diagnostic purposes, but all of the other body fluids may be chosen under appropriate circumstances.

The concentration of an enzyme in plasma or serum cannot be measured readily in terms of μg per liter because it is difficult to determine in a direct manner the mass of a specific enzyme per unit volume of fluid. Instead, the rate at which the enzyme catalyzes a reaction with a substrate is measured under specific conditions. The analytical measurement may involve the decrease in substrate concentration with time, the increase in a reaction product, or the increase or decrease of a cofactor such as NADH (reduced nicotinamide adenine dinucleotide). If the substrate is present in a sufficiently high concentration to saturate the enzyme (occupy all binding sites), then the concentration of the reaction product depends directly upon the concentration of the enzyme and the time period of the reaction.

The activity of an enzyme may be measured by two different approaches: fixed time (two-point) assay and continuous monitoring (kinetic assay).

Fixed Time (Two-Point) Assay

The first enzyme tests commonly employed in the clinical laboratory (amylase, lipase, alkaline and acid phosphatase) utilized a fixed time for the reaction and expressed the enzyme activity as the amount of substrate transformed by a specified volume of serum under the particular conditions of the test. Many of the incubation times were too long (24 hours in the lipase method of Cherry and Crandall,[4] and 1 hour in the alkaline phosphatase method of Bodansky[5]). As a consequence, there was some enzyme inactivation (partial denaturation), product inhibition, and suboptimal concentration of substrate after a time. The rate of enzyme activity decreased with time after an early period of linearity.

The greater stability and sensitivity of modern spectrophotometers have improved the performance of two-point assays by making shorter incubation times possible. In such systems, however, it is necessary to establish the limits of enzyme activity in which zero order kinetics are followed. Preliminary experiments with different time intervals and with various aliquots of serum sample are essential for the selection of the conditions that produce linearity.

In some clinical situations, the activities of particular serum enzymes may be so large that the substrate concentration in the two-point assay may become exhausted before the test is completed, or so reduced in concentration that the enzyme's binding sites are no longer saturated. In either case, the measured activity is no longer directly proportional to enzyme concentration or to time and a falsely low value is obtained. The test has to be repeated while using either a much shorter time period or with a smaller serum sample.

Zero order kinetics are also not followed if the serum to be tested has a lag phase, a time period during which the enzyme activity builds up to its linear, zero order rate. A lag phase may be produced by the enzyme binding some metabolic substrate in serum or an impurity in the added substrate that is slowly broken down. The lag phase is a direct result of transitory, competitive inhibition and may be avoided by instituting a short preincubation period before adding the desired substrate. There also may be a lag phase in linked enzyme reactions when it takes a little time for the indicator reaction to reach a maximum rate. Lag phase caused by impurities in the substrate can be eliminated only by substrate purification or by obtaining another source of substrate that does not contain the impurity. Figure 6.3 illustrates an enzyme reaction with a lag phase, segment *ab* in the curve. In this example, the true enzyme activity in a 5-minute, two-point assay is from minute 2 to minute 7, with an activity of segment *bd*. The activity measured from time zero to 5 minutes would be falsely low because of the lag phase. Also, in this example, the substrate no longer saturates the enzyme after 11 minutes, so zero order kinetics are followed only in segment *be*. A measurement of activity in a 15-minute period after the lag phase would be falsely low as indicated by the segment *fg*.

Continuous Monitoring (Kinetic Assay)

The *rate of the enzyme reaction as a function of time* is measured by incubating serum under specific reaction conditions and measuring the rate of

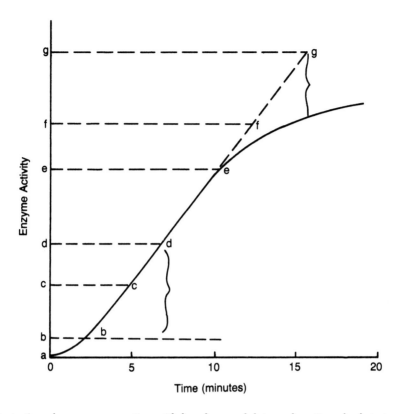

Fig. 6.3. Illustration of an enzyme reaction with lag phase and, later, exhaustion of substrate.

change in substrate, cofactor, or product concentration. Because the reaction time is usually short (a matter of some seconds or up to a few minutes), there is little danger of enzyme inactivation. Furthermore, continuous monitoring permits multiple readings for the determination of the rate. If the substrate concentration should fall below the optimum because of high enzyme activity, a warning is provided by the falling reaction rate; no such warning is obtained in a fixed time reaction. Continuous monitoring is used most commonly with those enzymes in which changes in NADH or NADPH are measured but can also be used for the determination of other enzyme activities (e.g., alkaline phosphatase) if a colored product is generated from a noncolored substrate. Many assays require the introduction of accessory, purified enzymes to produce a coupled system that utilizes or generates NADH, the substance whose change in concentration is measured by a decreasing or increasing absorbance at 340 nm. It is imperative to add the accessory enzymes in excess so that the rate limiting factor is the concentration (activity) of the enzyme of interest. A preincubation period is usually necessary so that extraneous reactions utilizing or producing NADH are completed before measurements are started on the enzyme of interest. Also, some serum enzymes have a lag phase, as described above in the two-point assay, and allowance for this lag period must be made before initiating measurements of the reaction rate. Some precautions also have to be taken with continuous monitoring methods in order to eliminate possible errors. Inspection of the chart recording of absorbance change with time allows one to check for linearity of the enzyme action and to see whether there is sufficient substrate for those serum samples with high enzyme activity. Some instruments provide a built-in check for linearity of the absorbance readings.

In the past, a great deal of confusion existed because enzyme activity was expressed in arbitrary units that were set by the person devising the method. Scientists have cooperated on an international scale to eliminate the use of arbitrary enzyme units and to substitute a uniform method for expressing enzyme activity. The International Unit (U) of enzyme activity was defined by the International Commission on Biochemical Nomenclature (1972) as the amount of an enzyme that will convert one μmole of substrate per minute in an assay system. The activity is usually expressed in terms of International Units per liter of serum (U/L). The Système Internationale introduced a confusing note, however, by designating the second as the basic unit of time; the S.I. unit of enzyme activity is called a *katal* and is defined as one mole of substrate transformed per *second*; 60 μ katal = one IU. 1 nkat = 16.7 U. The measurement of time in seconds, minutes and hours is worldwide; since the International Unit (μmol/min) is widely accepted in the U.S.A., we shall adhere to this unit of enzyme activity in this book. Since the activity of an enzyme is dependent upon a host of factors, such as pH, temperature, the concentrations of substrate, cofactors (if any), salts, and buffers, it is important to define the optimum conditions for a test and to stick with them. Normal values for a particular enzyme test may vary from laboratory to laboratory because of small or large differences in methodology, even though all of the laboratories may be expressing the results in International Units.

The historical nomenclature of enzymes was confusing because some names were arbitrarily chosen and others were constructed by adding the suffix *ase* to the substrate acted upon (e.g., protein*ase*, ester*ase*). Sometimes there were several common names for the same enzyme because they were discovered independently in different areas of the world. Now this confusion has been eliminated because the Enzyme Commission of the International Union of Biochemistry (IUB) has provided a systematic nomenclature* and code number for all enzymes.[6]

---

*There are no official, international abbreviations for the enzymes. The following nonstandard abbreviations are used in this book:

ACP, acid phosphatase
ALT (GPT), alanine aminotransferase
ALP, alkaline phosphatase
AST (GOT), aspartate aminotransferase
ALD, aldolase
CK (CPK), creatine phosphotransferase (phosphokinase), creatine kinase
CK-BB (CK$_1$), the BB (brain) isoenzyme of CK
CK-MB (CK$_2$), the MB (heart) isoenzyme of CK
CK-MM (CK$_3$), the MM (muscle) isoenzyme of CK
GD, glutamate dehydrogenase
GGT (GGTP), gamma glutamyl transpeptidase
GPD, glucose-6-phosphate dehydrogenase
LD (LDH), lactate dehydrogenase
LD$_1$, the most rapidly migrating LD isoenzyme (heart)
LD$_2$, the second most rapidly migrating LD isoenzyme
MD, malic dehydrogenase
NTP, 5'-nucleotidase
OCT, ornithine carbamoyl transferase

Since the measurement of an enzyme level makes use of its catalytic activity, its assay becomes a sensitive means for diagnosing disease. Tiny changes can be detected because each enzyme molecule is recycled over and over again as it acts upon the substrate. Although a few enzyme tests have been employed in the clinical laboratories for more than 50 years, the past 25 years have seen a tremendous increase in their use and scope. The discovery by LaDue, Wroblewski, and Karmen[7] in 1954 that the serum glutamic-oxaloacetic trans-aminase, or GOT (now called aspartate aminotransferase, AST), activity in serum was increased within 24 hours after a myocardial infarct, initiated the modern era of clinical enzymology. Many different enzymes were investigated in an attempt to find early indicators of specific tissue injury. The serum patterns of several different enzymes that are obtained after a suspected organ injury are usually more revealing than the activity of a single enzyme; this finding has led to the multiplicity of enzyme tests.

A further refinement in clinical enzymology has been the measurement of the different *isoenzymes* of a particular enzyme. Some enzymes exist in mul-timolecular forms (isoenzymes) that have similar catalytic activity but different biochemical and immunologic properties; they can be separated by electro-phoresis, which demonstrates a difference in charge on the various isoenzymes, by differences in adsorption properties, or by reaction with specific antibodies. Current theory ascribes the formation of isoenzymes to differences in compo-sition of the polypeptide chains comprising the enzyme. This is elaborated in more detail in the discussion of the CK and LD isoenzymes (pp. 187, 199).

Some of the enzyme tests that are currently being performed widely in clinical chemistry laboratories are described below and summarized in Table 6.1. This table lists not only the common or trivial name of the enzyme but also its common abbreviation, its systematic name, the Enzyme Commission (EC) code number, and some of the situations in which the tests are used for clinical diagnosis or patient management.

CLINICAL USAGE OF ENZYMES

Diagnosis of Myocardial Infarction (MI)

A myocardial infarct is a necrotic area in the heart caused by a deficient blood flow to that area as the result of a clot in a coronary vessel and/or narrowing of the vessel lumen by atheromatous plaques. When the cardiac cells in the necrotic area die, their intracellular enzymes diffuse out of the cell into tissue fluid and end up in plasma. Since it is not always possible to make a definitive diagnosis of myocardial infarction by an electrocardiogram, appro-priate enzyme tests are extremely helpful for this purpose.[3,8–10]

The enzyme tests that have proven to be most helpful in the diagnosis of myocardial infarction are listed in Table 6.2. These are creatine kinase (CK), aspartate aminotransferase (AST or GOT), lactate dehydrogenase (LD or LDH), isoenzyme CK-MB, and the isoenzymes of LD. All of the isoenzymes of LD have to be separated in order to obtain the absolute values of LD_1 and LD_2 as well as the ratio, LD_1/LD_2. Other intracellular enzymes have also been suggested for diagnostic purposes, but the preceding grouping provides the most infor-mation for the least cost. As shown in Table 6.2 and Figure 6.4, some of the enzyme activities increase early after an infarct (CK and CK-MB), some appear

TABLE 6.1

Some Serum Enzymes of Clinical Interest

COMMON NAME (ABBREVIATION)	SYSTEMATIC NAME (IUB)	EC* CODE NUMBER	DIAGNOSTIC PURPOSE
Aldolase (ALD)	Fructose 1,6-diphosphate: D-glyceraldehyde-3-phosphate lyase	4.1.2.13	Muscle disorders
Alkaline phosphatase (ALP)	Orthophosphoric monoester phosphohydrolase	3.1.3.1	Bone and liver disorders
Acid phosphatase (ACP)	Orthophosphoric monoester phosphohydrolase	3.1.3.2	Metastasizing cancer of the prostate
Amylase	1,4-α-D-Glucan glucanohydrolase	3.2.1.1	Acute pancreatitis
Creatine kinase (CK)	ATP:creatine phosphotransferase	2.7.3.2	Myocardial infarction, muscle disease
γ-glutamyl transpeptidase (GGT)	γ-glutamyl transferase	2.3.2.2	Liver disease
Glutamic-oxaloacetic transaminase (GOT)	L-aspartate:2-oxoglutarate aminotransferase (AST)	2.6.1.1	Myocardial infarction, liver disease, muscle disease
Glutamic-pyruvic transaminase (GPT)	L-alanine:2-oxoglutarate aminotransferase (ALT)	2.6.1.2	Liver disease
Guanase or guanine deaminase (GDS)	Guanine aminohydrolase	3.5.4.3	Liver disease
Lactate dehydrogenase (LDH or LD)	L-lactate:NAD+ oxidoreductase	1.1.1.27	Myocardial infarction, liver disease, malignancies
Lipase	Glycerol ester hydrolase	3.1.1.3	Acute pancreatitis
5'-nucleotidase (NTP)	5'-ribonucleotide phosphohydrolase	3.1.3.5	Liver disease
Ornithine-carbamoyl transferase (OCT)	Carbamoyl phosphate:L-ornithine carbamoyltransferase	2.1.3.3	Liver disease
Pseudocholinesterase	Acylcholine acylhydrolase	3.1.1.8	Exposure to organophosphate insecticides

*Enzyme commission

a little later (AST), and some increase even later and remain elevated for prolonged periods (LD, LD_1, and LD_2). Each enzyme has its own particular time course when the serum activity of the enzyme is plotted against time after the myocardial infarct (Fig. 6.4). Since the laboratory has no control over when the patient may elect to see the physician or when the enzyme tests are ordered, it is necessary to have some tests available that can help to diagnose a myocardial infarction in a time period that may vary from 4 hours to 10 days.

Creatine Kinase (CK)

CK is an enzyme that catalyzes the reversible reaction:

$$\text{Reaction 1: Creatine phosphate} + \text{ADP} \underset{Mg^{2+}}{\overset{CK}{\longleftrightarrow}} \text{Creatine} + \text{ATP}$$

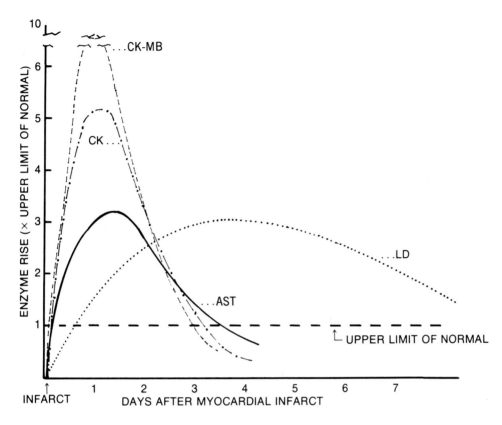

Fig. 6.4. Typical rise in serum enzyme activities following a myocardial infarction.

TABLE 6.2
Enzymes in the Diagnosis of Myocardial Infarction

ENZYME		ENZYME ACTIVITY INCREASE			
NAME	ABBREVIATION	ADULT UPPER LIMIT OF NORMAL (ULN) (Units/L)	USUAL RISE ($\times$ ULN)	TIME FOR MAXIMUM RISE* (Hr.)	TIME FOR RETURN TO NORMAL (Days)
Creatine kinase	CK	100	5–8	24	2–3
Isoenzyme MB	CK-MB	6	5–15	24	2–3
Aspartate					
aminotransferase	AST	25	3–5	24–48	4–6
Lactate					
dehydrogenase	LD	290	2–4	48–72	7–12
Isoenzyme #1	LD_1	100 or 34%	2–4†	48–72†	6–12
Isoenzyme #2	LD_2	115 or 40%	1.1–3†	48–72†	6–10

*Time after occurrence of infarct as manifested by chest pain.
†The LD-isoenzyme indication of a possible MI is an abnormal increase in the total amount of LD and an increase in LD_1 so that the ratio, LD_1/LD_2 is greater than 1 (the *flipped* LD ratio). The flipped ratio is usually not maintained as long as an abnormally elevated LD_1.

It is present in high concentration in skeletal muscle, cardiac muscle, thyroid, prostate, and brain; it is present only in small amounts in liver, kidney, lung, and other tissues. Hence, an increase in serum CK activity must be ascribed primarily to damage to striated muscle (skeletal or cardiac) and in rare cases, to brain. Differentiation between these various diseases can frequently be made upon clinical grounds, but there are situations when this is not possible. Measurement of the CK-MB isoenzyme helps to solve the problem (p. 188).

As seen in Figure 6.4, the serum CK activity begins to rise 3 to 6 hours after a myocardial infarction and reaches a maximum around 24 hours. The maximum rise in serum for a patient is usually about 5 to 8 × ULN (upper limit of normal), but in severe cases, this may reach 10 to 20 × ULN. In general, the higher the rise, the worse is the prognosis. The CK activity in serum may return to normal levels in 3 or 4 days; a subsequent increase indicates an extension of the infarct.

REFERENCE. Rosano, T.G., Clayson, K.J., and Strandjord, P.E.: Evaluation of adenosine 5′-monophosphate and fluoride as adenylate kinase inhibitors in the creatine kinase assay. Clin. Chem. 22, 1078, 1976.

PRINCIPLE. The enzyme, CK, contains sulfhydryl groups that may become oxidized on storage, with decreasing enzyme activity. The serum enzyme may be stabilized by adding sulfhydryl compounds, such as dithioerythritol or N-acetyl cysteine.

The stabilized CK reacts with creatine phosphate in the presence of ADP and Mg^{2+} as cofactors to produce creatine and ATP, as shown in reaction 1:

The rate of formation of ATP is determined by coupling reaction 1 with the following two reactions:

Reaction 2: ATP + glucose $\xleftrightarrow{\text{hexokinase}}$ ADP + glucose-6-phosphate

Reaction 3: Glucose-6-phosphate + $NADP^+$ $\xleftrightarrow{\text{GPD}}$ 6-phosphogluconate + NADPH + H^+

By fixing the conditions so that the auxiliary enzymes, substrates and the cofactors other than ATP are not rate limiting, the CK activity is followed by measuring the rate of increase in the absorbance at 340 nm as $NADP^+$ is reduced to NADPH.

REAGENTS AND MATERIALS

1. Stock Imidazole Buffer, 0.7 mol/L, pH 6.6 at 25° C. Dissolve 47.7 g anhydrous imidazole (Sigma) in about 800 ml water. Adjust pH to 6.6 with approximately 20 ml glacial acetic acid and make up to 1 L volume with water. Store in a glass bottle at 4° C. The solution is stable unless it becomes contaminated with microorganisms.

2. Working Imidazole Buffer, 0.14 mol/L, pH 6.6 at 25° C. Dilute 200 ml stock imidazole solution and 0.1 ml Triton X-100 (Harleco) to 1 L with water and check pH. Store at 4° C in a glass bottle.

3. Glucose-ADP Reagent. 160 mmol/L in glucose, 120 mmol/L in magnesium acetate, and 16 mmol/L in adenosine diphosphate (ADP). Dissolve 7.208 g glucose (MCB), 6.435 g $MgAc_2 \cdot 4H_2O$ (MCB), and 1.977 g ADP in 200 ml imidazole working buffer. Adjust the pH to 6.6 while adding 0.5 molar

NaOH with stirring; the Mg(OH)$_2$ which may precipitate temporarily while the base is being added, should completely redissolve. Dilute to 250 ml volume with the buffer. Store frozen in 5 ml aliquots; the solution is stable for at least 2 months at $-10°$ C.

4. Dithioerythritol (DTE), 400 mmol/L. Dissolve 3.086 g of DTE (Sigma) in approximately 40 ml of working imidazole buffer. Adjust pH to 6.6 if necessary. Dilute to 50 ml volume with the buffer. Freeze in 1.1 ml aliquots at $-10°$ C; they are stable for at least 4 weeks.

5. Nicotinamide Adenine Dinucleotide Phosphate (NADP), 60 mmol/L. Dissolve 2.458 g NADP (Sigma)* in a mixture of 40 ml of working imidazole buffer and 4.9 ml molar NaOH. Adjust the pH to 6.6 with molar NaOH and dilute to 50 ml volume with buffer. Freeze in 1.1 ml aliquots at $-10°$ C; they are stable for at least 4 weeks.

6. Creatine Phosphate, 100 mmol/L in sodium fluoride, 100 mmol/L. Dissolve 0.210 g NaF in approximately 40 ml working imidazole buffer. When it is completely dissolved, add 1.681 g creatine phosphate (CalBiochem-Behring) (mol. wt 336). When the creatine phosphate is dissolved, adjust the pH to 6.6 at 25° C with 2 mol/L acetic acid. Dilute to 50 ml volume with working imidazole buffer and store at 4° C in an amber bottle. The solution is stable for at least 2 weeks. Filter through a 0.45 micron filter apparatus.

7. Hexokinase (HK) from yeast (EC 2.7.1.1).
   a. Stock hexokinase (P-L Biochemicals) is a 50% (v/v) glycerol solution containing approximately 4500 U/ml of chromatographically pure enzyme (200 U/mg protein).
   b. Working HK $\geq$ 225 U/ml at 30° C. Add 0.5 ml of stock HK to 9.5 ml working imidazole buffer and store at 4° C; it is stable for at least 1 week at this temperature.

8. Glucose-6-phosphate Dehydrogenase (GPD) from yeast (EC 1.1.1.49).
   a. Stock GPD (P-L Biochemicals) in 50% (v/v) glycerol solution containing approximately 1700 U/ml of enzyme (200 U/mg of protein).
   b. Working GPD = 136 U/ml at 30° C: Add 1.0 ml of stock GPD to 11.5 ml of working imidazole buffer and store at 4° C, a temperature at which it is stable for at least 2 weeks.

9. Working Reaction Mixture: Thaw one tube each of the glucose-ADP reagent (#3), DTE (#4), and NADP (#5). Combine 5 ml of #3, 1.0 ml #4, and 1.0 ml of #5 in a 25-ml Erlenmeyer flask. To this, add 1.0 ml of working HK solution (#7b), 1.0 ml of working GPD solution (#8b), and 1 ml of working buffer (#2). Filter through a 0.45 micron filter. Prepare on the day of use and keep in an ice bath. The mixture is stable for 24 hours at 4° C.

10. NaOH, 1 mol/L: Dissolve 40 g NaOH pellets in water and dilute to 1 L.

11. Acetic Acid (HAc), 2 mol/L: Dilute 115 ml concentrated HAc to 1 L with water.

12. A stable spectrophotometer suitable for kinetic analysis.

PROCEDURE. The concentrations and volumes of reagents have been designed for optimal measurement of CK activity. Many kinetic analyzers may be used if the concentration of reactants in the final reaction mixture is maintained (see Note 1).

---

*The water of crystallization may vary from batch to batch. This affects the formula weight and hence the amount to be weighed for a specific concentration.

1. Turn on power to the instrument for at least 30 minutes before starting the assay in order to maintain a stable temperature of 30° C in the photometer cuvets.
2. Follow manufacturer's instructions for operation of instrument for CK assay. The following operations take place.
   a. 10 μl of serum are transferred with 90 μl of water wash-out to a tube. To the same tube, add 50 μl of working reaction mixture (#9).
   b. The mixture is stirred and preincubated for 6 minutes in order to use up any endogenous substrates that may be present.
   c. The CK reaction is started by the addition of 50 μl of creatine phosphate (#6). The final reaction mixture is stirred and an aliquot is transferred to a flow cell (cuvet) at 30° C.
   d. After a delay of 120 seconds to allow for lag time in the reaction, the absorbance change in the reaction mixture is integrated and averaged for 10 to 30 seconds at 340 nm. Microprocessors convert the absorbance readings to International Units/liter and print out the results.
   e. Repeat all samples that are nonlinear, have depleted substrate or have values greater than 800 U/L.
3. Calculation:

$$\text{CK activity, U/L} = \Delta A/\text{min} \times 1/6.22 \times \frac{0.200 \text{ ml}}{0.010 \text{ ml}} \times 1000$$

$$= \Delta A/\text{min} \times 3215$$

where $\Delta A/\text{min}$ is the absorbance change per minute, and 6.22 is the absorptivity of NADPH, mmol/L; the factor 1000 changes it to μmol/L, 0.200 ml is the total volume in the cuvet, and 0.010 ml is the volume of serum.
Notes:
1. The concentrations of reagents in the final reaction mixture are the following:

Imidazole buffer	70 mmol/L
Glucose	20 mmol/L
Magnesium acetate	15 mmol/L
ADP	2 mmol/L
NaF	25 mmol/L
NADP	1.5 mmol/L
Dithioerythritol	10 mmol/L
Hexokinase	5.6 U/ml at 30° C
GPD	3.4 U/mol at 30° C
Creatine phosphate	25 mmol/L

Reagent concentrations and volumes may be changed as long as the concentration in the final reaction mixture is the same as given above. This allows some flexibility in adjusting to different instrumentation systems.
2. CK activity is inhibited by uric acid, but this inhibition is reversed by the addition of a sulfhydryl compound (dithioerythritol).
3. NaF is added to the reaction mixture in order to inhibit the utilization of ADP by myokinase, which may be present in serum. NaF does not affect

CK activity but greatly inhibits the myokinase activity (MK): 2 ADP $\xrightarrow{MK}$ ATP + AMP.

4. The enzymes, HK and GPD, must be in excess so that the CK enzyme is rate limiting.

REFERENCE VALUES. Normal values for CK range from 10 to 100 U/L with the method described. Strenuous work or exercise 48 hours prior to CK measurement may increase the value; muscle trauma has a similar effect. CK activity is usually higher in males than in females because of the greater muscle mass and physical activity.

INCREASED ACTIVITY. Since CK is located primarily in skeletal muscle, myocardium, and brain, the serum activity increases after damage to these tissues and is not usually affected by the pathologic conditions in other organs.

1. *In damage to heart tissue.* As discussed earlier in this chapter, there is a sharp but transient rise in CK activity following a myocardial infarction; the degree of increase varies with the extent of the tissue damage. CK is the first enzyme to appear in serum in higher concentration following myocardial infarction and is probably the first to return to normal levels if there is no further coronary damage. The serum CK may be increased in some cases of coronary insufficiency without myocardial infarction.[3,8–10] The simultaneous determination of the CK-MB isoenzyme and LD isoenzymes will help to make the diagnosis, as described later.

2. *In damage to skeletal muscle.* The serum CK activity may rise to high levels following injury to skeletal muscles. Some of the causes may be trauma, muscular dystrophy, massage of chest during a heart attack, an intramuscular injection, or even strenuous exercise. The serum activity parallels the amount of muscle tissue involved. In prolonged shock, the CK enzyme also leaves the ischemic muscle cells and appears in the serum.

3. *In brain damage.* The CK levels in serum are increased in brain injury only when there is some damage to the blood brain barrier; the rise is in the BB fraction. Damage to the blood brain barrier may be caused by trauma, infection, stroke, or severe oxygen deficiency.

Creatine Kinase Isoenzymes

Enzymes are proteins assembled from polypeptide subunits. Creatine kinase is a dimer that consists of two subunits, M and B; the M subunit predominates in skeletal muscle and the B subunit in brain. There are three isoenzymes separated by electrophoresis that behave as dimers: CK-BB (CK_1), composed of two B chains; CK-MB (CK_2), which is made up of one M and one B chain; and CK-MM (CK_3), which is comprised of two M chains. The BB isoenzyme is the most negatively charged and migrates furthest toward the positive pole (anode), with the mobility of prealbumin. MB is of intermediate negative charge and is the middle band (mobility of β-globulin). MM has the least negative charge and migrates slightly toward the cathode similar to the migration of gamma globulin proteins.

TISSUE DISTRIBUTION. The MM isoenzyme is found primarily in skeletal and cardiac muscles but low activity exists in lung and kidney. Cardiac muscle cells contain a mixture of the CK-MM and CK-MB isoenzymes; the major portion is MM but the MB content is considerable and may comprise from 15 to 20% of the total CK activity. By contrast, skeletal muscle CK consists of approximately 99% MM fraction and only about 1% of MB. The BB isoenzyme is present in brain tissue, gastrointestinal and genitourinary tracts (colon, prostate, uterus), with lower activity in thyroid and lung.

SERUM CK ISOENZYMES. The predominant CK isoenzyme in the serum of normal individuals is the MM fraction, which comprises 94 to 98% of the total. The MB isoenzyme may be present up to 6% of the total but is usually only 2 to 4% (1 to 4 U). In normal serum, the BB isoenzyme is undetectable by electrophoretic methods but it may increase appreciably in women immediately postpartum, in patients with cardiovascular accidents (stroke), acute renal disease, adenocarcinomas of the prostate or other tissues, severe hypoxia (oxygen lack), and brain injury that damages the blood brain barrier.

The most important diagnostic use for CK isoenzymes is for the diagnosis of myocardial infarction (MI). Following a moderate to severe MI, the MB isoenzyme rises rapidly, reaches a maximum within 24 hours, and then falls rapidly (Fig. 6.4). Its relative increase in serum is greater than for total CK but it returns to normal values a little earlier than the latter. After a small MI, the MB isoenzyme may become elevated even though the total CK remains within normal limits. When dealing with a possible MI, it is important to consider the percentage increase in CK-MB as well as its activity in units because trauma to skeletal muscle results in an increase in both total CK and the MB isoenzyme; the MB activity as percent of total activity, however, is less than 3% in muscle trauma and greater than 6% in MI. The assay for CK-MB is an important diagnostic aid in patients suspected of having had an MI. Further information on CK isoenzymes in the diagnosis or assessment of MI appears in several excellent reviews.[9,10]

PRINCIPLE. The three isoenzymes of CK can be separated by electrophoresis, selective adsorption on columns, and by radioimmunoassay (RIA) for either the M or B polypeptide unit of CK (see p. 253 for discussion of RIA principles). Commercial kits are available for all three methods. The electrophoretic method will be described first because it is a more widely used technique than column chromatography or RIA and yields results in a shorter time.

ELECTROPHORETIC SEPARATION OF CK ISOENZYMES

REFERENCE. Modified from Procedure #23, May, 1981 (Helena Laboratories).

Principle. The CK isoenzymes in serum or cerebrospinal fluid are separated by electrophoresis at pH 8.8 on cellulose acetate. The strip after electrophoresis is overlaid with a cellulose acetate strip that has been soaked with CK substrate and accessory enzymes and reagents shown in reactions no. 2 and no. 3; the "sandwich" is incubated, the strips separated and dried. The dry electrophoresis strips are examined under an ultraviolet light that makes the CK isoenzyme bands fluoresce. For quantitation, the strips are scanned under a beam of ultraviolet light and the fluorescence is recorded and measured by a scanning densitometer.

Reagents and Materials
1. Tris-barbital-sodium barbital electrophoresis buffer: Add 2500 ml water to the contents of an Electra HR packet of preweighed buffer (Helena Laboratories). Check that the pH is 8.8. The buffer is stable indefinitely when it is stored at 4° C. A sodium barbital buffer, 0.03 mol/L, pH 8.8,

may be substituted. Dissolve 6.21 g sodium barbital in 700 ml water, adjust pH to 8.8 with acetic acid, and dilute to 1 L.
2. CK reagent (Sclavo) or CK-US Isoenzyme Reagent (Helena Laboratories). Add 3 ml Sclavo buffer to the reagent vial. The reconstituted reagent is stable for 2 days at 4° C.
3. Titan III Iso-Flur plates with mylar backing (Helena Laboratories).
4. Electrophoresis chamber and power supply (Helena Laboratories).
5. Scanning Densitometer with fluorescent and visible capabilities (Helena ChemScan, Helena Laboratories).

Procedure
1. Soak the Titan III Iso-Flur plate for 30 minutes at room temperature. Mark one strip (mylar side) for every seven specimens. Place these and an equal number of unmarked strips in the soaking rack and immerse slowly in the tris-barbital buffer.
2. Heat the development weight at 37° C by placing it on top of the heating box with both the fan and heater switches turned on.
3. Pour about 100 ml electrophoresis buffer (reagent 1) in the outer chambers of the electrophoresis cell. Wet two disposable wicks in buffer and drape one over each support bridge. They must remain in contact with buffer along one entire edge. Cover the cell.
4. Load 10 μl serum into seven of the eight wells in the sample plate. A control containing all three isoenzymes is placed in the eighth well.
5. Remove a marked plate from the buffer. Blot it well. Place it on the aligning plate with the longer edge on the line marked "This end is cathode."
6. Prime the applicator by depressing it in the wells of the sample plate 2 to 3 times and blotting on blotter paper. Repeat.
7. Apply the specimens onto the strip twice while keeping the applicator depressed for 5 to 10 seconds. Note which end is specimen 1.
8. Place the strip in the chamber with the application site nearest the cathode. Make sure that good contact is maintained with the wicks during the electrophoresis.
9. Replace the lid on the chamber and apply a constant voltage of 300 V for 11 minutes. An amperage of 5 to 7 ma per plate should be measured at the beginning of the run.
10. Remove a second cellulose acetate plate from the buffer and blot it well. Pour 1.0 ml substrate (reagent 2) on it and allow it to soak until the run is completed (5–8 minutes).
11. When the electrophoresis is completed, remove the plates, without tilting, and blot the wet edges.
12. Sandwich the plate with substrate to the electrophoresis strip and squeegee out the excess substrate by applying pressure in both directions to the plate.
13. Incubate the strips between blotter paper and under the development weight for 25 minutes.
14. Separate the strips and dry them by placing them underneath the heating block for about 5 minutes.
15. Scan the dried strips visually for fluorescent bands by placing them in the Photo Vubox or by use of an ultraviolet lamp (long wave, 366 nm gives the greatest sensitivity) in a darkened room. With practice, it is easy

to pick up abnormal MB and BB bands. All serums have a dim fluorescent spot that migrates with albumin between the BB and MB bands and that is colored slightly yellow.

16. Quantitatively scan the strips in an Auto Scanner Flur-Vis (Helena Laboratories) attached to a Quant II-T computer assembly. Follow the manufacturer's directions. The calculation of the percentage of each isoenzyme fraction is similar to that of the protein fractions in protein electrophoresis (p. 164).

   a. CK-MB relative percentage $= \dfrac{\text{MB band integral count}}{\text{total count}} \times 100$

   b. CK-BB relative percentage $= \dfrac{\text{BB band integral count}}{\text{total count}} \times 100$

If electrophoresis alone is used for isoenzyme separation, the activity of each isoenzyme band in International Units is obtained by multiplying the total CK activity by the relative percentage of each fraction.

If electrophoresis is used in conjunction with a chromatographic separation of the CK isoenzymes (see p. 191), multiply the activity of the eluate (contains both MB and BB fractions) by the relative percentage of the MB and BB fractions found by electrophoresis.

Control and/or Acceptable Limits. A mixture of human brain and heart homogenates that contains the three isoenzymes of CK may be used as a qualitative control for migration in an electrophoretic field. For a quantitative control, Beckman I.D. Zone CK Iso ELP Control (Beckman Instruments) may be used for each strip run. The CK-MB and CK-BB percentages must be within the acceptable range on each strip used for calculation of isoenzymes in patient serums.

Reference Values. The upper limit for CK-MB activity $= 6$ U/L. It usually is less than 6% of the total CK. The upper limit for CK-BB activity is not measurable by electrophoresis; normal individuals usually have no detectable BB isoenzyme.

CHROMATOGRAPHIC SEPARATION OF CK ISOENZYMES. Chromatographic separation of CK isoenzymes followed by quantitation of the enzymatic activity of the isolated isoenzymes is more sensitive and precise than electrophoretic methods. Many different techniques exist but we shall refer only to some of the widely used, well-tested ones.[11–13] In one of these methods,[12] the MB and the BB isoenzymes appear in the same eluate and the tacit assumption is made that the activity can be ascribed to the MB fraction, since BB activity is usually negligible. There are sufficient exceptions, however, to warrant a qualitative evaluation of each serum by electrophoresis to verify this assumption. In the other two methods,[11,13] the MB fraction appears in a separate eluate from the BB, so its activity can be assayed directly and without ambiguity.

Principle. The chromatographic separation of CK isoenzymes is based upon their charge differences. When serum is filtered through a DEAE-Sephadex A50 column (Pharmacia) under conditions of low ionic strength, pH 8.0, the CK-MB and -BB isoenzymes are adsorbed to the gel; CK-MM carries no negative charge and therefore flows through the column and is washed out. After washing the column to completely remove the MM isoenzyme, the MB and BB isoenzymes may be eluted together by increasing the ionic strength (NaCl concentration) of the wash buffer[12] or may be eluted separately[11,13] by proper adjustment of ionic strength and pH. The activity of CK in the various eluates or in a specific eluate is determined in a manner identical with that for total CK except that a larger aliquot must be used because the volume of eluate is much greater than that of the original sample.

Commercial chromatographic kits are available for the determination of CK isoenzymes.

IMMUNOTECHNIQUES FOR DETERMINATION OF CK ISOENZYMES. Various techniques based upon the binding of antibodies to the M or B subunit of CK have been devised. The assay of the MB isoenzyme by immunoinhibition of the M chain was introduced first but it was neither sensitive nor precise. The method involved the measurement of total CK activity before and after treating serum with an anti-M antibody. The antibody bound the M polypeptides and eliminated 100% of the activity of the MM isoenzyme and 50% of the MB fraction. To obtain MB activity, one has to multiply by 2 the activity obtained after antibody treatment and assume that the BB fraction is zero, an assumption that is not warranted in many hospital situations.

A more sensitive and promising assay is radioimmunoassay (see p. 253 for discussion of RIA principle) for the MB isoenzyme.[14] Some RIA methods determine the *amount of isoenzyme protein* present in serum but not the enzymatic activity. A commercial kit that appears to be promising, Isomune-CK (Roche Diagnostics), utilizes two separate tubes for the assay. In the first tube, anti-M antibody binds to all M subunits and inhibits 100% of the MM fraction and 50% of the MB; BB activity is unaffected. An anti-gamma globulin antibody is added to the second tube after the addition of anti-M antibody. The second antibody binds to and precipitates all molecules that have bound the anti-M antibody. The MM and MB fractions form a pellet after centrifugation, and the CK activity in the supernatant is derived from the BB isoenzyme. Upon subtracting the CK activity in tube #2 from that in tube #1, one-half the activity of CK-MB is obtained. Multiplication × 2 provides the activity of the CK-MB isoenzyme in International Units.

Reference Values. The CK-MB band in normal serum usually is not visible when separated by electrophoresis. Normal serum may have up to 6 U/L by chromatographic methods or 6 to 18 μg/L of MB protein by RIA.[18]

Increased Activity. The MB isoenzyme starts to increase within 4 hours after an MI and reaches a maximum within 24 hours. The maximum rise may range from 10 U to as high as 400 U, depending upon the severity of the infarct.

The MB isoenzyme always is elevated within 48 hours of an MI, but it can also be elevated with coronary insufficiency. When taken together with the LD isoenzymes, the finding of a flipped ratio ($LD_1 > LD_2$) plus an elevated MB isoenzyme is a positive indication of a myocardial infarction.[15] A diagnosis is always made with more assurance when the pattern of key enzyme changes is taken into consideration (see Fig. 6.4).

The BB isoenzyme is not detectable by electrophoresis or chromatography in the serum of healthy individuals but traces can be found by RIA. The BB fraction rises to electrophoretically measurable levels in the conditions mentioned on page 188.

Aminotransferases (Transaminases)

SERUM ASPARTATE AMINOTRANSFERASE, AST (TRANSAMINASE, GOT). Aspartate aminotransferase (AST) is found in practically every tissue of the body, including red blood cells. It is in particularly high concentration in cardiac muscle and liver, intermediate in skeletal muscle and kidney, and in much lower concentrations in the other tissues. The discovery by LaDue, Wroblewski, and Karmen[7] that the serum AST (GOT) concentration increased shortly after the occurrence of a myocardial infarction greatly accelerated the search for enzymes in serum as indicators of specific tissue damage. The measurement of the serum AST level is helpful for the diagnosis and following of cases of myocardial infarction, hepatocellular disease, and skeletal muscle disorders.

REFERENCE. Bergmeyer, H.U., Scheibe, P., and Wahlefeld, A.W.: Optimization of methods for aspartate aminotransferase and alanine aminotransferase. Clin. Chem. 24, 58–73, 1978.

Principle. The enzyme, aspartate aminotransferase, reversibly transfers an amino group from aspartate to α-ketoglutarate (oxoglutarate) according to Reaction 4 and forms oxaloacetate in the process. This reaction is coupled with that of malate dehydrogenase, in which the oxaloacetate is reduced to malate as NADH is simultaneously oxidized to NAD[+]; the decrease in absorbance at 340 nm is measured.

$$\text{Reaction 4: Aspartate } + \text{ } \alpha\text{-ketoglutarate } \xrightarrow{\text{AST}} \text{ glutamate } + \text{ oxaloacetate}$$

$$\text{Reaction 5: Oxaloacetate } + \text{ NADH } + \text{ H}^+ \xrightarrow{\text{MD}} \text{ malate } + \text{ NAD}^+$$

Preincubation of the serum sample with lactate dehydrogenase prior to adding the α-ketoglutarate substrate is necessary for the removal of any endogenous pyruvate that may be present. LD also utilizes NADH in converting pyruvate to lactate, but the supply of endogenous pyruvate is soon exhausted.

Reagents and Materials

1. Tris/L-aspartate (Tris 97 mmol/L, pH 7.8 at 30° C, L-aspartate 252 mmol/L). Dissolve 1.18 g Tris (hydroxymethyl) aminomethane and 3.35 g L-aspartic acid in 80 ml water; adjust pH to 7.8 with NaOH, 5 mol/L. Make up to 100.0 ml volume with water.
2. Tris/HCl Buffer (Tris 97 mmol/L, pH 7.8 at 30° C). Dissolve 1.18 g Tris buffer in 80 ml water; adjust pH to 7.8 at 30° C with HCl, 5 mol/L. Make up to 100.0 ml volume with water.
3. Pyridoxal Phosphate (6.3 mmol/L in 97 mmol/L Tris buffer, pH 7.8). Dissolve 16.7 mg pyridoxal phosphate in 10.0 ml Tris/HCl buffer (solution #2).
4. Reduced β-NADH (11.3 mmol/L in 97 mmol/L Tris buffer, pH 7.8). Dissolve 16 mg NADH-Na$_2$ (or an amount equivalent to this as determined by water of hydration or other contaminants) in 2.0 ml Tris/HCl buffer (solution #2).
5. Malate Dehydrogenase/Lactate Dehydrogenase (activities 75,600 U/L, each). Mix the enzyme solutions in glycerol according to their catalytic concentrations and adjust with glycerol, 50% (v/v), to give the above concentrations.
6. Reagent Mixture for overall aspartate aminotransferase reaction. Mix 100.0 ml Tris/L-aspartate (solution #1) with 2.0 ml pyridoxal phosphate (solution #3), 2.0 ml NADH (solution #4) and 1.0 ml enzyme (solution #5).

7. Reagent Mixture for individual sample blank. Mix 100.0 ml Tris/HCl buffer (solution #2) with 2.0 ml pyridoxal phosphate (solution #3), 2.0 ml NADH (solution #4), and 1.0 ml enzyme (solution #5).
8. α-Ketoglutarate, 144 mmol/L. Dissolve 273 mg α-ketoglutaric acid, disodium salt in 10.0 ml water. Adjust to pH 7.8 at 30° C with hydrochloric acid, 5 mmol/L.

Stability of Solutions. Store all solutions in stoppered containers in a refrigerator at 0° to 4° C. Prepare the reagent mixtures (solutions #6 and #7) fresh each day. Prepare the pyridoxal phosphate, the NADH, and the α-ketoglutarate (solutions #3, #4, and #8) every second week. The Tris/L-aspartate, the Tris/HCl and the enzymes (solutions #1, #2, and #5) are stable for at least 6 months. Deterioration of these latter three solutions is usually due to bacterial contamination. This can be prevented by addition of 8 mmol/L sodium azide. Store solutions #3, #4, #6, and #8 in dark bottles.

Procedure

1. Pipet into cuvets 2.00 ml of reagent mixture #6 containing Tris buffer, L-aspartate, NADH, pyridoxal phosphate, and the 2 supplementary enzymes, MD and LD.
2. Add 0.20 ml serum, mix, and incubate at 30° C for 10 minutes.
3. Start the AST reaction by the addition of 0.20 ml of α-ketoglutarate (solution #8).
4. Mix and record the change in absorbance at 340 nm, preferably by a kinetic rate analyzer instrument. With some instruments with microprocessors, the reaction time can be as short as 9 seconds. The time should be adjusted to the capability of the instrument used. Where the reaction rate is obtained by plotting the absorbance change with time on a recorder, follow the absorbance change long enough to obtain a straight line for at least 2 minutes.
5. Calculation:

$$\text{AST activity, U/L} = \Delta A/\text{min} \times 1/6.22 \times \frac{2.40 \text{ ml}}{0.20 \text{ ml}} \times 1000$$

$$= \Delta A/\text{min} \times 1929$$

where $\Delta A/\text{min}$ is the absorbance change per minute, 6.22 is the mmol/L absorptivity of NADH at 340 nm, 2.40 ml is the total volume in the cuvet, 0.20 ml is the volume of serum, and 1000 is the factor for converting mmoles to μmoles.

Notes:

1. The volumes proposed by the IFCC committee can be scaled down proportionately in order to conserve serum and reagents. Reagent concentrations and volumes may be changed so long as the final concentrations of the reactants are the same as given below:

Tris buffer	80 mmol/L
L-aspartate	200 mmol/L
α-ketoglutarate	12 mmol/L
Pyridoxal phosphate	0.10 mmol/L

NADH	0.18 mmol/L
Malate dehydrogenase	420 U/L*
Lactate dehydrogenase	600 U/L*

The use of many different automated enzyme analyzers may be accommodated in this manner.

2. See footnote, p. 185 for precautions to take concerning the water of crystallization of the various reagents and coenzymes when making up the solutions.

3. Blood specimens should be drawn in a manner to avoid hemolysis. Erythrocyte AST falsely elevates the serum AST activity if the blood is hemolyzed.

4. NADH powder should be protected from moisture and stored at 0° C or lower.

Reference Values. The normal concentration of serum AST (GOT) is 6 to 25 U/L.

Increased Activity. The serum activity of AST is increased after myocardial infarction, in liver disorders, in trauma to or in diseases affecting skeletal muscle, after a renal infarct, and in various hemolytic conditions.

Myocardial Infarction. The serum activity of AST begins to rise about 6 to 12 hours after myocardial infarction and usually reaches its maximum value in about 24 to 48 hours. It usually returns to normal by the fourth to sixth day after the infarct. This pattern is shown in Table 6.2 and Figure 6.4. The increase in activity is not as great as for CK, nor does it rise as early after the infarct. It is a much less specific indication of myocardial infarction than the rise in CK, since so many other conditions, e.g., liver, muscle, or hemolytic disease, can cause a rise in serum AST. Prolonged myocardial ischemia may be accompanied by a rise in serum AST. Congestive heart failure also is associated with an increased serum activity of AST because of the hepatic ischemia and anoxia that is produced.

In Hepatic Disorders. See pages 241 to 243 and references (3, 8).

In Diseases of Skeletal Muscle. See the section on diagnosis of muscle disorders later in this chapter and reference 3.

SERUM ALANINE AMINOTRANSFERASE, ALT (TRANSAMINASE, GPT). The enzyme that transfers an amino group from the amino acid alanine to α-ketoglutarate is alanine aminotransferase (ALT).

The concentration of ALT in tissues is not nearly as great as for AST. It is present in moderately high concentration in liver, but is low in cardiac and skeletal muscles and other tissues. Its use for clinical purposes is primarily for the diagnosis of liver disease (see Chap. 7) and to resolve some ambiguous increases in serum AST in cases of suspected myocardial infarction, when the CK test or isoenzyme tests for LD and CK-MB are not available. When both AST and ALT are elevated in serum, the liver is the primary source of the enzymes (liver ischemia because of congestive heart failure or other sources of liver cell injury). If the serum AST is elevated while the ALT remains within normal limits in a case of suspected myocardial infarction, the results are

---

*Enzyme activities are expressed in International Units (μmol/minute) and not in SI units (μmol/second) as in the original publication.

compatible with myocardial infarction. The role of ALT (GPT) in clinical enzymology is reviewed in several books.[3,8]

REFERENCE. Bergmeyer, H.U., Scheibe, P., and Wahlefeld, A.W.: Optimization of methods for aspartate aminotransferase and alanine aminotransferase. Clin. Chem. 24, 58–73, 1978.

Principle. ALT reversibly transfers the amino group from alanine to α-ketoglutarate, forming pyruvate and glutamate (Reaction 6).

The rate of formation of pyruvate is determined by coupling the ALT reaction with that of lactate dehydrogenase (LD), which converts the pyruvate to lactate (Reaction 7); the decrease in absorbance at 340 nm is measured as NADH is oxidized to NAD⁺.

$$\text{Reaction 6: Alanine} + \alpha\text{-ketoglutarate} \xrightarrow{\text{ALT}} \text{glutamate} + \text{pyruvate}$$

$$\text{Reaction 7: Pyruvate} + \text{NADH} + \text{H}^+ \xrightarrow{\text{LD}} \text{lactate} + \text{NAD}^+$$

Reagents and Materials

1. Tris (hydroxymethyl) Aminomethane, stock solution, 1.0 mol/L. Dissolve 121.1 g Tris base (Schwarz/Mann) in water and dilute to 1 L. Solution is stable at room temperature for at least 6 months.
2. Triton X-100 (Harleco).
3. Tris Buffer, 0.16 mol/L, pH 7.54 at 25° C ± 0.02 and pH 7.4 at 30° C. Dilute 160 ml of Tris stock solution with about 800 ml water. Adjust pH to 7.54 ± 0.02 at 25° C by the addition of approximately 5 ml of concentrated HCl. Dilute to 1 L volume with water. Add 0.1 ml of Triton X-100. Filter through an apparatus containing a 0.45 micron pore size filter (Millipore). The solution is stable at 4° C indefinitely.
4. L-alanine, 1.066 mol/L in 0.16 mol/L Tris buffer, pH 7.54 at 25° C. Dissolve 23.74 g L-alanine in about 150 ml water. Add 40 ml of Tris stock solution (solution #1). The pH of the mixture is about 8.4. Adjust the pH to 7.54 ± 0.02 at 25° C by the addition of concentrated HCl. Dilute to 250 ml volume with water and add 25 μl Triton X-100. Filter through a 0.45 micron filter apparatus (Millipore). Solution is stable at 4° C for 1 month.
5. Reduced Nicotinamide Adenine Dinucleotide, NADH (Boehringer-Mannheim), 2.4 mmol/L in Tris buffer, pH 7.54 at 25° C. The concentration is based upon 100% purity; since the analysis of commercially supplied NADH is about 85% purity, the concentration prepared is 2.4 × 100/85 or 2.82 mmol/L (2 mg/ml). Dissolve 2 mg of NADH per ml of Tris buffer (solution #3). Prepare only sufficient solution to make up Reagent A (solution #7).
6. LD Stock Solution, BMC (Boehringer-Mannheim): The preparation is a concentrate of LD from hog muscle in 50% glycerol, with an activity of 5500 U/ml.
7. Reagent A: To 24 ml of L-alanine (solution #4), add 8 ml of NADH (solution #5) and 120 μl of LD Stock (solution #6).
8. Reagent B: Dissolve 5.26 g α-ketoglutaric acid and 35.64 g L-alanine in about 250 ml water. Add 80 ml of Tris stock solution (solution #1) which brings the pH to about 7.4. Adjust the pH to 7.54 ± 0.02 at 25° C with

0.1 mol/L NaOH. Dilute to 500 ml with water and add 50 μl Triton X-100. Filter through a 0.45 micron pore size filter (Millipore).
9. Kinetic analyzer.

Procedure
1. Use an automatic pipetter to transfer 10 μl serum plus a 90 μl water washout to a cuvet. Add 50 μl reagent A and preincubate for 6 minutes at 30° C.
2. At the end of 6 minutes, add 50 μl reagent B, mix, wait 30 seconds, and measure the decrease in absorbance of the NADH with time at 340 nm. The above volumes and concentrations may be modified to accommodate a particular instrument provided that you adhere to the final concentrations of reactants in the mixture.

$$\text{ALT activity, U/L} = \Delta A/\text{min} \times 1/6.22 \times 0.200 \text{ ml}/0.010 \text{ ml} \times 1000$$

$$= \Delta A/\text{min} \times 3{,}215.$$

where ΔA/min is the absorbance change per minute, 6.22 is the mmol/L absorptivity of NADH at 340 nm., 0.200 ml is the total volume in the cuvet, 0.010 ml is the volume of serum sample, and 1000 is the factor to convert millimolar absorptivity to micromolar.

Notes:
1. The concentrations of reagents in the final reaction mixture are the following:

Tris buffer, pH 7.4 at 30° C	80 mmol/L
L-alanine	400 mmol/L
α-ketoglutarate	18 mmol/L
NADH	0.15 mmol/L
LD	5,100 U/L

2. The amount of water of crystallization in the various reagents must be taken into account when preparing the solution. See CK, page 185.
3. NADH powder should be exposed to moist air as little as possible and should be stored below 0° C with a desiccant.
4. The LD enzyme must be added in excess so that the activity of ALT is the rate-limiting factor.
5. The blood specimen should be drawn and handled carefully to avoid hemolysis. Hemolysis falsely elevates the ALT.
6. Serum ALT is not too stable and should be kept at 4° C for short periods or stored frozen at −20° C or lower for longer periods.

Reference Values. The normal range for serum ALT is 3 to 30 U/L.

Increased Activity. The serum activity of ALT is increased in a variety of hepatic disorders (see pages 241–243).

Lactate Dehydrogenase, LD, (LDH)

Lactate dehydrogenase reversibly catalyzes the oxidation of lactate to pyruvate by transferring hydrogen from lactate to the cofactor, NAD⁺, according to Reaction 8.

$$\text{Reaction 8: CH}_3 - \underset{\text{lactate}}{\text{CH}} - \text{C} - \text{O}^- + \text{NAD}^+ \xrightarrow{\text{LD}} \text{CH}_3 - \underset{\text{pyruvate}}{\text{C}} - \text{C} - \text{O}^- + \text{NADH} + \text{H}^+$$

LD is distributed widely in tissues and is present in high concentration in liver, cardiac muscle, kidney, skeletal muscle, erythrocytes, and other tissues.

The measurement of the serum concentration of LD has proven to be useful in the diagnosis of myocardial infarction. The LD enzyme activity in serum does not rise as much as CK or AST after myocardial infarction, but it does remain elevated for a much longer period of time. This is quite important when the patient does not see a physician for 3 or 4 days following an infarct. In hepatocellular disease, the serum activity of LD rises, but the measurement of this enzyme is much less useful than that of AST or ALT because the test is less sensitive. The role of LD in clinical enzymology is discussed in several books.[3,8]

REFERENCE. Quistorff, H., and Clayson, K.J.: Summary of studies of optimal conditions for serum LD assay. Chemistry Division, Department of Laboratory Medicine, University of Washington, 1973.

PRINCIPLE. In the conditions of this assay, the enzyme LD converts pyruvate to lactate while oxidizing NADH to NAD^+. The rate of decrease in absorbance of NADH at 340 nm is proportional to LD activity.

REAGENTS AND MATERIALS

1. Tris (hydroxymethyl) Aminomethane, 1 mol/L. Dissolve 121.1 g Tris Base (Schwarz/Mann) in water and dilute to 1 L volume. Stable at room temperature for at least 6 months.
2. Triton X-100 (Harleco).
3. Tris Buffer, 0.16 mol/L, pH 7.54 ± 0.02 at 25° C and pH 7.4 at 30° C. Dilute 160 ml of 1 mol/L Tris with 800 ml water. Adjust pH to 7.54 ± 0.02 at 25° C by the addition of approximately 5 ml concentrated HCl. Dilute to 1 L with water. Add 0.1 ml of Triton X-100 and filter through an apparatus containing a 0.45 micron pore size filter (Millipore). Solution is stable indefinitely at 4° C.
4. Reduced Nicotinamide Adenine Dinucleotide, NADH (Boehringer-Mannheim).
   a. Purchase 1-g bottles and divide the contents into aliquots sufficient for one batch of LD assays (approximately 10 mg and 25 mg aliquots). The aliquots are stored in small vials in individual desiccators (large test tubes, half-filled with a desiccant such as Drierite, which can be tightly stoppered) at −5° C.
   b. Reagent A. 0.6 mmol/L NADH in Tris buffer, 0.16 mol/L, pH 7.54 at 25° C. The concentration is based upon 100% purity of the NADH. Since commercial NADH is about 85% pure, compensate by making the solution 0.6 × 100/85 = 0.71 mmol/L or 0.5 mg/ml. Weigh out 10 mg NADH in a vial to the nearest 0.1 mg. Immediately add sufficient Tris buffer (solution #3) to make the concentration of NADH in buffer equal to 0.5 mg/ml. Filter through 0.45 micron pore filter (Millipore). Bergmeyer et al. report that the solution is stable for 2 weeks at 4° C. (See AST method, p. 192).
5. Reagent B. Pyruvate 4 mmol/L in Tris buffer (solution #3). Dissolve 0.022 g sodium pyruvate (Boehringer-Mannheim) in solution #3 and dilute to 50 ml volume with the buffer. The solution is stable for 10 days at 4° C. Filter an appropriate volume on the day of use through a 0.45 micron pore size filter (Millipore).

6. Kinetic analyzer. (See p. 185, #12).

PROCEDURE

1. Use an automatic pipetter to transfer 10 μl serum plus a 90 μl water washout to a cuvet. Add 50 μl of reagent A (solution #4) and preincubate for 6 minutes.
2. At the end of 6 minutes, add 50 μl of reagent B (solution #5) and stir. After a delay of 30 seconds after the addition of reagent B, read the absorbance for 30 to 90 seconds and record the results in International Units.
3. Calculation:

$$\text{LD activity, U/L} = \Delta A/\text{min} \times 1/6.22 \times 0.200 \text{ ml}/0.010 \text{ ml} \times 1000$$

$$= \Delta A/\text{min} \times 3215$$

where $\Delta A/\text{min}$ is the absorbance change per minute, 6.22 is the mmol/L absorptivity of NADH at 340 nm, 0.200 ml is the total volume in the cuvet, 0.010 ml is the volume of serum sample, and 1000 is the factor to convert millimolar absorptivity to micromolar.

Notes:

1. The concentrations of reagents in the final reaction mixture are the following:

Tris buffer, pH 7.4 at 30° C	80 mmol/L
Pyruvate	1.0 mmol/L
NADH (compensated for impurities)	0.15 mmol/L

   The ratios of activity of the different LD isoenzymes vary independently of the pyruvate concentration; the activities of LD_1 and LD_5 are identical per unit weight of enzyme when the pyruvate concentration is 1.0 mmol/L.

   Reagent concentrations and volumes may be altered to accommodate a particular analytical system as long as the concentrations in the final reaction mixture are the same. Larger aliquots should be taken if the method is performed manually, and the absorbance changes should be recorded in less than 2 minutes.

2. The purity of the Tris buffer is critical for the LD assay. Material from Schwarz/Mann is satisfactory, but some other brands give lower LD results.
3. The "purity" of the NADH can be checked by measuring the absorbance of a solution of NADH of known concentration and comparing it to the expected. Pure NADH at a concentration of 0.100 mmol/L should have an absorbance of 0.622 at 340 nm when the light path is 1 cm.

REFERENCE VALUES. The normal range of serum LD activity varies from 125 to 290 U/L.

INCREASED CONCENTRATION. The serum LD activity is increased in a wide variety of disorders because it is so widely distributed in tissues.[3,8] The principal clinical uses of the LD test are the following:

1. *In myocardial infarction.* Serum LD activity increases after myocardial infarction, but the rise occurs later than that for CK or AST and is of lesser intensity (Table 6.2). Its great value in the diagnosis of myocardial infarction lies in the prolongation of its increased activity; it may remain elevated for 7 to 10 days, long after the CK and AST levels have returned to normal (Fig. 6.4). The isoenzymes of LD also have an important role

in the diagnosis of myocardial infarction and will be described in the following sections.

2. *In other diseases.* Serum LD activity is increased in liver disease, but other enzymes are more sensitive and specific for liver disorders (Chap. 7). The serum activity is also increased following muscle trauma, renal infarct, hemolytic diseases, and pernicious anemia. Hemolyzed blood specimens will have artifactually elevated LD activities owing to LD enzymes coming from the ruptured red blood cells; the same is true if the serum is allowed to stand too long upon the clot.

LD Isoenzymes

The LD enzyme is composed of four subunits. There are two different poly-peptide chains: an M type typical of skeletal muscle and an H type from cardiac muscle. Five different isoenzymes are possible from this combination: HHHH (LD_1 from heart muscle, red blood cells, or kidney), MHHH (LD_2, also in heart), MMHH (LD_3 in lung and other tissues), MMMH (LD_4 in many tissues), and MMMM (LD_5, primarily in skeletal muscle and liver). LD_1 migrates farthest toward the anode and is called the fastest fraction; the others migrate more slowly, with LD_5 moving slightly toward the cathode as does gamma globulin.

REFERENCE. A modification of Rosalki, S.B.: Standardization of isoenzyme assays with special reference to lactate dehydrogenase isoenzyme electrophoresis. Clin. Biochem. 7, 29, 1974.

PRINCIPLE. The LD isoenzymes in serum are separated by electrophoresis on cellulose acetate in a manner similar to that for serum proteins. The membrane containing the separated isoenzyme bands is then overlaid on an agar gel containing lactate substrate, buffer, NAD^+, nitroblue tetrazolium, and phenazine methosulfate. As each LD isoenzyme band oxidizes lactate to pyruvate, the NADH formed transfers electrons to phenazine methosulfate, which reduces nitroblue tetrazolium to an insoluble formazan dye upon the isoenzyme band. After washing and drying the membrane, the intensity of the dye deposited on the isoenzyme bands is quantitated by densitometry.

REAGENTS AND MATERIALS

1. Electrophoresis Buffer, 0.025 mol/L barbital, Tris 0.050 mol/L, citrate 0.012 mol/L, pH 8.2 at 37° C. Dissolve 4.60 g barbital (Fisher), 6.05 g Tris (hy-droxymethyl) aminomethane (Schwarz/Mann) and 2.52 g citric acid in about 900 ml water. Adjust the pH to 8.5 at 25° C by the addition of 2 molar NAOH and make up to 1 L volume with water. Store in refrigerator.

2. Membrane Buffer. Dissolve 200 mg bovine gamma globulin (BGII, Sigma) in 200 ml of electrophoresis buffer.

3. Incubation Buffer. Dissolve 28.81 g Tris (Schwarz/Mann), 16.1 g sodium pyrophosphate (Mallinckrodt) ($Na_4P_2O_7 \cdot 10H_2O$), and 107 mg EDTA (eth-ylenedinitrilo) tetraacetic acid disodium salt $\cdot 2H_2O$, (Mallinckrodt) in about 600 ml water. Adjust the pH to 8.8 at 25° C (equivalent to pH 8.4 at 37° C) and add water to 700 ml volume. The solution is added to 300 ml agar solution in step 4 to give a final concentration of Tris, 0.10 mol/L, pyro-phosphate, 0.015 mol/L, EDTA 0.12 mmol/L, pH 8.8 at 25° C.

4. Buffered Agar Gel, 1% (w/v), pH 8.8 at 25° C. Dissolve 10 g Noble agar (or agarose) in 300 ml boiling water. When the temperature has dropped to 70° C, add 700 ml of the incubation buffer (#3) warmed to 70° C. Mix

well and transfer 6.1 ml aliquots to test tubes, stopper well, and store at 4° C.

5. L(+)-lactate, lithium (CalBiochem) 450 mmol/L. Dissolve 2.2 g L(+)-lithium lactate in water and make up to 50 ml volume. Solution is stable for 3 weeks at 4° C.

6. Phenazine methosulfate (PMS, Sigma) 2.0 mg/ml (6.5 mmol/L): Dissolve 20 mg PMS in 10 ml water. Wrap in aluminum foil to protect it from light. Solution is stable for 1 week at 4° C.

7. Nitroblue tetrazolium (NBT, Sigma) 2.0 mg/ml (2.44 mmol/L): Dissolve 10 mg NBT in 5 ml water. Stable for 2 weeks at 4° C in brown bottle.

8. Nicotinamide adenine dinucleotide (NAD, Boehringer-Mannheim) 6 mg/ml, (9 mmol/L): Prepare a 2 weeks' supply at one time, half of which can be stored for 1 week at 4° C while the second half should be stored frozen at −10° C in 1 ml aliquots in small glass tubes. For example, weigh out 120 mg of NAD^+, dissolve in 20 ml water, and divide and store the solution as described.

9. Buffered Agar Substrate (staining medium) for the LD isoenzyme reaction. The substrate plates are usually prepared during the electrophoresis and are described in step 1a–d under *Staining*. The final concentrations of the various constituents in the incubation plate are the following:

L(+)-lactate   12 mg/ml (125 mmol/L)
NAD^+ 0.60 mg/ml (0.9 mmol/L)
PMS   0.020 mg/ml (0.07 mmol/L)
NBT   0.20 mg/ml (0.25 mmol/L)

   Buffer constituents: Tris 0.10 mol/L, pyrophosphate 0.015 mol/L and EDTA, 0.12 mmol/L.

10. Fixing Solution, 0.5 mol/L, HNO_3. Dilute 30 ml concentrated HNO_3 to 1 L volume with water.

11. Microzone Electrophoresis Apparatus (Beckman Instruments).

12. Cellulose Acetate Membranes (Sartorius, Beckman, but others can be used).
PROCEDURE
Electrophoresis
1. Fill the Microzone electrophoresis cell to the mark with cold electrophoresis buffer.
2. Pour a little membrane buffer in a rinse pan and gently lay a cellulose acetate membrane on the surface, shiny side up, and allow to wet evenly from the bottom surface. Do not trap air under the membrane. When the membrane is evenly wet and there are no speckled areas indicating air bubbles, submerge it completely.
3. Remove the wet membrane, blot lightly and place it in the membrane holder, shiny side up. Position it in the electrophoresis chamber. The membrane must not dry and should be evenly stretched on the holder.
4. Attach cell cover properly.
5. Apply the serum samples through the transverse trough closest to the cathode end. Apply that volume of serum that is equivalent to 600 to 1000 U/L of LD in one application, i.e., if serum LD activity is 200 U/L, make 3 applications; if it is 2000 U/L, dilute the serum with an equal volume of saline and make a single application.

6. Transfer the electrophoresis chamber to a pan of ice and subject it to electrophoresis for 20 minutes at 7 ma constant current. The initial starting voltage should be 350 to 400 v as a cross-check.

Staining

1. Prepare the agar substrate-staining medium while the electrophoresis is progressing.

    a. Place a tube of buffered agar gel in a beaker of boiling water. Turn the overhead lights off.

    b. As soon as the agar melts, transfer the tube to a beaker of water maintained at 65° C.

    c. To a small beaker, transfer 1.0 ml of NTB, 2.0 ml of lithium lactate, 1.0 ml of NAD$^+$, and 0.1 ml of PMS solutions. Mix with a stirring rod and quickly add the warm (65° C) buffered agar solution.

    d. Quickly mix and pour into a plastic staining box, approximately 6.5 × 9 cm; tilt so that the liquid spreads evenly and then place on a level surface. Protect it from the light with aluminum foil. After the agar has solidified, wrap the box tightly in foil and refrigerate until 15 minutes before use. Prepare fresh daily.

2. After electrophoresis, remove the membrane from the chamber and cut it to fit the substrate gel.

3. In subdued light, place the strip, shiny face downward, on the gel surface. It is best done by starting at one end and gradually lowering the other end so that no bubbles are trapped. Force any trapped bubbles to one side with tweezers.

4. Wrap the box in aluminum foil and float in a 37° water bath for 45 minutes.

Fixing

1. Remove the membrane and place in the dilute HNO$_3$ solution for 2 minutes while agitating occasionally. This clears the background.

2. Rinse the membrane in running tap water for 20 seconds.

Drying

1. Blot the wet membrane with blotting paper.

2. Lay the membrane *shiny side down* on a clean glass plate and fix in place with cellulose tape.

3. Dry the membrane for 10 to 15 minutes in an oven with blower at about 40° C.

4. Cool the plate and cut the tape along the edge of the membrane with a razor blade.

5. Gently lift the membrane from the plate, starting at one corner. If it sticks, try to free the membrane by gently applying a razor blade to the point where it sticks.

6. Place in a transparent, plastic envelope and scan in a densitometer, with the shiny side of the membrane facing the light source.

Notes:

1. The substrate-staining medium must be protected from light because phenazine methosulfate is light-sensitive. The background becomes dark if the membrane is exposed to light while on the substrate gel or before fixing.

2. The electrophoresis membrane has to be handled rapidly during the process of removing from the chamber and placing on the substrate gel. If it should dry, the enzymes become denatured.
3. The glass plate must be clean or else the membrane may stick to it.
4. If there is insufficient time to scan the membrane, it may be kept overnight in a drawer, protected from the light.
5. Always run a control serum on each membrane.

REFERENCE VALUES. The serum LD isoenzyme composition of normal individuals as % of total LD and as units is the following:

LD_1	20–34%	100 U/L
LD_2	32–40%	115 U/L
LD_3	17–23%	65 U/L
LD_4	3–13%	40 U/L
LD_5	4–12%	35 U/L

CLINICAL APPLICATIONS. The LD isoenzyme determination may be useful in the following clinical situations (see Table 6.3):

1. *In myocardial infarction.* The LD isoenzymes from heart (LD_1 and LD_2) increase in serum within 24 to 48 hours after an infarct. LD_1 and LD_2 are the two fastest LD isoenzymes. The upper limit of normal for LD_1 is 100 U/L or 34% of the total LD; the value for LD_2 is 115 U/L or 40% of the total. In a myocardial infarction, the ratio of LD_1/LD_2 becomes *flipped,* i.e., greater than one. Since this isoenzyme pattern (the flipped ratio) may occur after a renal infarct and in hemolytic situations, these possibilities must be ruled out upon other grounds. If the time interval is right, obtaining a CK isoenzyme pattern will resolve the problem. The combination of an elevated CK-MB isoenzyme and a flipped LD isoenzyme ratio in a

TABLE 6.3

Summary of Serum LDH Isoenzyme Patterns in Some Important Clinical Conditions

Myocardial infarction	Moderate elevation of LDH_1; slight elevation of LDH_2
Pulmonary infarction	Moderate elevation of LDH_2, LDH_3; slight elevation of LDH_4
Acute hepatitis	Marked elevation of LDH_5; moderate elevation of LDH_4
Arthritis and joint effusions	Elevation of LDH_5
Muscular dystrophy	Elevation of LDH_1, LDH_2, LDH_3
Dermatomyositis	Elevation of LDH_5
Sickle cell anemia	Moderate elevation of LDH_1, LDH_2
Megaloblastic anemia	Marked elevation of LDH_1
Agnogenic myeloid metaplasia	Elevation of LDH_2, LDH_3; mild elevation of LDH_4
Granulocytic leukemia	Elevation of LDH_2; slight elevation of LDH_3
Essential myoglobinuria	Moderate to severe elevation of LDH_5; moderately elevated LDH_3, LDH_2
Intravascular hemolysis	Marked elevation of LDH_1; and moderate elevation of LDH_2, LDH_3
Muscular dystrophy— muscle biopsy	LDH_5 markedly decreased compared to normal muscle
Arthritis with joint effusions— synovial fluid	Elevation of LDH_5
Hemolyzed RBC	Elevation of LDH_1, LDH_2
Destruction of lymphocytes	Elevation of LDH_3, LDH_4, LDH_5
Aplastic anemia	LDH_1, LDH_2, LDH_3, LDH_4, LDH_5 *all* elevated due to destruction of both RBC and WBC

From Nerenberg, S.T.: Electrophoretic Screening Procedures. Philadelphia, Lea & Febiger, 1973, p. 100.

patient suspected of having a myocardial infarct makes the diagnosis certain. This combination never occurs in coronary insufficiency without a myocardial infarct.[15]

2. *In liver disease.* See page 245 for its use in helping to distinguish between viral hepatitis and infectious mononucleosis.[16]

3. *In pulmonary infarction.* In this condition, LD_3 frequently becomes a prominent band.

Diagnosis of Hepatic Disease

See pages 239 to 245 for a discussion of the various enzymes employed in the diagnosis and management of liver disease. These are AST, ALT, LD, ALP, NTP, GGT, and OCT.

Diagnosis of Bone Disease

Alkaline Phosphatase[3,8,17]

Alkaline phosphatases are a group of enzymes that split off a terminal phosphate group from an organic phosphate ester in alkaline solution. Their optimum pH is usually around pH 10, but this varies with the particular substrate and isoenzyme. Alkaline phosphatase (ALP) is widely distributed in the body and is present in high concentration in bone (osteoblasts, the cells of growing bone), intestinal mucosa, and renal tubule cells and in lower concentration in the liver (it is highest in the biliary tree), leukocytes, and placenta.

ALP ISOENZYMES.[3,19] Isoenzymes of ALP that are separated by acrylamide gel electrophoresis have been identified in the following tissues: liver, bone, intestinal mucosa, placenta, and bile. Physical and chemical properties sometimes are used for distinguishing between the various isoenzymes. For example, the placental enzyme is heat-stable at 65° C but not the others; at 55° C, the liver isoenzyme is moderately heat-stable but the bone enzyme is heat-labile. This property is sometimes used to differentiate between bone and liver disease when the total ALP is elevated. The intestinal ALP isoenzyme is inhibited by L-phenylalanine. Normal serum ALP consists of a mixture of isoenzymes derived primarily from liver, intestines, and bone, with the one from liver predominating except under conditions of rapid skeletal growth.

REFERENCE. McComb, R.B., and Bowers, G.N., Jr.: A study of optimum buffer conditions for measuring alkaline phosphatase activity in human serum. Clin. Chem. *18*, 97, 1972.

Principle. Serum alkaline phosphatase (ALP) catalyzes the hydrolysis of p-nitrophenylphosphate to p-nitrophenylate ion and phosphate. The substrate is colorless, but the p-nitrophenylate resonates to a quinoid form in alkaline solution and strongly absorbs light at 404 nm; the reaction is followed by continuous monitoring. A buffer, such as 2-amino-2-methyl-l-propanol, is used to act as a phosphate acceptor (transphosphorylation) and speed up the reaction. Mg^{2+}, as well as Zn^{2+}, are activators of ALP.

Reagents

1. Stock Buffer (50% w/v). Dilute 500 g 2-amino-2-methyl-l-propanol (2A2M1P) to 1000 ml volume with water. The solution is stable, but store at 4° C.

2. 2A2M1P Buffer, 0.89 mol/L, pH 10.33 at 30° C. Transfer 160 ml of stock buffer to a 1000-ml volumetric flask, add 500 ml water, and 160 ml 1.0

mol/L HCl. Adjust pH to 10.33 at 30° C if necessary. Protect from atmospheric CO_2.

3. MgCl$_2$, 1.5 mmol/L. Dissolve 300 mg MgCl$_2$·6H$_2$O in water and make up to 1000 ml volume.
4. p-Nitrophenylphosphate, 210 mmol/L in 1.5 mmol/L MgCl$_2$. For each ml of substrate desired, dissolve 61 mg of disodium p-nitrophenylphosphate ·6H$_2$O in 1 ml of MgCl$_2$ solution. Substrate should be prepared freshly or stored for no longer than one week at 4° C.
5. Stock Standard, p-nitrophenol, 1.0 mmol/L. Dissolve 139.1 mg of high purity p-nitrophenol in water and make up to 1000 ml volume. It is stable for several months when protected from the light.
6. Working Standard, p-nitrophenol, 0.04 mmol/L in 2A2M1P buffer, pH 10.33. Note: The final pH of the reaction mixture is 10.30. Transfer 10.0 ml of stock standard to a 250-ml volumetric flask and dilute to the mark with 2A2M1P buffer (0.90 mmol/L).

Procedure

1. Add 0.10 ml serum to 2.7 ml buffer in a cuvet and place in a 30° C water bath for at least 5 minutes.
2. Initiate the reaction by adding 0.20 ml of substrate previously warmed to 30° C and mix. Transfer to 30° C reading compartment of spectrophotometer.
3. Record the absorbance at 404 nm for at least 2 minutes against a substrate blank (2.8 ml buffer + 0.20 ml substrate).
4. Calculate the change in absorbance per minute (ΔA/min) from the recording.

$$U/L = \frac{\Delta A/min}{18.75} \times \frac{3.0}{0.10} \times 1000$$

$$= \Delta A/min \times 1600$$

where 18.75 = the absorbance of 1 μmol/L of p-nitrophenol in 2A2M1P buffer at 404 nm in a 1 cm light path, 3.0 is the volume in the cuvet, 0.1 is the sample volume, and 1000 is the factor converting ml to 1 L. The activity at 37° C is approximately 1.32 that measured at 30° C.

Reference Values

Adults = 20 to 105 U/L
Infants and children
    0–3 months = 70 to 220 U/L
    3 mo–10 years = 60 to 150 U/L
    10 yr–puberty = 60 to 260 U/L

The increased normal values for growing children reflect the increased osteoblastic activity that occurs during periods of rapid skeletal growth.

Increased Activity. Serum ALP activity is raised in all *bone disorders* accompanied by *increased osteoblastic activity.* This includes Paget's disease (osteitis deformans), osteoblastic tumors with metastases, hyperparathyroidism when there is mobilization of Ca and P from bone, rickets, and osteomalacia.

The activity of the enzyme, 5'-nucleotidase (NTP) is normal under these circumstances (see p. 244). Serum ALP activity is also increased in liver disease, particularly in *disorders of the hepatic biliary tree* (see p. 240), and during the *third trimester of pregnancy* owing to the elaboration of a placental isoenzyme of ALP that is absorbed into the maternal bloodstream.

The predominant serum ALP isoenzyme is the bone fraction in bone diseases, the liver and (sometimes) bile fractions in liver ailments, and the placental type in late pregnancy. Atypical isoenzymes may appear in some malignancies.

Decreased Activity. Low levels of ALP are found in a rare congenital defect, *hypophosphatasemia*, in dwarfs because of depressed osteoblastic activity, in hypothyroidism, and in pernicious anemia. In the two latter conditions, deficiencies of thyroid hormone and vitamin B_{12}, respectively, are responsible for the lowered serum ALP activity.

5'-Nucleotidase

The enzyme 5'-nucleotidase (NTP) is useful in interpreting an elevated alkaline phosphatase result of unknown etiology.[3,8] NTP is a phosphatase enzyme that splits off the phosphate from 5'-adenosine monophosphate. Its serum concentration is elevated in liver disorders, but not in bone diseases. A high ALP activity with a normal NTP activity indicates that the ALP is coming from a tissue other than liver and presumably from bone if this is the alternative. If both activities are elevated, the high ALP can be ascribed to a liver problem and not to bone (see p. 244).

Diagnosis of Muscle Disorders[3,8]

The three serum enzymes used most frequently for this purpose, in order of their general usefulness, are creatine kinase (CK), aspartate aminotransferase (AST), and aldolase (ALD). These enzymes are elevated in all types of progressive muscular dystrophy, but the greatest relative rise in the CK enzyme occurs in the *Duchenne* type of muscular dystrophy. The CK serum activity may be increased as much as 50 × ULN early in the disease, while the activity of AST and ALD may be 10 and 6 × ULN, respectively. The activities of the enzymes in serum become progressively lower with the duration of the disease because of the decreased muscle mass.

Raised serum enzyme levels are detectable in virtually all neurogenic muscle atrophies and are temporarily raised following muscle trauma, surgery when muscles are cut, and intramuscular injections when long-lasting preparations are employed.

Methods for CK and AST are described on pages 184 and 192, respectively, of this chapter. The following is a description of aldolase (ALD).

Aldolase

The enzyme adolase, which is listed in Table 6.1, converts fructose-1,6-diphosphate into two triose phosphate esters, dihydroxyacetone phosphate and glyceraldehyde-3-phosphate. This is an early step in the glycolysis of glucose.

REFERENCE. Pinto, P.V.C., Kaplan, A., and Van Dreal, P.A.: Aldolase: II. Spectrophotometric determination using an ultraviolet procedure. Clin. Chem. *15*, 349, 1969.

PRINCIPLE. Aldolase (ALD) converts fructose-1,6-diphosphate into equimolar quantities of two trioses, glyceraldehyde-3-phosphate (GAP) and dihydroxy-acetone phosphate (DAP), as shown below in Reaction 9. This reaction is coupled with Reactions 10 and 11 by adding an excess of the two enzymes, triosephosphate isomerase (TPI, EC 5.3.1.1) and glycerol-1-phosphate dehydrogenase (GD EC 1.1.1.8). When these subsidiary enzymes are added in excess, the activity of aldolase becomes rate limiting and is proportional to the decrease in absorbance at 340 nm. Two moles of NADH are oxidized to NAD$^+$ per mole of substrate hydrolyzed by ALD. The three reactions are the following:

Reaction 9: FDP $\xrightarrow{\text{ALD}}$ GAP + DAP

Reaction 10: GAP $\xrightarrow{\text{TPI}}$ DAP

Reaction 11: DAP + NADH + H$^+$ $\xrightarrow{\text{GD}}$ G1P + NAD$^+$

where FDP = fructose-1,6-diphosphate, GAP = glyceraldehyde-3-phosphate, DAP = dihydroxyacetone phosphate, G1P = glycerol-1-phosphate. The enzyme abbreviations have been described.

Diagnosis of Acute Pancreatitis[4,8]

It is difficult to establish the diagnosis of acute pancreatitis without the assistance of laboratory tests. The patient usually complains of an intense pain in the upper abdomen that could be caused by several different disorders. The two tests most commonly used for diagnostic purposes are those of serum amylase and serum lipase; the measurement of the amount of amylase excreted into the urine per hour also provides useful information.

Serum Amylase (EC 3.2.1.1; 1,4-α-D glucan glucanohydrolase)

Starch is the storage form of carbohydrate in plants, and glycogen is the form of storage in animals. Both are polymers of glucose molecules that are linked together in a chain as condensation takes place between the hydroxyl groups on carbons 1 and 4 of adjacent glucose molecules. Water is split off, and the linkage is called an α-1,4-glycosidic bond. An occasional branch point in the starch or glycogen molecule is produced by forming an α-1,6-glycosidic bond.

In humans, the amylases are enzymes that randomly split the α-1,4-glycosidic bonds on the starch chain, breaking it down into much smaller units. The end products are a mixture of glucose, maltose, and dextrins (small polymers containing the branch points).

Amylases are secreted by the salivary and pancreatic glands into their respective juices, which enter the gastrointestinal tract. These enzymes are important for the digestion of ingested starches, but the amylase from the pancreas plays the major role, since the salivary amylase soon becomes inactive in the acid condition prevailing in the stomach. There are several isoenzymes of both pancreatic and salivary amylase. The amylase that is normally present in serum is derived from both pancreas and salivary glands.[20] The activity of serum amylase rises following an obstruction to the flow of fluid from either the salivary or pancreatic glands, but the elevation is usually much greater when the outflow from the latter gland is blocked. Acute pancreatitis is caused by blockage of the pancreatic ducts or by direct injury to the pancreatic tissue by toxins, inflammation, trauma, or by impaired blood flow to the pancreas. The inflammation and autodigestion by pancreatic enzymes that accompany pancreatic injury usually result in an obstruction to the flow of pancreatic juice into the intestine.

Several different approaches are used for the determination of serum amylase activity.[21] The amyloclastic methods measure the disappearance of starch substrate; saccharogenic methods measure the reducing sugars (glucose and maltose) that are produced as a result of enzymatic action; dye methods measure the absorbance of soluble dye that is split from an insoluble amylose-dye substrate. Promising methods are being developed that permit continuous monitoring.[21] They are based upon coupling p-nitrophenol in an alpha-glycosidic linkage to a glucose or maltose polymer. The amylase splits off the p-nitrophenol, whose absorbance can be continuously monitored at 404 nm. Results obtained on patients by this method must be compared to the large body of results obtained by conventional methods before judgment can be passed upon its suitability for clinical use. A commercial kit is available (Pantrak, E.K., CalBiochem-Behring Corp.)

Amyloclastic Method

REFERENCE. Rice, E.W.: Improved spectrophotometric determination of amylase with a new stable starch substrate. Clin. Chem. 5, 592, 1959 as modified by Lowry, J., Delaney, C.J., Kenny, M.A., Yoshida, L., Legaz, M.E., and Clayson, K.J.: Ongoing studies of amylase methodology (1972–1975). Seattle, Chemistry Division, Department of Laboratory Medicine, University of Washington.

PRINCIPLE. A buffered starch solution is incubated with serum for 10 minutes at 37° C in a 2-point assay. The initial starch concentration and that present at the end of 10 minutes' incubation are measured by the addition of molecular iodine, which forms a deep blue color with linear starch chains; the blue color, therefore, measures the amount of unhydrolyzed starch remaining at the end of the assay. Amylase is activated by Cl^-, which is contained in the substrate solution. The enzyme action is stopped by the addition of iodine-EDTA solution, as the EDTA chelates Ca^{2+}, an ion required for amylase activity.

REAGENTS AND MATERIALS

1. Tris (hydroxymethyl) aminomethane (Schwarz/Mann), 1.0 mol/L. Dissolve 121.1 g of Tris base in water and dilute to 1 L.
2. NaCl, 1.0 mol/L. Dissolve 58.4 g NaCl in water and dilute to 1 L.
3. HNO_3, 1 mol/L. Dilute 61 ml of concentrated HNO_3 to 1 L with water.
4. Starch/Buffer Solution, containing 1.0 g Lintner soluble starch, 200 mmoles NaCl, and 15 mmoles sodium azide per liter of Tris buffer, 80 mmol/L and pH 7.4 at 25° C. Make a paste of 1.00 g Lintner soluble starch with about 10 ml water in a liter beaker. Add 80 ml of Tris buffer, 200 ml of the NaCl solution and about 110 ml of water. Bring to a boil while stirring. Then cool to room temperature and add 1.0 g sodium azide as a preservative. (Note: The sodium azide will decompose if added to a hot solution.) Adjust the pH to 7.4 at 25° C by the addition of 1 molar nitric acid. Transfer to a liter volumetric flask and make up to volume.
5. Stock Iodine-EDTA (0.05 molar I_2 and 1 mmolar EDTA). Dissolve 24 g KI in 250 ml H_2O in a 1000-ml volumetric flask. Add 0.37 g disodium EDTA

and stir until dissolved. Add 10.4 g I$_2$ and stir until dissolved. Make up to 1000 ml volume with water and store in brown bottle.

6. Working I$_2$-EDTA. Dilute the stock solution 1:100 with water as needed.
7. Test Tubes, 19 × 150 mm.

PROCEDURE

1. Accurately pipet 1.0 ml of starch/buffer solution in each of two test tubes for every specimen to be run. One is for specimen and the other for specimen blank. Treat control sera in the same manner as specimens. One additional tube is prepared for a reagent blank.
2. Place all tubes in a 37° C water bath for 5 minutes to come up to temperature.
3. At exactly 15- or 30-second intervals, add 25 μl of appropriate serum to specimen tubes. Incubate for 10 minutes.
4. At the end of exactly 10 minutes of incubation, stop the reaction by adding 15 ml iodine-EDTA solution to each tube in the same 15- or 30-second sequence. Cover each tube with Parafilm and mix by inverting about 5 times.
5. Add 15 ml of the iodine-EDTA solution to each of the specimen blank tubes and the reagent blank tube. Add 25 μl serum to each specimen blank and 25 μl water to the reagent blank. Cover with Parafilm and invert 5 times.
6. After 3 minutes, read and record the absorbance of each tube against water at 600 nm.

Note: High specimens, exceeding 12 U/ml have to be repeated. If acute pancreatitis is suspected, time can be saved in the original analysis by setting up one tube with 10 μl of serum in addition to the usual 25 μl one. If the activity is high, use the value obtained with the 10 μl sample, and if below 12 U/ml, use the result from the 25 μl tube.

7. Calculation:

$$\text{Amylase activity in U/ml} = \frac{A_B - A_U}{A_B} \times \frac{6.2}{10} \times \frac{1}{0.025}$$

$$= \frac{A_B - A_U}{A_B} \times 24.8$$

where A$_B$ = absorbance of specimen blank, A$_U$ = absorbance of unknown, 6.2 is μmol starch per ml, 10 is the incubation time in minutes, and 0.025 is the serum volume in ml. The factor for converting 1 mg of starch to μmoles is derived by dividing the weight of starch in 1 ml solution (1.0 mg) by the effective molecular weight of the starch monomer, which is 162.

REFERENCE VALUES. The normal concentration of serum amylase varies from 0.8 to 3 U/ml.

INCREASED ACTIVITY. Serum amylase activity is raised considerably in acute pancreatitis, obstruction of the pancreatic ducts, and mildly in obstruction of the parotid (salivary) gland. The rise in serum amylase activity after obstruction of the pancreatic ducts, whether by a stone, inflammation, or compression of the common bile duct by a cancer of the head of the pancreas, is rapid and temporary. It usually reaches a maximum value in about 24 hours, which may be from 6 to 10 × ULN, with a return to normal in 2 or 3 days. The increase in serum amylase activity caused by a stone in the parotid duct or by the disease, *mumps*, usually is less than 4 × ULN.

Serum amylase is rapidly cleared by the kidney, so measurement of urinary amylase is a valuable adjunct to the serum test and is described below. Some types of renal damage may be accompanied by a mildly elevated level of serum amylase because of impaired excretion.

DECREASED ACTIVITY. A decreased concentration of serum amylase may be found in acute or chronic hepatocellular damage, but this is not a sensitive liver function test.

Serum Amylase Isoenzymes

Several isoenzymes of salivary amylase and several different ones from pancreas have been demonstrated in serum. Not all isoenzymes are present in the serum of a normal person, with the most prominent peaks being that of P2 (the intermediate migrating pancreas band), and S1 (the slowest migrating salivary band). The salivary isoenzymes migrate farther toward the anode than those from the pancreas. In a series of normal persons and those with pancreatitis and a variety of other diseases, Legaz and Kenny[20] found that none of the 25 normal blood donors had the P3 isoenzyme band, but that 40 of 40 patients with acute pancreatitis had a prominent P3 band. None of the 85 patients with bowel obstruction, facial trauma, or other non-pancreatic diseases had the P3 isoenzyme band in their serum. The only patients in addition to those with pancreatic disease who had the P3 band in their serum were about 37% of patients with severe renal disease. For details of this procedure, see reference 20.

PRINCIPLE. Serum is separated by electrophoresis, using cellulose acetate as the support medium. After separation, the membrane is laid on an agarose gel containing a starch-dye substrate and is incubated for 20 minutes. Dark-colored bands appear in areas in which the isoenzymes have migrated. The membrane is air-dried, and the absorbance of the various bands is read in a recording densitometer.

Urine Amylase

Since amylase has a molecular weight below 50,000 daltons, it is readily excreted by the kidney. Accordingly, the urinary excretion of amylase is high in patients with acute pancreatitis. The increased excretion of amylase persists longer than the elevation in serum amylase activity, and can help to establish the diagnosis of acute pancreatitis. The urinary amylase may be elevated for 7 to 10 days, whereas the serum amylase returns to normal in 2 or 3 days after an attack.

METHOD. The method is the same as for serum amylase. The concentration of the enzyme in urine is roughly the same per ml as for serum.

PROCEDURE. Collect an accurately timed urine specimen for 1 to 2 hours and measure the enzyme activity in the same manner as for serum.

CALCULATION. Calculate the enzyme activity/ml urine in the same manner as for serum. Multiply this values times the urine volume (ml) per hour.

REFERENCE VALUES. The normal excretion of amylase is up to 400 U/hr.

INCREASED EXCRETION. The amount of amylase excreted per hour may become high in acute pancreatitis, as much as 5 to 10 × ULN. The elevation may persist for as long as a week after the serum amylase has returned to normal. Other conditions that increase the serum amylase also result in an elevated urinary amylase excretion.

DECREASED EXCRETION. The test for urinary amylase is meaningless in acute or chronic renal disease because, with a decreased glomerular filtration rate (as shown by a decreased creatinine clearance), the clearance of amylase will also be decreased. Low values may also be obtained when there is damage to liver cells because of the lowered level of serum amylase. The reporting of results as a ratio of amylase clearance to creatinine clearance may be misleading.

Serum Lipase (Glycerol ester hydrolase, EC 3.1.1.3)

Lipase is an enzyme that hydrolyzes emulsified triglycerides. The ester bonds at carbon atoms 1 and 3 of glycerol are preferentially split, liberating two moles of long-chain fatty acid and one mole of 2-acylmonoglyceride per mole of triglyceride attacked. The reaction with triolein is shown below:

| TRIOLEIN | GLYCERYL MONOOLEATE | OLEIC ACID |

R stands for $CH_3 - (CH_2)_7 - CH = CH - (CH_2)_7 -$

The pancreas is the principal organ for the production of lipase, which is secreted in the pancreatic juice along with the other digestive enzymes. Some lipases may be found in gastric and intestinal mucosa, but they play no significant role in the digestion of fat.

Serum normally contains a low activity of lipase, but the tissue source of this enzyme has not been clearly identified. Like serum amylase, the activity of serum lipase is rapidly elevated in acute pancreatitis but remains elevated for a longer period of time than amylase. The test for serum lipase, however, has not been as easy to perform technically, nor has it been as reproducible as the serum amylase test, so it is not used as much. It usually takes too much time to be employed as an emergency test.

REFERENCE. Bandi, Z., and Kenny, M.A.: A sensitive and accurate assay of lipase activity in serum. Clin. Chem. *20,* 880, 1974.

PRINCIPLE. Purified olive oil is emulsified with gum acacia. An aliquot of the emulsion is incubated with serum at pH 7.8 at 30° C for 90 minutes. The liberated fatty acids are extracted with heptane, then redissolved in isopropanol-heptane mixture after evaporation of the original heptane solution to dryness. The fatty acids are titrated with 0.05 mol/L alcoholic KOH, using thymol blue as an indicator.

REAGENTS

1. Gum Acacia Solution. Dissolve 80 g gum acacia and 2 g sodium benzoate in 1 L of deionized water. Filter the solution into a glass-stoppered bottle and store at 4° C.

2. Purified Olive Oil. To 300 ml olive oil add 60 g alumina and stir with a magnetic stirrer for 1 hour. Allow the alumina to settle and filter through a Whatman No. 1 filter paper, using a Buchner funnel to facilitate filtration. If an aspirator is used as the source of vacuum, a water trap should be used to prevent the water from backing up into the olive oil when the water pressure drops. The purity of the olive oil should be checked by thin layer chromatography. Apply 10 μl of 5% (v/v) solution of purified olive oil in hexane to a 20 × 20 cm silica gel G plate, 0.3 mm in thickness, and develop the plate with a mixture of hexane/ether/acetic acid, 60:40:1 (by volume). The various lipid classes are detected by spraying the plate with 50% sulfuric acid and then heating it to 110° C. The olive oil is sufficiently pure if it contains not more than trace amounts of fatty acids, mono-, and diglycerides.

    Another way to determine the purity of the olive oil is to measure the amount of free fatty acids it contains. To 2.5 ml purified olive oil, add 6 ml titration solvent, and titrate with 0.05 mol/L KOH. Repeat the purification if more than 0.1 ml of the base is needed for the titration.

3. Olive Oil Emulsion. To 100 ml gum acacia solution in a blender add 100 ml purified olive oil and mix for a total time of 10 minutes at maximum speed. Prevent the mixture from getting warmer than 30° C by placing it in a refrigerator when this temperature is reached. When the emulsion is chilled, continue the mixing. Store in a refrigerator. Do not freeze. Discard if blank is more than 0.3 ml of 0.05 mol/L KOH. If the emulsion is not used for more than 10 days, it should be checked for free fatty acid content before it is used for lipase assay. Take 5 ml of the emulsion and titrate it with 0.05 mol/L KOH (aqueous) to pH 10.5 by pH meter. Discard the emulsion if titration value is more than 0.3 ml.

4. Sodium Taurocholate, 18.5 mmol/L. Dissolve 1.00 g sodium taurocholate in water to make 100 ml solution.

5. Buffer Solution, 1 mol/L, Tris buffer, pH 7.8. Dissolve 12.11 g of Tris (hydroxymethyl) aminomethane in 50 ml deionized water and adjust to pH 7.8 with concentrated HCl using a pH meter. Allow the solution to reach room temperature and adjust the pH again to 7.8 with 1 mol/L HCl. Make up the volume to 100 ml.

6. Palmitic Acid Standard, 50 mmol/L. Dissolve 1.275 g palmitic acid in chloroform and make up to 100 ml volume. Place 2 ml of this solution (100 μmol) in a 200 × 25 mm test tube with Teflon-lined screw cap. Evaporate the chloroform under a stream of filtered air or nitrogen in a water bath at about 40 to 60° C. Close the tubes with screw caps and store in a convenient place. The dry palmitic acid is stable.

7. Extraction Solvent. Mix 800 ml isopropyl alcohol, 200 ml heptane, and 20 ml 1 mol/L sulfuric acid. Keep mixture in a well-stoppered bottle.

8. 0.5 mol/L H_2SO_4. To about 500 ml water add 27 ml concentrated sulfuric acid and make up the volume to 1 L with water.

9. Indicator-Titration Solvent. Dissolve 200 mg thymol blue in 120 ml 95% ethanol. When indicator dissolves, add 900 ml benzene, and titrate to an orange-yellow color with 0.05 mol/L KOH in ethanol.

10. Alcoholic KOH (0.05 mol/L). 2.80 g KOH is dissolved in 800 ml of ethanol and made up to 1000 ml volume with this solvent.

PROCEDURE

Preincubation

1. Rehomogenize the emulsion until it reaches 35° C. The emulsion should be kept in the homogenizer and stored in a refrigerator at 4° C.

2. Set up 200 × 25 mm screw cap tubes as follows (examine each tube; if the lip of the tube is damaged, discard it):
   a. Standard: To the test tubes containing 100 μmol palmitic acid (#6), add 0.5 ml of buffer (#5), 5 ml olive oil emulsion (#3) and 1 ml taurocholate solution (#4).
   b. Test and Control: Add 0.5 ml buffer, 5 ml olive oil emulsion, and 1 ml taurocholate solution.
   c. Blank for Test and Control: Add 0.5 ml buffer, 5 ml olive oil emulsion, and 1 ml taurocholate solution. Cap tubes with screw caps and mix well. Incubate at 30° C for 10 minutes.

Incubation with Enzyme

3. Add 0.5 ml serum to control and tests at 30 second intervals.

4. Cap tubes with screw caps and mix well with vortex.

5. Incubate at 30° C for 90 minutes.

Extraction

6. Add 30 ml extraction solvent to each tube (30-second intervals). Add 0.5 ml control serum to standard. Add 0.5 ml serum to the blank of test and control.

7. Cap tubes and shake up and down vigorously for 10 seconds. Allow tubes to stand for 5 minutes at room temperature. The processing of the samples may be stopped after the extraction solvent has been added and may be restarted at any convenient time within 3 days.

8. To each tube add 18 ml heptane and 12 ml water. Shake and centrifuge for 10 minutes at low speed. From the upper (heptane) layer of each tube pipet 25.0 ml into a 100 × 25 mm test tube. One should be careful not to take any of the lower layer (water phase) or to touch the inner wall of the tube with the tip of the pipet. The water phase is strongly acidic; if any of the water layer enters the pipet, expel all the heptane back into the tube, recentrifuge, and repeat the pipetting with a *new* pipet.

Titration

9. Evaporate the 25 ml heptane under a stream of filtered air or nitrogen in a water bath at about 40 to 60° C. To the lipid residue, add 10 ml titration solvent containing the indicator. The color of the indicator should remain yellow when added to the lipid extract of the incubation medium. If the indicator changes to red, it shows the presence of acid contamination from the lower phase of the extraction mixture, which will give falsely high results.

10. Place a fluorescent light source at the left side of a microburet and titrate

the fatty acids to a stable green color with 0.05 mol/L KOH. Stirring is accomplished by bubbling CO_2-free nitrogen through the titration mixture. The titration should be done in a well-ventilated hood because the nitrogen stream will carry benzene from the titration medium into the surrounding atmosphere.

11. Calculation:

$$\text{Lipase in U/L} = \frac{V_U - V_{UB}}{V_S - V_{SB}} \times 100 \times \frac{2000}{90} = \frac{V_U - V_{UB}}{V_S - V_{SB}} \times 2222$$

Where V_U = ml KOH to titrate the unknown, V_S = ml KOH to titrate the standard, V_{UB} = ml KOH to titrate the blank of the unknown, V_{SB} = ml KOH to titrate the blank of the standard, the factor 100 is the number of μmoles of standard in 2 ml, and the factor 2000/90 converts the volume of serum to one liter and the time of incubation to one minute.

If a specimen should have an activity exceeding 1500 U/L, dilute one volume of serum with four volumes of 0.85% (w/v) NaCl and repeat the test.

REFERENCE VALUES. The normal range for serum lipase extends from 28 to 200 U/L.

INCREASED CONCENTRATION. Serum lipase activity is elevated in acute pancreatitis and may reach 10 to 40 ULN, depending upon the method of assay and the time at which the sample was collected. The serum lipase activity usually reaches its maximum at 72 to 96 hours after an attack of acute pancreatitis and declines more slowly than does the amylase activity.

Diagnosis of Metastasizing Cancer of the Prostate[3,22]

Cancer of the prostate is a common disease that primarily affects elderly men. It is difficult to diagnose in the early stages because of the lack of signs and symptoms; most of the time the diagnosis is made only after metastasis has occurred.

The prostate gland is rich in acid phosphatase (ACP), which is present in lower concentration in many tissues (spleen, kidney, liver, bone, blood platelets and others). ACP acts optimally below pH 6.0 to split off a phosphate group from a variety of organic phosphate esters. The ACP activity of normal serum is derived from some of the above tissues but primarily from blood platelets. Normal serum has a low activity of ACP but in metastasizing carcinoma of the prostate, its activity increases greatly and may rise to 3 to 15 × ULN. The carcinoma has to metastasize, i.e., to invade blood capillaries, lymph channels, and other tissues, before the elevation in the serum level of acid phosphatase occurs; a discrete prostatic cancer that has not penetrated beyond the capsule does *not* cause this rise in serum ACP.

Because erythrocytes and blood platelets also contain an acid phosphatase, it is essential to distinguish between the ACP derived from these sources during the clotting of the blood specimen and that coming from the prostate. Two different techniques may be employed to assist in identifying the serum ACP derived from prostatic tissue. The first is to use a substrate that the ACP from prostate splits more readily than does the ACP from platelets and erythrocytes; sodium thymolphthalein monophosphate and α-naphthylphosphate are such substrates. The second technique is to measure the ACP activity before and after adding tartrate to the mixture. Tartrate greatly inhibits the ACP from prostate, but is much less inhibitory for the ACP from erythrocytes or platelets. A combination of both techniques is considered to be the most satisfactory when employing α-naphthylphosphate as the substrate.

Serum Acid Phosphatase (ACP)

REFERENCES. Babson, A.L., and Phillips, G.E.: An improved acid phosphatase procedure. Clin. Chim. Acta 13, 264, 1966.

Berger, L., and Rudolph, G.G.: Alkaline and acid phosphatase. *In* Standard Methods of Clinical Chemistry, Vol. 5, edited by S. Meites. New York, Academic Press, 211, 1965.

PRINCIPLE. The ACP in serum splits α-naphthylphosphate into equimolar parts of α-naphthol and phosphate. The α-naphthol is quantitated by coupling it with a diazonium salt to form a highly colored azo dye whose absorbance is measured. The hydrolysis is carried out in the absence and presence of tartrate. The enzyme activity inhibited by the tartrate is presumed to be derived from the prostate, but this is not always the case. The ACP from erythrocytes and blood platelets is somewhat inhibited by tartrate, depending upon the substrate used, but that coming from liver or kidney is not appreciably affected. The "prostatic" or tartrate-inhibitable ACP is calculated by subtracting the activity of ACP in tartrate from the total activity.

REAGENTS

1. Citrate Buffer (0.07 mol/L, pH 5.6). Dissolve 14.7 g citric acid monohydrate in about 500 ml of water. Add 200 ml 1 mol/L NaOH and adjust the pH to 5.6 with NaOH or HCl. Dilute to 1 liter and store at 4° C.
2. Buffered Substrate (2.7 mmol/L). Dissolve 33 mg α-naphthylphosphate (monosodium, monohydrogen, MWt = 246) in 50 ml of citrate buffer. It is stable for one week in the refrigerator. Adjust to pH 5.6.
3. Diazo Reagent. Dissolve 40 mg Fast Red B Salt (Diazotized 5-nitro-o-anisidine, Sigma No. F1125 is satisfactory) in 100 ml 0.1 mol/L HCl. The reagent will remain stable for one week in the refrigerator.
4. 0.1 mol/L NaOH. Dilute a solution of concentrated NaOH to 0.1 mol/L with water.
5. Stock Standard Naphthol, equivalent to enzyme activity of 200 μmol/min. Since the incubation period is 30 minutes, a stock standard solution of 200 U must contain 200 × 30 = 6000 μmoles per liter. Dissolve 86.5 mg α-naphthol in 10 ml absolute ethanol and dilute to 100 ml with water. Store in a refrigerator.
6. Working Standards in 5% Bovine Serum Albumin (BSA).

μmol/L	Units	ml Water	ml Stock Standard
0	0	2.5	0
60	2	2.0	0.5
300	10	0	2.5

Dilute each to 50 ml volume with 5% BSA, aliquot, and store in freezer.
7. 5% BSA Stock. Dissolve 7.5 g BSA in 140 ml sterile normal saline. Adjust pH to 6.0 with 0.1 mol/L NaOH and make up to 150 ml volume with saline solution in a graduated cylinder.

8. Tartrate Buffer (0.6 mol/L, pH 5.6). Suspend 9.0 g crystalline L-tartaric acid in about 95 ml water. Add clear, saturated NaOH (about 19 mol/L) slowly while stirring until the pH reaches 5.2, making certain that all of the tartaric acid has dissolved. Dilute to 100 ml with water. Store in a glass bottle in the refrigerator with about 1 ml chloroform as a preservative.

PROCEDURE

1. Set up three 19-mm tubes for each unknown serum, labeled "Total," "Tartrate," and "Blank." Pipet 1.0 ml buffered substrate into each tube.
2. Set up two tubes for the control serum and two tubes for each standard (2 and 10 U). One of each pair of tubes is for total, and the other is for the blank activities. Transfer 1.0 ml substrate (#2) to each of these tubes.
3. To all tubes labeled "Tartrate," add 50 µl tartrate buffer (#8). Bring all tubes to 37° C in the water bath.
4. To all tubes labeled "Total" and "Tartrate" add 0.2 ml of the appropriate specimen (patient's serum, control serum or standard) at 30-second intervals and incubate for exactly 30 minutes. Incubate the "Blank" tubes also.
5. At the end of 30 minutes, remove the "Total" and "Tartrate" tubes from the bath at 30-second intervals and add 1.0 ml diazo reagent (#3), mix, and add immediately 5.0 ml NaOH solution (#4) and mix again.
6. To the "Blank" tubes at the end of 30 minutes' incubation, add 0.2 ml serum plus 1.0 ml diazo reagent in rapid order, mix, add the NaOH, and mix again.
7. Read the absorbance of all tubes in a spectrophotometer against a water blank at 600 nm.
8. Calculation: Calculate the ACP activities of "Total" and "Tartrate" tubes for each patient. *Total* activity only is measured for the standards and control serum.

$$\text{Total ACP activity} = \frac{A_u - A_{ub}}{A_s - A_{sb}} \times C$$

where A_u = absorbance of the serum specimen, A_{ub} = absorbance of the serum blank, A_s = absorbance of the standard, A_{sb} = absorbance of the standard blank, and C = concentration of the standard.

A similar calculation is made for the tubes marked "Tartrate," using the same standards as above. *Prostatic ACP* is the tartrate-inhibitable ACP activity, or *Total ACP — Tartrate ACP.*

REFERENCE VALUES. Total ACP: 0.2 to 1.8 U/L; Tartrate Inhibitable: 0 to 0.6 U/L.

INCREASED CONCENTRATION. Increased activity of the tartrate-inhibitable ACP is characteristically found in metastasizing carcinoma of the prostate. Occasionally, elevated activities of the tartrate-inhibitable ACP may be found in Gaucher's disease or some bone diseases (Paget's disease or female breast cancer that has metastasized to bone), and these obviously are not derived from the prostate. Massage of the prostate increases ACP activity for 1 or 2 days.

DECREASED CONCENTRATION. No physiologic significance is attached to a low serum ACP activity.

ACP Isoenzymes

There are a number of ACP isoenzymes that can be separated electrophoretically; several originate in the prostate gland. The different prostatic isoenzymes vary in their content of sialic acid (an acetylated amino-sugar acid), and since this group is charged, the isoenzymes migrate to different spots in an electric field. Various immunotechniques, such as RIA and counter immunoelectrophoresis (see Chap. 8 for principles of these techniques), for measurement of prostatic isoenzymes in serum are being tested clinically because of their greater ability to detect low enzyme concentrations (about 1,000-fold less than by conventional ACP measurement). The aim is to diagnose cancer of the prostate before metastasis has occurred. Many commercial kits are available for this purpose that are more specific (fewer false positives) but they lack sensitivity because they produce too many false negatives. If the technological progress continues, the measurement of ACP prostatic enzymes may become a valuable screening test for the population at risk for prostatic cancer.

Investigation of Genetic Disorders

Each genetic disease is characterized by a deficiency in at least one enzyme system. Demonstration of the deficiency in certain situations can be made by the finding of an absent or greatly decreased enzyme activity in particular tissues or cells. Carriers of the trait usually demonstrate a moderately lowered concentration of the enzyme. This type of enzyme analysis is usually carried out in special laboratories dedicated to the study of genetic disorders because of the special techniques involved. Mention will be made below of a few of the enzymes that are measured for this purpose, but details of the methodology will not be given.

Pseudocholinesterase (acylcholine acylhydrolase, EC 3.1.1.8)

Patients undergoing surgery frequently are injected with a short-acting muscle relaxant, succinylcholine. An enzyme in plasma, pseudocholinesterase, rapidly hydrolyzes the succinylcholine so that its action in the body is short-lived. Some people have a genetic deficiency of this enzyme,[3,23] and when injected with succinylcholine, they fail to inactivate the drug and may be subjected to its action for as long as 2 to 3 hours instead of the usual 2 minutes. The respiratory muscles may be so relaxed by the drug that breathing is inadequate and the patient's life may be endangered. This situation can be avoided by screening all patients beforehand for decreased activity of pseudocholinesterase, using succinylcholine as the substrate.

Measurement of the enzyme is also useful in following patients who have been exposed to organophosphates, a common insecticide used in certain farm areas. Enzyme inhibition is proportional to the amount of exposure to the insecticide.

Glucose-6-phosphate Dehydrogenase, GPD
(D-glucose-6-phosphate: NADP oxidoreductase, EC 1.1.1.49)

The enzyme, GPD, catalyzes the oxidation of glucose-6-phosphate as the first step in the pentosephosphate pathway. The enzyme is present in many types of cells, but it is particularly important in the red blood cell because the NADPH that is generated is a necessary ingredient for other enzyme systems in the erythrocyte to prevent the accumulation of methemoglobin. Some people have a genetic deficiency of GPD, and those who have 30% or less of the normal amount of the enzyme are susceptible to acute hemolysis of the red blood cells if exposed to certain drugs, such as antimalaria drugs (primaquine), some sulfonamides, quinine, and others. Screening methods are available for detecting a deficiency of GPD in red cells. An excellent review of the subject, with a discussion of various methodologies, has been written by Beutler.[24]

Galactose-1-phosphate Uridyl Transferase (EC 2.7.7.12)

The disease galactosemia is caused by the hereditary deficiency of the enzyme galactose-1-phosphate uridyl transferase, which is necessary for the conversion of ingested galactose to glucose. This conversion must take place before galactose can be utilized for energy. In the absence of the enzyme, continued ingestion of galactose (a component of lactose, the sugar in milk) causes a buildup of galactose-1-phosphate in cells which produces the typical symptoms of galactosemia: cataract formation, enlarged liver and spleen, and mental retardation. If the deficiency is discovered early enough, the infant can be saved by removing all sources of galactose from the diet. This is accomplished by substituting an artificial milk devoid of lactose for natural milk.

Infants suspected of having galactosemia can be checked by analyzing their red blood cells for the presence of the enzyme, galactose-1-phosphate uridyl transferase. The subject is well reviewed by Segal.[25]

Enzymes as Reagents or Labels

In addition to the measurement of serum enzyme activity as an aid to the diagnosis and management of disease, enzymes are widely utilized in the clinical chemistry laboratory as reagents highly specific for particular chemical constituents of serum. A few examples should suffice:

1. *Urease* in the determination of serum urea nitrogen (BUN). When serum is incubated with the enzyme urease (usually prepared from jack bean meal), urea is the only compound found in the body that is split by this enzyme, the reaction producing NH_4^+ and CO_2. The NH_4^+ may be analyzed in a variety of ways, and its concentration is proportional to the urea concentration.

2. *Glucose oxidase* in the determination of serum glucose concentration. Glucose oxidase reacts specifically with glucose $+ O_2$ to produce gluconic acid $+ H_2O_2$. The glucose concentration may be determined from the rate of O_2 consumption or by the amount of H_2O_2 produced. The latter measurement may be made by chemical methods or by decomposing the H_2O_2 with catalase or peroxidase and coupling this reaction with a hydrogen donor (o-tolidine, dianisidine) which becomes colored (see p. 313).

3. *Uricase* in the determination of uric acid. The enzyme converts uric acid into allantoin, CO_2, and H_2O_2. Since uric acid absorbs light at 293 nm and allantoin does not, the uric acid concentration is proportional to the decrease in UV absorbance at 293 nm. Other variants of the method measure the amount of H_2O_2 produced by employing peroxidase and a suitable chromogen as in the glucose oxidase method for glucose.

Much more recently, enzymes have been introduced into the clinical chemistry laboratory as a means of labeling antigens, and thereby specifically determining the serum concentration of drugs or other compounds difficult to analyze by other means. This is elaborated further in Chapter 8. Suffice it to say that the use of an enzyme label provides a means for amplification of a chemical reaction; it greatly increases the sensitivity of detection of constituents in serum by several orders of magnitude. It approaches radioimmunoassay in sensitivity and specificity, but is performed without the use of radioactive compounds or counting equipment.

REFERENCES

1. Lehninger, A.L.: Biochemistry: The Molecular Basis of Cell Structure and Function, 2nd ed. New York, Worth Publishers, 1975, p. 183.
2. Toporek, M.: Basic Chemistry of Life, 2nd ed. New York, Appleton-Century-Crofts, 1975, p. 266.
3. Wilkinson, J.H.: The Principles and Practices of Diagnostic Enzymology. Chicago, Year Book Medical Publishers, 1976.
4. Cherry, I.S., and Crandall, L.A., Jr.: The specificity of pancreatic lipase: Its appearance in the blood after pancreatic injury. Am. J. Physiol. *100*, 266, 1932.
5. Bodansky, A.: Phosphatase studies. II Determination of serum phosphatase. Factors influencing the accuracy of the determination. J. Biol. Chem. *101*, 93, 1933.
6. Dixon, M., and Webb, E.C.: Enzymes, 3rd ed. New York, Academic Press, 1979, p. 684.
7. LaDue, J.S., Wroblewski, F., and Karmen, A.: Serum glutamic oxaloacetic transaminase activity in human acute transmural myocardial infarction. Science *120*, 497, 1954.
8. Coodley, E.L. (editor): Diagnostic Enzymology. Philadelphia, Lea & Febiger, 1970.
9. Roberts, R., and Sobel, B.E.: Creatine kinase isoenzymes in the assessment of heart disease. Am. Heart J. *95*, 521, 1978.
10. Lott, J.A., and Stang, J.M.: Serum enzymes and isoenzymes in the diagnosis and differential diagnosis of myocardial ischemia and necrosis. Clin. Chem. *26*, 1241, 1980.
11. Mercer, D.W., and Varat, M.A.: Detection of cardiac-specific creatine kinase isoenzyme in sera with normal or slightly increased total creatine kinase activity. Clin. Chem. *21*, 1088, 1975.
12. Yasmineh, W.G., and Hanson, N.Q.: Electrophoresis on cellulose acetate and chromatography on DEAE-Sephadex A-50 compared in the estimation of creatine kinase isoenzymes. Clin. Chem. *21*, 381, 1975.
13. Klein, B., Foreman, J.A., Jeunelot, C.L., and Sheehan, J.E.: Separation of creatine kinase isoenzymes by ion-exchange column chromatography. Clin. Chem. *23*, 504, 1977.
14. Roberts, R., Sobel, B.E., and Parker, C.W.: Immunologic detection of myocardial infarction with a radioimmunoassay for MB creatine kinase. Clin. Chim. Acta *83*, 141, 1978.
15. Galen, R.S.: The enzyme diagnosis of myocardial infarction. Hum. Pathol. *6*, 141, 1975.
16. Meeker, D., Clayson, K.J., and Strandjord, P.E.: Differential diagnosis of acute hepatocellular injury in infectious mononucleosis. Clin. Res. *21*, 519, 1973.
17. McComb, R.B., Bowers, G.N., Jr., and Posen, S.: Alkaline Phosphatase. New York, Plenum Press, 1979.
18. Homburger, H.A., and Jacob, G.L.: Creatine kinase radioimmunoassay and isoenzyme electrophoresis compared in the diagnosis of acute myocardial infarction. Clin. Chem. *26*, 861, 1980.
19. Moss, D.W.: Alkaline phosphatase isoenzymes. Clin. Chem. *28*, 2007, 1982.
20. Legaz, M.E., and Kenny, M.A.: Electrophoretic amylase fractionation as an aid in diagnosis of pancreatic disease. Clin. Chem. *22*, 57, 1976.
21. Kaufman, R.A., and Tietz, N.W.: Recent advances in measurement of amylase activity—a comparative study. Clin. Chem. *26*, 846, 1980.
22. Bodansky, O.: Acid phosphatase. Adv. Clin. Chem. *15*, 43, 1972.
23. Lehmann, H., and Liddell, J.: The cholinesterase variants. *In* The Metabolic Basis of Inherited Disease, 3rd ed., edited by J.B. Stanbury, J.B. Wyngaarden, and D.S. Fredrickson. New York, McGraw-Hill, 1972, p. 1730.
24. Beutler, E.: Glucose-6-phosphate dehydrogenase deficiency. *In* The Metabolic Basis of Inherited Disease, 4th ed., edited by J.B. Stanbury, J.B. Wyngaarden, and D.S. Fredrickson. New York, McGraw-Hill, 1978, p. 1430.
25. Segal, S.: Disorders of galactose metabolism. *In* Metabolic Basis of Inherited Disease, 4th ed., edited by J.B. Stanbury, J.B. Wyngaarden, and D.S. Fredrickson. New York, McGraw-Hill, 1978, p. 160.

Chapter 7

THE LIVER AND TESTS
OF HEPATIC FUNCTION

The liver is an important organ that participates in numerous metabolic functions and serves additionally as a secretory-excretory organ. The bile that flows from the liver or gallbladder into the small intestine not only is essential for the digestion of fats and absorption of the fat-soluble vitamins but also serves as an excretory mechanism for certain pigments and waste products. Total loss of the liver usually results in death within 24 hours.

PHYSIOLOGICAL ROLE OF THE LIVER

The liver is a large organ that lies just under the diaphragm on the right side of the body. It has a tremendous blood supply and weighs about 1500 g. Like all organs, the liver receives oxygenated blood pumped from the heart in an artery that branches and subdivides to arterioles and finally perfuses the tissue as capillaries. The arterial exchange of O_2 for CO_2 between blood and tissue cells and the exchange of nutrients and waste products take place in the capillary beds. The capillary branches combine to form venules and then veins that finally return the blood from the organ to the heart.

The hepatic artery and hepatic vein serve this function for the liver, but the liver is unique in that it has a second great venous blood supply coursing through it. The portal vein, which brings blood from all parts of the gastrointestinal tract (stomach, small and large intestines, pancreas, and spleen), flows directly from these organs to the liver where it terminates in capillary-like vessels called sinusoids; the blood from the sinusoids finally leaves the liver by merging with the blood of the hepatic vein. The great physiological significance of the portal flow to the liver is that *all nutrients arising from the digestion of food in the gastrointestinal tract,* with the exception of fats, *pass through the liver first* before entering the general circulation for transmission to the rest of the body. This nutrient-rich blood supply provides the liver with a high concentration of various substrates that enables it to carry out the many metabolic functions that characterize this organ as the "metabolic factory" of the body. Popper and Schaffner describe the structure and function of the liver in their book.[1]

The liver is also unique in its fine anatomic structure or architecture. In essence, it is characterized by a series of plates of hepatic cells, one layer thick, in contact with the bloodstream on one side and the bile canaliculi (bile channels) on the other side. Thus, each hepatic cell (hepatocyte) has a large surface area in contact with both a nutrient-intake system from the sinusoids (portal vein "capillaries") and an outlet system, the bile canaliculi, which carries away

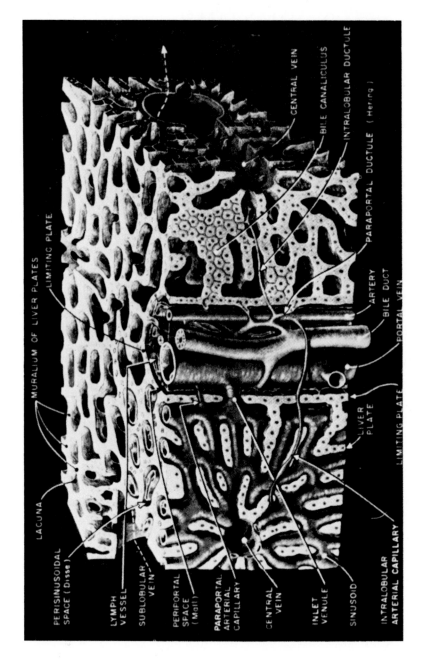

Fig. 7.1.   Three-dimensional representation of internal liver structure. (From H. Elias and J.C. Sherrick. Morphology of the Liver. New York, Academic Press, 1969, p. 135.)

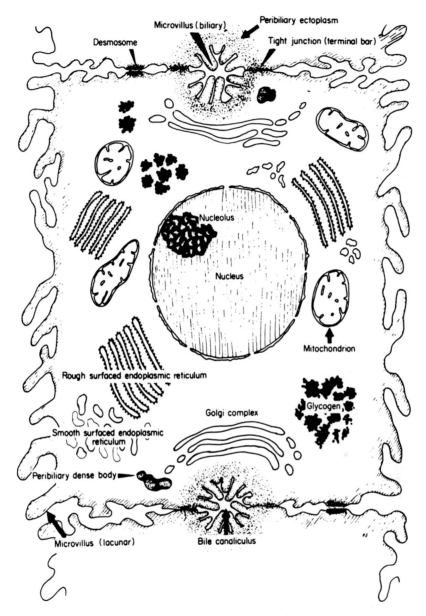

Fig. 7.2. Diagram of generalized liver cell. (From H. Elias and J.C. Sherrick. Morphology of the Liver. New York, Academic Press, 1969, p. 56.)

the secretions and excretions from the hepatocytes. Bile is a viscous fluid that is produced in this process, and it consists of a solution of bile salts, bile pigments, cholesterol, and other materials. The bile canaliculi combine to form bile ducts that carry the bile secretions into the small intestine. The anatomic arrangement of the liver provides maximum efficiency for receiving materials to be utilized or processed as well as for the removal or excretion of undesirable products.[1] The liver architecture is illustrated in Figure 7.1, and a liver cell is shown in Figure 7.2.

The liver performs many important functions dealing with or affecting metabolism, excretion of various substances, protection, circulation, and blood coagulation.

Metabolic Functions

Some of the diverse metabolic functions carried out by the liver are exclusive to this organ. The activities involve many phases of carbohydrate, protein, and lipid metabolism, as well as the various interconversions from one category to another and the production of intermediary forms.

CARBOHYDRATE METABOLISM. Glycogen is synthesized from glucose coming from ingested carbohydrates and is stored in the liver during the temporary periods of carbohydrate excess. During periods of low plasma glucose or of need for glucose because of no intake (fasting) or of excess utilization (exercise or hard physical work), the stored glycogen is converted to glucose in an attempt to maintain a constant blood glucose concentration. Amino acids and fatty acids may also be transformed into glucose by the liver during periods of great need. In addition, other hexoses are converted into glucose by hepatic cells.

PROTEIN SYNTHESIS. Almost all of the plasma proteins, with the exception of the gamma globulins, are synthesized in the liver. The plasma proteins originating in the liver include albumin, lipoproteins, many specific carrier proteins, and some of the proteins involved in the blood coagulation process. This latter group includes prothrombin, fibrinogen, Factors V, VII, and X from the extrinsic clotting system, and Factor IX from the intrinsic clotting system.

LIPID METABOLISM. Plasma lipoproteins, the carriers of plasma lipids, are synthesized in the liver as well as the endogenous portion of the primary lipid components—cholesterol, phospholipids, and triglycerides. The liver esterifies some cholesterol to form cholesterol esters. Cholesterol is also degraded into bile acids by liver cells and secreted into bile. The bile acids are essential emulsifiers of fat in the small intestine; fat digestion and absorption (p. 327) do not occur in its absence (occlusion of the bile ducts or cholestasis).

OTHER SYNTHESES OR TRANSFORMATIONS. A large number of other metabolic transformations too numerous to mention here occur in the liver. These include various transaminations (conversion of a keto acid to an amino acid, with concomitant production of a new keto acid), production of ketone bodies (acetoacetic acid and β-hydroxybutyric acid) from acetyl CoA, conversion of some amino acids to glucose and vice versa, conversion of lactic acid to glucose (or glycogen), synthesis of fatty acids from acetyl CoA, and many other reactions of importance.

Storage Functions

The liver is the primary storage site for glycogen, vitamins A, D, and B_{12}. Significant amounts of iron are stored in the reticuloendothelial cells of the liver, as well as in the bone marrow and spleen.

Excretory Functions

The liver secretes bile, with its content of pigments (primarily bilirubin esters), bile salts (cholic acid and its various conjugates), cholesterol, and other substances extracted from blood into bile (some dyes, heavy metals, enzymes), into the small intestine by way of the bile ducts. Many of these materials are waste products, but the bile salts are detergents that are essential for the emulsification of ingested fats as the first step in their digestion.

Protective Functions

The liver helps protect the body from various foreign or dangerous materials by phagocytic action and detoxification reactions.

PHAGOCYTIC ACTION. The liver contains a large number of phagocytic cells, called Kupffer cells, which line the sinusoids and are active in removing foreign materials from the blood. The Kupffer cells are part of the reticuloendothelial system and act as scavengers.

DETOXIFICATION. Many noxious or comparatively insoluble compounds are converted to other forms that are either less toxic or become water-soluble and therefore excretable by the kidney. The conversion to a less toxic form may involve esterification, methylation, oxidation, reduction, or other changes. Ammonia, a very toxic substance arising in the large intestine through bacterial action upon amino acids, is carried to the liver by the portal vein and converted into the innocuous compound, urea, by the hepatocytes. Esterification with glucuronic acid is a common mechanism for converting insoluble materials into water-soluble compounds that can be excreted by the kidney. The very insoluble pigment, bilirubin, is esterified to a diglucuronide, which is then excreted into bile. Likewise, many of the steroid hormones are esterified and excreted through the kidney, thereby shortening their active stays in the body. A portion of these materials may be converted into sulfate esters and excreted.

Circulatory Functions

The liver plays a role in immunologic defense through its reticuloendothelial system (Kupffer cells), helps to regulate blood volume by serving as a blood storage area, and is a means for mixing the blood from the portal system with that of the systemic circulation.

Function in Blood Coagulation

As mentioned under protein synthesis, certain proteins essential for the coagulation of blood are synthesized solely by hepatic cells. These are fibrinogen, prothrombin, and Factors V, VII, IX, and X. The latter four have short half-lives, turn over quite rapidly, and thus may quickly become limiting for the coagulation process in the presence of severe hepatic pathology. Prothrombin and Factors VII, IX, and X require the presence of vitamin K, a fat-soluble vitamin, for their synthesis.

HEPATIC DYSFUNCTION

Alterations in hepatic cell activity as a result of disease may affect some of the functions or activities of the liver to a greater or lesser degree and help to serve as an indicator of liver pathology. Estimation of the presence or absence of hepatic dysfunction is complicated by the large functional reserve of the liver and its power to regenerate rapidly. Under experimental conditions with rats, as much as 80 to 85% of the liver must be removed before certain laboratory tests for liver function become abnormal. Regeneration of new liver tissue is rapid. It is impossible to quantitate the amount of hepatic tissue damaged by a disease process; a diffuse minimal involvement of the liver may produce a more grossly abnormal laboratory test than a focal necrosis.

In view of the large number of activities engaged in by the liver, it is not surprising that numerous tests have been introduced to test a particular hepatic function. Most such tests are nonspecific, since diseases other than those involving the liver may produce similar changes. All of them present difficulty in interpretation because of the large functional hepatic reserve.

The large number of liver tests that has been proposed is evidence of the complexity of the problem and illustrates the lack of consensus as to which tests are best suited for the purpose. The reasons for requesting liver function tests are threefold: (1) for diagnosis, to see whether there is impairment of the hepatobiliary system; (2) for differentiation, if there is hepatic impairment, to discover whether the problem consists primarily of damage to hepatic cells or is the result of an obstruction to the biliary system; (3) for prognosis, to estimate the extent of the damage and the probable outcome with appropriate therapy. The ideal test is specific for hepatic disease, sensitive (capable of detecting early or miminal hepatic disease), and selective (able to detect specific types of liver disorders), but such a test does not exist. A battery or panel of liver function tests is usually employed, but the composition of the panel may vary according to the medical needs, that is, whether used for diagnosis, differentiation of liver disease, or for prognosis. In certain circumstances, special tests may be employed.

Bile Pigment Metabolism

Bilirubin

Bilirubin, the principal pigment in bile, is derived from the breakdown of hemoglobin when senescent red blood cells are phagocytized. The erythrocytes are loaded with hemoglobin, a complex molecule containing 4 heme groups (ferroprotoporphyrin) attached to a protein, globin (see Fig. 7.3); the erythrocytes survive in the body for about 120 days before the aged cells are engulfed and digested by phagocytic cells. These scavenger cells of the reticuloendothelial (R-E) system that destroy the old red blood cells are located primarily in the spleen, liver and bone marrow.

The degradation of the hemoglobin molecule, shown schematically in Figure 7.3, leads to the formation of bilirubin. About 80% of the bilirubin formed daily is derived from the breakdown of the senescent erythrocytes; the remainder comes from the degradation in bone marrow of immature red blood cells and from the destruction of other heme-containing proteins (myoglobin, catalase, cytochromes). Thus, about 6 to 6.5 g of hemoglobin in aged red blood cells are broken down daily in an adult to form about 220 mg of bilirubin; another 50 or 60 mg of bilirubin originate from other sources. The stepwise breakdown of hemoglobin and the fate of the product, bilirubin, are shown schematically in Figure 7.4.

Step 1. The exact mechanism is not well understood, but the net effect is the splitting off of the protein globin, which may be reutilized or hydrolyzed to amino acids that join the amino acid pool for recycling. The porphyrin ring of the heme molecule is broken open, with loss of one of the methene groups connecting the 4 pyrrole rings. The resulting open-chain tetrapyrrole loses its

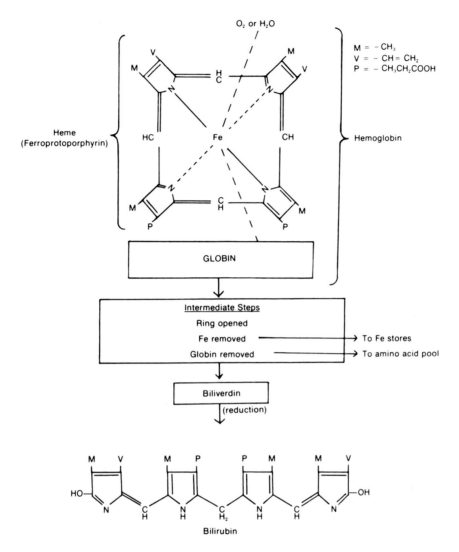

Fig. 7.3. Degradation of hemoglobin. The heme portion of hemoglobin is cyclic and contains four substituted pyrrole rings connected by methene groups; Fe^{2+} is bound inside the pyrrole ring. Hemoglobin is converted to bilirubin by a series of steps that include disruption of the cyclic pyrrole chain, removal of Fe and of globin, and reduction of one methene group.

iron, which becomes bound to a protein (ferritin) where it is stored until utilized again for the synthesis of new heme compounds in the bone marrow. The resulting pigment is reduced to form bilirubin, a reddish-yellow waste product that must be excreted.

Step 2. Bilirubin leaves the reticuloendothelial cell and is solubilized in plasma by firmly binding to the protein, albumin.

Step 3. Upon reaching the liver sinusoids, the bilirubin-albumin complex is transferred into the hepatocyte by an active process. The bilirubin is detached from its albumin carrier and transported inside the hepatic cell network to microsomes in rough endoplasmic reticulum.

Step 4. Esterification (conjugation) of bilirubin takes place in the endo-

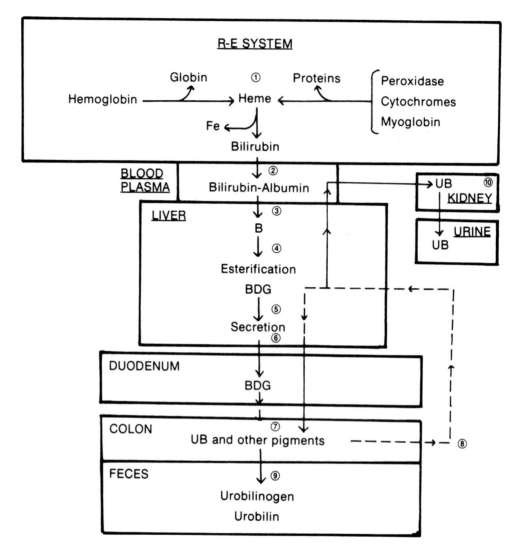

Fig. 7.4. Schematic illustration of formation and excretion of bile pigments. The following abbreviations are used: R−E = reticuloendothelial; B = bilirubin; BDG = bilirubin diglucuronide; UB = urobilinogen. The uptake of bilirubin in the liver is from a sinusoid into a hepatocyte; esterification takes place on a microsome. The BDG is excreted into bile canaliculi, which merge into bile ducts.

plasmic reticulum (Fig. 7.2). An enzyme, uridyldiphosphate glucuronyl transferase (UDPG), transfers a glucuronic acid molecule to each of the two propionic acid side chains in bilirubin, converting bilirubin into a diglucuronide ester. The product, bilirubin diglucuronide (BDG), is frequently referred to as conjugated bilirubin; in this book, it will be called esterified bilirubin. A monoester, bilirubin monoglucuronide, was thought to be present also in plasma, but this seems to be a complex containing equimolar amounts of nonesterified bilirubin and the diglucuronide.

Step 5. The bilirubin diglucuronide, which is water soluble, is secreted from the hepatic cell by a transport mechanism into the bile canaliculi as a step in the formation of bile (Fig. 7.2).

Step 6. The bilirubin diglucuronide, along with the rest of the bile, passes into larger channels, the bile ducts. Bile is produced continuously by the liver, but between periods of digestion of foods, it is stored temporarily in the gallbladder where it is concentrated by the absorption of water. Under proper stimulation (food in the stomach or hormones that stimulate the flow of bile), bile from the gallbladder and that coming directly from the liver flow through the common bile duct into the small intestine.

Step 7. As bilirubin diglucuronide approaches the colon (large intestine), it is exposed to the action of bacteria whose enzymes cleave off the glucuronic acid moieties and reduce (hydrogenate) the bilirubin molecule. The reduction products are known collectively as urobilinogen.

Step 8. A portion of the urobilinogen is absorbed into the portal blood, which flows to the liver. The healthy liver cells remove almost all of the urobilinogen as it passes through and re-excrete it into the bile. The small portion of urobilinogen that escapes excretion by the hepatocyte (about 1% of the colon urobilinogen content) reaches the peripheral circulation where it is excreted by the kidney into the urine (Step 10).

Step 9. The urobilinogen that is not absorbed in the colon becomes partially oxidized to urobilin and other brownish pigments that are excreted in the feces.

In summary, about 250 to 300 mg of bilirubin are produced daily in normal, healthy adults. A normally functioning liver is required to eliminate this amount of bilirubin from the body, a process requiring the prior esterification of bilirubin to form the water-soluble diglucuronide. Practically all of it is eliminated in the feces in the form of pyrrole fragments and pigments (dipyrroles, urobilinogen, urobilin); a small amount of the colorless product, urobilinogen, is eliminated in the urine. A low concentration of bilirubin is found in normal plasma, most of which is nonesterified (free), but firmly bound to albumin. A small percentage of the total bilirubin exists in normal plasma as the diglucuronide. There are analytical tests for measuring the concentration of total bilirubin and esterified bilirubin in serum and urobilinogen in urine.

Disorders of Bile Pigment Metabolism

Analytical tests for the measurement of concentrations of total and esterified bilirubin in serum have been available for a long time. The concentration of the nonesterified (unconjugated, indirect-reacting) bilirubin is obtained by difference. The concentration of total bilirubin in the serum of normal adults ranges from 0.1 to 1.0 mg/dl (1 to 10 mg/L or 1.7 to 17.0 μmol/L). The concentration of the esterified bilirubin may be up to 0.3 mg/dl (5.0 μmol/L). When the bilirubin concentration in the blood rises, the pigment begins to be deposited in the sclera of the eyes and in the skin. This becomes evident to an experienced observer when the serum bilirubin reaches a concentration of 2.5 mg/dl or greater. This yellowish pigmentation in the skin or sclera is known as jaundice, or *icterus;* when visible jaundice is present, the serum bilirubin concentration must of necessity be above the upper limit of normal.

Since most liver diseases and several nonhepatic disorders are accompanied by jaundice, the differential diagnosis of jaundice plays an important role in elucidating hepatic dysfunction. An elevated level of serum bilirubin may be produced by one or more of the following four processes involved in the metabolism of bilirubin, as listed below and shown in Figure 7.4.

AN EXCESSIVE LOAD OF BILIRUBIN PRESENTED TO THE LIVER (Fig. 7.4, Step 2). This occurs in hemolytic disease or in any process of excessive erythrocyte destruction. With chronic hemolysis, more bilirubin may be produced and presented to the liver for excretion than the liver can handle. This problem of *overproduction with retention is prehepatic* in origin. The serum bilirubin elevation in chronic hemolytic anemia is mild and usually is below 5 mg/dl because the liver can excrete much more than the normal load of this pigment. The bilirubin fraction that is increased is the nonesterified bilirubin that is albumin-bound (Fig. 7.4, Step 2); the esterified fraction remains at normal levels. No bilirubin is detected in the urine because there has been no rise in the esterified fraction.

During the catabolism of this increased load of bilirubin, more bilirubin is esterified in the liver (Fig. 7.4, Step 4) and excreted into the gut (Step 5) where it is converted into urobilinogen (Step 7). As a result of the formation of excessive amounts of urobilinogen and other pigments, fecal urobilinogen and urobilin are greatly increased (Step 9), while urine urobilinogen (Step 10) is moderately increased. These changes are tabulated in Table 7.1.

DEFECTIVE TRANSPORT INTO THE HEPATOCYTE. An active transport system is required for the transportation of the bilirubin bound to plasma albumin through the hepatic cell membrane to the microsomes (Step 3). A congenital malfunction of this transport system produces a mild jaundice (serum bilirubin below 2 mg/dl) that is harmless and usually is detected accidentally during routine screening tests for blood chemistry or in monitoring of serum bilirubin concentrations during other illnesses. This transport defect is known as *Gilbert's disease.*

IMPAIRMENT IN THE ESTERIFICATION OF BILIRUBIN

Physiologic Jaundice of the Newborn. The UDPG transferase enzyme system responsible for converting bilirubin into the diglucuronide (Fig. 7.4, Step 4) is not fully developed at birth. In the full-term infant, it takes several days before

TABLE 7.1
Results Frequently Found with Liver Function Tests in Various Liver Disorders[a]

| | TYPES OF JAUNDICE | | | | | | |
| | PREHEPATIC | | INTRAHEPATIC | | | | EXTRAHEPATIC |
COMMON LABORATORY TESTS	HDN*	Hemo-lytic	Acute Injury	Chronic Injury	Chole-stasis	Cirrhosis	Obstruction (Cholestasis)
Bilirubin—Total	↑↑↑	↑	↑↑	→↑	↑↑↑	↑	↑↑↑
Bilirubin—Esterified	→	→	↑↑	→↑	↑↑↑	↑	↑↑↑
Urine Bilirubin	0	0	↑↑↑	↑	↑↑↑	↑	↑↑↑
Urine Urobilinogen	0	↑	↑	↑	→↑	→↑	↓↓
Fecal Urobilinogen	0	↑↑↑	↓	↓	↓	→↓	↓↓
Albumin			→	↓	→	↓↓	→
γ-globulin			↑[b]	↑↑[b]	→	↑↑	
Prothrombin Time			↑↑	↑	↑[c]	→	↑[c]
ALP			↑	↑	↑↑	→	↑↑↑
AST			↑↑↑↑	↑	↑↑	↑	↑↑
ALT			↑↑↑↑	↑	↑	→↑	↑

[a] See text for situations where these tests are most useful.
[b] γ-globulin is increased in viral hepatitis but not in a hepatitis caused by drugs or chemical toxins.
[c] Reversed 24 hours after injection of soluble vitamin K preparations.
*Hemolytic disease of the newborn

the enzyme is produced in sufficient quantity to esterify the bilirubin presented to it. The serum bilirubin may rise as high as 8 mg/dl in the normal full-term infant by the third to the sixth day of life as a result of the enzyme immaturity before falling to normal adult levels; this condition is known as physiologic jaundice of the newborn. This process is aggravated in premature infants who have to wait a longer time for the generation of the esterifying enzyme (UDPG transferase). The serum bilirubin concentration may climb as high as 15 mg/dl in premature infants in the absence of any disease process.

Hemolytic Disease of the Newborn (HDN). This disorder is caused by either an Rh or an ABO system incompatibility, which greatly intensifies the jaundice usually encountered in the newborn, because of the increased amount of bilirubin produced from the phagocytized red blood cells. Hemolytic disease of the newborn occurs, for example, when the erythrocytes of a fetus with Rh positive blood type become coated with maternal antibodies to Rh protein and are destroyed by the infant's reticuloendothelial system. Since the immature liver cannot esterify the bilirubin, a step necessary for its excretion, the concentration of bilirubin in the plasma rises rapidly. Plasma albumin has a limited capacity for binding bilirubin, and when primary binding sites are saturated, the bilirubin becomes less tightly bound to secondary sites. When the blood circulates to the brain, the loosely bound bilirubin is partitioned between the lipid covering of brain cells and the plasma. It enters the brain cells and causes irreversible damage to the basal ganglia. Nuclear staining of brain cells by bilirubin is known as kernicterus. Many infants who survive the kernicterus become spastic and may also suffer from mental retardation.

The ABO type of HDN may occur when the mother is blood type O and the baby is either type A or type B, even though they are Rh compatible. The mother, being type O, has antibodies to both type A and B red blood cells. Since her IgG antibodies can cross the placenta, the fetal red cells are vulnerable to attack. Fortunately, the antibodies of the ABO type of hemolytic disease of the newborn are much weaker than those of the Rh type and do much less damage. The jaundice produced in the newborn is much less severe, and the plasma bilirubin rises more slowly and crests at lower levels. The need for an exchange transfusion is much less frequent in this type of HDN.

The critical concentration of serum bilirubin for possible brain damage in newborns is around 20 mg/dl, although this may vary, depending upon the concentration of serum albumin, the administration of certain drugs, and other factors. An exchange transfusion is usually performed in order to lower the concentration of circulating bilirubin; this may have to be repeated several times in some cases.

It is a crucial responsibility for the laboratory to accurately perform and report the bilirubin analysis in cases of hemolytic disease of the newborn. The physician depends heavily upon the laboratory values in deciding whether an exchange transfusion is necessary.

The rise in serum bilirubin levels in physiologic jaundice of the newborn and in hemolytic disease of the newborn is in the nonesterified portion because of the immaturity of the transferase enzyme system. The first sign of production of the enzyme is an increase in the plasma concentration of the esterified fraction.

Congenital Deficiency of the UDPG Transferase Enzyme System. Some children are born with a defect in the transferase enzyme, which leads to a severe jaundice of the type seen in hemolytic disease of the newborn. The disease, known as the Crigler-Najjar syndrome, is rare but severe; over 50% of the reported patients with this disease died within 1 year of birth, and about half of the survivors suffered from irreversible brain damage.

DISTURBANCES IN EXCRETION OF BILIRUBIN (CHOLESTASIS). The largest percentage of patients with jaundice is found in this category. In this large group of disorders, the elevated concentration of serum bilirubin consists of both the esterified and nonesterified fractions, since bilirubin is able to enter the hepatic cell and become esterified. Because of the impairment in the excretory transport system or an obstruction in the biliary tree impeding the excretion of bile into the intestine (Fig. 7.4, Steps 5 and 6), the concentration of esterified bilirubin builds up in the hepatocyte. Consequently, some of the esterified bilirubin is regurgitated into the bloodstream by way of the sinusoids. A second consequence of the increased amounts of bilirubin diglucuronide in the hepatic cells is the slowdown of the transport system that transfers the bilirubin-albumin complex of plasma into the hepatocyte. The net result is a hyperbilirubinemia, with the esterified bilirubin constituting 50 to 70% of the total. The test for urine bilirubin is usually positive in the early stages of the disease because of the elevated level of bilirubin diglucuronide in plasma that can pass through the renal glomerular membrane (Table 7.1).

With respect to the bilirubin degradation products, less bilirubin is reaching the gut because of the obstruction, so fewer pigments are produced. The feces may vary in color from pale yellow or brown to clay-colored, depending upon the completeness of the obstruction or impairment of the excretory transport process (Table 7.1). With complete cholestasis, the urinary urobilinogen falls to zero because no bile and hence no bilirubin reaches the intestine for later conversion to urobilinogen. In hepatocellular disease, the urinary excretion of urobilinogen may be increased despite a reduction in the amount of bilirubin that may reach the gut. This seeming paradox is explained by the *failure of the damaged hepatic cells to remove the reabsorbed* urobilinogen from the portal vein blood; the urobilinogen in the blood then becomes available for excretion by the kidney.

When the bile is completely excluded from the gut by an obstruction, there is a gross failure to digest and absorb ingested fats because of the absence of bile salts to serve as emulsifiers; this produces a steatorrhea—pale, fat, bulky stools.

The various types of disorders responsible for faulty or limited excretion of bilirubin by the liver may be divided into two groups: (1) those of hepatic origin (intrahepatic) and (2) those originating beyond the liver (posthepatic).

Intrahepatic. The most common disorders are diffuse hepatocellular damage and liver cell destruction. These may be caused by viral hepatitis, hepatitis produced by toxins (drugs, chemicals such as phosphorus or organic arsenicals, chlorinated solvents such as carbon tetrachloride, and chloroform), cirrhosis, or intrahepatic cholestasis (as in hepatic edema). A congenital condition also exists in which there is a disorder in the transport system for excreting the bilirubin diglucuronide into the bile canaliculi; (Fig. 7.4, Step 5); this is called the Dubin-Johnson syndrome (Fig. 7.4, Step 5).

All of the intrahepatic problems must be treated by medical management; surgery would be harmful to the patient.

Posthepatic. These include all types of extrahepatic cholestasis or obstruction of the flow of bile into the intestine. The obstruction may be caused by stones in the bile ducts or gallbladder, by carcinoma of the head of the pancreas, by other tumors that obstruct the common bile duct, or by strictures of the common bile duct. Problems related to obstruction by stones, benign tumors, and strictures are definitely alleviated or cured by surgery, whereas with malignancies, a great deal depends upon the type of malignancy, its size and distribution, and the time of discovery (whether before or after metastasis). Since surgical intervention will be of benefit to patients with certain types of jaundice but harmful to others, it is extremely important to make the correct differential diagnosis of jaundice. The various laboratory tests that assist in making this distinction will be discussed later in this chapter after the full complement of tests has been described.

TESTS OF HEPATIC FUNCTION

A number of tests of serum bilirubin, urine bilirubin, excretory function, serum proteins, and enzymes have proven to be useful in the handling of patients with suspected liver disease.

Serum Bilirubin

Two forms of bilirubin exist in plasma, both of which are bound to albumin. These are (1) bilirubin as such (nonesterified, unconjugated, indirect reacting) and (2) bilirubin diglucuronide (BDG, esterified, conjugated, direct reacting). Routine analytical procedures exist for the determination of *total* bilirubin (esterified + nonesterified) and for the measurement of bilirubin diglucuronide alone. The nonesterified fraction is obtained by difference. The measurement of both entities is necessary for the differential diagnosis of hemolytic disease and is quite useful in monitoring the course of hemolytic disease of the newborn (HDN). In following the course of liver disease, the analysis of total bilirubin usually suffices for the measurement of bile pigment in serum.

The classic method for quantitating bilirubin is by converting it into an azo dye and measuring its absorbance at a specific wavelength. The diazo reaction splits both bilirubin diglucuronide and bilirubin in the middle to form two dipyrroles, with loss of the methene group at the scission point. Each dipyrrole forms an azo dye by coupling with the diazonium salt; the resulting azobilirubins have essentially the same spectral absorbance values.

Although many diazonium salts could be used, the one most commonly employed is the one prepared from sulfanilic acid. The latter is diazotized with sodium nitrite in acid solution and then coupled in aqueous solution with bilirubin diglucuronide. Since the nonesterified bilirubin is tightly bound to albumin, it requires the presence of a polar solvent (methanol, ethanol) or some accelerator substances that promote the displacement of bilirubin from albumin and speed up the diazo coupling.

The azo dye is red in neutral solution and blue in strong alkali. The original Malloy and Evelyn method[18] and its various modifications employed methanol as an accelerator and have been used widely for many decades. They are not as accurate as the Jendrassik and Grof method at low concentrations of bilirubin (near the upper limit of normal) and are affected by the protein matrix.

Manual Determination

REFERENCE. Koch, T.R., and Doumas, B.T.: Bilirubin, total and conjugated, modified Jendrassik-Grof method. In Selected Methods of Clinical Chemistry, Vol. 9, edited by W.R. Faulkner and S. Meites. Washington, D.C., American Association for Clinical Chemistry, 1982, p. 113.

PRINCIPLE

Total Bilirubin: Serum is mixed with a slightly acidic solution of caffeine and sodium benzoate, compounds that accelerate the diazotization of nonesterified bilirubin. Diazotized sulfanilic acid cleaves both forms of bilirubin (esterified and nonesterified) and forms an azo dye with the dipyrroles. The addition of alkaline tartrate (pH 13) converts azobilirubin to a blue color whose absorbance is measured at 600 nm. The solution appears green because of the reagent's yellow color.

Esterified (Conjugated) Bilirubin: The serum is mixed directly with the diazo reagent in the absence of accelerators. Only esterified bilirubin reacts within 5 minutes under these conditions. After 5 minutes, ascorbic acid is added to decompose excess diazo salt and stop the reaction. The solution is made alkaline, and the absorbance measured at 600 nm.

REAGENTS

1. *Caffeine reagent:* Dissolve 56 g of anhydrous sodium acetate, 56 g of sodium benzoate and 1 g of EDTA in about 600 ml water. Add 38 g caffeine, mix well, and make up to 1 L volume. Store at room temperature in a polyethylene bottle. It is stable for at least 6 months.

2. *Sulfanilic acid, 5.0 g/L:* Dissolve 5.0 g sulfanilic acid and 15 ml concentrated HCl in about 600 ml water and make up to 1.0 L with water. Solution is stable in a polyethylene bottle for at least 6 months.

3. *Sodium nitrite, 5.0 g/L:* Dissolve 0.5 g $NaNO_2$ in water and make up to 100 ml volume. Store in borosilicate bottle at 4° C. It is stable for at least 2 weeks and usually much longer. Discard when it begins to turn yellow.

4. *Diazo reagent:* Combine and mix 0.1 ml $NaNO_2$ with 4.0 ml sulfanilic acid; keep these proportions if a larger volume is needed. Use within 5 hours.

5. *Alkaline tartrate:* Dissolve 75 g NaOH and 263 g sodium tartrate·$2H_2O$ in water and dilute to 1 L. You may substitute 323 g sodium potassium tartrate·$4H_2O$ for the sodium tartrate. Stable for at least 6 months in a polyethylene bottle.

6. *Ascorbic acid, 40 g/L:* Dissolve 0.2 g ascorbic acid in 5.0 ml water. Prepare fresh daily.

7. *Bilirubin Standard.* Bilirubin standards can be prepared according to the directions of Doumas et al.[2] Twenty mg of Bureau of Standards bilirubin are weighed to 0.05 mg and suspended in 1.0 ml dimethyl sulfoxide (DMSO) in a 100-ml volumetric flask. After dispersing and wetting in the DMSO, add 2.0 ml 0.1 mol/L Na_2CO_3. Add 80 ml of a solution containing 4 g bovine serum albumin (BSA) per 100 ml, adjusted to pH 7.4. Two ml 0.1 mol/L HCl are added, and the solution is made up to volume with BSA solution. This provides a stock standard solution of 20.0 mg/dl. Shield

from light and store 0.1 ml aliquots, preferably at $-70°$ C in the deep freeze; standards are stable for several months at $-20°$ C, longer at $-70°$ C. A standard curve can be prepared by using the stock solution and appropriate dilutions with BSA solution. Some commercial control sera are available.

PROCEDURE

Total Bilirubin

1. Label tubes for Test and Blank for each serum, control and standard.
2. Pipet 50 μl of serum (test, control, or standard) to each tube.
3. Transfer 0.5 ml caffeine reagent to each tube.
4. Add 0.25 ml freshly prepared diazo reagent to each test tube, mix, and let stand for 10 minutes.
5. While waiting, transfer 0.25 ml sulfanilic acid solution to each Blank tube and mix.
6. After the Test has reacted for 10 minutes, add 0.5 ml alkaline tartrate to every tube. Cover the tubes with Parafilm and mix thoroughly by inversion.
7. Set the spectrophotometer to zero absorbance with water at 600 nm.
8. First, read all the Blank solutions and then the Test solutions, if using a flow-through cuvet, in order to minimize carryover. The azobilirubin color is stable in alkali for at least 30 minutes.
9. Calculation.

Subtract each Blank absorbance from the Test to obtain ΔA.

$$\text{Total bilirubin, mg/dl} = \Delta A_u / \Delta A_s \times C,$$

where ΔA_u and ΔA_s are the respective corrected absorbances of Test and Standard, and C is the concentration of the Standard.

Esterified Bilirubin

1. Pipet 50 μl of serum (test, control, or standard) to tubes labeled Test and Blank. Add 0.25 ml of 50 mmol/L HCl to each tube.
2. Add 0.25 ml diazo reagent to Test tubes and mix. After 5 minutes, add 25 μl of ascorbic acid to each tube and mix well. This decomposes the diazo salt and stops the reaction.
3. To the Blank tubes, add 0.25 ml sulfanilic acid and 25 μl ascorbic acid.
4. Pipet 0.5 ml alkaline tartrate into each tube, mix well, and read absorbance at 600 nm against water as a reference as described for total bilirubin.
5. *Calculation:*

Esterified bilirubin, mg/dl $= \Delta A_u / \Delta A_s \times C$

Nonesterified Bilirubin = Total − Esterified

Notes:

1. When performing bilirubin tests on patients with hemolytic disease of the newborn, it is advisable to use a control serum or standard with a concentration close to 20 mg/dl because this is near the critical concentration at which decisions are made concerning exchange transfusions.
2. Bilirubin is altered by ultraviolet light in such a way that it no longer is able to form azobilirubin. It is important to protect serum samples and standards from direct sunlight or ultraviolet light.

REFERENCE VALUES

	Age	Term	Premature
Newborns:	0–24 hr	up to 6 mg/dl	up to 6 mg/dl
(See adult values for > 1 month)	24–48 hr	6–7 mg/dl	6–8 mg/dl
	3–5 days	4–6 mg/dl	10–12 mg/dl
Adults:	Total: up to 1.1 mg/dl (19 μmol/L) Esterified: up to 0.3 mg/dl (5 μmol/L)		

The elevated bilirubin concentration in the normal newborn and in those with hemolytic disease of the newborn is in the nonesterified fraction.

INCREASED CONCENTRATION

Total bilirubin. Total bilirubin is increased mildly in chronic hemolytic disease (below 5 mg/dl), moderately to severely (10 to 30 mg/dl) in hepatocellular disease, and markedly in cholestasis (internal or external obstruction to bile flow, where the concentration could vary from 10 to 50 or 60 mg/dl).

Esterified bilirubin. This fraction is increased in both hepatocellular disease and cholestasis.

Nonesterified bilirubin. The increase in bilirubin concentration in hemolytic disease, including HDN, is almost entirely in this fraction. It also accounts for about 30 to 50% of the bilirubin rise in hepatocellular disease or cholestasis.

Decreased concentration. Of no clinical significance.

Automated Determination

REFERENCE. For continuous flow methodology, the method of Jendrassik and Grof[3], as modified by Gambino, S.R., and Schreiber, H., Technicon Symposium paper #54, 1964.

Principle. Total bilirubin is converted to azobilirubin with diazotized sulfanilic acid, using an acetate solution of caffeine and sodium benzoate as an accelerator. Reagent details and the flow diagram may be obtained from the Technicon Instrument Corp. as method #N85 I/II. Gambino[4] has also described the Jendrassik and Grof procedure as a manual method.

Direct Spectrophotometric Determination

Total bilirubin in the serum of newborns may be estimated directly from the spectral absorbance of bilirubin at 455 nm. This test cannot be done with adult serum because carotene or other pigments that absorb at this wavelength may be present. These pigments are not present in the sera of newborns but hemolysis gives false elevations in the bilirubin concentration because oxyhemoglobin also absorbs some light at 455 nm. The effect of hemolysis can be avoided by measuring the absorbance at 455 and 575 nm and correcting for hemoglobin as described below.

REFERENCE. White, D., Haidar, G.A., and Reinhold, J.G.: Spectrophotometric measurement of bilirubin concentrations in the serum of the newborn by use of a microcapillary method. Clin. Chem. 4, 211, 1958.

PRINCIPLE. The absorbance of a diluted serum is measured at two wavelengths and a correction is made for the presence of oxyhemoglobin if any should be present. The first wavelength, 455 nm, is approximately the point of maximum absorbance of bilirubin but also is a measure of the hemoglobin concentration. At 575 nm, the second wavelength, the absorbance of hemoglobin is essentially the same as at 455 nm, while bilirubin has zero absorbance. The correction for hemoglobin is made by subtracting A_{575} from A_{455}.

REAGENTS AND MATERIALS

1. A narrow band-pass spectrophotometer.
2. Phosphate buffer, 0.067 mol/L, pH 7.4. Dissolve 7.65 g disodium hydrogen phosphate, Na_2HPO_4, and 1.74 g monobasic potassium phosphate, KH_2PO_4, in water and make up to 1 L. Check pH and adjust to 7.4 if necessary.
3. Bilirubin Standard, 10 and 20.0 mg/dl. Prepare as indicated on page 232.

PROCEDURE

1. Pipet 20 μl of patient serums and standards, respectively, into 1.0 ml phosphate buffer in cuvets.
2. Mix and read absorbances at 455 nm and 575 nm against phosphate buffer.

CALCULATION

$$\text{Bilirubin, mg/dl} = \frac{(A_{455} - A_{575}) \text{ Unknown}}{(A_{455} - A_{575}) \text{ Standard}} \times \text{Concentration Standard}$$

In S.I., bilirubin, μmol/L = mg/dl $\times$ 17.1

Some bilirubinometers available commercially (Advanced Instruments; American Optical) are dedicated to the measurement of serum bilirubin at two different wavelengths simultaneously and automatically make the correction for the second wavelength, giving a direct readout of the corrected bilirubin concentration.

Urine Bilirubin

Bilirubin is not detectable by conventional methods in the urine of a normal, healthy person. When found to be present in the urine, it indicates some pathologic condition of the liver or biliary system. The bile pigment found in the urine in these conditions is bilirubin diglucuronide, the water-soluble ester of bilirubin.

BY TABLET. Bilirubin can be detected by the classic procedure of converting it into the dye, azobilirubin. This conversion may be done with the commercially available Ictotest tablets (Ames) containing a diazotized salt in an appropriate buffer. Five drops of urine are placed on an asbestos-cellulose mat. If bilirubin is present, it is adsorbed on the surface of the mat. A tablet containing the diazotized salt and a mixture of $NaHCO_3$, salicylic acid, and boric acid is placed on the wet spot. Two drops of water are flowed down the tablet to the mat. If bilirubin is present, a blue to purple color forms in the mat within 30 seconds. A red or orange color is read as a negative test because it means that a urinary compound other than bilirubin has been converted to an azo dye.

BY DIPSTICK. Bilirubin in urine may also be detected by any of the several multitest dipsticks (Ames, Bio-Dynamics). The diazotized salt in the stick is different from that of the tablet and consequently, it is of a different color. The presence of any bilirubin in the urine is abnormal.

BY OXIDATION SPOT TEST. This test is a Watson and Hawkinson modification of the Harrison test.[5] A thick filter paper is soaked in saturated barium chloride solution, dried, and cut into small strips. The lower half of a strip is inserted into a urine sample and removed. A drop of Fouchet's reagent (1 g ferric chloride dissolved in 100 ml solution of 250 g/L trichloroacetic acid) is applied at the boundary between the wet and dry areas of the strip. The presence of bilirubin is indicated by a green or blue color. Bilirubin is oxidized by Fe^{3+} to biliverdin (green) or to oxidation products of biliverdin (blue). The intensity of the color varies directly with the concentration of bilirubin present and is graded arbitrarily from zero to $4+$.

Urine Urobilinogen

Urobilinogen is a colorless compound derived from bilirubin that has been excreted in the bile and partially hydrogenated by bacteria in the intestines. It is partially reabsorbed into the portal vein, almost completely extracted from the blood by normal liver cells, and re-excreted into the bile. The small portion that is not taken up by the hepatocytes is excreted by the kidney as urobilinogen. Normally, about 1 to 4 mg/24 hr are excreted in the urine. The excretion of urinary urobilinogen is elevated in hemolytic disease (excess production from bilirubin), in hepatocellular liver disease (decreased removal by hepatocytes), and in congestive heart failure (impaired circulation to liver).

BY DIPSTICK. A dipstick (Ames; Bio-Dynamics) that is impregnated with p-dimethylaminobenzaldehyde and an acid buffer turns red in the presence of urobilinogen. Porphobilinogen reacts in a similar fashion. Comparison with a color chart will indicate whether the urobilinogen is present in normal amounts or in excess.

SEMIQUANTITATIVE DETERMINATION

REFERENCE. Henry, R.J., Jacobs, S.L., and Berkman, S.: Studies on the determination of the bile pigments. III Standardization of the determination of urobilinogen as urobilinogen-aldehyde. Clin. Chem. 7, 231, 1961.

Principle. The intensity of color produced by reacting the urobilinogen in a 2-hour urine collection (from 1 or 2 P.M. to 3 or 4 P.M.) with Ehrlich's aldehyde reagent is measured spectrophotometrically and quantitated by comparing it with a standard phenolsulfonphthalein (PSP) solution. Ascorbic acid is added as a reducing agent to stabilize the urobilinogen-aldehyde complex. A saturated sodium acetate solution is added to increase the pH and minimize the production of a colored product with indole and skatole, compounds that may be produced in the colon by bacterial action, absorbed, and excreted into the urine. The method is considered to be only semiquantitative, since pure urobilinogen cannot be obtained as a standard. PSP is used as a standard, and the absorbance of a 0.20 mg/dl solution of dye is equivalent to that of 0.346 mg/dl of urobilinogen combined as the aldehyde. The concentration is expressed in Ehrlich units, where 1 Ehrlich unit has the color equivalence of 1 mg urobilinogen.

Reagents
1. Ehrlich's Reagent. Dissolve 0.7 g p-dimethylaminobenzaldehyde in 150 ml concentrated HCl. Add 100 ml water and mix.
2. Sodium Acetate, saturated. Saturate about 800 ml water with NaAc, anhydrous or NaAc·3 H_2O. Some undissolved reagent should be present.

3. Ascorbic Acid Powder.
4. PSP Stock Standard. Dissolve 20.0 mg phenolsulfonphthalein (phenol red) in 100 ml of approximately 0.01 mol/L NaOH.
5. Working Standard. Dilute the stock standard 1:100 with 0.01 mol/L NaOH. This standard (0.2 mg/dl of PSP) has the absorbance equivalency of a 0.346 mg/dl solution of urobilinogen when reacted with Ehrlich's aldehyde.

Procedure

1. Measure the volume of the 2-hour urine sample.
2. Bilirubin interferes if more than a trace is present; therefore, test for its presence. If necessary, remove the bilirubin by adding 2 ml 100 g/L $BaCl_2$ to 8 ml urine, mix, and filter. Use the filtrate, but multiply the final result by 1.25 to correct for the dilution.
3. To 10 ml urine (or filtrate), add approximately 100 mg ascorbic acid and mix.
4. Transfer 1.5 ml aliquots to "Blank" and "Unknown" tubes, respectively.
5. To the "Unknown" tube, add 1.5 ml Ehrlich's aldehyde reagent, mix, and quickly add 3 ml saturated NaAc solution.
6. For the blank, first mix 1 volume of Ehrlich's aldehyde reagent with 2 volumes of saturated NaAc solution. Then transfer 4.5 ml of the mixture to the blank tube. By raising the pH in this manner, the production of a pink color by indole or skatole is minimized.
7. Read the absorbances of "Unknown" and "Blank" against water at 562 nm, within 5 minutes. Measure the absorbance of working standard against water.
8. Calculation:

$$\text{Ehrlich U/dl} = \frac{A_u - A_b}{A_s} \times 0.346 \times \frac{6.0}{1.5} = \frac{A_u - A_b}{A_s} \times 1.38$$

$$\text{Ehrlich U/2 hr.} = \frac{A_u - A_b}{A_s} \times 1.38 \times \frac{\text{vol in ml}}{100}$$

Reference Values. The usual amount excreted is 0.1 to 1.0 Ehrlich units/ 2 hr.

Fecal Urobilinogen

Fecal urobilinogen is decreased in hepatocellular disease as well as in obstructions of the biliary tree. Visual inspection of the feces usually suffices to detect decreased urobilinogen because the stools become pale or clay-colored with decreasing amounts of pigment. The original source of the brownish fecal pigments is the bilirubin that undergoes bacterial action in the gut.

Dye Tests of Excretory Function

In addition to the excretion of bilirubin, the liver is capable of excreting various dyes or drugs by the same excretory pathway. Bromsulphophthalein (BSP) is one such dye used as an excretory test. It has been replaced by enzyme assays that provide better clinical information without the risk because BSP injection is a potential danger for some patients; it has caused anaphylactic shock in allergic patients, and the dye solution may produce tissue necrosis if it does not enter the vein. Indocyanine green is a less allergenic dye to use for a pigment excretory test if one should be necessary. It is useless to perform such a test on a *jaundiced* patient, however; the dye excretion will be abnormally low if the patient is unable to excrete bilirubin adequately.

Tests Based upon Abnormalities of Serum Proteins

The liver is the sole source of synthesis of most of the proteins in plasma except for the gamma globulins, which are produced in lymphocytes and plasma cells. The only protein analyses that provide useful information concerning liver disease, however, are those of albumin, total globulins, the globulin fractions as obtained by electrophoresis, the clotting factors measured by the one-stage prothrombin time, and α-fetoprotein, which is usually elevated in primary hepatic carcinoma. Knowing the concentration of *total* serum proteins is of little diagnostic help because there is no characteristic or predictable change. The proteins synthesized in the liver are usually decreased in hepatocellular disease, but the immunoglobulins are increased in viral hepatitis and in chronic liver infections. Short-term obstruction of the biliary system usually does not affect the protein concentration.

Thus, depending upon the type and duration of hepatic disease, the serum total protein concentration may be normal, increased, or decreased. The determination of serum total protein is carried out, however, because this value is needed for the calculation of the globulin concentration when albumin is measured directly (p. 159). Fractionation by electrophoresis also requires the estimation of total protein in order to express all concentrations as g/dl or g/L.

ALBUMIN. Albumin has a half-life in the body of approximately 17 days, which means that its concentration in plasma would decrease at the rate of about 3% per day if hepatic synthesis of albumin were completely stopped. Since complete stoppage of synthesis is seldom the case, the drop in albumin concentration proceeds somewhat slowly in hepatocellular disease. Therefore, it is a good test for following chronic hepatocellular damage, but is not suited for early detection of hepatic dysfunction.

See p. 159 for the method for measuring the concentration of serum albumin. The normal concentration of serum albumin varies from 3.5 to 5.2 g/dl. In parenchymal liver disease (damage to hepatocytes), the decrease in the serum albumin concentration varies directly with the *degree and duration* of the disease. The prognosis is poor when the hypoalbuminemia is severe and prolonged. A hypoalbuminemia is a common finding in liver cirrhosis. The serum albumin concentration is usually not affected by obstruction of the biliary system unless the condition has been prolonged so that parenchymal damage also occurs.

TOTAL GLOBULINS. The immunoglobulins (γ-globulins) in serum are increased in infections and particularly in some viral diseases, such as infectious hepatitis. The concentration of total globulins increases when there is a rise in the γ-globulins because the increase in the latter more than compensates for a possible fall in the α₁-globulins. It is far better, however, to have a complete protein fractionation and see what is happening to the γ-globulin fraction than to rely on the total globulin concentration.

GAMMA GLOBULINS. Refer to page 166 for a discussion of the polyclonal (γG, γA, and γM) rise in the γ-globulins that occurs in viral hepatitis and chronic infections. A typical electrophoretic pattern obtained in this disease is shown in Figure 5.2. Normal γ-globulin concentrations vary from 0.7 to 1.5 g/dl, but these may triple in a severe viral hepatitis. The γ-globulin values are also high in the disease sarcoidosis, when this is accompanied by granulomas in the liver, and in liver cirrhosis. The protein electrophoresis pattern in cirrhosis usually shows a bridging between the β- and γ-globulin bands.

ALPHA GLOBULINS. Serum α_1-globulins are usually decreased along with albumin in parenchymal liver disease, but this decrease is not of great diagnostic value. Changes in α_2- and β-globulins are also not significant.

CLOTTING FACTORS. Various proteins that participate in blood coagulation are synthesized in the liver: fibrinogen, prothrombin (Factor II), and Factors V, VII, IX, and X. Vitamin K is essential for the hepatic synthesis of prothrombin and Factors VII, IX, and X. The one-stage test of the plasma prothrombin time requires the presence of prothrombin plus Factors V, VII, and X. If any one of these factors is below a critical concentration, the prothrombin time is prolonged. The plasma half-life of Factor VII is short, being less than 6 hours, so the plasma concentration of this factor is rapidly depleted when there is damage to hepatocytes.

The measurement of the plasma prothrombin time is usually performed in the hematology laboratory or in a coagulation division, if one exists, so details of measurement are considered beyond the scope of this book. For those who may be interested in the performance of this test, see reference #6.

Interpretation. An increased one stage, prothrombin time indicates the failure of hepatic synthesis of one or more of the aforementioned clotting factors. This could be caused either by a deficiency of vitamin K, a fat-soluble vitamin that requires the presence of bile salts for its absorption, or by damage to parenchymal cells. To resolve the dilemma, a water-soluble preparation of vitamin K is injected into the patient, and the test for prothrombin time is repeated 24 hours later. The prothrombin time will be restored to normal if the fault was the failure of bile salts to reach the intestine (cholestasis or obstruction). In the case of parenchymal damage, the prothrombin time will still be prolonged despite the injection of vitamin K. This test helps to differentiate the causes of jaundice. As with all tests of liver function, there is an overlapping of results when there are elements of parenchymal damage and cholestasis in the same patient.

Selected Enzyme Tests

A discussion of the general use of enzyme tests in the diagnosis and management of various diseases appears in Chapter 6. Only those useful in liver disease will be considered in this chapter (see Table 7.2).

The measurement of the serum activities of a number of enzymes is helpful as an adjunct in the diagnosis and management of patients with suspected liver disease. Numerous enzyme tests have been proposed for this purpose, but with many there is a great deal of overlap in the information they provide. A rise in the serum activities of some enzymes may be characteristic for some types of injury to the hepatobiliary system, but others may be much less specific because of their wide distribution in other tissues. In the interest of utility and economy, one has to select those tests that are most suitable for a particular

TABLE 7.2

Upper Limits of Normal (ULN) of Various Serum Enzymes Providing Information
in Hepatobiliary Disease

ENZYME		CONVENTIONAL ABBREVIATION	UPPER LIMIT OF NORMAL (ULN) IN U/L
COMMON NAME	INTERNATIONAL NAME		
Alkaline phosphatase	Same	ALP	Adult 105
Glutamic-oxaloacetic transaminase	Aspartate aminotransferase	AST (GOT)*	25
Glutamic-pyruvic transaminase	Alanine aminotransferase	ALT (GPT)*	30
γ-glutamyl transpeptidase	γ-glutamyl transferase	GGT	Male 35 Female 30
Lactate dehydrogenase	Same	LD (LDH)*	290
5′-nucleotidase	Same	NTP	7
Ornithine carbamoyl transferase	Same	OCT	6

*Old abbreviation

diagnostic or management situation and must correlate the results with those obtained with other tests of hepatic function.

The common enzyme tests that are available in most laboratories for assisting in the diagnosis and management of liver disease are alkaline phosphatase (ALP), aspartate aminotransferase (AST or GOT), and alanine aminotransferase (ALT or GPT). In special cases, larger laboratories may also measure the activities of 5′-nucleotidase (NTP), gamma glutamyl transferase (GGT), ornithine carbamoyl transferase (OCT), and the isoenzymes of lactate dehydrogenase (LD) or alkaline phosphatase. The latter group of enzymes, which are measured only in larger laboratory centers, will be discussed along with the more common enzyme tests, but only a reference is given to methodology instead of procedural details (p. 243); the procedures for the common enzyme tests are described in Chapter 6. Many other enzyme tests will not be mentioned because they offer nothing unique that at least one of the above-mentioned enzyme tests does not provide.

SERUM ALKALINE PHOSPHATASE (ALP). The determination of serum alkaline phosphatase activity was the first enzyme test to be employed in the differential diagnosis of jaundice. It is still a useful procedure when carried out in conjunction with other hepatic tests, despite the proliferation in the number of other enzyme tests available.[7-9]

Reference Values. The normal range of alkaline phosphatase activity in serum as performed by the method described on page 203 is 20 to 105 U/L for adults. Refer to page 204 for values in different age groups.

Increased Activity. The enzyme ALP is widely distributed in many tissues, including the osteoblasts (the bone-building cells in bone), cells lining the sinusoids, and bile canaliculi in the liver. Thus, one always has to exclude the possibility of osteoblastic activity when utilizing the ALP test in liver disease. The activity of serum ALP rises both in hepatocellular disease and cholestasis, but the rise is usually greater in the latter condition.

The increase in ALP activity is usually great in space-occupying lesions of the liver, such as carcinoma, amebic abscess, amyloidosis, and granulomatous lesions (sarcoidosis, tuberculosis of the liver). Thus, in about 75% of patients with a primary hepatitis, the rise in serum ALP activity is less than 2.5 × the upper limit of normal (ULN). In contrast to this, the increase in ALP activity exceeds 2.5 × ULN in about 75% of the patients with obstruction of the bile duct. The diagnostic value of the test increases when taken in conjunction with a serum protein fractionation and other tests.

AMINOTRANSFERASES, AST (GOT) AND ALT (GPT). The reversible reactions of the two common aminotransferases, or transaminases, as they were formerly called, are described on pages 192 and 195. The two aminotransferases of clinical importance are aspartate aminotransferase, (AST) and alanine aminotransferase (ALT). The liver is a rich source of both of these aminotransferases, and their measurements have been of value in the diagnosis of liver disorders.[7-9]

Reference Values. The normal values for serum AST and ALT as performed by the methods described on pages 192 to 196 are 6 to 25 U/L and 3 to 30 U/L, respectively (Table 7.2).

Increased Activity. The transaminases are present in normal serum at a low concentration as manifested by a low serum activity, but when tissue cells containing large amounts of these enzymes are injured or killed, the enzymes diffuse into the bloodstream where a temporary high degree of enzyme activity occurs. The degree of activity depends, of course, upon the extent of the tissue damage, the prior concentration of the enzyme in the tissue, and the time course following the tissue injury.

Because the serum levels of both aminotransferases tend to rise or fall more or less together with hepatic cell damage, no real purpose is served in ordering both tests. The rise in serum ALT activity in acute hepatic cell injury is usually a little greater than that of AST, but the determination of either test would provide the diagnostic information in most cases.

Normal values for the aspartate aminotransferase and values encountered in various liver diseases are shown in Table 7.3. In general, the serum aminotransferase activity may rise to 100 × ULN in a case of severe viral hepatitis (Fig. 7.5) or of acute toxicity with a drug that severely damages liver tissue (e.g., ingestion of chloroform or carbon tetrachloride, phosphorus compounds). The result depends, of course, upon obtaining the blood specimen at the peak of the diffusion of the enzyme from severely damaged hepatic cells into the bloodstream. Since the laboratory and the physician have no control over the time when a patient may seek medical aid, the peak occurrence of the aminotransferase activity in serum may have passed by the time the specimen is obtained, although the activity may still be greatly elevated. Each enzyme has

TABLE 7.3
Typical Range of AST in Various Liver Diseases

DISEASE STATE	RANGE OF AST ACTIVITY × UPPER LIMIT OF NORMAL (ULN)
Normal adults	0.3–1.0
Viral hepatitis—nonicteric phase	1.5–8
Viral hepatitis—acute stage	12–100
Acute poisoning (CHCl$_3$ or CCl$_4$)	12–125
Cholestasis, extrahepatic or intrahepatic	1–8
Primary or metastatic hepatic carcinoma	1–8
Alcoholic cirrhosis	1.3–6

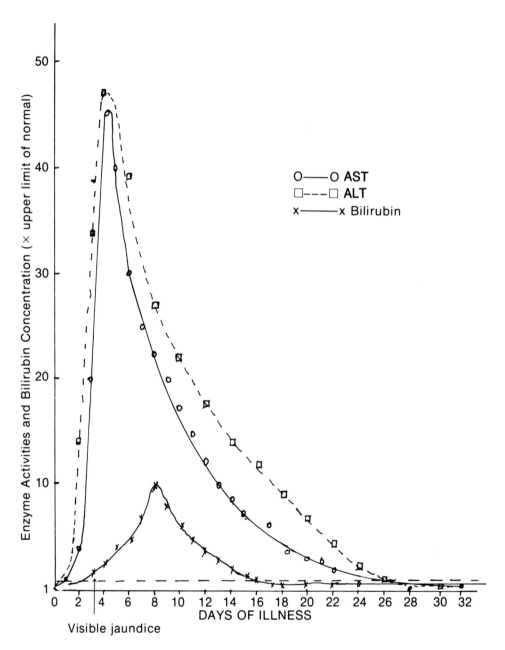

Fig. 7.5. Example of bilirubin concentrations and serum aspartate (AST) and alanine (ALT) aminotransferase activities in serum of patient with acute case of viral hepatitis. Note that the rise in transaminase activities precedes the jaundice.

its own particular time course in serum after tissue injury. The rise in serum aminotransferase activity in viral hepatitis begins early in the disease, frequently before jaundice is visible, and reaches a maximum during the acute stage of the disease when destruction of hepatic cells is at its height. The peak rise in aminotransferase following an acute chemical toxicity, such as after the ingestion of carbon tetrachloride in a suicide attempt, may occur within 24 hours and decline rapidly by the second day. The time factor must be taken into consideration in any interpretation.

Moderate elevations (1 to 8 or 9 × ULN) of serum aminotransferase activity usually occur in cholestasis, cirrhosis, hepatic tumors, infections, and infectious mononucleosis. This enzyme test is of value in the differentiation of these conditions from those causing acute hepatocellular injury. Some of the usual findings are summarized in Table 7.3.

Less Common Enzyme Tests for Special Situations

Most cases of liver disease can be diagnosed by the judicious application of the common enzyme tests used in conjunction with the other tests of hepatic function, the patient's history, physical examination, and other studies, such as gallbladder visualization when appropriate, liver biopsy, or a radioactive liver scan. There may always be a few situations, however, in which the diagnosis may still be in doubt. The special enzyme tests described in this section are those which may be helpful in specific situations.

γ-GLUTAMYL TRANSFERASE (GGT). This enzyme, also referred to as γ-glutamyl transpeptidase (GGTP), transfers the γ-glutamyl portion of a peptide to another peptide, amino acid, or water. Glutathione is the common substrate of this enzyme in the body; the enzyme participates in amino acid transport by transferring the γ-glutamyl portion of glutathione to other amino acids, enabling them to cross cell membranes more readily. The enzyme is present in relatively high concentration in kidney, pancreas, liver, and prostate. GGT is a sensitive indicator of liver disease.[9,10]

Reference Values. The normal values as performed by the method of Rosalki and Tarlow[15] are 3 to 35 U/L for males and 3 to 30 U/L for females.

Increased Activity. Serum GGT activity is usually elevated in both hepatocellular and obstructive liver diseases. The enzyme is not absolutely specific for disorders of the hepatobiliary tree, but its specificity surpasses that of the other enzymes commonly used in the diagnosis of hepatic disorders. Thus, the serum activity of GGT is not elevated in bone disease, as is ALP, nor in muscle disease or hemolytic conditions, as is AST (GOT). The estimation of serum GGT is most helpful in the following situations:

1. *In detecting hepatic injury caused by alcoholism:* Serum GGT is considered to be the enzyme of choice when investigating possible alcoholism.
2. *Hepatic metastasis in the anicteric patient:* Serum GGT frequently is elevated in hepatic metastasis, but since its activity may also rise in the presence of other malignancies, this test is useful in conjunction with tests of ALP or NTP, which are less sensitive but a little more specific.[17]
3. *In the management of patients with infectious hepatitis:* If GGT is monitored serially during the course of the disease, a return of serum GGT activity to normal is an excellent prognosis; a later rise in GGT activity indicates an exacerbation of the disease. GGT measurement is of greater

value in this situation than either AST or ALT, although the estimation of NTP may be equally good.

4. *In chronic obstruction of the bile ducts:* Usually an elevated serum ALP activity provides this information, but if there is any question of a concomitant bone disease, which also elevates serum ALP, the determination of serum GGT or NTP will resolve the problem, since these enzyme tests are more specific for liver disease and are not affected by the proliferation of osteoblasts.

5'-NUCLEOTIDASE (NTP). The enzyme 5'-nucleotidase (NTP) specifically hydrolyzes the phosphate from 5'-adenosine mononucleotide (AMP). It is present as a microsomal enzyme in liver as well as in other tissues, and its activity in serum is increased in hepatobiliary disease. Bodansky and Schwartz presented a review of the enzyme.[16]

Reference Values. Normal values for tests performed by the method of Clayson, Quast, and Strandjord[11] range from 0.4 to 7.0 U/L.

Increased Activity. The serum activity of NTP is increased in various types of hepatic disorders. Its measurement is of value in the following situations:

1. *In ascertaining whether an increase in ALP is caused by osteoblastic activity or liver disease:* Skeletal disorders usually do not cause an increase in serum NTP, but liver diseases of various types do.[12] This type of differentiation is frequently necessary in growing children who usually have an elevated ALP.

2. *In the diagnosis of hepatic metastases in the anicteric patient:* Serum GGT, NTP, and ALP are usually elevated in this situation. The GGT enzyme test is the most sensitive, but the NTP test would be confirmatory.

3. *In the management of patients with infectious hepatitis:* As mentioned under the GGT test, the activities of both GGT and NTP rise in infectious hepatitis and fall with recovery; a secondary rise indicates an exacerbation. Tests of GGT and NTP are sensitive and give the same information.

ORNITHINE CARBAMOYL TRANSFERASE (OCT). The enzyme OCT transfers a carbamoyl group from carbamoyl phosphate to ornithine as an early step leading to urea formation. This reaction takes place in the liver where the enzyme is located. Aside from low concentrations of OCT in the small intestine, the enzyme is virtually absent from all other tissues. This fact insures that a rise in serum OCT activity is highly specific for hepatic injury.

Reference Values. The normal range of serum OCT activity when performed by the method of Strandjord and Clayson[13] is 0 to 6 U/L.

Increased Activity. The determination of serum OCT activity is the method of choice for the diagnosis of acute, intermittent obstruction of the common bile duct. It is the most sensitive as well as the most specific test. The test can also be used to differentiate between hepatocellular injury and injury to other organs, since the enzyme is scarcely detectable in any organ except liver.[7,12]

ISOENZYMES. As described on page 199, some enzymes exist in multimolecular forms that can be separated from each other by electrophoresis. These isoenzymes can be quantitated by incubating the electrophoretic strip in appropriate substrate mixtures.

Since the isoenzyme patterns for a particular enzyme may vary from organ to organ, the serum isoenzyme pattern may reveal damage to a specific organ or help to differentiate between possible injury to one of two organs under consideration. Thus, the determination of the serum isoenzymes of alkaline phosphatase may aid in differentiating between liver or bone disease as the primary reason for the presence of an elevated serum alkaline phosphatase activity. The isoenzyme of alkaline phosphatase derived from bone migrates in an electrical field somewhat more slowly than the isoenzyme found in the liver.

In a similar manner, the determination of the isoenzymes of lactate dehydrogenase (p. 199) may help to differentiate between a viral hepatitis and hepatocellular injury secondary to infectious mononucleosis. When the activity units of LD_2 are plotted against the activity units of LD_5, the isoenzyme that predominates in liver, it is found that in viral hepatitis the activity of LD_5 rises rapidly, while that of LD_2 (in high concentration in heart and some other tissues) is scarcely elevated. In infectious mononucleosis with hepatic damage, the activity of both isoenzymes is increased, so that it is possible to separate the patients with mononucleosis from those with hepatitis.[14]

REFERENCES

1. Popper, H., and Schaffner, F.: Liver: Structure and Function. New York, McGraw-Hill, 1957, p. 7.
2. Doumas, B.T., Perry, B.W., Sasse, E.A., and Straumfjord, J.V., Jr.: Standardization in bilirubin assays: Evaluation of selected methods and stability of bilirubin solutions. Clin. Chem. *19*, 984, 1973.
3. Jendrassik, L., and Grof, P.: Vereinfachte photometrische Methoden zur Bestimmung des Blutbilirubins. Biochem. Zeit. *297*, 81, 1938.
4. Gambino, S.R.: Bilirubin (Modified Jendrassik and Grof)—Provisional. *In* Standard Methods of Clinical Chemistry, Vol. 5, edited by S. Meites. New York, Academic Press, 1965, p. 55.
5. Watson, C.J., and Hawkinson, V.: Semiquantitative estimation of bilirubin in the urine by means of the barium-strip modification of Harrison's test. J. Lab. Clin. Med. *31*, 914, 1946.
6. Miale, J.B.: Laboratory Medicine: Hematology, 4th ed. St. Louis, C.V. Mosby, 1972, p. 1271.
7. Coodley, E.L. (Editor): Diagnostic Enzymology. Philadelphia, Lea & Febiger, 1970.
8. Batsakis, J.G., and Briere, R.O.: Interpretive Enzymology. Springfield, IL, Charles C Thomas, 1967.
9. Wilkinson, J.H.: The Principles and Practices of Diagnostic Enzymology. Chicago, Year Book Medical Publishers, 1976.
10. Rosalki, S.B.: Gamma-glutamyl transpeptidase. Adv. Clin. Chem. *17*, 53, 1975.
11. Clayson, K.J., Quast, N.M., and Strandjord, P.E.: Studies of optimal conditions for human serum and aorta 5'-nucleotidase activity. Chemistry Division, Department of Laboratory Medicine, University of Minnesota, 1969.
12. Strandjord, P.E., and Clayson, K.J.: Clinical enzymology in the evaluation of heart and liver disease. *In* Enzymology in the Practice of Laboratory Medicine, edited by P. Blume and E.F. Freier. New York, Academic Press, 1974, p. 381.
13. Clayson, K.J., Fine, J.S., and Strandjord, P.E.: A more sensitive automated method for determination of ornithine carbamoyltransferase activity in human serum. Clin. Chem. *21*, 754, 1975.
14. Meeker, D., Clayson, K.J., and Strandjord, P.E.: Differential diagnosis of acute hepatocellular injury in infectious mononucleosis. Clin. Res. *21*:519, 1973.
15. Rosalki, S.B., and Tarlow, D.: Optimized determination of γ-glutamyl transferase by reaction-rate analysis. Clin. Chem. *20*, 1121, 1974.
16. Bodansky, O., and Schwartz, M.K.: 5'-nucleotidase. *In* Advances in Clinical Chemistry, Vol. 11, edited by O. Bodansky and C.P. Stewart. New York, Academic Press, 1968, p. 278.
17. Kim, N.K., Yasmineh, W.G., Freier, E.F., Goldman, A.I., and Theologides, A.: Value of alkaline phosphatase, 5'-nucleotidase, γ-glutamyltransferase and glutamate dehydrogenase activity measurements (single and combined) in serum in diagnosis of metastasis to the liver. Clin. Chem. *23*, 2034, 1977.
18. Malloy, H.T., and Evelyn, K.A.: The determination of bilirubin with the photoelectric colorimeter. J. Biol. Chem. *119*, 481, 1937.

Chapter 8

IMMUNOCHEMICAL TECHNIQUES

A discussion of immunochemical techniques must of necessity introduce some new terms that require definition for proper understanding. The most important new terms are the following:

Immunochemical

1. *Antibody:* An immunoglobulin (p. 154) produced by the plasma cells of a host in response to the introduction of a foreign macromolecule, usually a protein. It has the capacity to react with (bind) the particular molecule that stimulated its production or with certain groups on the surface of that molecule.
2. *Antigen:* A molecule that is able to react with an antibody. Proteins of molecular weight greater than 4000 usually induce the formation of antibodies when injected into a host of a different species, but antigens of lower molecular weight do not possess this property unless bound to a protein.
3. *Hapten:* A small molecule with the capacity to bind to a specific antibody. In this respect, it is an antigen, but it cannot induce the formation of antibodies when injected into a host *unless it is covalently bound to a foreign protein.* A hapten is *not* an immunogen.
4. *Immunogen:* A molecule that is able to induce antibody formation when injected into a host.
5. *Label:* A distinctive substance (element, compound, or reactive group) attached to a molecule that readily distinguishes it from neighboring molecules by some physical or chemical characteristic. The label may be a radionuclide, an enzyme, a fluorophor, a chromogen, or some other group.

Radiochemical

1. *Beta particle:* A high-speed electron; one of three forms of radioactivity.
2. *Gamma ray:* X-ray of short wavelength (<0.1 nm), high-energy electromagnetic radiation; one of three forms of radioactivity.
3. *Isotope:* One of two or more atoms having the same atomic number but different atomic weights. Isotopes have identical chemical properties but slightly different physical properties.
4. *Radioactivity:* The emission of alpha, beta or gamma rays as an unstable (radioactive) chemical element decays; it is converted into another element in the process.
5. *Radionuclide:* A natural or artificially created isotope of a chemical element having an unstable nucleus that decays, emitting alpha, beta, or gamma rays.
6. *Radiolabel:* The binding of a radioactive isotope to a compound.

7. *Scintillation fluid:* A liquid that emits light when exposed to beta particles. The radioactivity of beta emitters is counted by submersing the emitter in scintillation fluid and measuring the light emitted by the scintillation fluid.

8. *Specific activity:* Radioactivity per unit weight of labeled compound.

9. *Counting:* Recording the disintegrations per minute of the radioactive elements. A gamma counter is used for those emitting gamma rays and a scintillation counter for beta emitters.

Chemical

1. *Chromogen:* A substance having the capacity to become colored in a particular environment, such as pH or redox potential.

2. *Fluorophor:* A compound that fluoresces (emits light) when irradiated with light of the proper wavelength.

As knowledge of the different plasma proteins increased, it became clinically important to quantitate specific proteins, when possible; this was particularly true for the classes of immunoglobulins. Chemical methods were inadequate for clinical purposes because of the large amount of plasma required and the long, laborious separation process. Progress was made by turning to immunochemical techniques.

When a foreign protein is injected into an animal, the plasma cells (B lymphocytes) react to this stimulus (antigen) by synthesizing an antibody that is complementary to some reactant groups on the surface of the antigen. There is a good "fit" as the reactant groups of the antigen are bound by the antibody (Fig. 8.1), similar to that of a key fitting a lock; antibodies

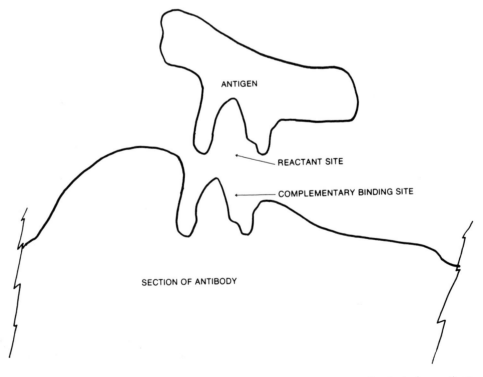

Fig. 8.1. Schematic illustration of an antibody binding site to a reactant group (hapten) of an antigen.

usually do not react with other antigens unless these antigens are closely related in structure. Antibodies are immunoglobulins and are specific in recognizing and binding a particular antigen. Small molecules (under 4000 daltons) by themselves are unable to generate antibodies when injected into animals but consistently do so when covalently bound to a foreign protein. The reactant group of such an immunogen is a hapten. Haptens do combine readily with the specific antibody. Antibodies to drugs, steroid and thyroid hormones, chemotherapeutic agents, and other biologically interesting compounds are produced by using these small molecules as haptens. Antibodies have such a high degree of specificity in recognizing and binding particular haptens or antigens in a complex mixture of serum proteins that they have become an invaluable analytical tool in the clinical laboratory. With modern techniques, it is possible to produce antibodies against practically any molecule of medical interest, from macromolecules (proteins and polypeptides) to such small molecules as drugs, amino acid derivatives, or anti-cancer chemicals.

Six immunochemical techniques are described in this chapter: immunodiffusion; immunoelectrophoresis; precipitin measurement by nephelometry; radioimmunoassay and competitive protein binding; enzyme immunoassay; and fluorescent immunoassay. The first two are used primarily for identification purposes and are qualitative or semiquantitative in nature. The latter four techniques are employed for quantitative analysis of many compounds of medical or biological interest. They are in the forefront of a rapidly expanding field of clinical chemistry.

IMMUNODIFFUSION

Double Diffusion in Agar

Antibody in solution interacts with its antigen quite rapidly, but the reaction complex is soluble at first. With the passage of time, the antigen-antibody complex continues to aggregate until visible particles (precipitin) are formed. Most antibodies are divalent (have two separate combining sites for antigen). Aggregation occurs as the divalent antibodies bridge separate antigens and form a lattice. The reaction is quite complex and is affected by the pH, temperature, ionic strength of the solution, time, and the ratio of antibody to antigen; a critical ratio is necessary for the formation of a precipitin. When either antigen or antibody is present in excess, no precipitation occurs.

Ouchterlony[1] initiated a technique for the identification of particular proteins in the complex mixture found in serum. He prepared agar gels about 3 mm thick in Petri dishes with well patterns of various types and designs. Antibody is usually placed in a central well, and antigens or antigen and serum are placed in the other wells. Since both antigens and antibodies diffuse through the gel until they finally meet, the technique is called "double diffusion." It is a qualitative procedure that identifies proteins.

Typical diffusion patterns and precipitation lines are illustrated in Figure 8.2. When antibody to an antigen is placed in the lower, central well (Fig. 8.2A) and its antigen is placed in each of the two upper wells, antigen and antibody diffuse out of their respective wells in concentric circles. In 1 or 2 days, the antibody meets the antigen and combines with it. Diffusion continues and the

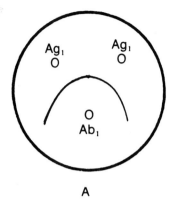
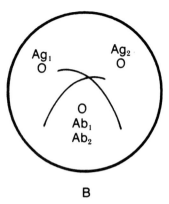

A   B

Fig. 8.2. Illustration of double diffusion in agar. A, The same antigen, Ag_1, is placed in each of the upper wells, and the antibody to Ag_1 is placed in the central well. The precipitation pattern is a smooth arc, a line of identity. B, This demonstrates the pattern of nonidentity, which develops when nonrelated antigens, Ag_1 and Ag_2, are placed in their respective wells and antibodies to each are placed in the central well.

process of aggregation of the antigen-antibody complex begins. A line of precipitation of the complex (precipitin) forms in an arc as the favorable ratio of antibody to antigen builds up. A line of precipitation can be seen when reading the plates against a black background with side lights. A continuous arc of precipitation is formed, as illustrated in Figure 8.2A, since the antigen in the left well is identical with the antigen in the right well; this smooth arc is called "a line of identity" and signifies that the antigens in the upper two wells are identical. Figure 8.2B illustrates the type of precipitation formed when one antigen (Ag_1) is in one well, another antigen (Ag_2) is in the well adjacent to it, and antibodies to each are in the central well. The pattern of crossed precipitation lines is one of "non-identity" and signifies nonrelated antigens.

The technique of double diffusion was modified to become more micro by using thinner gels. Double diffusion techniques are widely used to identify particular proteins in serum or other body fluids. Many companies have kits available that contain everything needed for testing (gels, antibodies, antigen standards). A kit is particularly useful in identifying the abnormal immunoglobulin in monoclonal gammopathies (p. 155) or for detecting Bence Jones protein in urine and classifying the type (whether kappa or lambda chain, p. 156).

Radial Immunodiffusion (RID)

Radial immunodiffusion is a single diffusion technique whereby the antigen diffuses in a gel containing antibody. Individual proteins or protein classes in serum may be quantitated by incorporating specific antibodies or class-specific antibodies in an agar or agarose gel containing a series of small wells at spaced intervals. Some wells are filled with measured amounts (5 to 10 μl) of solutions of the purified protein at known concentration; others are filled with patient serum. The antigen diffuses out of the wells, reacts with the antibody, and forms a diffuse precipitation zone around the well. This process continues until an equilibrium is reached in 2 or 3 days. The logarithm of the antigen concentration varies with the diameter of the precipitation zone in some techniques

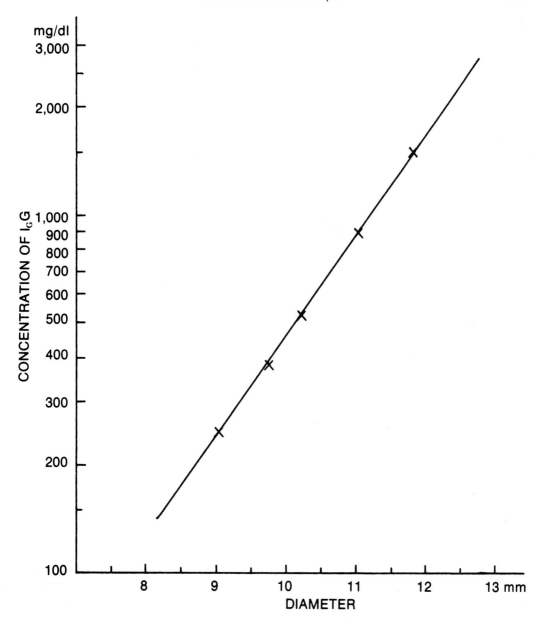

Fig. 8.3. Calibration curve for IgG by radial immunodiffusion. The IgG concentration is plotted on semilog paper against the diffusion ring diameter.

and with the square of the diameter in others involving shorter time intervals (4 to 20 hours). Berne[2] finds that antigen concentration is directly related to the diameter when equilibrium is reached (2 to 4 days). When the shorter incubation times are employed, a standard curve is prepared by plotting on semilog paper the concentration of the purified protein (on log scale) versus the diameter of the diffusion zones (Fig. 8.3). Radial immunodiffusion kits are available commercially for quantitating a number of individual proteins or classes of proteins present in serum, cerebrospinal fluid, urine, or other fluids.

The precision of RID methods is not high, so they are being supplanted by more precise immunotechniques.

IMMUNOELECTROPHORESIS (IEP)

A development that was particularly good for separating about 20 to 40 different proteins in serum at one time was immunoelectrophoresis.[3] In immunoelectrophoresis, an electrophoretic separation of the protein mixture takes place on a suitable support medium (agar, agarose, cellulose acetate) in conventional manner (see p. 161). Antibody solution is then placed in a long, narrow trough (for agar or agarose films) parallel to the direction of electrophoresis, and incubation proceeds as in the double diffusion technique. The proteins diffuse outward from their position after electrophoresis, and the antibodies diffuse outward from the trough. (When using cellulose acetate as a support medium, a long, narrow strip of cellulose acetate is wet with the antibody solution and laid on the membrane parallel to the electrophoretic direction.) The antibodies meet their specific antigens and precipitate them in an arc as shown in Figure 8.4. Over 20 identifiable precipitation lines can be identified in human serum when using an antihuman serum preparation in the trough. If a single antibody such as an anti-IgG solution is used in the trough, the only precipitation arc is that of the IgG-antibody complex.

Laurell[4] modified the immunoelectrophoretic technique by performing the electrophoresis of the serum proteins in agarose gels containing specific antibodies. As the electrophoresis continues, the proteins migrate in the electric field. The antigen reacts with the antibody and the complex precipitates further and further from the starting point as the electrophoretic migration continues until the supply of the particular antigen (protein) is exhausted (completely precipitated). The precipitation pattern is in the form of a rocket and the length of the rocket is proportional to the protein concentration.* By running a series of standards with pure protein, quantitation is simple. This technique is known as electroimmunoassay, or the rocket technique, to differentiate it from immunoelectrophoresis, which is qualitative in nature.

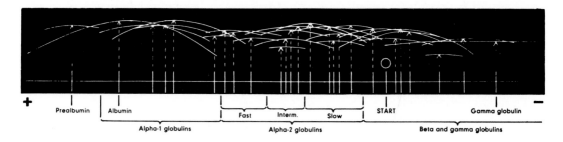

Fig. 8.4. Diagrammatic representation of an immunoelectrophoretic pattern of a normal human serum. (Courtesy of Hirschfeld, J.: Immunoelectrophoresis-procedure and application to the study of group-specific variations in sera. Science Tools, 7, 18, 1960, published by LKB-Produkter AB, Stockholm.)

---

*A rocket kit is available from Worthington Diagnostics to quantitate the following serum proteins: IgA, IgG, IgM, albumin, alpha-1-antitrypsin, C3, transferrin, and haptoglobin.

IMMUNONEPHELOMETRY

The quantitative analysis of antigen-antibody complexes is not confined to radial immunodiffusion and measurement of precipitation zones. An advance in technique was made by measuring the light scattering of the precipitin by nephelometry. When conditions are optimized, the analysis takes much less time and can be automated. The technique is compatible with automation by continuous flow, fast centrifugal analyzers, and nephelometers designed specifically for this purpose.[6,12] A laser beam increases the sensitivity of a nephelometer. The procedure is not sufficiently sensitive to measure the serum concentration of polypeptide hormones but is being used successfully for quantitating many serum proteins, such as albumin, immunoglobulins, transferrin, haptoglobin, some complement components, and many others. The review by Savory[12] covers the principles, practices, and instrumentation of immunonephelometry.

RADIOIMMUNOASSAY (RIA) AND COMPETITIVE PROTEIN BINDING (CPB)[7–10]

A great technical advance in the quantitation of individual serum or plasma proteins was made when Yalow and Berson[7] devised a radioimmunoassay for the measurement of the insulin concentration in plasma. The technique was far more sensitive and specific than any method in existence at the time. It was particularly well suited for the analysis of peptide hormones in plasma; picogram quantities $(10^{-12}$ g) of some hormones can be measured in a plasma protein mixture containing about a billion times as much of other proteins. Its specificity depends upon the well-known ability of an antibody to recognize and bind only its antigen in a mixture of many different proteins. The unique feature of radioimmunoassay was the utilization of a small amount of radiolabeled antigen to compete with the plasma or serum antigen for binding to a limited amount of antibody. The free antigen was separated from that bound to antibody and the radioactivity was counted; this provided a sensitive means of estimating the unknown antigen (non-radioactive).

The unique features of the RIA insulin assay were (1) the utilization of a specific antibody to react with insulin molecules in a mixture of serum proteins and (2) the competition between serum insulin and a trace amount of radiolabeled insulin for antibody binding sites. The quantitation was performed by counting radioactivity after separation of the antibody-bound labeled insulin from the unbound, an exquisitely sensitive method. The higher the specific activity of the labeled antigen, the more sensitive is the assay. RIA assays have been expanded to include practically all compounds of medical interest.

Principle of RIA[9]

Two items must be prepared or purchased for an RIA test: a high potency antibody against the purified antigen or hapten, and a radiolabeled antigen or hapten of high specific activity. For the latter, a radioactive element (commonly ^{125}I, but sometimes ^{14}C or ^{3}H) is covalently bound to the antigen by a gentle chemical procedure. The principles are the following:

1. A limited amount of specific antibody (sufficient to bind 40–60% of the radiolabeled antigen) is incubated with a mixture of the serum antigen plus trace amounts of radiolabeled antigen.

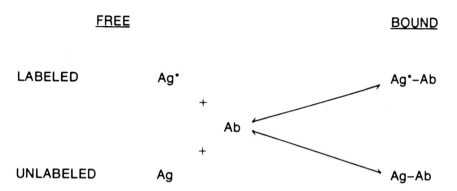

Fig. 8.5. Illustration of the competition between radioactively labeled* and unlabeled antigen for binding to a limited amount of antibody.

2. The radiolabeled and unlabeled antigens compete for binding sites on the antibody (Fig. 8.5).
3. When the equilibrium is reached, the bound antigen-antibody complex (labeled plus unlabeled) is separated from the unbound (free) antigen, and the radioactivity is counted. The ratio of antibody-bound, labeled antigen to total labeled antigen (B/T) decreases as the concentration of unlabeled antigen increases.
4. The percent bound is calculated: $\%B = B/T \times 100$.
5. A series of known antigen concentrations are run in the same batch as the unknowns and the percent bound calculated. A standard curve is prepared by plotting %B against antigen concentration; the concentrations of the unknowns are obtained from the standard curve (Fig. 8.6).

   Equilibrium in the antigen-antibody binding need not be complete since the conditions are the same for both standards and unknowns. It is essential to run a standard curve with every batch of unknowns because slight changes in conditions (temperature, pH, amount of antibody, or radioactive label) may alter the standard curve. Refer to the article by Kenny[10] for the various problems of RIA quality control.

Aspects of RIA Technique

TECHNOLOGICAL. When the RIA technique was introduced into the clinical laboratory, it was unique in its specificity and in its sensitivity. It utilized the great specificity of antibodies and combined it with the sensitivity of quantitating radioactive labels (the counting of radioactive disintegrations per minute). RIA no longer has a monopoly on the analytical use of antibodies; as described later in this chapter, enzyme immunoassay and fluorescent immunoassay also employ antibodies as specific analytical tools. The amplification power of enzymes and the sensitivity of fluorescence measurement enable the two latter techniques to reach a degree of sensitivity that is adequate for most of the assays hitherto performed by RIA.

It is theoretically possible but usually not practical for a clinical laboratory to generate its own antibodies and radiolabel the antigens with ^{125}I. It is more cost-effective to purchase the requisite materials from reliable companies with good quality control of their products.

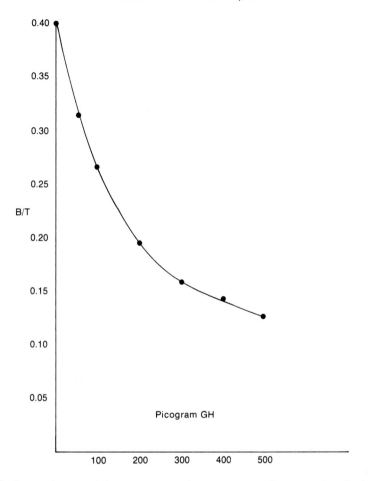

Fig. 8.6. Standard curve for growth hormone (GH). The percent Bound (%B) is plotted against picograms per ml of GH.

Many different techniques exist for separating the bound antigen-antibody complex from the free antigen. Since the antibody is a gamma-globulin, the complex may be precipitated by agents that generally aggregate globulins, such as ammonium sulfate or polyethylene glycol. The free antigen may also be adsorbed to dextran-coated charcoal or other adsorbents or separated by gel filtration. In some techniques, a second antibody, an anti-gamma-globulin from a different animal species, is added; the second antibody combines with and precipitates the first antibody, including the antibody-antigen complex, because it forms such a large lattice. In some techniques, solid phase separation is used; the antibody is combined with or adsorbed to a solid surface, such as the wall of a plastic test tube, and the free antigen is decanted from that bound by the antibody attached to the wall. Each company has its own preferred method for separation.

RIA has a temporary advantage over other immunoassay techniques in that there are more methods available for analysis because it has been in the field longer. For certain analytes, it may also be a little more sensitive. The disadvantages of RIA,[16] however, are that it poses a health hazard (exposure to radiation), presents serious radioactive waste disposal problems, requires a license, and is subjected to government regulation, involves more costly instrumentation, takes longer to perform, and is a heterogeneous system that requires the separation of bound label from the unbound; the reagents are less stable and some isotopes have a short half-life.

VARIETY OF TESTS AVAILABLE. RIA tests have been introduced for many different compounds of interest to the clinical chemistry laboratory. These include all of the protein and polypetide hormones; many individual plasma proteins (the different classes of immunoglobulins, albumin, transferrin, ferritin, haptoglobin, and many others); some proteins considered to be cancer markers (carcinoembryonic antigen, alpha fetoprotein, prostatic acid phosphatase isoenzyme, and others); Australian antigen (for hepatitis); various isoenzymes (e.g., of CK, acid phosphatase, and others); and for many low molecular weight compounds that can serve as haptens when linked to a foreign protein [steroid hormones, T_3 and T_4, drugs of abuse (opiate family, barbiturates, and others), all types of therapeutic drugs and medications, vitamins (B_{12}, folic acid, 1,25-dihydroxy-D_3), and many other biologic or therapeutic materials].

COMPETITIVE PROTEIN BINDING. The *competitive protein binding technique* is closely related in principle to the radioimmunoassay. The only difference is the use of a specific binding or carrier protein derived from tissues or plasma instead of an antibody. The technique is particularly applicable for the analysis of hormones that have specific carrier proteins or bind tightly to particular target organs. The competitive protein binding test for thyroxine (T_4) employs thyroid-binding globulin; a test for cortisol uses cortisol-binding globulin. The specific binding protein is incubated with a mixture of plasma or serum and the radiolabeled hormone. The labeled and unlabeled hormones compete for binding sites; the ratioactivity is counted after separation of the bound hormone from the free. The concentration is read from a standard curve prepared in the same manner as for RIA.

ENZYME IMMUNOASSAY (EIA)[13,15–17]

The technique of enzyme immunoassay is similar in principle to that of radioimmunoassay except that the antigen or hapten is labeled with a stable enzyme instead of a radioactive element.

In most instances, the enzyme-hapten compound is enzymatically active; the enzymatic activity is severely reduced when it becomes bound to antibody (Fig. 8.7). The loss of activity may be explained by steric hindrance (the antibody blocks the binding of substrate to enzyme) or by a change in conformation of the enzyme so that the substrate no longer "fits." Enzyme immunoassay is just as selective as radioimmunoassay because it utilizes the specificity of antibodies for recognizing their antigens or haptens; it depends upon the competition between enzyme-labeled and unlabeled antigen or hapten for binding to a limited amount of antibody. It differs from RIA in several respects[16]: (1) it eliminates the hazards of radioactivity, avoids dealing with government regulations and licensing entailed by working with radionuclides; (2) it uses conventional laboratory instruments only since quantitation is accomplished by means of a spectrophotometer; (3) the compounds are stable and the shelf life is long; (4) results are obtained in a much shorter time than in RIA, within 5 to 10 minutes compared to 1 to 5 days for RIA. The enzymatic methods are

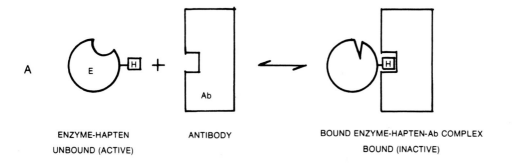

ENZYME-HAPTEN   ANTIBODY   BOUND ENZYME-HAPTEN-Ab COMPLEX

UNBOUND (ACTIVE)   BOUND (INACTIVE)

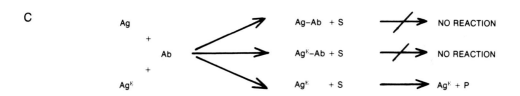

ENZYME-HAPTEN + SUBSTRATE ⇌ ENZYME-HAPTEN—SUBSTRATE ⇌ ENZYME-HAPTEN + PRODUCTS

Fig. 8.7. Illustration of the principles of EMIT. A, Change in conformation when the enzyme-hapten molecule is bound by antibody. The substrate is denied access to its binding site in the enzyme. B, Ability of the enzyme-hapten to combine with and split its substrate. C, Summation of the events that occur when antigen (hapten) and enzyme-hapten (in excess) are added to a mixture of antibody to the hapten and the specific substrate for the enzyme.

much more sensitive than the usual chemical methods because the enzymes are continually recycled to react with substrate. The sensitivity can be amplified by extending the time of incubation with the substrate.

Many different enzymes have been employed as the label.[15] The characteristics that govern the choice are stability, cost, and generation of a suitable chromogen or fluorophor that can be quantitated easily. A few of the enzymes that are commonly employed are glucose-6-phosphate dehydrogenase (generates NADH), alkaline phosphatase (splits p-nitrophenylphosphate into a colored compound), and β-galactosidase (splits substrate into a fluorescent product).

Enzyme immunoassays may be of two general types, homogeneous and heterogeneous systems, which we will describe.

Homogeneous System, Enzyme-Multiplied Immunoassay Technique (EMIT)[13]

The distinguishing characteristic of a homogeneous immunoassay system is the elimination of the necessity for separating the bound from the free hapten. The EMIT technique, employing a homogeneous system, was first devised for the quantitation of drugs and hormones in biological fluids. A specimen sample of serum or other body fluid is mixed with a solution containing antibody, enzyme-hapten complex, and buffered substrate. The mixture is incubated at 37° C in a cuvet for a short time (frequently 2 minutes) and the absorbance change is measured. Many of the enzyme labels generate NADH so the absorbance change at 340 nm is recorded. The free enzyme-antigen molecule is enzymatically active while that bound to antibody has greatly reduced activity (Fig. 8.7). The concentration of analyte (drug or hormone) is obtained from a standard curve derived from analyzing various concentrations of pure antigen in the same manner as patient samples and plotting the concentration of antigen against absorbance. A standard curve should be run with each batch of samples.

Heterogeneous System, Enzyme-Linked Immunoabsorbent (ELISA)[17]

In the ELISA system, the antibody is always bound to a solid surface (surface of test tube, resin beads, or to a membrane) but there are several variants in carrying out the test. The ELISA system has been used with great success to identify viruses and antibodies against various viruses, bacterial products, and parasites. The system is beginning to be used for quantitation of some drugs and constituents of interest in clinical chemistry (Immunotech). For drug analysis, the antibody is bound to a membrane. The hapten is linked to an enzyme, and the hapten in serum competes with the enzyme-hapten for binding to the immobilized antibody. The antibody-bound haptens must be separated from the unbound (often by decantation) and the enzymatic activity of the free hapten-enzyme determined by incubation with substrate. The concentration is obtained from a standard curve.

FLUORESCENCE IMMUNOASSAY (FIA)[14]

The principles of FIA are similar to those of RIA and enzyme immunoassay, the only difference being in the label. In FIA, the label has the capacity to fluoresce or else an enzyme-labeled hapten liberates a fluorophor upon incubation with its substrate. In all cases, the labeled hapten competes with the serum hapten for binding to a limited amount of antibody. The homogeneous systems require no separation of bound label from the unbound but the heterogeneous systems do. The concentration of hapten in both systems is also obtained from a standard curve. There is a further description of FIA on page 385, because this technique is mainly used for drug analysis at present. The EMIT system is incorporating FIA in its forthcoming generation of tests, which are almost identical to the present test system except that they will quantitate the NADH by fluorescence rather than spectrophotometry; fluorescence is more sensitive.

REFERENCES

1. Ouchterlony, O.: Handbook of immunodiffusion and immunoelectrophoresis. Ann Arbor, MI, Ann Arbor Science Publishers, 1968, p. 16.
2. Berne, B.H.: Differing methology and equations used in quantitating immunoglobulins by radial

immunodiffusion, a comparative evaluation of reported and commercial techniques. Clin. Chem. *20*, 61, 1974.

3. Grabar, P.: The immuno-electrophoretic method of analysis. *In* Immunoelectrophoretic Analysis, edited by P. Grabar and P. Burtin. Amsterdam. Elsevier Publishing Co., 1964, p. 3.
4. Laurell, C-B.: Quantitative estimation of proteins by electrophoresis in agarose gel containing antibodies. Anal. Biochem. *15*, 45, 1966.
5. Killingsworth, L.M., Buffone, G.J., Sonawane, M.B., and Lunsford, G.C.: Optimizing nephelometric measurement of specific serum proteins: Evaluation of 3 diluents. Clin. Chem. *20*, 1548, 1974.
6. Buffone, G.J., Savory, J., and Cross, R.E.: Use of a laser-equipped centrifugal analyzer for kinetic measurement of serum IgG. Clin. Chem. *20*, 1320, 1974.
7. Yalow, R.S., and Berson, S.A.: Assay of plasma insulin in human subjects by immunological methods. Nature *184*, 1648, 1959.
8. Berson, S.A., and Yalow, R.S.: General principles of radioimmunoassay. Clin. Chim. Acta *22*, 51, 1968.
9. Oxley, D.K.: Radioimmunoassay. *In* Interpretations in Therapeutic Drug Monitoring, edited by D.M. Baer and W.R. Dito. Chicago, American Society of Clinical Pathologists, 1981, p. 220.
10. Kenny, M.A.: Practical aspects of quality control for radioimmunoassays. Am. J. Med. Tech. *42*, 318, 1976.
11. Langan, J., and Clapp, J.J. (editors): Ligand Assay. New York, Masson Publishing, 1981.
12. Savory, J.: Nephelometric immunoassay techniques. *In* Ligand Assay, edited by J. Langan and J.J. Clapp. New York, Masson Publishing, 1981, p. 257.
13. Schneider, R.S.: Recent advances in homogeneous enzyme immunoassay. *In* Ligand Assay, edited by J. Langan and J.J. Clapp. New York, Masson Publishing, 1981, p. 151.
14. Ullman, E.F.: Recent advances in fluorescence immunoassay techniques. *In* Ligand Assay, edited by J. Langan and J.J. Clapp. New York, Masson Publishing, 1981, p. 113.
15. Wisdom, G.B.: Enzyme-immunoassay. Clin. Chem. *22*, 1243, 1976.
16. Schall, R.F., Jr., and Tenoso, H.J.: Alternatives to radioimmunoassay: Labels and methods. Clin. Chem. *27*, 1157, 1981.
17. Voller, A., Bartlett, A., and Bidwell, D.E.: Enzyme immunoassays with special reference to ELISA techniques. J. Clin. Path. *31*, 507, 1978.

Chapter 9

ENDOCRINOLOGY

HORMONE AND ENDOCRINE SYSTEM

The endocrine system consists of a number of ductless glands that produce highly active chemical regulators called hormones. These chemical messengers are secreted into the bloodstream and carried to target organs elsewhere where they act upon specific receptor cells.[1,6] The protein and peptide hormones (hypothalamus, anterior and posterior pituitary, pancreas, parathyroid) are soluble in plasma, but the steroids (from adrenal cortex and gonads) and the iodinated thyronine derivatives (from the thyroid) are poorly soluble; they require specific plasma transport proteins.

Functions of Hormones

Hormones regulate metabolism and assist in perpetuation of the species by influencing a wide variety of biologic systems to perform the following functions:

1. To maintain a constant internal environment in the body fluids (homeostasis).
2. To regulate the growth and development of the body as a whole.
3. To promote sexual maturation, maintain sexual rhythms, and facilitate the reproductive process.
4. To regulate energy production, stabilize the metabolic rate.
5. To help the body adjust to stressful or emergency situations.

The individual hormones, however, vary considerably in the extent of their functions; the parathyroid hormone, for example, is concerned with the maintenance of a constant plasma concentration of calcium ion (p. 348), while the thyroid hormone regulates the basal metabolic rate of all cells in the body and, in conjunction with growth hormone, is also essential for proper growth and development.

Functional Types

There are three functional types of hormones: releasing factors[7] from the hypothalamus, which stimulate the secretion of some anterior pituitary hormones, tropic hormones that stimulate the growth and activity of other endocrine glands, and the nontropic or effector hormones, which are secreted by all of the endocrine glands other than the anterior pituitary and hypothalamus. The target cells of the effector hormones are the particular tissue cells upon which these hormones exert their metabolic effects; releasing factors act only

261

TABLE 9.1
Types, Sources, and Action of Some Human Hormones*

CHEMICAL TYPE	HORMONE	ENDOCRINE GLAND	TARGET TISSUE	PRINCIPAL ACTION
Glycoproteins (2 chains, α and β)	Thyrotropin (TSH)	Anterior pituitary	Thyroid	Formation and secretion of thyroxine
	Follicle stimulating (FSH)	Anterior pituitary	Female—ovary Male—testis	Follicular growth, estradiol secretion Spermatogenesis
	Luteinizing (LH)	Anterior pituitary	Female—ovary Male—testis	Luteinization of follicles; with FSH → ovulation Secretion of testosterone
	Chorionic gonadotropin (hCG)	Placenta	Ovary	Growth of corpus luteum, maintenance of pregnancy
Protein & long-chain polypeptides	Growth (Somatotropin) (GH)	Anterior pituitary	Whole body	Growth of bone & muscle
	Adrenocorticotropin (ACTH)	Anterior pituitary	Adrenal cortex	Formation & secretion of cortisol
	Prolactin (lactogenic) (PRL)	Anterior pituitary	Breasts	Growth of breasts and stimulation of milk secretion
	Insulin	Pancreas, β islet cells	Muscle, liver, adipose cells	Lowers blood glucose concentration; lipogenesis
	Glucagon	Pancreas, α cells	Liver	Glycogenolysis
	Parathyroid hormone (PTH)	Parathyroid	Bone Kidney	↑ Plasma Ca by ↑ bone resorption ↓ Plasma P by ↑ renal P excretion
	Calcitonin	Thyroid	Bone	↓ Plasma Ca by ↓ bone resorption
	Gastrin	Stomach	Acid-secreting stomach cells	↑ Secretion of gastric HCl
	Secretin	Intestine	Pancreas & stomach	↑ Pancreatic H_2O & HCO_3^- secretion ↑ Pepsin secretion in stomach
	Cholecystokinin (Pancreozymin)	Intestine	Pancreas & gall bladder	↑ Secretion of enzymes by pancreas ↑ Gall bladder contraction
	Erythropoietin	Kidney	Stem cells	↑ Formation of red blood cells
	Renin	Kidney	Angiotensinogen	↑ Blood pressure (formation of Angiotensin I → Angiotensin II)
	Releasing Factors (RF): Growth hormone (GHRF)	Hypothalamus	Anterior pituitary	Release of growth hormone (GH)
	Corticotropin (CRF)	Hypothalamus	Anterior pituitary	Release of ACTH

		Source	Target tissue	Action
Short-chain polypeptides†	Releasing Factors (RF): Thyrotropin (TRF)	Hypothalamus	Anterior pituitary	Release of TSH
	FSH (FSHRF)	Hypothalamus	Anterior pituitary	Release of FSH
	LH (LHRF)	Hypothalamus	Anterior pituitary	Release of LH
	PRL (PRLRF)	Hypothalamus	Anterior pituitary	Release of PRL
	ADH (vasopressin)	Posterior pituitary	Arterioles / Kidney tubules	↑ Blood pressure / ↑ Water reabsorption
	Oxytocin	Posterior pituitary	Uterine muscle	Contraction of uterus / Initiates milk secretion
	Somatostatin (GHIF)	D cells, pancreatic islets	Anterior pituitary	Inhibition of GH release
Steroids	Glucocorticoids (Cortisol)	Adrenal cortex	All tissues	↑ Gluconeogenesis, ↑ liver glycogen, ↑ protein catabolism, ↓ inflammatory tissue reactions
	Mineralocorticoids (aldosterone)	Adrenal cortex	Kidney tubules	↑ Na reabsorption, ↑ K excretion
	Androgens	Adrenal cortex	Tissues responsible for masculinization	An excess induces virilization in females and immature males
	Estrogens (estradiol)	Ovary	Female accessory sex organs	Promote secondary sex characteristics; maintain pregnancy
	Progesterone	Ovary (plus placenta during pregnancy)	Uterus	Prepares uterus for ovum implantation; suppresses menstruation
	Testosterone	Testis	Male accessory sex organs	Promotes secondary sex characteristics
	$1,25$-dihydroxy-D_3	Kidney	Intestine, bone, kidney	↑ Ca absorption, ↑ bone resorption, ↓ Ca excretion (?)
Amino acid derivatives	Thyroxine (T_4)	Thyroid	All tissues	↑ Metabolic rate
	Triiodothyronine (T_3)	Thyroid	All tissues	↑ Metabolic rate
	Epinephrine Norepinephrine	Adrenal medulla	Sympathetic nerve endings, liver, heart, blood vessels	Sympathetic neurotransmitters; ↑ glycogenolysis, ↑ lipolysis; ↑ heart rate and ↑ blood pressure
	Dopamine (PRLIF)	Hypothalamus	Anterior pituitary	Inhibition of PRL release

*The list is not complete. Some hormones that are not definitely established or that do not have significant application in clinical chemistry have been omitted.

†Less than 15 amino acids in the chain.

upon the anterior pituitary, and tropic hormones act only upon other endocrine glands.[2]

The tropic hormones of the pituitary that act upon specific endocrine glands are thyrotropin (TSH), a hormone that causes the thyroid gland to produce thyroxine, adrenocorticotropin (ACTH), which stimulates the adrenal cortex to produce cortisol, and the gonadotropins (FSH and LH), which stimulate the ovaries to produce estrogen and the testes to produce testosterone, respectively (Table 9.1).

The effector hormones exert their particular effects upon nonendocrine tissues. Thus, gonadal hormones (testosterone in males, estrogens in females) affect a wide variety of tissues during puberty to produce the typical secondary sex characteristics for each sex; thyroid hormones affect the metabolic rates of all tissues; growth hormone stimulates skeletal growth and affects many metabolic processes.

Chemical Types

Hormones are of five general chemical types: glycoproteins composed of two dissimilar chains; proteins or long-chain polypeptides; short-chain polypeptides; steroids; and amino acid derivatives (Table 9.1). Half (TSH, FSH, LH) of the anterior pituitary hormones are glycoproteins; the rest are long-chain polypeptides. The posterior pituitary hormones, ADH (vasopressin) and oxytocin, and all of the hypothalamus releasing factors are short-chain peptides. Hormones from the adrenal cortex and the gonads are steroids, whereas epinephrine and norepinephrine from the adrenal medulla are amines derived from the amino acid tyrosine. The thyroid hormones are iodinated derivatives of tyrosine. Some of the protein hormones may be stored in their glands as prohormones, inactive compounds of larger molecular weight; they are converted to the active hormone when proteases cleave off a portion of the polypeptide chain, as in the case of insulin. The growth hormone exists as a preprohormone from which growth hormone and other polypeptides with hormonal activity may be split off; one fragment mobilizes fatty acids from the fat depots.

The production and secretion of hormones by an endocrine gland may be initiated by one or more of five different types of signals,[6] which include the following:

1. Stimulation of the cerebral cortex by thoughts, emotions, stress, circadian (periodic) rhythms that may be daily, monthly, or seasonal, and by chemical transmitters active at nerve junctions, such as norepinephrine, dopamine, acetylcholine, and serotonin that cause release of hypothalamic or adrenal medullary hormones.
2. A change in the plasma concentration of particular ions or compounds (secretion of parathyroid hormone when the plasma concentration of Ca^{2+} is low or of insulin when the plasma level of glucose is high).
3. Secretion of tropic hormones (e.g., ACTH and TSH cause the adrenal and thyroid glands to produce and secrete cortisol and thyroxine, respectively).
4. Variation in blood osmolality or volume (secretion of antidiuretic hormone, ADH, by the posterior pituitary when the plasma osmolality increases and renin by the kidney in response to low blood volume).

5. Release of hormones in the gastrointestinal tract in the presence of various foods (food in the stomach stimulates gastrin secretion by the stomach).

Hormonal Regulation by the Hypothalamus

The control and regulation of the hormones that exert such profound effects upon so many tissue cells are very complicated. A schematic representation appears in Figure 9.1. Primary control is believed to rest in the hypothalamus,

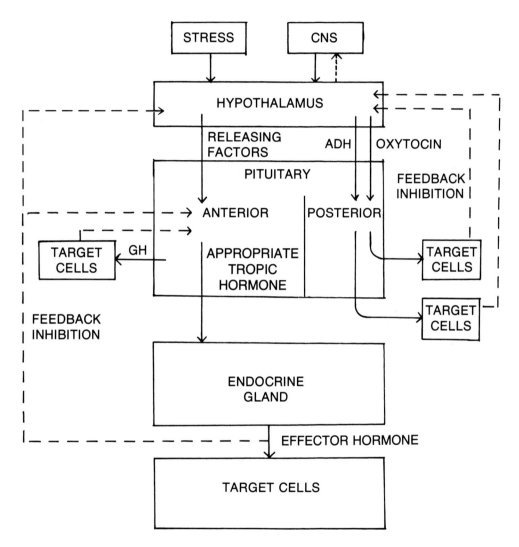

Fig. 9.1. Schematic representation of the control of hormone secretion. The hypothalamus may be stimulated by the central nervous system (CNS) or by stress to secrete releasing factors (GHRF, CRF, TRF, LHRF, or FSHRF), ADH, or oxytocin. GHRF stimulates the anterior pituitary to secrete growth hormone, which acts directly upon target cells. Other releasing factors stimulate secretion of tropic hormones that induce adrenal cortex, thyroid, or gonads to secrete their respective hormones, which then act upon target cells. Hormones ADH and oxytocin travel down the pituitary stalk into the posterior pituitary where they are secreted. Broken lines represent feedback control of secretion of hypothalamic or anterior pituitary hormones; some hormones (TSH) may directly inhibit release of tropic factor by the anterior pituitary. See Table 9.1. for the meaning of abbreviations.

a small gland and nerve center situated just adjacent to the pituitary. It is connected to the latter by the pituitary stalk through which blood flows from the hypothalamus to both pituitary lobes and through which nerve fibers pass to the posterior lobe.

The hypothalamus may be acted upon by the central nervous system or by stress to secrete one or more of a group of releasing factors. The releasing factors are polypeptides that stimulate the anterior pituitary to elaborate and secrete the appropriate tropic hormone. This results in a great amplification of the original signal as the tropic hormone activates its target gland to produce and secrete a specific hormone. Thus, each endocrine gland in the chain pours out many molecules for each stimulus received and this produces a cascade effect. The hypothalamus also contains peptides that inhibit the pituitary release of hormones; for example, *somatostatin* inhibits the pituitary output of growth hormone. This provides further "fine-tuning" control of hormonal concentrations in blood.

Hormonal Regulation by the Anterior Pituitary

The pituitary gland is composed of two distinct parts, the anterior lobe or adenohypophysis and the posterior lobe or neurohypophysis. The anterior lobe of the pituitary is the master gland of the endocrine system because it secretes tropic hormones which regulate the growth and hormonal production of other endocrine glands, as shown in Table 9.1. The tropic hormones are ACTH, TSH, FSH, and LH, which respectively act upon the adrenal cortex, thyroid, and gonads to produce and secrete their particular hormones. An effector hormone, somatotropin or growth hormone (GH), is also secreted by the anterior pituitary in response to the GH releasing factor. GH acts upon the long bones and muscles to promote skeletal growth, but it also has many metabolic effects (p. 306). GH secretion is stimulated by hypoglycemia as well as by the GH releasing factor.

The hypothalamus does not make releasing factors for the hormones of the posterior pituitary or for any endocrine gland other than the anterior lobe of the pituitary. The hypothalamus does synthesize ADH and oxytocin, two hormones that travel through nerve fibers in the pituitary stalk to the posterior lobe where they are stored and from which they are secreted.

The effector hormones produced by the endocrine glands exert a feedback inhibition either upon the hypothalamus to retard or block further secretion of releasing factors or upon the anterior pituitary to inhibit the secretion of tropins. This mechanism completes the control loop, as indicated schematically in Figure 9.1. For a cyclical event, such as the menstrual cycle, there is a positive feedback control between estradiol and the LH releasing factor so that a regular periodic pattern is established. The central nervous system also affects hormonal secretions during periods of nervous excitement, anxiety, or stress; the menstrual cycle is frequently disturbed in such situations.

Hormones act in different ways upon their target cells. The protein and peptide hormones and the catecholamines (epinephrine and norepinephrine) become attached to target cell membranes and activate adenyl cyclase to produce cyclic AMP (c-AMP). Cyclic AMP activates protein kinases, enzymes that activate phosphorylases; liver phosphorylase, for example, converts glycogen to glucose. Hormonal action upon the cell membrane is usually short-lived because of rapid conversion to inactive metabolites by enzymes.

The chemical nature of the steroid and thyroid hormones is such that they are able to enter their target cells and penetrate to the cell nuclei. They react with nuclear DNA in such a way as to repress some genes, derepress others, or uncover new genes to make new messenger RNA's with instructions for the synthesis of new proteins. Thus, these hormones have relatively long-term effects and indirectly affect many metabolic systems. Special binding proteins in the cytoplasm aid in the transfer of the hormone to the nucleus. The events at the cellular level are beginning to be clarified.

Since the main function of a hormone is to exert a fine regulatory control over metabolic processes or growth and development, it is obvious that either the overproduction or the underproduction of various hormones will lead to abnormalities. It is not a simple problem, however, to ascertain whether an endocrine dysfunction is caused by a malfunction of the endocrine gland in question, by a deficiency of a tropic hormone (pituitary dysfunction), by a deficiency of releasing factors (hypothalamus or feedback control dysfunction), by a defect in the target cell, or by other factors involving transport proteins and autoimmune disease. Pinpointing the pathologic process responsible for an endocrine abnormality is further complicated by the problem of ectopic hormones. Ectopic hormones are hormones produced elsewhere in the body than in the customary endocrine gland. Many types of malignancies produce ectopic hormones, which, of course, do not respond to the normal regulatory processes of the body. Ectopic production of ACTH is discussed on page 282.

Clinical chemistry laboratories are able to measure the plasma concentration of many hormones by various immunotechniques (see p. 253). The availability of more dependable antibodies and commercial kits has helped to make this possible. It is costly, however, to perform these assays when the demand for them is low because the ratio of nonpaid (standards and controls) to paid tests (patient samples) is high when the test volume is low; also, proprietary assay materials are expensive. The only details of hormone methodologies presented in this chapter are those performed by chemical techniques. The principles of immunotechniques are discussed on pages 253 to 258; for each particular kit, the manufacturer's directions have to be followed.

THYROID HORMONES

The thyroid gland is a small gland situated in the neck, wrapped around the trachea just below the larynx (voice box). It secretes the hormones that increase the metabolic rate and oxygen consumption and that are necessary for proper growth and development. The thyroid also secretes calcitonin, a hormone that participates in the regulation of plasma Ca^{2+} concentration by inhibiting bone resorption (see discussion of the parathyroid on p. 296).

The circulating thyroid hormones, thyroxine (3,5,3',5'-tetraiodothyronine, T_4) and 3,5,3'-triiodothyronine (T_3), are iodinated derivatives of the amino acid, tyrosine, and their structures are shown in Figure 9.2. Their immediate precursors are monoiodotyrosine (MIT) and diiodotyrosine (DIT), but these latter compounds do not appear to any significant extent in plasma except under rare pathologic conditions. Neither MIT nor DIT has any hormonal activity.

The various events occurring in the synthesis and release of thyroid hormones are depicted in Figure 9.3. In step 1, the entry of inorganic iodide ion into the thyroid cell is facilitated by an enzyme system that can be inhibited by anions

Thyroxine (T$_4$)
(3,5,3′,5′-Tetraiodothyronine)

3,5,3′-Triiodothyronine (T$_3$)

Fig. 9.2. Structures of the thyroid hormones, T$_4$ and T$_3$. The parent molecule with no iodine atoms is thyronine. T$_4$ and T$_3$ are tetraiodo- and triiodothyronine, respectively. Reverse T$_3$ (rT$_3$) is not shown, but its iodine molecules are in the 3,3′,5′ position of triiodothyronine.

such as ClO$_4^-$, SCN$^-$, or high concentrations of I$^-$. The I$^-$ is trapped once it enters the thyroid gland and does not diffuse out readily. In step 2, the I$^-$ is oxidized with the help of a peroxidase enzyme system to an active state, presumably I$_2$, which prepares it for step 3. The activated iodine reacts in step 3 with tyrosyl residues in the protein thyroglobulin, a large glycoprotein present in the thyroid cells. The products are MIT and DIT residues attached to thyroglobulin, with the iodine in the 3-position in MIT and in the 3,5-positions in DIT. The MIT and DIT residues are oxidatively coupled in step 4 by an enzyme system to form T$_3$ and T$_4$ residues attached to thyroglobulin; two DIT residues are condensed to form T$_4$, whereas the coupling of MIT with DIT yields T$_3$.

Although T$_4$ is the principal iodinated hormone secreted by the thyroid, peripheral tissues convert T$_4$ to T$_3$, the active hormonal entity that enters tissue cells. About 70% of the plasma T$_3$ is derived from the tissue deiodination of T$_4$, while the rest is of thyroidal origin. A reverse T$_3$ (rT$_3$), in which the iodine molecules are in the 3,3′,5′ position instead of the 3,5,3′ position of T$_3$, is formed to a minor extent during this deiodination process. Reverse T$_3$ has no hormonal activity.

The thyroglobulin with its mixture of iodinated tyrosyl and thyronine residues is stored in the thyroid follicles until the signals for release are given. In step 5, some of the stored, iodinated thyroglobulin is degraded by a cellular protease to yield a mixture of free MIT, DIT, T$_3$, and T$_4$. MIT and DIT are enzymatically deiodinated, and the I$^-$ is recycled within the gland (step 6). T$_3$ and T$_4$ diffuse through the thyroid cell wall into plasma where they are quickly bound by several different proteins (step 7).

Both T$_4$ and T$_3$ are poorly soluble in plasma and are transported primarily by a thyroid-binding globulin (TBG); albumin carries about 15% of the total circulating amounts of each hormone, and a prealbumin binds a like percentage of T$_4$ but a smaller one of T$_3$. The plasma concentration of free (unbound) hormone is quite low because of the high affinities of the binding proteins for T$_4$ and T$_3$. The ratio of total T$_4$ to free T$_4$ is about 2000:1 with a corresponding figure of 200:1 for T$_3$; the free hormone is in equilibrium with the bound. Only the *free hormone can enter cells and exert its hormonal action* by reacting with nuclear DNA (p. 267). The plasma distribution of thyroid hormones is approximately 97% T$_4$, 2% T$_3$ and <1% rT$_3$.

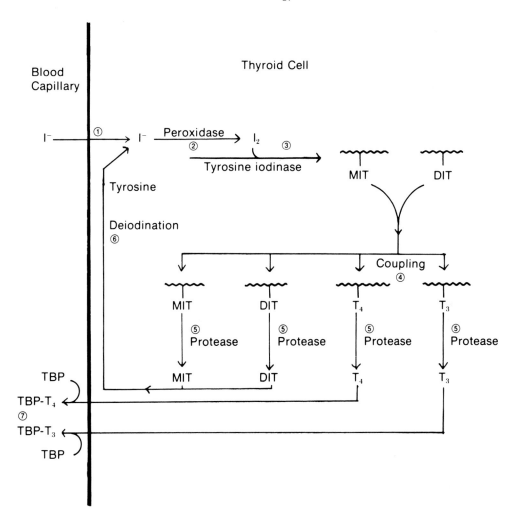

Fig. 9.3. Schematic representation of the production of thyroid hormones. MIT = monoiodotyrosine, DIT = diiodotyrosine; T_3 = 3,5,3'-triiodothyronine; T_4 = thyroxine (3,5,3',5'-tetraiodothyronine); TBP = T_3- and T_4-binding proteins; wavy line represents the thyroglobulin molecule. Defects may occur at any of the numbered steps; these are discussed in the text.

Despite the greater plasma concentration of T_4, T_3 is considered to be the actual active hormone of the thyroid gland. T_3 is distributed throughout the extracellular water, but T_4 is confined to the plasma; the daily turnover of hormone is much more rapid for T_3 than for T_4. When thyroid hormone is administered therapeutically, the physiologic effects occur sooner after administration of T_3 than after T_4.

Evaluation of Thyroid Function

The evaluation of thyroid status is not a simple procedure because it does not depend solely upon the measurement of circulating thyroid hormones.[8] First, there must be assurance that the diet is adequate in iodine. Then, the concentration of thyroid-binding proteins and their relative degree of saturation with T_4, the actual concentrations of free (unbound) T_4 and T_3, the status of the hypothalamus and anterior pituitary with their output of releasing factor (TRF) and thyroid stimulating hormone (TSH), respectively, are all factors that must be sorted out and evaluated. A further complication in malignant exophthalmos (Graves' disease) is the presence in plasma of a long-acting thyroid stimulator (LATS), that is not of pituitary origin. It is measured by bioassay only.

The assessment of thyroid function by laboratory methods has improved tremendously since 1950 to 1955. Before that time, the measurement of the basal metabolic rate (rate of oxygen consumption during a 15-minute period while fasting and at bedrest) was the principal test employed. When the test was well done, gross dysfunction of the thyroid could be documented but there were many variables and too much imprecision for diagnosis of moderate thyroid dysfunction.

The next advance, the measurement of PBI[14], represented the first analytical attempt at estimation of thyroid hormonal iodine. It was possible to separate the serum inorganic iodide from the iodine bound to T_3 and T_4 by precipitating the serum proteins and measuring the iodine in the protein precipitate. Estimation of hormonal iodine was later improved by quantifying the iodine in serum T_4 after adsorption of the hormone to an anionic resin column[15] (T_4 by column). Both methods have been supplanted, however, by various immunotechniques that are commercially available and that have the advantages of fewer interfering substances, higher precision, use of smaller serum samples, and a more rapid performance.

Total T_4 and T_3 by RIA or Enzyme Immunoassay

The assessment of thyroid function by measuring the concentration of serum T_4 is based upon the assumption that serum T_4 reflects T_3 levels and that the serum concentration of each rises or falls in parallel fashion in disease states. This is true most of the time but there are sufficient exceptions that the measurement of both hormones is indicated in some instances. A rise in the concentration of serum T_3 may be the first abnormality in some cases of hyperthyroidism and, in rare cases, may be the only abnormality.

The most widely used methods for the measurement of serum T_4 and T_3 are those of radioimmunoassay (RIA) and enzyme immunoassay. The RIA and enzyme immunoassays are sensitive, are not affected by iodine compounds in serum, and are readily available as commercial kits. Only the principles of these two techniques will be given because the procedural details vary with each manufacturer.

PRINCIPLE OF RIA TECHNIQUE. The principle is the same for both the T_4 and T_3 assay. The only differences are (1) the antibody must be specific for the hormone to be assayed and not cross-react with the other and (2) the pure ^{125}I-labeled hormone has to correspond to the hormone being assayed. There are various technical procedures for releasing hormones from binding proteins or for separating antibody-bound antigen from the unbound (p. 255) but only one for each is described in the following steps for the measurement of T_4 or T_3:

1. *Releasing the hormone in serum from its binding proteins:* Transfer a serum aliquot along with barbital buffer, pH 8.6, to a test tube containing bovine γ-globulin (2 g/L) and 8-anilino-1-naphthalene sulfonic acid (600 mg/L). The latter compound displaces both T_4 and T_3 from all binding

proteins. The γ-globulin is included to make a larger pellet when the proteins are precipitated in step 3.

2. *Incubation with labeled hormone and specific antibody:* Add a known amount of ^{125}I-labeled hormone and appropriate antibody to the above tube. Mix and incubate at 37° C. The serum hormone and the radiolabeled hormone compete for antibody-binding sites; the higher the concentration of serum T_4 in the T_4 assay, the less ^{125}I-T_4 is bound to the antibody against T_4. The antiserum has to be titered to provide the optimum concentration (to bind 50–60% of antigen) prior to use.

3. *Separation of antibody-bound hormone from the free:* The above mixture is cooled in ice and the hormones bound to antibody are precipitated by the addition of cold polyethylene glycol (200 g/L), a protein-denaturing agent. Upon centrifugation, the precipitated antibody and other proteins form a pellet and are separated from the supernatant by decantation.

4. *Counting the antibody-bound radioactivity.* The ^{125}I in the pellet is counted in a gamma counter. The percent bound (%B) of ^{125}I-hormone $=$

$$\frac{\text{Counts in sample}}{\text{Total counts in } ^{125}\text{I-hormone aliquot}} \times 100$$

5. *Converting %B to hormone concentration from Standard Curve.* A standard curve is prepared with each batch of tests and the %B of ^{125}I-hormone is plotted against the concentration of pure hormone (μg/dl for T_4, ng/dl for T_3). The %B of patients' samples are converted to concentration from the curve.

REFERENCE VALUES (Table 9.2). In RIA and enzyme immunoassay techniques, the concentrations of T_4 and T_3 are expressed as *amount of hormone* rather than *hormonal iodine* per deciliter or liter of serum.* The reference values vary somewhat with the methodology and the purity of the standards used, but in our laboratory the reference values for T_4 are 4.1 to 11.3 μg/dl of hormone (41 − 113 μg/L) or 53 to 147 nmol/L; the mean is 7.4 μg/dl. The corresponding

TABLE 9.2
Some Reference Values* for Total and Free T_4 and T_3

| THYROID STATUS | T_4 | | T_3 | |
	TOTAL ng/dl	FREE ng/dl	TOTAL† ng/dl	FREE ng/dl
Euthyroid	4,100–11,300 (7,400)	5.4 ± 1 0.046%–0.054% of total	65–199 (132)	0.39–0.87 or 0.46% of total T_3
Hyperthyroid	13,000–31,000	19 ± 6	200–1900 (600)	
Hypothyroid	<6000	1.7 ± 0.7	5–130 (30)	

*Expressed in terms of ng of the entire molecule per dl serum and *not* of iodine.
†Mean values in parentheses.

---

*1 nmol T_4 = 777 ng = 508 ng T_4 iodine; 1 nmol T_3 = 651 ng = 381 ng T_3 iodine.

values for total T_3 are 65 to 199 ng/dl of hormone, with a mean of 132 ng/dl. This is equivalent to 1.0 to 3.1 nmol/L.

INCREASED CONCENTRATION. The serum concentrations of both T_4 and T_3 are usually elevated in hyperthyroidism, with the few exceptions noted on page 270 in which only the T_3 may be abnormal. Both T_4 and T_3 levels are increased in conditions that raise the plasma concentration of TBG (pregnancy, taking of birth control pills or estrogens). There is no increase in the basal metabolic rate, however, because the extra T_4 and T_3 are bound to the newly synthesized TBG as equilibrium between the free and bound hormones is maintained. The person is euthyroid because the level of *free hormone*, the biologically active entity, remains essentially unchanged.

DECREASED CONCENTRATION. The serum concentrations of T_4 and T_3 are usually decreased in hypothyroidism but there may be some overlap with the reference range in mild cases. Their concentrations are also lowered when the TBG level has been decreased by disease (nephrosis, in which there is loss of TBG and the other binding proteins into the urine, and in liver disease, in which there is decreased synthesis of TBG), by medications that reduce the synthesis of TBG (androgens, anabolic steroids) or compete with T_4 and T_3 for binding sites (aspirin, diphenylhydantoin, tolbutamide and others). The lowering of plasma T_4 and T_3 concentration accompanying a decrease in TBG concentration has no effect upon the basal metabolic rate.

PRINCIPLE OF ENZYME IMMUNOASSAY. Enzyme immunoassay eliminates the use of radioactivity and counting equipment. It does require, however, a good automatic pipetter-diluter, an automated system for transferring and incubating reagents with samples for a precise time at 37° C, and for a spectrophotometer capable of recording absorbance changes at 340 nm. The EMIT thyroxine assay is available only from Syva.

The EMIT system is a homogeneous enzyme immunoassay technique that utilizes the following principles:

1. T_4 is chemically bonded to the enzyme, malate dehydrogenase (MD). The T_4-MD has greatly reduced catalytic activity, an exception to the usual EMIT hapten-enzyme complex. In all of their drug assays, the hapten-enzyme complex is enzymatically active; the enzymatic activity is greatly reduced when the drug-enzyme becomes bound to antibody (p. 258).

2. Serum samples are first mixed with a sodium hydroxide solution in order to break the binding of patients' T_4 with TBG and the other binding proteins.

3. The treated serum is mixed with a buffered solution of NADH, antibody against T_4 and finally, with a mixture of T_4-MD and MD substrate. Patient T_4 and T_4-MD compete for antibody binding sites; the higher the patient T_4, the less T_4-MD is attached to antibody, and vice versa.

4. The mixture is incubated at 37° C for a set time as in any two-point enzyme assay and the increase in absorbance at 340 nm is recorded. The unbound T_4-MD has little enzyme activity, but the T_4-MD bound to antibody is quite active; NAD^+ is converted to NADH when malate is dehydrogenated by the T_4-enzyme complex.

5. A standard curve is run with pure thyroxine and the absorbance is plotted against the concentration of T_4. There is an inverse relation between serum

T_4 (or Standard T_4) concentration and absorbance because only the T_4-MD bound to antibody has appreciable activity.

REFERENCE VALUES. The reference range by this system is close to that obtained by RIA, but each laboratory should prepare its own range. The interpretation of increased or decreased serum T_4 concentration is the same as described for RIA.

T_3 Uptake and Free Thyroid Index (FTI).

The importance of the amounts and saturation of thyroid-binding proteins with T_4 was appreciated from the days when the PBI test was initiated. The problem of interpretation of PBI values, and later of T_4, became more acute as more and more women began to use birth control pills containing estrogens; these pills increased the concentration of TBG and consequently of T_3 and T_4 without altering the metabolic rate. The T_3 uptake test was introduced to provide an indirect estimate of the TBG. It measures the uptake of *excess* ^{125}I-T_3 *from an adsorbent* after saturating TBG with the labeled hormone.

PRINCIPLE

1. *Incubation of serum with a known amount of* ^{125}I-T_3 *with serum in an alkaline buffer* enables the labeled T_3 to attach to all unoccupied binding sites of TBG. The alkaline buffer prevents the binding of label by albumin and prealbumin.
2. The unbound ^{125}I-T_3 is adsorbed to the adsorbent (resin or silicate) and physically separated from the label bound to TBG (by centrifugation when using silicate).
3. *The radioactivity in the pellet (the label taken up by the silicate; i.e., not bound to TBG)* is counted in a gamma counter after centrifugation.
4. The T_3 uptake $= \dfrac{\text{Test counts}}{\text{Total counts}} \times 100.$
5. A serum pool of known (calibrated) uptake is provided for adjustment or calibration of test results.

REFERENCE VALUES. In euthyroid subjects who do not have renal or liver disease and are not taking medications that increase or decrease TBG, the range is from 35 to 45%, using this kit.*

INCREASED VALUES. The T_3 uptake is increased in *hyperthyroidism* (>45%) because more of the TBG binding sites are occupied by the patients' T_3 and T_4, which leaves more of the labeled-T_3 available for adsorption by the secondary binder (silicate). In like manner, the uptake is elevated in all situations in which there is a *decrease in TBG concentration*, whether through loss by the kidney in renal disease or decreased synthesis by the liver that occurs in hepatocellular disease or after the administration of certain steroids. *Drugs that lower the serum T_3 and T_4 concentration by competitive binding to TBG increase the T_3 uptake because the labeled T_3 cannot displace the drugs from their binding to TBG; more of the labeled T_3 is taken up by the silicate.*

DECREASED VALUES. The T_3 uptake is decreased in *hypothyroidism* (<35%) because there is a lowered hormonal output by the thyroid and consequently, a less saturated TBG. The TBG binds more of the added label and *less is taken*

---

*NML Laboratories, Inc.

up *by the silicate.* Pregnancy and estrogens, which are present in birth control pills, increase the plasma concentration of TBG. More binding sites are available for accepting the ^{125}I-T_3, so less of the label appears in the silicate.

Since the T_3 uptake test varies inversely with the T_3 and T_4 concentration in a wide variety of conditions in which thyroid function is normal, and directly with the T_3 and T_4 concentration in hypothyroidism and hyperthyroidism, it is much more useful to use this test in conjunction with the T_4 measurement. The free thyroid index (FTI) has become the best way of clarifying this relation. FTI = T_3 uptake × T_4 concentration. The reference values for FTI depend upon the methodologies employed. The main relationships are the following:

1. Normal FTI: In euthyroid subjects, pregnancy, women taking estrogens (birth control pills), in nephrosis, hepatitis, and in persons taking drugs that elevate T_3 uptake.
2. Increased FTI: In hyperthyroidism.
3. Decreased FTI: In hypothyroidism.

Free T_4 and Free T_3

The effective thyroid hormones are free T_4 and T_3, since these are the ones that act upon tissue cells; the T_4 and T_3 bound to plasma proteins function as a reservoir of circulating hormones and are in equilibrium with the free forms. Free T_4 and free T_3 in serum can be separated from the protein-bound hormones by dialysis or other separation techniques and then determined by radioimmunoassay.[9] The levels of free T_4 and T_3 correlate much better with the thyroid status than total T_4 or T_3 and are not affected by organic iodide contaminants in serum or by abnormal TBG concentration. Some common reference values are shown in Table 9.2. The serum total T_4 is about 7400 ng/dl compared to 132 ng/dl for total T_3. Free T_4 comprises only 0.046% of the total T_4, and its concentration is about 5 ng/dl. The concentration of free T_3 is extremely low, approximately 0.6 ng/dl (600 pg/dl).

TSH

Serum TSH levels are elevated in hypothyroidism primary to the thyroid gland because of the absence of the negative feedback control. Hypothyroidism in newborns can also be detected by the presence of an elevated TSH. The reference values for TSH in euthyroids vary with the different RIA methodologies employed but appear to be in the range of 0 to 3 μU/ml when using a standardized human TSH product.[9,13] Values from 5 to 500 μU/ml may be found in hypothyroidism, with most of them above 30 μU/ml.

Thyroid Releasing Factors (TRF)

Injection of TRF and measurement of the output of TSH and T_3 may have some value as a test for indicating combined pituitary and thyroid function or for separating hypothalamic from pituitary disease.

Guidelines for Use of Thyroid Tests

When there is clinical evidence of hyperthyroidism:

1. Elevated serum concentration of T_4 is found in most cases, which confirms the diagnosis.

2. If T_4 should be normal, measure serum T_3 and Free T_4 Index (FTI). $\uparrow T_3$ indicates T_3 hyperthyroidism while $\uparrow$ FTI signifies a decreased plasma concentration of TBG. A free T_4 assay may be substituted for the FTI.

When there is clinical evidence of hypothyroidism:
1. $\downarrow T_4$ is found in many cases.
2. $\uparrow$ TSH confirms the diagnosis.
3. If T_4 is normal and TSH elevated, an FTI is indicated for evaluation of the TBG concentration.
4. In a small percentage of cases, it may be necessary to measure free T_4 or T_3; sometimes an rT_3 assay may help to clarify an ambiguous situation.

Screening of newborns for hypothyroidism: In some states, it is customary to measure serum T_4 and to check all low or questionable values with a TSH test; a $\uparrow$ TSH confirms a diagnosis of hypothyroidism. The TSH assay may be performed first in other states and elevated values confirmed by a $\downarrow T_4$ concentration. It is important to make the diagnosis early and institute treatment with T_3 or T_4 before there is arrested development.

ANTERIOR PITUITARY HORMONES[6,10]

The anterior pituitary gland controls the hormonal outputs of some of the other endocrine glands by its secretion of tropic hormones, which is discussed in the preceding section. The secretion of tropic factors is in part regulated by feedback control loops, depending upon the concentration of circulating, primary hormone and, in part, by releasing factors emanating from the hypothalamus.

The concentration of circulating anterior pituitary hormones is low except under certain pathologic conditions and is best measured by sensitive RIA methods. Prior to the introduction of RIA techniques, bioassay was the only methodology available, but this was too unreliable and insensitive to measure plasma concentrations. A brief summary of the clinical significance of pituitary hormones is given below; commercial kits are available for the RIA measurement of all pituitary hormones of clinical interest.

Growth Hormone (GH, Somatotropin)

Growth hormone promotes protein synthesis when injected into young animals and stimulates skeletal growth. In humans, a deficiency of GH in children causes dwarfism and an excess, gigantism. A hypersecretion of GH occurring in adulthood after closure of the long bones results in large, gross-featured persons, a condition known as acromegaly.

The plasma concentration of GH during the day is low in humans and is around 5 ng/ml. There is a diurnal variation in concentration, with the greatest rise occurring early in sleep, around 11 P.M. or midnight, which may be eight- or tenfold higher than the basal level during the day. Plasma GH is usually elevated in acromegalics and measurement of it can assist in the diagnosis, especially in the early stages.

Growth hormone has many metabolic effects that are antagonistic to those of insulin (p. 306). Hypoglycemia stimulates the pituitary to secrete GH. Production of hypoglycemia by means of an insulin injection has been used as a stimulation test for GH secretion, with measurement of the latter by RIA methods, but it is dangerous and has been replaced by the intravenous injection of arginine, which also activates secretion of GH. A GH stimulation test is employed in investigating dwarfism of children in order to ascertain whether or not there is a deficient output of GH. When there is a hormone deficiency, growth can be induced by the injection of GH obtained from human pituitaries.

Thyroid Stimulating Hormone (TSH, Thyrotropin)

The significance and measurement of TSH by RIA is discussed in the preceding thyroid section. TSH is used for the detection or confirmation of hypothyroidism in newborns (screening test) and sometimes for confirming the diagnosis of hypothyroidism in adults when the serum T_4 result is borderline or normal; the plasma TSH concentration is elevated in hypothyroidism because its output is triggered by $\downarrow T_4$. In rare cases, the test may be needed to differentiate between primary hypothyroidism and that secondary to pituitary ($\downarrow$ TSH output) or hypothalamus ($\downarrow$ thyroid releasing factor) problems.

Adrenocorticotropin (ACTH)

ACTH is a tropic hormone, a long-chain polypeptide that binds to cells of the adrenal cortex and influences their activities. The net effect is to stimulate the formation of adrenal steroids by increasing the synthesis of pregnenolone from cholesterol (Fig. 9.5). A second effect of ACTH is to promote the growth of adrenal cortical tissue by stimulating cortical RNA to synthesize adrenal proteins.

The concentration of ACTH in plasma is highest between 6 and 8 A.M. and lowest in the evening between 6 and 11 P.M.; the plasma concentration in a normal person does not exceed 50 pg/ml at its peak, whereas the basal level may be close to 5 pg/ml. High plasma levels of ACTH are found in three pathologic conditions: (1) in primary adrenal cortical deficiency; (2) in patients with Cushing's disease (hyperactivity of the adrenal cortex caused by excess ACTH secretion by the pituitary); (3) and in patients with ectopic tumors that produce ACTH. These conditions are discussed in the section dealing with the adrenal cortex (p. 282).

Gonadotropins (FSH, LH)

FSH and LH are glycoproteins secreted by the anterior pituitary that are necessary for the proper maturation and function of the gonads in both men and women. These hormones induce growth of the gonads and secretion of gonadal hormones and are necessary for the reproduction process (development of mature ova in females and of spermatozoa in males). Until recently, the only methods available for measurement in urine were relatively insensitive, imprecise bioassay methods. RIA methods have been introduced that are sufficiently sensitive to measure the circulating hormone concentrations. The purity of the hormones utilized for producing the antibodies is an important factor, since antibodies to impure FSH cross-react extensively with LH.

FSH and LH are present in the plasma of both males and females at all ages. A small rise occurs at puberty in both sexes, but a great increase in the concentration of plasma LH and FSH takes place in women after the menopause and remains elevated for the remainder of their lives. In ovulating females, the concentration of both LH and FSH rises sharply from the basal level just prior

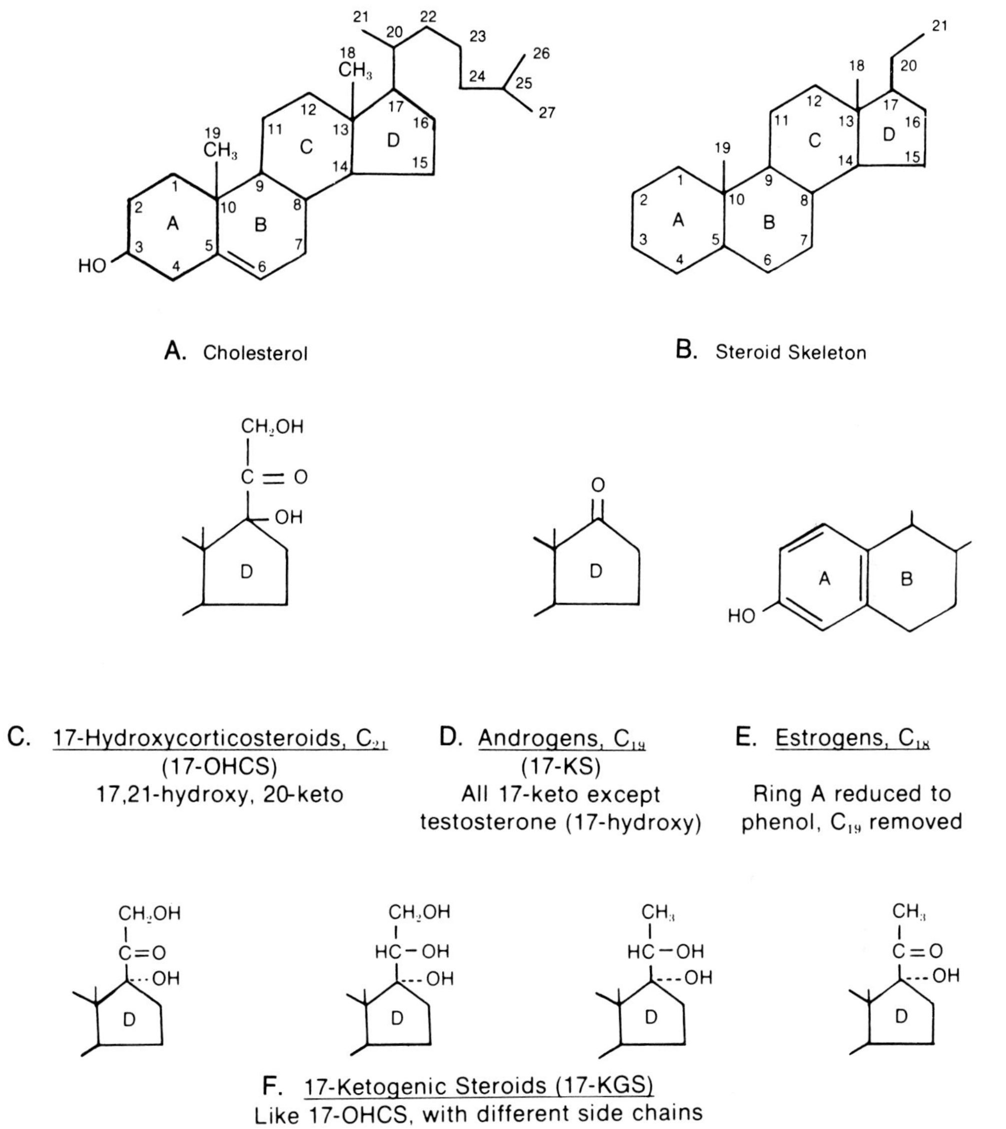

A. Cholesterol

B. Steroid Skeleton

C. 17-Hydroxycorticosteroids, C$_{21}$
(17-OHCS)
17,21-hydroxy, 20-keto

D. Androgens, C$_{19}$
(17-KS)
All 17-keto except
testosterone (17-hydroxy)

E. Estrogens, C$_{18}$
Ring A reduced to
phenol, C$_{19}$ removed

F. 17-Ketogenic Steroids (17-KGS)
Like 17-OHCS, with different side chains

Fig. 9.4. Summary of configuration of steroids.

to ovulation and then it rapidly falls. The clinical uses of the test are discussed in the section on gonads (p. 286).

Other Anterior Pituitary Hormones

The anterior pituitary also secretes prolactin (PRL), a hormone derived from the same parent molecule as GH. They have many amino acid sequences in common. Prolactin participates with gonadal steroids in breast growth during pregnancy. After parturition, it stimulates milk secretory activity, although the *initiation* of milk flow is induced by one of the posterior pituitary hormones, oxytocin. Plasma prolactin can be measured by RIA. Its concentration may be elevated in the following situations:

1. *By pituitary tumors.* Many women during pregnancy have an enlarged pituitary gland that may cause pressure headaches or visual difficulties (pressure upon the optic chiasm). The possibility of a tumor also has to be considered and investigated by measurement of plasma PRL.
2. *Inappropriate production of breast milk* (galactorrhea) by women who are not nursing a child postpartum. Some medications may elevate the plasma PRL. Reference values for PRL are
   Male <20 ng/ml
   Female <30 ng/ml

There are two melanocyte-stimulating hormones (MSH), both short-chain polypeptides whose amino acid sequence is closely related to that of ACTH. MSH stimulates melanocytes to produce pigments. Measurement of MSH is of little clinical significance at the present time.

POSTERIOR PITUITARY HORMONES[11]

The posterior pituitary stores and secretes two closely related peptide hormones, ADH (vasopressin) and oxytocin. Both are composed of eight amino acids arranged in a five-membered ring that contains cystine with its disulfide bridge, and a three-membered tail. Six of the eight amino acids of ADH and oxytocin are identical in their respective positions in the hormones.

The primary physiologic function of ADH is to increase the reabsorption of water by the renal tubules (p. 112) when the plasma osmolality becomes elevated; the hormone also has an effect upon blood pressure. Deficiency of ADH is associated with the disease diabetes insipidus, which is characterized by the passage of large volumes of dilute urine. Because diagnosis is made upon clinical evidence, there is no need to measure the hormone concentration in blood. The purified hormone may be used therapeutically in those with an ADH deficiency.

Oxytocin, the other posterior pituitary hormone, is a potent stimulant for the contraction of smooth muscle. Clinically, it is sometimes used to induce labor by its promotion of uterine contractions. Oxytocin also stimulates the ejection of milk from the mammary glands. No particular medical need is served at present by measuring the concentration of circulating oxytocin.

ADRENAL CORTEX HORMONES[4,12]

The adrenal cortex produces a large number of steroid hormones derived from cholesterol and with the ring structure of the latter molecule. Cholesterol is a waxlike lipid found in all cells; its structure is shown in Figure 9.4 (also see p. 326). The synthesis of the cortical hormones proceeds by a series of enzymatic steps involving cleavage of part of the side chain of cholesterol, dehydrogenation (conversion of the hydroxyl group to a keto group or formation of a double bond in one of the rings), hydroxylation in a few favored locations, and other actions. An outline of the steps involved in the conversion of cholesterol to an adrenal steroid appears in most biochemistry and endocrinology textbooks and will not be presented here.

The different types of steroid hormones synthesized by the adrenal cortex (Fig. 9.4) are glucocorticoids (cortisol primarily), which affect protein and carbohydrate metabolism, mineralocorticoids (aldosterone is the most potent one), which affect electrolytes, androgens (masculinizing hormones), and minute

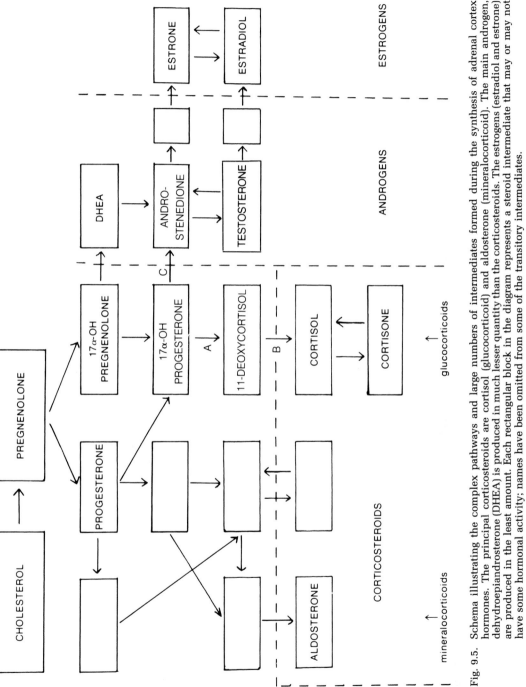

Fig. 9.5. Schema illustrating the complex pathways and large numbers of intermediates formed during the synthesis of adrenal cortex hormones. The principal corticosteroids are cortisol (glucocorticoid) and aldosterone (mineralocorticoid). The main androgen, dehydroepiandrosterone (DHEA) is produced in much lesser quantity than the corticosteroids. The estrogens (estradiol and estrone) are produced in the least amount. Each rectangular block in the diagram represents a steroid intermediate that may or may not have some hormonal activity; names have been omitted from some of the transitory intermediates.

amounts of estrogens (female sex hormones). An abbreviated schema (Fig. 9.5) illustrates the multiple and interrelated pathways for production of the various adrenal hormones.

The main hormonal product of the adrenal cortex in both males and females is the glucocorticoid, cortisol. The physiologically important mineralocorticoid, aldosterone, is present in plasma at approximately one-thousandth the concentration of cortisol. The most potent male sex hormone, testosterone, is secreted primarily by the testis and is mainly responsible for the production of the male secondary sex characteristics, but the adrenal cortex does produce in small amount some weaker androgens (androstenedione and dehydroepiandrosterone). An abnormally high production of androgens, however, may cause precocious sexual development in immature males and virilization in females. Estradiol, the most potent female sex hormone, is secreted primarily by the ovary, but the adrenal cortex does secrete minute amounts of this hormone. A change in the proportions or activities of some of the key enzymes in the adrenal cortex, however, can lead to a serious hormonal imbalance (overproduction of some hormones and underproduction of others). This may be caused by disease processes, radiation damage, genetic defects, or by drugs that inhibit some of the key enzymes.

Corticosteroids

Cortisol

Cortisol, the principal glucocorticoid, acts upon target cells by penetration and transport to the cell nucleus, binding to DNA and altering the transcription of RNA. Accordingly, cortisol alters various metabolic processes: it accelerates the enzymatic breakdown of muscle proteins and conversion of their amino acids into glucose; fat in adipose tissue is mobilized for energy purposes; the uptake of glucose by muscle is inhibited and thus, cortisol acts as an insulin antagonist (p. 306); and it reduces cellular reaction to inflammatory agents and lessens the immune response by inhibiting antibody formation.

ACTH stimulates the adrenocortical production and secretion of cortisol and to a lesser extent, androgens, but has no effect upon that of aldosterone. Cortisol is the only adrenal hormone to inhibit the anterior pituitary secretion of ACTH by negative feedback. The secretion of ACTH in turn is promoted by corticotropin releasing factor (CRF) produced by the hypothalamus; sustained ACTH production may also result from prolonged stress. The secretion of the mineralocorticoid aldosterone, by contrast, is activated by nerve impulses from volume receptors in the kidney when the blood volume falls and by plasma angiotensin, a polypeptide that raises blood pressure (p. 115).

The daily cortisol secretion of the adrenal gland is approximately 25 mg, but the usual concentration of the hormone in plasma varies from 6 to 25 μg/dl. The plasma concentration of cortisol closely follows the diurnal variation pattern of ACTH and is highest in the early morning hours and lowest at night. Cortisol is transported in plasma in three ways: 85% is transported by corticosteroid-binding α-globulin (CBG, transcortin), 10% is bound to albumin, and approximately 5% is free (not bound to proteins). Cortisol and its oxidation product, cortisone, are inactivated in the liver by two separate reduction processes (hydrogenation) to tetrahydro-derivatives, which are rapidly conjugated with glucuronic acid and excreted in the urine.

Not more than 1% of the total cortisol synthesized daily is excreted as such in the urine. About 30 to 50% appears in the form of glucuronide conjugates of tetrahydro-derivatives of cortisol and cortisone. All of these compounds contain the dihydroxyacetone group in the side chain (Fig. 9.4) and are known as 17-hydroxycorticosteroids (17-OHCS). They can be determined chemically by the Porter-Silber reaction,[16] or by kits using RIA; the latter methods have largely supplanted the former.

The determination of plasma cortisol is the most useful test for the diagnosis of the majority of adrenocortical disorders, both hyper- and hypofunction. The assay may be performed by competitive protein binding or by RIA. Prior to these sensitive methods, *plasma* cortisol was measured fluorometrically by the Porter-Silber reaction.[17]

In earlier days, however, it was easier to measure the concentration of cortisol metabolites *in urine* because of the higher concentrations. Two different chemical methods were used for this purpose: the measurement of 17-ketogenic steroids (17-KGS, which are sometimes called 17-oxogenic steroids) and 17-hydroxysteroids (17-OHCS). In the first method,[18] all compounds with the configuration of Figure 9.4F, are converted to a 17-ketosteroid (Fig. 9.4D) by reduction of a urinary extract with sodium borohydride (to convert cortisol precursors to cortisol), followed by oxidation to 17-KS with sodium metaperiodate. The 17-KS are reacted with m-dinitrobenzene in alcoholic KOH (Zimmerman reaction) and the absorbance measured. The 17-OHCS include cortisol, cortisone, and their tetrahydro-derivatives (Fig. 9.4C). The major portion of the daily cortisol production appears in the urine as 17-OHCS, primarily as glucuronides. The only other steroid of importance that the 17-OHCS method measures is 11-deoxycortisol, a steroid that normally is not present at high concentration in the urine. 11-Deoxycortisol is formed and excreted in high concentration, however, when there is a congenital deficiency of a key enzyme, 11β-hydroxylase, because Path B, Figure 9.5, is blocked. The deficiency of this enzyme results in a lowered to absent production of cortisol and an outpouring of ACTH because of the failure of feedback control. The stimulated adrenal cortex increases its synthesis of other steroids, including 11-deoxycortisol and androstenedione, an androgen. The female subject suffers from virilization. 11-Deoxycortisol is also excreted in large quantities if a drug (metyrapone) that inhibits the enzyme 11β-hydroxylase is administered because the conversion to cortisol is blocked (Path B, Fig. 9.5). Injection of this drug is a test of the pituitary's ability to secrete ACTH when the plasma level of cortisol is forced to drop.

The chemical analysis of urinary 17-OHCS or of 17-ketogenic steroids does not give a true estimate of plasma cortisol concentration or output because they may arise from other steroids similar in structure. These tests have been replaced by direct measurement of plasma cortisol. The principles of the most common techniques follow.

PLASMA CORTISOL METHODOLOGIES. Heparinized blood is collected around 8:00 A.M. (near peak time for plasma cortisol concentration) in all methods.

By Fluorescence.[17] This method has been supplanted by more specific ones that are easier to perform. The principal steps in this procedure are the following:

1. *Extraction of cortisol* plus other steroids from plasma into dichloromethane.
2. *Removal of various interfering steroids* by washing the extract with alkali.
3. Evaporation of solvent and *removal of some non-cortisol steroids* by dissolving the residue in water, followed by extraction with carbon tetrachloride. The CCl_4 layer is discarded.
4. *Extraction of cortisol from the aqueous residue with dichloromethane,* removal of water traces from the solvent by adding powdered Na_2SO_4.
5. *Production of fluorescence* by adding an H_2SO_4-ethanol mixture to an aliquot of the dichloromethane extract from step 4 and passing an activating light (470 nm) through the upper layer. *The fluorescence is measured at 520 nm and compared with that of standard solutions.*

By Competitive Protein Binding. Cortisol is relatively insoluble in plasma and is transported by a cortisol-binding globulin (transcortin), in a manner analogous to the binding and transport of T_4 by TBG. The binding affinity is high. The principles for cortisol determination by competitive protein binding are similar to those described for T_4.

1. Cortisol is removed from plasma transcortin by extraction.
2. The plasma cortisol, radiolabeled cortisol (commonly labeled with an ^{125}I-derivative because it is cheaper to use a gamma-counter for ^{125}I than a beta-counter for tritiated [3H]-cortisol) and serum transcortin are incubated together. The plasma- and labeled cortisol compete for binding sites.
3. The bound cortisol is separated from the unbound and counted.
4. The plasma cortisol concentration is obtained from a standard curve in which %B is plotted against cortisol concentration. Kits are available for this procedure.

By RIA.

1. Some kits avoid the extraction step in the RIA procedure by heat or by adjusting the pH of the plasma to dissociate the cortisol from transcortin.
2. ^{125}I-ligand-cortisol (e.g., ^{125}I-histamine coupled to cortisol) and a limiting amount of antibody are added and incubated with the buffered serum.
3. The antibody-bound cortisol is separated from the unbound and the radioactivity is counted in a gamma-counter.
4. The cortisol concentration is obtained from a standard curve of %B vs. cortisol concentration.

Reference Values. The morning plasma cortisol concentration in normal adults has a mean value of approximately 15 μg/dl, with a range of 6 to 25 μg/dl.

Increased Concentration. An elevation of plasma cortisol produces Cushing's Syndrome, which can be caused by four classic conditions: bilateral adrenal hyperplasia, thought to be due to increased corticotropin releasing factor (CRF) from the hypothalamus; autonomous (uncontrolled by pituitary) production of cortisol by a cancer of the adrenal cortex; overproduction of ACTH by a basophilic neoplasm of the anterior pituitary; and an ectopic carcinoma elsewhere that produces ACTH, e.g., some carcinomas of the lung. Plasma cortisol concentration is elevated in all of the conditions that elevate the 17-OHCS excretion except for those that increase the production of 11-deoxycortisol (drugs that inhibit the enzyme, 11β-hydroxylase).

Several tests help to differentiate between the various conditions causing an elevation in plasma cortisol. The concentration of the plasma ACTH level is low in adrenal neoplasia and elevated in pituitary tumors and ectopic carcinomas producing ACTH. Also, the administration of large doses of dexamethasone to suppress the pituitary secretion of ACTH does not cause a fall in plasma cortisol concentration when there is an adrenal neoplasm because ACTH secretion is already suppressed by the high level of cortisol.

The elevated plasma cortisol concentration caused by a pituitary neoplasm does not show as great a diurnal variation as in normals; the plasma cortisol concentration late at night does not fall below 8 μg/dl in those patients. Large doses of dexamethasone suppress the secretion of ACTH and cortisol but not small doses.

Ectopic production of ACTH by carcinomas usually is accompanied by high plasma levels of ACTH that are not suppressed by high doses of dexamethasone.

Decreased Concentration. Plasma cortisol levels are decreased in Addison's disease (destruction of adrenal cortical tissue by infective agents, autoimmune disease or other causes), in hyposecretion of ACTH by the pituitary, and in some genetic defects when there is a deficiency of an adrenal enzyme necessary for cortisol synthesis.

The urinary excretion of 17-OHCS and 17-KGS is increased in all of the above conditions that raise the concentration of plasma cortisol, as well as in the adrenogenital syndrome in which the production of androgens is elevated. The excretion of 17-OHCS and 17-KGS is low in situations that depress the production of cortisol, except when there is a deficiency of the enzyme, 11β-hydroxylase or 21-hydroxylase (p. 284). With either of these deficiencies, cortisol cannot be formed and its precursors are diverted into the formation of androgens. Reference values are listed in Appendix 1, page 403.

URINE FREE CORTISOL. Most of the urine tests for cortisol derivatives (17-OHCS and 17-KGS) have been replaced by the assay of cortisol in plasma or urine by RIA. The cortisol in urine reflects the plasma level of free cortisol (not bound to transcortin) because only the unbound hormone can be excreted. The RIA test for urine cortisol is similar to that described for plasma after a preliminary extraction and gel filtration step to eliminate interfering steroids. It is necessary to add cortisol labeled with tritium (^{3}H-cortisol) to the urine before the extraction step, however, in order to correct for losses in the extraction and gel filtration steps.

Adrenal Androgens[4,12]

The distinguishing features of the adrenal androgens and their metabolites are shown in Figure 9.4D.

The most potent androgen is testosterone, but the weaker ones, dehydroepiandrosterone and androstenedione, are converted by peripheral tissues into testosterone. They are C_{19} compounds and all are 17-ketosteroids (17-KS) except testosterone, which has a hydroxyl instead of a ketone group at C_{17}. Very little testosterone, the male sex hormone, is secreted as such by the adrenal cortex; the testis is the primary site for its production, but peripheral tissues convert a small portion of the androgens to testosterone.

ADRENOGENITAL SYNDROME. The adrenogenital syndrome is characterized by virilization in females and precocious puberty in young males and is caused by the adrenocortical overproduction of androgens. Congenital defects in either of two key enzymes in the cortisol synthesis pathway are usually responsible for this disorder. The most common defect is a deficiency in the enzyme that hydroxylates the steroid nucleus at C_{21} (Fig. 9.5, step A). The pathway to cortisol is blocked, so the intermediates flood the androgen pathway (Fig. 9.5, step C).

The other enzymatic defect consists of a deficiency of the enzyme 11β-hydrolase, which converts 11-deoxycortisol to cortisol (Fig. 9.5, step B). Again, the intermediates are largely transformed into androgens. The lowered plasma cortisol induces the secretion of ACTH; the net result is adrenal hyperplasia with increased production of androgens from cortisol precursors. Some adrenal tumors also produce increased secretion of androgens.

17-KETOSTEROIDS (17-KS). The 17-KS in the urine of females are derived almost exclusively from the adrenal gland, but in adult males, only 70% originates from this gland. The chemical measurement of 17-KS in urine was formerly the main laboratory test for investigating the adrenogenital syndrome. This method, however, has been superseded by the direct measurement of plasma cortisol and testosterone by RIA or CPB techniques. The testosterone concentration is elevated in most cases while the cortisol concentration provides information about possible blocks in the synthetic pathway. The basis for the 17-KS method[19] follows.

Principle. The urinary 17-KS are excreted as water-soluble conjugates of sulfate and glucuronide. The conjugates are hydrolyzed by boiling an aliquot in HCl, and the free steroids are selectively extracted into a petroleum ether-benzene mixture. The use of a nonpolar solvent mixture leaves in the aqueous layer many potential interfering steroids, including estrone. The 17-KS in the extract are then reacted with m-dinitrobenzene in alkaline solution (Zimmerman reaction) to give a red color whose absorbance is measured at 520 nm.

Reference values are 8 to 20 mg/d for the adult male, 5 to 15 mg/d for the adult female, and less than 2 mg/d for a child under 5 years.

Increased values are found in adrenal hyperplasia, adrenal carcinoma, in uncontrolled ACTH secretion whether caused by a pituitary basophilic tumor or an ectopic carcinoma secreting ACTH, and in testicular tumors. Decreased amounts of 17-KS are excreted in primary hypofunction of the adrenals, in hypofunction secondary to impaired function of the anterior pituitary, in adenoma of one adrenal that suppresses the contralateral production of DHEA, and in decreased gonadal function.

Aldosterone

Aldosterone is by far the most potent mineralocorticoid secreted by the adrenal cortex, even though several corticosteroids, including the glucocorticoids, have some mineralocorticoid activity. Like cortisol, it is a C_{21} compound and has a 11-hydroxy group, but it does not have a hydroxyl at C_{17} and differs from all of the other steroids by the presence of an aldehyde group at C_{18}, which is in equilibrium with the hemiacetal form. Aldosterone increases the plasma concentration of Na^+ by increasing Na^+ reabsorption in the renal tubules; the plasma K^+ concentration falls because aldosterone promotes the renal excretion of this ion (see Chap. 4).

Aldosterone is present in plasma at extremely low concentrations and is quite difficult to measure accurately. A normal person who has been in the upright position for several hours has a plasma aldosterone concentration between 5 and 20 ng/dl. If the subject is recumbent for several hours, the plasma aldosterone level is much lower and may be only 10 to 40% of that in the upright position. The liver converts aldosterone into a glucuronide and a tetrahydroglucuronide, which pass into the urine; the kidney also inactivates aldosterone by transformation into water-soluble glucuronide.

Hyperaldosteronism has been found in patients with some adrenal tumors. These patients usually have elevated serum Na concentration, lowered K, and hypertension. When measured, there is an increased plasma aldosterone concentration and an increased excretion of this hormone in the urine.

FEMALE SEX HORMONES

Ovarian Hormones

The female gonad, or ovary, has a double function; it not only produces and secretes the female sex hormones, but it is the site of production and maturation of the ova. One mature ovum is released approximately once every 4 or 5 weeks by a nonpregnant woman during the years between the onset of menstruation and the menopause.

The reproductive system of females is far more complicated than in males because of the cyclical events that take place during the menstrual cycle and the even greater changes that occur during a pregnancy. Two different chemical types of steroid hormones are produced and secreted by the ovary in nonpregnant females. During pregnancy, the same hormones are produced by the ovary, but in different proportions. The placenta also makes the hormones that are necessary for the maintenance of pregnancy.

The first group of female sex hormones, the estrogens, originate in the ovarian follicles (and also in the placenta during pregnancy). The estrogens participate in the menstrual cycle and are essential for the development and maintenance of the reproductive organs and secondary sex characteristics.

The second group comprises progesterone and its metabolites (progestational hormones), which are formed in the corpus luteum, the body that develops from the ruptured ovarian follicle. Progesterone is secreted after ovulation, stimulates the uterus to undergo changes that prepare it for implantation of the fertilized ovum, and suppresses ovulation and secretion of pituitary LH. If pregnancy occurs, the secretion of progesterone by the corpus luteum and also by the placenta, suppresses menstruation for the duration of the pregnancy.

The estrogens are C_{18} steroids, in contrast to the androgens, which are C_{19}; another difference is that ring A of the estrogen steroid is phenolic (has 3 double bonds) as shown in Figure 9.6. Progesterone is a C_{21} compound and chemically is more closely related to the adrenal steroids; as a matter of fact, progesterone is an intermediate in the production of adrenal steroids (Figs. 9.5 and 9.6).

Estrogens

The principal and most potent estrogen, estradiol, is partially converted in the body to estrone, a much weaker hormone.

There are RIA kit methods for the measurement of plasma levels of estradiol,

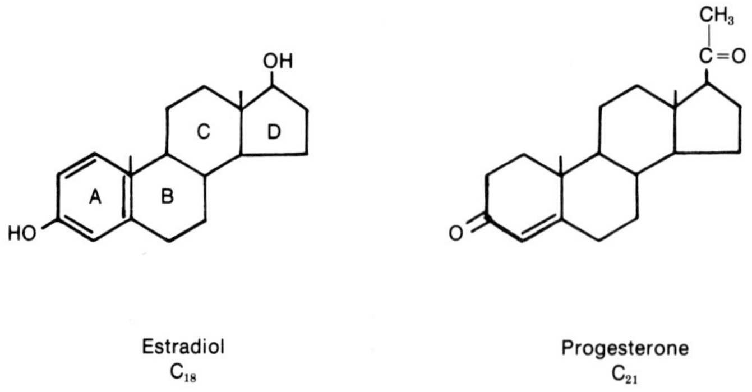

Fig. 9.6. Types of steroids produced by the ovary (and placenta). Ring A of estradiol is a phenolic ring that makes the hormone slightly acidic.

estrone, and estriol; the principles for determination are similar to those for cortisol. Plasma estradiol levels are valuable for the investigation of women with menstrual difficulties, since estradiol is an indicator of ovarian function. More information can be gained if the pituitary tropic hormones, FSH and LH, are measured at the same time in order to ascertain whether the problem is of pituitary or ovarian origin. The plasma estrone concentration is an indicator of estradiol production, since it is an end product arising from estradiol metabolism. The concentration of estrogens in the plasma of normal, nonpregnant women is in the pg/ml range and varies with the stage of the menstrual cycle.

ESTRIOL (E_3). Estriol has no hormonal activity. It is produced in relatively large quantities during the last trimester of pregnancy by the placental conversion of fetal adrenal steroids. Thus, its concentration in the urine or plasma of pregnant women provides some indication of fetal well being (placenta-fetus viability). Serial measurement of serum or urine estriol during late pregnancy is an important test in obstetrics because a sudden drop in estriol concentration or output is a danger signal of fetal-placental dysfunction.

Reference Values for E_3. The concentration of serum E_3 in the third trimester of pregnancy varies from 5 to 40 ng/ml for unconjugated E_3 and 40 to 500 ng/ml for total E_3. Because of the wide individual variations in serum E_3 concentration, measurements are made on three successive days at approximately the same time (to reduce the effect of diurnal variation) and the average taken. If a subsequent E_3 value decreases from the mean by more than 30%, it indicates danger to fetal well-being. Penney and Klenke[24] report that the serum concentration of *unconjugated E_3* is a more reliable indicator than *total E_3*.

Urinary estriol can be measured by gas chromatography, fluorometry, or colorimetry, but all of the methods are long and tedious, requiring many extractions and purification steps. Of the three techniques, gas chromatography is the most precise but it requires an additional step of preparing a volatile derivative. Reference is given for a colorimetric method[21] but many laboratories prefer the measurement of *serum* estriol by RIA because results can be obtained the same day.

The principles for the colorimetric determination of estriol follow.

Principle. Urinary estrogen conjugates are hydrolyzed by heating acidified urine and extracting the steroids into ethyl ether. Impurities are removed by a mild alkaline wash, and after evaporation of the ether and extraction with a less polar organic solvent, the estrogens (phenolic steroids) are separated from all of the other steroids. Since the estrogens are weakly acidic, they are extracted from the petroleum ether-benzene mixture with sodium carbonate. The Na_2CO_3 solution is made less alkaline (pH lowered to <10) by the addition of $NaHCO_3$ and the estrogens are transferred to ethyl ether by shaking with this solvent. After evaporation of the ether, the estrogens are dissolved in ethanol and heated with sulfuric acid to produce a yellow color (Kober reaction). A second heating with hydroquinone produces a pink color whose absorbance is measured at 512.5 nm. An Allen correction (correction for background) is made by subtracting from the peak absorbance the average absorbance at ± 30 nm from the peak.

Reference Values. The urinary estriol excretion is high in late pregnancy but quite variable from woman to woman. Whereas the excretion of estriol in the twentieth week of a normal pregnancy averages about 4 mg/day (2 to 7 mg range), the mean value jumps to 13 mg/day by the thirty-second week, 18 mg/day by the thirty-sixth, and 26 mg/day by the fortieth; the range for 95% of the women is from 14 to 44 mg/day (49–154 µmol/day) at term.

Decreased Excretion. The major portion of the estriol excreted by pregnant women originates as androgens in the fetal adrenal cortex, which are converted to estriol by placental enzymes. A sudden drop in estriol excretion is an indication of fetal jeopardy; something has gone wrong with either the fetus or the placenta. Serial determinations of urinary estriol provide a good indication of fetal risk in late pregnancy.

Increased Excretion. Usually of no clinical significance but frequently is increased when there are twins.

Placental Hormones

The placenta in the pregnant female serves as an endocrine organ in addition to its anatomical function of providing nutrients to the developing embryo and removing its waste products. In the human, the placental hormones are identical with those produced by the ovarian follicle and corpus luteum except for chorionic gonadotropin (hCG) and lactogen (hPL, similar to pituitary prolactin and growth hormone).

Human Chorionic Gonadotropin (hCG)

hCG is a glycoprotein hormone that resembles pituitary LH in that they have identical α-chains but different β-chains; they have similar hormonal action in stimulating the corpus luteum to produce progesterone; progesterone helps to maintain the pregnancy by preventing menstruation. The formation of hCG by the chorion of the developing placenta begins shortly after implantation of the fertilized egg in the uterine wall; its concentration in plasma and urine rises steadily from the first week after conception for the next 10 to 12 weeks. The early presence of hCG in urine became the basis for a pregnancy test.

The first laboratory pregnancy test was the bioassay of Ascheim and Zondek. An aliquot of urine was injected into virgin, female rabbits; the presence of hCG (and hence, pregnancy) was indicated by the presence of an enlarged uterus and large corpora lutea in the injected animal. The technology for pregnancy testing has improved markedly so it may be carried out rapidly and conveniently by simple slide tests that are available commercially. The principle of the pregnancy test that we use (Pregnosis, Roche Diagnostics) follows:

Principle of a Latex Agglutination Inhibition Slide Test for Pregnancy

1. A drop of urine is mixed on a slide with one drop of rabbit antiserum against hCG. The antibody binds to antigen if at least 2.0 ± 0.5 U/ml of hCG are present in the urine. Some women, but not all, have this urinary concentration of hCG as early as 5 days after the first missed period.

2. Mix the urine with the antibody for 30 seconds and then add one drop of antigen (a suspension of latex polymer particles that have been chemically bonded to hCG).

3. Mix for 1 to 2 minutes and observe for agglutination in direct lighting.

4. Interpretation: Absence of agglutination within 2 minutes is a positive test for pregnancy. hCG in the urine has bound sufficient antibody so that there are too few sites left to bind the antigen on the coated latex particles. The latex polymer does not agglutinate in the absence of antibody binding. Agglutination denotes a negative test for pregnancy because the antibody binds the hCG of the latex particles in the absence of competing hCG in urine. Negative tests should be repeated within 1 to 2 weeks to make certain of the results or an RIA test carried out if there are clinical reasons for suspecting an ectopic pregnancy. The latter test is more sensitive and usually resolves borderline results.

The serum RIA assay for hCG is indicated in the following situations: to settle an ambiguous pregnancy test; to pick up early an ectopic pregnancy (hCG is lower than in a placental pregnancy); and to indicate the presence of a hydatidiform mole or choriocarcinoma ($\uparrow \uparrow \uparrow$ hCG) by quantitation of hCG.

The presence of hCG in the serum of males with a testicular nodule frequently serves as a tumor marker; the presence of β-hCG is even more convincing because it is specific for hCG. About 40 to 60% of males with testicular cancer have high serum concentrations of hCG or the β-subunit.

Serum hCG Reference Values. The serum concentration of hCG in males and ovulating females varies from 0 to 10 mU/ml.

Increased Concentration. A value greater than 30 mU/ml confirms a positive pregnancy test. hCG levels are usually very high (up to 1,000,000 mU/ml) in moles (hydatidiform) and tumors that secrete hCG.

Serum β-hCG Concentration. Antibodies to hCG cross-react to some extent with LH because they both have the same α-chain. Antibody to β-hCG does not react with LH. The serum concentration of β-hCG in normal, nonpregnant individuals is less than 3 mU/ml.

MALE SEX HORMONES

The male gonads are the testes. Like the ovary, they too have a double function: to produce and secrete the male hormone, testosterone, and to produce the spermatozoa which are essential for fertilization of the ovum in the reproductive process. The pituitary gonadotropin, LH, stimulates interstitial cells in the testis to produce testosterone, and FSH promotes spermatogenesis by the germinal cells.

Practically all of the circulating testosterone in males is derived from the testis; the contribution of the adrenal cortex is negligible. Hence, measurement of the plasma testosterone concentration is a good way of studying hypogonadism and hypergonadism in males. Of course, the role of the pituitary still has to be assessed in order to determine whether an abnormality is primary to the testis or secondary because of an LH deficiency or excess. Plasma testosterone levels in normal adult males range from 350 to 1200 ng/dl (3.5 to 12 μg/L or 9.4 to 49 nmol/L), and 30 to 90 ng/dl in females.

Plasma testosterone levels are much lower in women, usually amounting to only 5% of those found in men, but they can still be measured by sensitive RIA or CPB methods; the principles are the same as for the cortisol assay. Kits are available. Testosterone in women may arise by tissue conversion of certain androgens in both the adrenal cortex and ovary.

INCREASED CONCENTRATION. Elevated plasma testosterone concentration in males may be caused by testicular carcinomas or by some abnormalities of pituitary gonadotropin. In females, plasma testosterone may be elevated in some cases of virilism or hirsutism.

DECREASED CONCENTRATION. Plasma testosterone concentration in males may be decreased in a variety of conditions directly affecting the testes, by pituitary failure, and in certain chromosomal abnormalities involving the sex chromosomes.

HORMONES OF THE ADRENAL MEDULLA[5,12]

The adrenal medulla originates in the developing fetus from cells of the sympathetic nervous system, and its hormones reflect this close relationship. In fact, one of its hormones, norepinephrine (noradrenaline), is also a neurotransmitter produced locally at nerve synapses beyond the ganglia, as well as in the adrenal medulla. The other medullary hormone, epinephrine (adrenaline) produces effects upon tissues and organs similar to that following stimulation of the sympathetic nervous system, such as increase in heart rate, increase in blood pressure, dilation of eye pupils, decrease in stomach and intestinal motility, induction of sweating, and erection of hair follicles. In addition, both hormones have some metabolic effects that promote the use of energy by the organism; the hormones raise the concentration of plasma glucose by inducing glycogenolysis (p. 306) and of free fatty acids by promoting lipolysis in adipose tissue. All of these effects, sympathetic-like and metabolic, are produced rapidly and help a mammal to meet an emergency situation that requires "fight or flight." These effects can be seen in both animals and humans in circumstances that elicit fear, anger, or aggression. The adrenal medulla is not essential for the life of man, but wild animals subject to predation probably would not survive as long without this gland because their emergency reactions would be handicapped.

Epinephrine and norepinephrine exert their rapid action upon tissues by binding to cell membranes and activating adenyl cyclase, which produces cyclic AMP within the cell. The second messenger, cyclic AMP, promotes the various metabolic and physiologic effects.

Epinephrine and norepinephrine are amines derived from the amino acid, tyrosine. Since they are related to catechol, shown in Figure 9.7, they are known as catecholamines. Tyrosine is first hydroxylated and then decarboxylated to form norepinephrine. Epinephrine is formed from norepinephrine by methy-

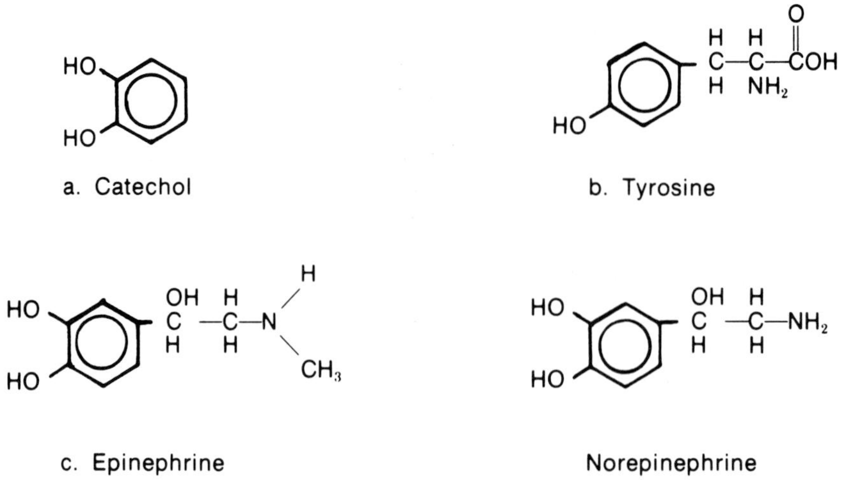

a. Catechol

b. Tyrosine

c. Epinephrine

Norepinephrine

Fig. 9.7. Structural formulas of (a) catechol, the parent compound after which the catecholamines are named, (b) tyrosine, the precursor of the adrenal medulla hormones, (c) epinephrine and (d) norepinephrine, the hormones of the adrenal medulla.

lation of the amino group. Epinephrine comprises 80 to 90% of the medullary hormones.

The medullary hormones are so potent physiolgically that only small amounts are needed to obtain their effects. Their action is transitory because the hormones are rapidly inactivated. The hormones are catabolized or inactivated by several different pathways, which involve O-methylation of the hydroxyl group on C_3 and/or oxidative deamination. Their principal metabolites are vanillylmandelic acid (VMA, 4-hydroxy-3-methoxymandelic acid), metanephrine, normetanephrine, and sometimes homovanillic acid. The *metanephrines* are formed by the methylation of the hydroxyl group (O-methylation) on C_3, which is in the meta position to the side chain (Fig. 9.7). VMA is derived from the oxidative deamination of the metanephrine side chain to a mandelic acid derivative. VMA can also arise from the oxidative deamination of epinephrine or norepinephrine, followed by O-methylation. It does not matter which reaction comes first; the final product is VMA except for the intermediate products that may be excreted before both reactions occur. Unchanged epinephrine and norepinephrine (catecholamines) in a 24-hour period usually account for about 1% of the medullary hormones that have been turned over, whereas VMA accounts for 75%, the metanephrines 10%, and the remainder are closely related metabolites.

The measurement of urinary catecholamines or their metabolites is of value in two clinical conditions: (1) in cases of unexplained high blood pressure (hypertension) in order to rule out or in the possible presence of a pheochromocytoma, a tumor of the adrenal medulla that is a cause of hypertension; and (2) to detect a neuroblastoma, a fatal malignancy in children in which there is excess production of norepinephrine by this cancer of the nervous system. Both of these conditions are relatively rare, but since hypertension is a common ailment, there are a fair number of requests to screen for the possibility of a pheochromocytoma as being the cause. It is important to detect one if it should be present because these adrenal tumors usually are benign and correctable by surgical removal of the gland. In screening for a possible pheochromocytoma, many laboratories measure the 24-hour urinary excretion of VMA, and if elevated, confirm by measuring the total catecholamines (epinephrine and norepinephrine) or by doing two-dimensional chromatography in order to rule out artifacts. These techniques also pick up most of the cases of neuroblastoma, but occasionally some patient will excrete more of the metanephrines, with the VMA being close to the upper limit of normal. Chemical methods for these three constituents are given. HPLC techniques may also be used.

Urinary VMA

REFERENCE. VMA-SKREEN instruction manual (Brinkmann-Eppendorf).

PRINCIPLE. A 24-hour urine sample is collected and preserved with 15 ml of 6 mol/L HCl. The VMA is extracted from a urine aliquot with ethyl acetate. This separates VMA from metanephrines and most metabolites that are not strongly acidic but further clean up is necessary because there may be many substances in urine that react with diazo salts. The VMA in the extract is adsorbed on a silica gel column. Interfering substances, such as vanillic acid and 4-hydroxy, 3-methoxy phenylethylene glycol are removed by washing. VMA is eluted with water, and converted into an azo dye by coupling with p-nitroaniline. The VMA-dye is further purified by extraction into ethyl acetate. The absorbance of the blue-purple VMA-dye is read at 570 nm. The VMA excretion/24 hours is calculated.

REAGENTS AND MATERIALS

1. Brinkmann Clini-Skreen "S" silica gel columns (Eppendorf-Brinkmann, Inc.).
2. VMA standard, 300 mg/L. Reconstitute the Brinkmann vial containing 6.0 mg VMA (4-hydroxy, 3-methoxy mandelic acid) with 20 ml water and 1 drop concentrated HCl. Stable for 3 months at 4° C.
3. HCl, 0.01 mol/L, pH 2.0. Dilute 0.42 ml concentrated HCl to 500 ml with water.
4. Ethyl acetate, reagent grade.
5. Sodium carbonate, 100 g/L. Dissolve 50 g Na_2CO_3 in water and make up to 500 ml volume. Stable for >1 year when tightly stoppered at 4° C.
6. Tetrabutylammonium hydroxide (TBAH, Brinkmann). Stable for 1 year at room temperature.
7. Washing Solution, ethyl acetate: ethanol, 1:1. Mix 500 ml ethyl acetate with 500 ml absolute ethanol.
8. Diazo reagent (Brinkmann). Reconstitute each vial with 5.5 mL water and 0.1 ml concentrated HCl just prior to use. Shake until dissolved.

PROCEDURE

1. Filter patient specimens through rapid filtration paper and check that the pH is below 5.
2. Transfer 3.0 ml of specimen, control and HCl solution (in that order) into appropriately labeled 50 ml extraction tubes.
3. Transfer 50 and 200 μl, respectively, of VMA Standard into separate extraction tubes containing 3.0 ml HCl solution. The Standard concentrations are equivalent to 5 and 20 mg/L urine.

4. Add approximately 1.5 g NaCl with a measuring thimble to each tube and mix well. The solution becomes saturated with NaCl.

5. Add 12 ml ethyl acetate to each tube and shake on a mechanical shaker for 5 minutes or mix on a vortex mixer for 30 seconds. Centrifuge for 5 minutes.

6. Transfer 5.0 ml of ethyl acetate layer (upper) into 16 × 100 mm tubes containing 0.1 ml glacial acetic acid. Mix well.

7. Place Clini-Skreen "S" columns in large test tubes. Pour specimens into separate columns, allow to drain by gravity, and discard the filtrate.

8. Rinse each tube with 10 ml washing solution and pour into drained columns. Discard wash solution.

9. Remove excess solvent from column by attaching aspirator to lower end and applying suction for at least 30 seconds (timed). Wipe off excess solvent from inside reservoir with a cotton swab in order to remove all interfering chromogens.

10. Place columns in 50 ml extraction tubes and elute each column with 20 ml water.

11. Reconstitute the diazo vial at this time. Swirl until dissolved.

12. Immediately add 1.0 ml 10% Na_2CO_3 solution to each eluate, mix, and then add 0.5 ml diazo reagent to each tube. Mix and let stand for 5 minutes.

13. Add 10 ml ethyl acetate to each tube, shake for 5 minutes and centrifuge for 5 minutes.

14. Transfer 5.0 ml of ethyl acetate layer (upper) to 16 × 100 mm test tubes containing 0.2 ml TBAH and mix well.

15. Read the absorbance against the reagent blank at 570 nm. The color is stable for 1 hour.

16. Calculation. Use the low Standard for most specimens and the high Standard only when you have a high result. Using the low Standard, VMA,

$$\text{mg/24 hours} = \frac{A_u}{A_s} \times 5.0 \text{ mg/L} \times \text{urine vol (L/24 hours)}.$$

Notes:

1. Dietary restrictions are not necessary because metabolites are removed.

2. The following drugs interfere with the test:
   Clofibrate (Atromid) lowers value.
   Nalidixic acid (Negram) increases value.
   Methyldopa (Aldomet) prevents formation of catecholamine.
   Azathioprine (Imuran) results in high blanks.
   L-Dopa (Levodopa) gives a pinkish blank and very high values.

REFERENCE VALUES. For adults, the excretion of VMA by this method extends from 2 to 8 mg/day or 10 to 40 μmol/day. The method gives results that are about the same as Pisano's.[22] Since the method is not specific, elevated excretion of VMA should be confirmed by two-dimensional chromatography or by analyzing also for catecholamines.

INCREASED EXCRETION. The daily excretion of VMA is usually increased in patients with pheochromocytomas and neuroblastomas, but some cases have been reported in which the VMA excretion is normal but that of epinephrine and norepinephrine, the metanephrines, or dopamine is elevated.

DECREASED EXCRETION. Clofibrate causes falsely low values.

Urinary Catecholamines

The catecholamines, epinephrine and norepinephrine, may be measured together as total catecholamines or obtained separately by oxidation of the catecholamines in the eluate at two different pH values; when oxidized at pH 6.5, both epinephrine and norepinephrine form fluorescent products, but only epinephrine does so at pH 2. The following directions are given for the analysis of *total* catecholamines; metanephrines can be measured by chemical treatment of the eluate. Tumors of the adrenal medulla secrete epinephrine primarily, whereas gangliomas (carcinomas involving the ganglia, neuroblastomas) produce norepinephrine.

The method given below for the determination of catecholamines utilizes a resin column for separation of the analytes. Commercial kits are available for those who do not wish to prepare the columns and reagents.

REFERENCE. Sandhu, R.S., and Freed, R.M.: Catecholamines and associated metabolites in human urine. *In* Standard Methods of Clinical Chemistry, Vol. 7, edited by G.R. Cooper. New York, Academic Press, 1972, p. 231.

PRINCIPLE. The pH of the urine is adjusted to pH 6.5, and the unconjugated catecholamines and metanephrines are adsorbed on a weak carboxylic cation exchange resin. Many impurities are washed out, and then the catecholamines are eluted with boric acid; the metanephrines remain on the column. Epinephrine and norepinephrine are oxidized by means of $K_3Fe(CN)_6$ at pH 6.5 to the adrenochromes, which rearrange to form the corresponding trihydroxyindoles that fluoresce when activated by light at 400 nm. The presence of ascorbate in alkaline solution inhibits the further oxidation of trihydroxyindole to nonfluorescent compounds. A urine blank is prepared by omitting the $K_3Fe(CN)_6$. The fluorescence of the trihydroxyindoles is measured at regularly timed intervals (exactly 10 minutes) after the addition of the alkaline ascorbate because the trihydroxyindole is stable for only a limited period of time.

REAGENTS AND MATERIALS

1. Acetic Acid, glacial.
2. HCl, 0.1 mol/L. Dilute concentrated HCl 1:120 with water.
3. NaOH, 5 mol/L. Dissolve 200 g NaOH in water and dilute to 1 L.
4. NaOH, 1 mol/L. Dilute some 5 mol/L NaOH 1:5 with water.
5. Potassium Ferricyanide $[K_3Fe(CN)_6]$, 8 mmol/L. Dissolve 2.5 g in a liter of water. Store at 4° C and prepare fresh monthly.
6. Ascorbic Acid, 0.1 mol/L. Prepare just before use by dissolving 0.2 g in 10 ml water.
7. Alkaline Ascorbate. Prepare just before use by adding 18 ml 5 mol/L NaOH to 2 ml ascorbic acid solution.
8. Boric Acid, H_3BO_3, 0.6 mol/L. Dissolve 40 g H_3BO_3 in water and make up to 1 L volume.
9. a. Catecholamine Stock Standard, norepinephrine hydrochloride salt. (CalBiochem or Sigma.) Prepare a solution containing 200 µg/ml free base in 0.1 mol/L HCl by dissolving 24.4 mg norepinephrine hydrochloride in 0.1 mol/L HCl and making up to 100 ml volume.
   b. Working Standard 2.0 µg/ml. Transfer 1.0 stock standard to a 100 ml volumetric flask, add 2 ml 0.1 mol/L HCl and make up to volume with water.

10. Polyethylene columns. (Whale Scientific)
11. Resin, Bio-Rex 70 (Bio-Rad) cation exchange resin, 200–400 mesh, sodium form. Suspend 200 g resin in water in a liter beaker and decant the fines. Repeat four more times. Adjust the suspension to pH 6.5 with glacial acetic acid and let stand overnight with occasional stirring. Decant and store in water, checking to see that the pH is 6.5.
12. Bromthymol Blue, 0.4 g/L. Dissolve 0.4 g bromthymol blue (Harleco) in water and make up to 1 L volume.
13. Preparation of Columns. Plug the bottom of the polyethylene column with a little glass wool and pour a slurry of resin into it, using the collared funnel that slips over the end of the column. Fill the column to about 5 cm of settled resin. Add a small layer of glass wool and wash column with 10 ml water.
14. Fluorometer with a primary (exciting) #405 filter (Turner), composite filter #7–51 (Corning), and Wratten #2C (Eastman) and secondary (emitting) filters of Wratten #65A and #8. The excitation wavelength for spectrofluorometers is 396 nm; the emission is measured at 506 nm.

PROCEDURE

1. *Urine Collection*
    a. Collect a 24-hour urine sample in a bottle containing 20 ml 6 mol/L HCl (concentrated HCl diluted with an equal volume of water). The final pH should be pH 3 or below. The acidified specimen should be refrigerated until assayed.
    b. Transfer 5.0 ml urine to a beaker and adjust to pH 6.5 by the cautious addition of 1 mol/L NaOH while stirring.
2. *Adsorption of Catecholamines and Unconjugated Metanephrines*
    a. Transfer sample to the resin column. The flow rate should not exceed 1 ml per minute.
    b. Wash three times with 10-ml portions of water and discard wash.
3. *Elution of Catecholamines.* Elute the catecholamines with 8 ml boric acid.
4. *Determination of Epinephrine + Norepinephrine*
    a. Set up a series of tubes marked "T" for unknown test, "T_i" for test + internal standard, "B" for urine blank, "S" for standard, and "B_s" for standard blank. An internal standard is added to the test samples in order to compensate for quenching. See Table 9.3 for the additions; the explanations are continued below.
    b. Transfer 1.0 ml of appropriate eluate to the T, T_i, and B tubes and 1.0 ml boric acid to the S and B_s tubes.
    c. Add 100 µl of catecholamine standard (0.20 µg) to each T_i tube and to the S tube.
    d. Add one drop bromthymol blue to all tubes, The T, T_i, and B tubes require additional boric acid solution to bring the pH to 6.5 (from yellow to green color), but the S and B_s tubes are acid and require dilute NH_4OH, 0.02 mol/L to change the color to green.
    e. Add 0.1 ml $K_3Fe(CN)_6$ to all tubes except the two blanks (B and B_s) and allow to stand for 3 minutes after mixing.
    f. Then add 1 ml alkaline ascorbate to all tubes and mix well.
    g. Add the $K_3Fe(CN)_6$ to the blank tubes and mix.
    h. Add water to all tubes to 5.0 ml volume.

TABLE 9.3
Protocol for Catecholamine Assay

OPERATION	TEST (T)	TEST + INTER-NAL STANDARD (T$_i$)	URINE BLANK (B)	STANDARD (S)	STANDARD BLANK (B$_s$)
Eluate, ml	1.0	1.0	1.0	—	—
Boric acid	—	—	—	1.0	1.0
Catecholamine working stand- ard, 2.0 μg/ml	—	100 μl	—	100 μl	—
Bromthymol blue (drops)	1	1	1	1	1
Adjust to pH 6.5 (green color)	Boric acid	Boric acid	Boric acid	NH$_4$OH	NH$_4$OH
K$_3$Fe (CN)$_6$, ml	0.1	0.1	—	0.1	—

Prepare alkaline ascorbate 3 minutes before proceeding to next step.

Alkaline ascorbate, ml	1.0	1.0	1.0	1.0	1.0
K$_3$Fe (CN)$_6$, ml	—	—	0.1	—	0.1
Add water to 5.0 ml volume for all tubes					

i. Read in fluorometer 10 minutes after the addition of alkaline ascorbate. Primary filter: 405 nm; secondary filters, Wratten #65 A and #8 (Eastman). Record the fluorescence after setting the instrument to zero with a water blank.

5. *Calculation.* Let F_u = fluorescence reading of tube T minus reading of its blank, B; F_s = fluorescence reading of standard minus its blank B$_s$; F_{si} = fluorescence reading of internal standard which = $F_{ti} - F_t$.

If $F_{si} \gtreqqless 0.9 \times F_s$, no correction for quenching is necessary. Otherwise multiply results by F_s/F_{si}.

$$\frac{F_u}{F_s} \times 0.20 \times \frac{8 \text{ ml}}{1 \text{ ml}} \times \frac{100 \text{ ml}}{5 \text{ ml}} = \frac{F_u}{F_s} \times 32 \text{ μg/dl}$$

$$\frac{F_u}{F_s} \times 32 \times \frac{TV \text{ (ml)}}{100 \text{ ml}} = \text{μg catecholamine/day}$$

REFERENCE VALUES. The daily excretion of epinephrine + norepinephrine by adults is usually less than 100 μg/day (<0.56 μmol/day) but this varies to some extent with muscular activity.

INCREASED EXCRETION. The same as described for VMA.

Urinary Metanephrines

The metanephrines are usually excreted as conjugates and must be hydrolyzed prior to analysis. The metanephrines are adsorbed upon the same weak cation exchange resin as the catecholamines but are eluted by a dilute solution of NH$_4$OH instead of boric acid. The metanephrines are then oxidized to vanillin and are determined spectrophotometrically. See the catecholamine reference for details, p. 293.

REFERENCE VALUES. 0.3 to 0.9 mg/day (1.5 to 4.5 μmol/day) for adults.

INCREASED EXCRETION. *See* VMA.

PARATHYROID HORMONE

Parathyrin (PTH)

The parathyroids are two pairs of small glands located close to the posterior surface of the thyroid gland. The glands produce a polypeptide hormone, parathyrin (PTH), which, together with calcitonin (a hormone produced in the thyroid gland) and 1,25-dihydroxy-D_3 from the kidney regulates the level of ionized calcium in plasma (p. 348). PTH is different from the other hormones in that it is not stored in the gland as granules; it is secreted directly into the plasma from the gland. Its secretion is directly controlled by the concentration of plasma Ca^{2+}. PTH has a short half-life in plasma of 20 to 30 minutes before it is inactivated, presumably by proteolysis (peptide chain cleavage).

PTH, in collaboration with 1,25-dihydroxy-D_3, raises the plasma Ca^{2+} by several mechanisms: (1) by promoting dissolution of bone; (2) by increasing the tubular reabsorption of Ca^{2+} in the kidney; and (3) by promoting the intestinal absorption of Ca^{2+}. At the same time, the plasma phosphate is lowered by the action of PTH in decreasing the tubular reabsorption of phosphate.

Increased plasma concentrations of PTH are found in primary hyperparathyroidism, usually caused by a benign adenoma; hyperplasia of the parathyroid in response to a chronically decreased plasma Ca^{2+} concentration (e.g., severe renal disease); and autonomous (uncontrolled) hypersecretion or ectopic PTH production from malignant tumors of non-endocrine origin (from some types of lung cancer).

A decreased plasma level of PTH is usually found in primary hypoparathyroidism caused by atrophy or damage to the glands and in hypoparathyroidism secondary to a persistent hypercalcemia.

The measurement of plasma PTH by RIA techniques is of diagnostic value in the above conditions. Circulating PTH exists as a mixture of the prohormone, the intact hormone, and fragments from both the carboxyterminal (C-PTH) and aminoterminal (N-PTH) ends of the molecule after splitting of PTH by proteases. The C-PTH fragment constitutes the major portion of circulating PTH in primary hyperparathyroidism even though it is biologically inactive; the intact molecule and N-PTH, the active forms of the hormone, are also present.

C-PTH or N-PTH can be measured separately by RIA techniques by use of the appropriate antibodies. The C-PTH assay is the method of choice for investigating primary hyperparathyroidism and is helpful in assessing the hypercalcemia of malignancies or in prolonged hypocalcemia. The N-PTH assay is most helpful for detecting hyperparathyroidism secondary to chronic renal disease, particularly in patients undergoing dialysis.[20] The investigation of these types of cases requires the simultaneous determination of total serum calcium and frequently that of ionized calcium (p. 351).

PTH Reference Range. 0.10 to 0.65 ng/ml when ionized calcium = 1.02 to 1.16 mmol/L

Calcitonin has the opposite effect of PTH upon calcium ion concentration, although by other mechanisms. When the concentration of Ca^{2+} is increased, the secretion of calcitonin inhibits the mobilization of calcium from bone by its inhibitory effect upon osteoclasts, the cells that dissolve bone. The half-life of calcitonin is shorter than that of PTH; calcitonin is also a polypeptide but is much smaller than PTH. Calcitonin can also be measured by RIA, but there is little clinical use for this determination at present.

HORMONES OF THE GASTROINTESTINAL TRACT

The gastrointestinal tract secretes several different hormones that stimulate particular exocrine glands to secrete their digestive juices in large quantities. One hormone of this group is gastrin, the most powerful stimulant for gastric acid secretion that is known; it is 1500 times as potent as histamine in this respect, the strongest acting drug that we have. Gastrin weakly stimulates the gastric cells to secrete some pepsin and intrinsic factor and causes the pancreas to increase it flow of digestive juices, but its primary action is to greatly increase the flow of gastric HCl.

Gastrin is produced primarily by cells of the gastric antrum (the end of the stomach adjoining the duodenum), but some gastrin is also produced by adjacent duodenal mucosal cells and by delta cells of the pancreatic islets. Some types of pancreatic islet tumors (non-β cell) produce large amounts of gastrin.

Gastrin is a relatively small peptide (17 amino acids in the chain), but a larger peptide ("big" gastrin) with the same physiologic activity as the smaller peptide and other gastrins, both larger and smaller, have been found. The concentration of gastrin in plasma can be measured by RIA methods.

Gastrin measurements are important clinically because overproduction of this hormone causes ulcers in the upper gastrointestinal tract. The most important clinical use of the gastrin test is to detect a condition known as the Zollinger-Ellison syndrome. In this disease, non-beta cell tumors of the pancreas produce large amounts of gastrin, and the patient suffers severely from ulcers; the plasma gastrin level may be grossly elevated.

The gastrointestinal tract generates other hormones, notably, cholecystokinin-pancreozymin and secretin. Cholecystokinin and pancreozymin were formerly thought to be two separate hormones, with cholecystokinin inducing contraction of the gallbladder and pancreozymin stimulating the secretion of pancreatic enzymes. They have turned out to be identical hormones secreted by the intestinal mucosa in response to the presence of food. Another intestinal hormone, secretin, increases the pancreatic secretion of water and HCO_3^-. These hormones are sometimes injected for the purpose of studying pancreatic response, but there is no clinical demand as yet for measurement of the plasma concentration of these hormones.

PANCREAS HORMONES

The pancreas is one of the few glands with both an exocrine and endocrine function. The pancreas plays an important role in the digestive process by secreting into the intestine by way of the common bile duct a juice that is rich in bicarbonate and a multiplicity of digestive enzymes. The bicarbonate in pancreatic juice neutralizes the acid in the fluid coming from the stomach, and different enzymes hydrolyze proteins, peptides, triglycerides, starch, disaccharides, esters, and other substances. The hormones insulin and glucagon are produced in the pancreatic islets of Langerhans and are secreted into the blood. Insulin is made by the beta cells of the islets and glucagon by the alpha cells.

Insulin

Insulin is a relatively small protein, molecular weight 6000, which is originally synthesized as a much larger, physiologically inactive protein, proinsulin. A section of the polypeptide chain of proinsulin is cleaved by proteolytic enzymes in the islets and eliminated, leaving the active insulin molecule stored in the islets as granules. A small amount of proinsulin, nevertheless, can be found in the bloodstream.

As described on page 305, insulin is a potent hormone that accelerates glucose uptake by tissues and promotes glycogen storage, triglyceride synthesis, and protein synthesis. It is best known as the antidiabetes hormone, since it corrects many of the metabolic defects in this disease, but insulin is unable to prevent the occurrence of defects in the renal and retinal capillaries that are typical of chronic diabetes. The diagnosis of diabetes is usually made upon the basis of history and blood glucose concentration in either the fasting state or after a glucose load (see Chap. 10) and not upon the basis of plasma insulin concentration.

Measurement of the plasma insulin concentration is of value in the diagnosis of adenoma of the pancreas (insulinoma); the insulin concentration is high in this disease. Insulin analysis is performed by radioimmunoassay. The plasma insulin concentration of normal individuals does not exceed 20 μU/ml (860 pg/ml). It is elevated in patients with an insulinoma and in those with excessive growth hormone secretion (gigantism in children, acromegaly in adults). Quantitation of plasma insulin is of no value in the diagnosis of diabetes mellitus.

Glucagon

Glucagon is also a polypeptide hormone produced in the pancreatic islets, but by different cells from those that make insulin. Glucagon has an effect upon plasma glucose similar to that of epinephrine; it increases the glucose concentration by inducing the rapid breakdown of stored liver glycogen (p. 306). Glucagon also stimulates the hydrolysis of triglycerides in adipose tissue to fatty acids and glycerol.

Measurement of plasma glucagon levels has no therapeutic value at this time. Glucagon is sometimes injected intravenously as a provocative test in certain diagnostic situations. In patients with insulinoma, the injection of glucagon produces hyperglycemia, followed by hypoglycemia; the hypoglycemia does not occur to a significant extent in individuals who do not have insulinoma. If glucagon is injected into patients with von-Gierke's disease (Type I glycogen storage disease), the plasma glucose concentration does not rise because the hepatic enzyme, glucose-6-phosphatase, is missing; the glucose is "locked" in the cell; this helps to confirm the diagnosis.

REFERENCES

1. Cryer, P.E.: Diagnostic Endocrinology, 2nd ed. New York, Oxford University Press, 1979.
2. *Ibid.*, p. 8.
3. *Ibid.*, p. 33.
4. *Ibid.*, p. 55.
5. *Ibid.*, p. 117.
6. Gornall, A.G. (editor): Applied Biochemistry of Clinical Disorders. Hagerstown, Harper and Row, 1980.
7. Hall, R., and Gomez-Pan, A.: The hypothalamic regulatory hormones and their applications.

In Advances in Clinical Chemistry, Vol. 18, edited by O. Bodansky and A.L. Latner. New York, Academic Press, 1976, p. 173.

8. Hershman, J.M. (editor): Endocrine Pathophysiology: A Patient Oriented Approach. 2nd ed., Philadelphia, Lea & Febiger, 1982.

9. Hershman, J.M., et al.: Thyroid disease. *In* Endocrine Pathophysiology: A Patient Oriented Approach, 2nd ed., edited by J.M. Hershman. Philadelphia, Lea & Febiger, 1982, p. 34.

10. Carlson, H.E.: Pituitary disease. *In* Endocrine Pathophysiology: A Patient Oriented Approach, 2nd ed., edited by J.M. Hershman. Philadelphia, Lea & Febiger, 1982, p. 7.

11. Weitzman, R.E., and Kleeman, C.R.: Disorders of water metabolism. *In* Endocrine Pathophysiology: A Patient Oriented Approach, 2nd ed., edited by J.E. Hershman. Philadelphia, Lea & Febiger, 1982, p. 254.

12. Wolfsen, A.R., Tuck, M., and Kaplan, S.A.: Adrenal disease. *In* Endocrine Pathophysiology: A Patient Oriented Approach, 2nd ed., edited by J.E. Hershman. Philadelphia, Lea & Febiger, 1982, p. 69.

13. Wellby, M.L.: The laboratory diagnosis of thyroid disorders. *In* Advances in Clinical Chemistry, Vol. 18, edited by O. Bodansky and A.L. Latner. New York, Academic Press, 1976, p. 103.

14. Foss, O.P.: Determination of protein-bound iodine in serum. *In* Standard Methods of Clinical Chemistry, Vol. 4, edited by D. Seligson. New York, Academic Press, 1963, p. 125.

15. Pileggi, V.J., and Kessler, G.: Determination of organic iodine compounds in serum. IV. A new nonincineration technic for serum thyroxine. Clin. Chem. 14, 339, 1968.

16. Silber, R.H.: Free and conjugated 17-hydroxycorticosteroids in urine. *In* Standard Methods of Clinical Chemistry, Vol. 4, edited by D. Seligson. New York, Academic Press, 1963, p. 113.

17. Smith, E.K., and Muehlbaecher, C.A.: A fluorometric method for plasma cortisol and transcortin. Clin. Chem. 15, 961, 1969.

18. Rutherford, E.R., and Nelson, D.H.: Determination of urinary 17-ketogenic steroids by means of sodium metaperiodate oxidation. J. Clin. Endocrinol. Metab. 23, 533, 1963.

19. Peterson, R.E.: Determination of urinary neutral 17-ketosteroids. *In* Standard Methods of Clinical Chemistry, Vol. 4, edited by D. Seligson. New York, Academic Press, 1963, p. 151.

20. Simon, M., and Cuan, J.: Diagnostic utility of C-terminal parathyrin measurements as compared with measurements of N-terminal parathyrin and calcium in serum. Clin. Chem. 26, 1672, 1980.

21. Hobkirk, R., and Metcalfe-Gibson, A.: Urinary estrogens. *In* Standard Methods of Clinical Chemistry, Vol. 4, edited by D. Seligson. New York, Academic Press, 1963, p. 65.

22. Pisano, J.J., Crout, J.R., and Abraham, D.: Determination of 3-methoxy-4-hydroxy mandelic acid in urine. Clin. Chim. Acta 7, 285, 1962.

23. Kenny, M.A.: Practical aspects of quality control for radioimmunoassays. Am. J. Med. Technol. 42, 318, 1976.

24. Penney, L.L., and Klenke, W.J.: Variability in unconjugated and total estriol in serum during normal third trimester pregnancy. Clin. Chem. 26, 1800, 1980.

Chapter 10

DISORDERS OF CARBOHYDRATE METABOLISM

Carbohydrate is the major component of the human diet in many areas of the world and is an important source of body energy. The capacity of the body to store carbohydrate is limited, however, and is confined to liver (can store up to 10% of its wet weight) and muscle (can store up to 0.5% of its wet weight); the total amount of stored carbohydrate is sufficient to fulfill the energy requirements of a person for only about one-half day.

All tissues can utilize glucose, the principal and almost exclusive carbohydrate circulating in blood, but under fasting conditions, only a few tissues depend entirely upon glucose as a source of energy. These are the brain, by far the most important glucose consumer, followed to a much lesser extent by red blood cells, platelets, leukocytes, and the kidney medulla. Other tissues readily oxidize fatty acids and ketones for energy purposes. As will be shown later in the chapter, plasma glucose levels are maintained during fasting by mobilization of muscle protein; the liver synthesizes glucose (gluconeogenesis) from certain amino acids (primarily alanine) derived from proteolysis. Liver gluconeogenesis is the mechanism during fasting for providing glucose to the tissues that are completely dependent upon glucose for viability.

INTERMEDIARY CARBOHYDRATE METABOLISM

Plasma glucose is derived from the hydrolysis of dietary starch and polysaccharides, from the conversion of other dietary hexoses into glucose by the liver, and from the synthesis of glucose from amino acids or pyruvate.

In times of glucose excess (elevated blood glucose, as after a meal), glucose is enzymatically polymerized in the liver to form glycogen (glycogenesis). When the blood glucose begins to drop, the glycogen is converted to glucose (glycogenolysis) by a different set of enzymes. Thus, independent mechanisms exist for regulating the blood glucose level by means of glycogenesis-glycogenolysis reactions. The liver is the main organ for the storage of excess carbohydrate as glycogen, and skeletal and heart muscles can store minor amounts of muscle glycogen. When the glycogen storage capacity is saturated, glucose is converted to fat by adipose cells where it is stored.

The energy stored in the glucose molecules is made available to the organism through several catabolic pathways that mainly generate ATP. The principal pathway of glucose oxidation consists of two phases, anaerobic and aerobic oxidation (oxidative phosphorylation), both of which are comprised of many different enzymatic steps. The anaerobic phase is better known as *glycolysis* and is illustrated in Figure 10.1. The glucose-6-P is converted through several steps to a triose phosphate and then to pyruvate. All of these reactions take place in the cytoplasm. Glycolysis can be reversed, i.e., pyruvate can be converted back to glucose-6-P, but by a partially different pathway.

When there is sufficient oxygen present, the aerobic phase of glucose utilization begins with the decarboxylation of pyruvate (Fig. 10.2). Pyruvate is

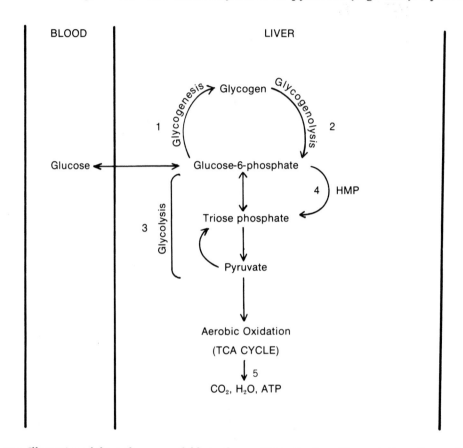

Fig. 10.1. Illustration of the pathways available to glucose-6-P in the liver. The anabolic pathway of glycogenesis is shown as pathway 1, and the regeneration of glucose-6-P by hydrolysis of glycogen in pathway 2. The catabolic pathways are glycolysis (3) and the hexose monophosphate shunt (HMP, 4). The catabolism of glucose-6-P is completed by aerobic oxidation (5).

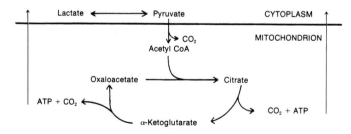

Fig. 10.2. Illustration of the aerobic phase of glucose oxidation. Pyruvate is converted to acetyl CoA in the mitochondrion where the acetyl CoA enters the tricarboxylic acid cycle (TCA). Only three of the many intermediate compounds of the TCA cycle are shown. Fifteen moles of ATP are produced per mole of pyruvate in the process of oxidative phosphorylation (30 moles ATP per mole glucose).

converted in the mitochondria to acetyl CoA that is oxidized to CO_2 and H_2O, to generate many molecules of ATP. The energy derived from the aerobic oxidation of glucose is about 19 times as great as that obtained by anaerobic glycolysis alone.

Acetyl CoA is the point where carbohydrate, protein, and fat metabolism intersect. Acetyl CoA is the two-carbon fragment derived from the catabolism of glucose, some amino acids, and fatty acids and then oxidized completely in the TCA cycle (see Figs. 10.2 and 10.3). Extrahepatic tissues utilize the acetyl CoA almost exclusively in the TCA cycle, but other pathways are possible in the liver. Under the combined conditions of depletion of carbohydrate and mobilization of fatty acids, as occurs in starvation and diabetes mellitus, two molecules of acetyl CoA may condense in liver mitochondria to form acetoacetyl CoA (Fig. 10.4). This latter compound then forms ketone bodies or is utilized for the synthesis of cholesterol (Fig. 10.5). This will be discussed further in the section on ketone bodies.

Glucose-6-P is also a pivotal point for three possible metabolic pathways, which are illustrated in Figure 10.1. In addition to the glycogenesis and the glycolytic pathways discussed above, glucose-6-P may proceed by the hexose monophosphate pathway (HMP) to form 1 mole of triose phosphate and 3 moles of CO_2 per mole of glucose-6-P. The HMP pathway is important as an energy

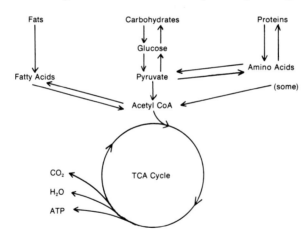

Fig. 10.3. Intersection of carbohydrate, protein, and fat metabolism at the level of pyruvate or acetyl CoA. This 2-carbon fragment is common to the catabolism of all three foodstuffs.

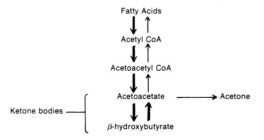

Fig. 10.4. Production of ketone bodies (acetoacetate, β-hydroxybutyrate, and acetone) during fasting states and in diabetes mellitus. Acetoacetate is converted to β-hydroxybutyrate by an enzyme; the conversion of acetoacetate to acetone is a spontaneous reaction that proceeds quite slowly.

source for erythrocytes that lack mitochondria and hence, cannot carry out oxidative phosphorylation.

Four pathways intersect at triose phosphate (Fig. 10.6). (1) Triose phosphate is produced by the glycolysis of glucose-6-P, but, since the enzymatic steps are reversible, triose phosphate in appropriate circumstances may be converted to glucose-6-P (gluconeogenesis); (2) triose phosphate is also produced by the HMP shunt; (3) it may be converted to glycerol in a side reaction that is reversible. The glycerol utilized for the synthesis of triglycerides from fatty acids and glycerol is produced by this reaction. Conversely, glycerol derived from

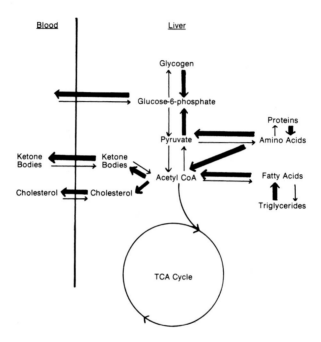

Fig. 10.5. Metabolic changes characteristic of diabetes mellitus. The overproduction of acetyl CoA, which comes from the accelerated catabolism of amino acids and fatty acids is diverted to the synthesis of ketone bodies and cholesterol instead of entering the TCA cycle; the supply of oxaloacetate in the TCA cycle is limited because of the depressed glucose catabolism.

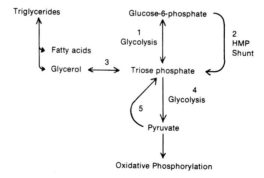

Fig. 10.6. Illustration of four pathways for the formation of triose phosphate: from glycolysis (1), from the HMP shunt (2), from glycerol (3), and from pyruvate by a separate enzymatic path (5). Triose phosphate may continue on the glycolytic pathway to pyruvate (4), be converted back to glucose-6-P (glucogenesis, reversal of 1), or be converted to glycerol (reversal of 3).

the hydrolysis of triglycerides enters the glycolytic pathway through conversion to triose phosphate. (4) Triose phosphate may be produced from pyruvate.

In spite of the multiple pathways possible for some of the key compounds in intermediary carbohydrate metabolism, the blood glucose concentration is kept remarkably constant in ordinary circumstances. This is made possible by a multiplicity of delicate control mechanisms for some of the key reactions, with the net result that some enzymatic reactions are inhibited while others are accelerated in order to maintain a relatively constant blood glucose concentration to meet the body's energy needs. Some of the control mechanisms work by feedback inhibition; others are under hormonal control. The operation of the TCA cycle is limited by the available oxygen supply and by the concentration of oxaloacetate, a compound derived from carbohydrate metabolism.

From a clinical viewpoint, most patients with disturbances in carbohydrate metabolism have some defect in their hormonal control, although some patients may be deficient in a particular enzyme.

KETONE BODIES. Ketone bodies are produced when excessive amounts of fatty acids are catabolized and the availability of glucose is limited (Figure 10.5). These conditions occur most commonly in diabetes mellitus and in prolonged fasting.

The three ketone bodies, acetoacetate, β-hydroxybutyrate, and acetone, are shown in Figure 10.4 as arising from the oxidation of fatty acids. When there is insufficient oxaloacetate to drive the TCA cycle, 2 moles of acetyl CoA join to form 1 mole of acetoacetyl CoA, which is then converted to acetoacetate. A portion of the latter compound is enzymatically reduced to β-hydroxybutyrate while a portion spontaneously decomposes to acetone and CO_2. The proportion of the three ketone bodies in blood and urine is variable.

Heart and skeletal muscle can utilize acetoacetate as a source of energy to some extent; the brain acquires a limited capacity for oxidizing ketone bodies after about 3 weeks of fasting, a protective measure to curtail muscle wasting (gluconeogenesis from muscle protein). When the production of ketone bodies exceeds the capacity of the tissues to use them, a condition known as *ketosis*, they are rapidly excreted in the urine. The excretion of ketones into the urine is called *ketonuria*, and an elevated level of ketones in the blood is called *ketonemia*. The overproduction of ketoacids causes acidosis or lowered blood pH.

HORMONAL REGULATION OF PLASMA GLUCOSE

Hormones play an important role in the regulation of the plasma glucose concentration. Insulin, the hormone produced by the pancreatic β cells, is the only one that lowers the concentration of plasma glucose. Hormones secreted by the anterior pituitary (growth hormone), adrenal medulla (epinephrine), adrenal cortex (cortisol), α-cells of the pancreas (glucagon), and thyroid (thyroxine) increase the plasma glucose concentration. The net result of the multiplicity of hormone effects, nervous system control, and feedback mechanisms is a very finely regulated system that keeps the plasma glucose concentration quite constant.

Decrease of Plasma Glucose Concentration

Insulin exerts a plasma glucose-lowering effect by several mechanisms: (1) by increasing the entry of glucose from plasma into muscle and adipose tissue cells; (2) by promoting liver glycogenesis· and thus converting glucose to glycogen; (3) by promoting glycolysis which accelerates glucose utilization; (4) by promoting lipid synthesis from glucose in adipose tissue; and (5) by promoting amino acid synthesis from glucose intermediates. In summary, relative excess of insulin in the blood produces a lowering of the plasma glucose concentration, and a relative deficiency is associated with an elevated plasma glucose concentration.

The rate of insulin secretion into the bloodstream is governed by the blood glucose concentration. When the plasma glucose concentration rises as it does after a meal, the β cells of the pancreatic islets are stimulated to secrete more insulin. Insulin secretion returns to the resting state when the plasma glucose concentration returns to normal. Conversely, a below normal plasma glucose concentration shuts off the secretion of insulin.

Some amino acids, such as leucine and arginine, lower the plasma glucose concentration by stimulating insulin secretion. Tolbutamide, a sulfonylurea, also induces insulin secretion and so is used as an oral, antidiabetic agent.

Increase of Plasma Glucose Concentration

GROWTH HORMONE. Growth hormone is produced by the anterior pituitary gland (p. 275). Its action is antagonistic to that of insulin. It raises the plasma glucose concentration by (1) inhibiting the entry of glucose into muscle cells, (2) inhibiting glycolysis, and (3) inhibiting the formation of triglycerides from glucose. The secretion of growth hormone is stimulated by a lowered plasma glucose concentration (hypoglycemia) which is just the opposite of the stimulus for insulin secretion.

EPINEPHRINE (ADRENALINE). Epinephrine and norepinephrine are the hormones secreted by the adrenal medulla (p. 289). They raise the plasma glucose concentration by causing the rapid breakdown of liver glycogen to glucose. The stimulus for epinephrine secretion is physical or emotional stress, which is neurogenic.

GLUCAGON. Glucagon is the hormone produced by the α-cells of the pancreas (p. 298). Glucagon secretion raises the plasma glucose concentration by promoting hepatic glycogenolysis in a manner similar to that of epinephrine.

Glucagon secretion is governed by the plasma glucose level but in an opposite manner to that of insulin. Glucagon output is stimulated by hypoglycemia and suppressed by hyperglycemia.

CORTISOL. Cortisol (hydrocortisone) as well as other 11-oxysteroids secreted by the adrenal cortex (p. 280) raise the plasma glucose concentration by (1) promoting gluconeogenesis from the breakdown of proteins (the conversion of amino acids to glucose), and (2) decreasing the entry of glucose into muscle cells; cortisol is an insulin antagonist. The stimulus for the increased output of cortisol is an increased secretion of ACTH, the adrenal cortical stimulating hormone of the anterior pituitary.

THYROXINE. Excess hormone from the thyroid gland promotes the conversion of liver glycogen to glucose and accelerates the absorption of glucose from the intestine during meals; its role in carbohydrate metabolism is minor.

DISORDERS OF CARBOHYDRATE METABOLISM

The various disorders in carbohydrate metabolism may be grouped into several categories that are dependent primarily upon laboratory findings. They are those that are associated with a raised plasma glucose concentration (hyperglycemia), a decreased plasma glucose concentration (hypoglycemia), and reducing sugars in the urine.

Diseases Associated with Hyperglycemia

Hyperglycemia is demonstrably harmful to the body when it is so high that the increased extracellular osmotic pressure causes cellular dehydration; coma can be produced by severe dehydration of brain cells. Other associated factors, such as acidosis and dehydration, may cause cellular damage and produce illness. A good example is uncontrolled diabetes mellitus. The patient may be found in a coma, with a blood glucose concentration exceeding 600 mg/dl; the accompanying acidosis, ketosis, dehydration, and electrolyte imbalance must be corrected as a lifesaving measure.

The potential pathological changes that may be ascribed to periodic, mild hyperglycemias in diabetics taking insulin are subtle but may become significant over a period of years. When the plasma glucose concentration is elevated over a period of time, normal hemoglobin A becomes glycosylated (HbA_{1c}); a nonenzymatic reaction occurs by which glucose covalently binds to the terminal valine of the hemoglobin A beta chain. This is demonstrable by column chromatography because the glycosylated hemoglobin has different adsorption and elution properties. This is used as a measure for ascertaining the effectiveness of blood glucose control in a diabetic because the concentration of HbA_{1c} varies directly with the elevation in plasma glucose.[7] The question arises: What other proteins, so far undiscovered, are also glycosylated in hyperglycemia? Could this be responsible for the long-term complications of diabetes mellitus described below? About 7 to 11% of a diabetic's hemoglobin is HbA_{1c} compared to 4 to 6% for a nondiabetic.

Diabetes Mellitus[1,2]

Diabetes mellitus is by far the most important disease affecting carbohydrate metabolism, but the disorder is by no means limited to this area; protein and fat metabolism are also affected. This is manifested by protein-wasting, with accelerated conversion of amino acids to glucose, and by increased catabolism of triglycerides, accompanied by overproduction of ketone bodies and cholesterol. The metabolic changes that are characteristic of uncontrolled diabetes are illustrated in Figures 10.4 and 10.5.

Many pathologic changes are associated with the chronic form of diabetes.[2] The most common complications are vascular lesions in capillaries and small veins leading to typical changes in the retina of the eye and to nephrosclerosis (kidney lesions), calcification of the large blood vessels, causing atherosclerosis ("hardening of the arteries") and coronary vascular disease, and neurologic defects (neuropathy). The blood capillary defect is a thickened basement membrane in all capillaries that hinders the exchange of substances and fluid. Blindness, renal disease, neuropathy and atherosclerosis are common occurrences in long-standing diabetes.

Diabetes is caused by a relative or absolute lack of insulin. An *absolute* deficiency of insulin occurs when there is degeneration of the pancreatic islet cells and insufficient synthesis of insulin to meet body needs. A *relative* deficiency of insulin may occur when antibodies to insulin are circulating or hormones that antagonize or oppose the action of insulin (growth hormone, ACTH, cortisol) are produced in excess.

The metabolic effects of insulin were discussed in the section dealing with the hormonal regulation of plasma glucose (p. 305). An insulin deficiency causes a reversal of all of the metabolic actions and leads to the following physiologic abnormalities:

1. *Hyperglycemia* is produced because the uptake of plasma glucose by muscle and fat cells is decreased; the breakdown of liver glycogen to glucose predominates; glycolysis is inhibited; the mobilization of protein from muscle cells, with conversion of amino acids to glucose is stimulated.

2. Ketosis (ketonemia) is produced as fatty acids are mobilized from the depot stores and converted to acetoacetyl CoA, as shown in Figure 10.4. The acetoacetyl CoA leads to the production of large quantities of ketone bodies and cholesterol, which diffuse from hepatic cells into the blood. Normally no ketone bodies are detected in plasma by conventional methods.

3. The fat mobilization and increased cholesterol synthesis lead to hyperlipidemia (increased concentration of blood lipids) and hypercholesterolemia.

4. The hyperglycemia leads to a large excretion of glucose in the urine (glucosuria or glycosuria) and to polyuria.

5. The presence of ketonemia leads to excretion of ketone bodies in the urine (ketonuria).

6. The high production and excretion of acetoacetate and β-hydroxybutyrate, which are moderately strong acids, lead to exhaustion of the blood buffer systems, with production of an acidosis. The blood pH falls.

7. The excretion into the urine of large amounts of glucose and ketone bodies produces an osmotic diuresis, with much loss of water and electrolytes.

Without treatment, the patient becomes severely acidotic and dehydrated and loses consciousness. Some patients in coma have been brought to the hospital with a blood pH as low as 6.9 and with a serum glucose concentration of greater than 1000 mg/dl.

DIAGNOSIS OF DIABETES MELLITUS. It has been customary to classify diabetes into four stages. The following test results are usually found in the various stages of the disease, and are described here in the nomenclature of the American Diabetes Association.[2]

Overt Diabetes (the severe case, with clear clinical symptoms and positive laboratory findings). The fasting serum glucose concentration is grossly elevated (usually exceeding 180 mg/dl), and glucose and ketone bodies are found in the urine, particularly several hours after a meal. A glucose tolerance test is not necessary for the diagnosis.

Chemical or Latent Diabetes. The fasting serum glucose concentration may be normal or mildly elevated (up to 120 mg/dl). No glucose or ketones are detectable in the urine. A loading test (postprandial glucose or oral glucose tolerance test) is required to demonstrate the abnormality. Usually the 2-hour postprandial serum glucose concentration exceeds 120 mg/dl, and the glucose tolerance test (GTT) is abnormal.

Suspected (Subclinical) Diabetes. The fasting serum glucose concentration is normal, and there is no glucosuria or ketonuria. The glucose tolerance test is normal except during pregnancy and conditions of stress. The diagnosis is usually made by a cortisone-glucose tolerance test, which simulates a stress situation.

Prediabetes. All laboratory tests are normal in this stage. The only individuals in this stage are those who have the gene for diabetes but who have not yet shown any of the metabolic abnormalities of diabetes.

Excess Growth Hormone

Excessive secretion of growth hormone may produce a diabetic-like reaction with respect to carbohydrate metabolism. The fasting serum glucose level may or may not be elevated, but the response to a glucose load is similar to that found in diabetes; postprandial serum glucose concentrations are mildly elevated, and the response to a glucose tolerance test is similar to that of a mild diabetic. The differential diagnosis may be made by measuring circulating growth hormone levels.

Excess Cortisol

Cortisol acts antagonistically to insulin with respect to carbohydrate metabolism and, when present in excess, causes a diabetic-like response to glucose loading tests. The plasma concentration of cortisol can be measured directly (p. 281) to detect the elevation.

Excess ACTH

Excessive amounts of ACTH, the adrenal cortical stimulating hormone, cause excessive secretion of cortisol, with the effects described above. Blood levels of ACTH can be measured.

Disorders Associated with Hypoglycemia[1]

Any condition in which the plasma glucose concentration falls below the lower limits of normal by greater than 2 standard deviations (below 60 mg/dl) is called *hypoglycemia*. Since the brain is dependent upon an adequate supply of glucose for its energy, the clinical symptoms of hypoglycemia resemble those of cerebral anoxia, which may include one or more of the following: faintness, weakness, dizziness, tremors, anxiety, hunger, palpitation of the heart, or "cold sweat"; there may even be mental confusion and motor incoordination. Consciousness is usually lost in an adult when the plasma glucose concentration falls below 40 mg/dl, but the rapidity of fall is also a factor. Widespread convulsions may accompany the coma or even precede it. Newborn infants are less sensitive to a decreased concentration of plasma glucose and may not go into convulsions until the plasma glucose concentration falls below 25 or 30 mg/dl.

The lower the plasma glucose level, the deeper is the coma for both adults and infants. The cortical centers are the first to be affected because they have the highest energy requirement in the brain, but with time, the lower centers are also affected; irreversible brain damage or death may occur if the hypoglycemia and coma persist too long. If the length of time in coma is short (less than 20 minutes), the intravenous injection of glucose usually restores consciousness immediately, with no permanent brain damage.

Note: It is imperative for a technologist who finds a serum glucose concentration in a patient at a hypoglycemic level to contact the attending physician at once by telephone and transmit the result because of the limited time for appropriate action.

There are many possible causes for hypoglycemia, of which the following are the principal ones:

Hormonal
1. *Insulin excess*
    a. Overdosage of insulin in a diabetic (too much insulin injected or failure of a diabetic to eat after usual dosage).
    b. Excessive secretion of insulin by the pancreas.
        (1) Pancreatic hyperplasia.
        (2) Islet cell tumor of the pancreas (insulinoma).
        (3) Insulin secretion induced by leucine.
        (4) Insulin secretion induced by sulfonylureas.
2. *Deficiency of cortisol (Addison's disease).* Fasting produces hypoglycemia.
3. *Deficiency of ACTH.* Fasting produces hypoglycemia.
4. *Deficiency of growth hormone.* Hypoglycemia may appear upon fasting.

Hepatic
1. *Depleted liver glycogen stores*
    a. Prolonged fasting or starvation.
    b. Severe hepatocellular damage.
    c. Acute drug toxicity.
2. *Failure to release liver glycogen (genetic defects)*
    a. Type 1 glycogen storage disease (von Gierke's disease), a deficiency of glucose-6-phosphatase in liver.
    b. Rarer glycogen storage diseases (types 3 and 8) in which glycogen-splitting enzymes are deficient in the liver.

Hereditary Disorders (Enzymatic Defects) with Reducing Sugars in the Urine
1. Galactosemia[3]: In this hereditary disease, galactose cannot be metabolized because an enzyme, galactose-1-P uridyl transferase is lacking or greatly reduced in activity. As a result, galactose-1-P piles up in cells and cannot be converted to glucose. Since galactose is a component of lactose, the sugar in milk, the disease is very serious in infants with this defect. In order to survive, they must receive an artificial milk containing no lactose. The diagnosis must be made early to avoid the crippling effects of galactose, which are hepatosplenomegaly (enlarged liver and spleen), cirrhosis of the liver, cataracts, and mental retardation. The infants fail to thrive and may die unless lactose is removed from the diet.

    The diagnosis is made by identifying galactose in the urine and confirmed by finding a deficiency of the enzyme galactose-1-P uridyl transferase in erythrocytes. The method for the detection of urine sugars will be given at the end of this chapter. It does not pay for a laboratory to set up the enzyme test unless an active screening program is in progress because the disease is rare.
2. Hereditary Fructose Intolerance.[4] This is a rare genetic disorder in which fructose-1-P accumulates in cells because the enzyme aldolase, which

converts fructose-1-P to triosephosphate, is lacking. The ingestion of fruit or sucrose (a disaccharide composed of glucose and fructose) produces vomiting, hypoglycemia, failure to thrive, and hepatomegaly. Children with this defect who learn to avoid fruit and products containing cane sugar survive. The laboratory findings are fructose in the urine and hypoglycemia after the administration of a test meal containing fructose.

3. There are several harmless hereditary conditions in which reducing sugars (sugars that reduce Cu^{2+} to Cu_2O in a hot, alkaline solution) appear in the urine. These are *essential fructosuria* and *essential pentosuria* in which fructose and xylulose, respectively, are the sugars excreted. In both conditions, the lack of a specific enzyme causes an increase in concentration of the particular sugar and its resultant excretion in urine. It is important to make the diagnosis so that the patient does not have to undergo repeated examinations throughout life for possible diabetes.

SERUM (PLASMA) GLUCOSE CONCENTRATION

Methods for the quantitative measurement of glucose in blood were introduced at the beginning of the century. The principal use of the test is for the diagnosis and management of diabetes, but it is absolutely essential for the detection and proper management of hypoglycemia, a condition encountered much less frequently. Until 15 or 20 years ago, the majority of the quantitative tests for glucose determination depended upon the oxidation of glucose by hot, alkaline copper solutions or solutions of potassium ferricyanide. These were supplanted by the orthotoluidine test and later by enzyme methods employing either glucose oxidase or hexokinase. Good discussions concerning glucose methodology have been written by Cooper[5] and by Quam, Westgard, and Carey.[6]

Various glucose methods are discussed below. No matter which method is employed, one must take precautions in the sample collection to prevent the utilization of glucose by leukocytes. The glucose loss upon standing in a warm room may be as high as 10 mg/dl per hour. The decrease in serum glucose concentration is negligible if the blood sample is kept cool and the serum separated from the clot within 0.5 hours of drawing. Otherwise, the addition to the collection tube of 2 mg sodium fluoride per ml blood to be collected will prevent glycolysis for 24 hours.

Chemical Methods

By Copper Reduction

The Somogyi-Nelson method[8] was the reference manual method for the serum analysis of "true" glucose for a long time but has been abandoned because it required too much handling and could not be automated readily. Its values do agree closely with those obtained by the o-toluidine and enzymatic assays.

The copper reduction method of Bittner and Manning[9] was widely used in continuous flow automated systems but has been supplanted by the more specific enzyme assays. The serum was dialyzed to separate the serum proteins from the smaller constituents. The dialysate stream was mixed with an alkaline solution of neocuproine (2,9-dimethyl-1,10-phenanthroline hydrochloride) and Cu^{2+} and heated to 95° C. The glucose is oxidized while the Cu^{2+} is reduced to Cu^+; the Cu^+ is complexed by the neocuproine to yield a soluble yellow-orange chromogen whose absorbance is measured at 454 nm. The results by this method are usually 2 to 10 mg/dl higher in normal persons than the true glucose value but may be much higher than this in patients with uremia or ketosis.

By Coupling with o-Toluidine, Manual and Automated

The condensation of glucose with o-toluidine to yield a soluble, green chromogen is widely used for the determination of glucose in various body fluids. The method is simple and rapid. There are few interfering substances, although the reagent reacts with other aldoses (sugars with an aldehyde end group). Galactose and mannose (aldohexoses) and the aldopentoses are usually in such low concentration in plasma that they cause no problems.

REFERENCE. Cooper, G.R., and McDaniel, V.: The determination of glucose by the ortho-toluidine method. Standard Methods of Clinical Chemistry, Vol. 6, edited by R.P. MacDonald. New York, Academic Press, 1970, p. 159.

PRINCIPLE. When heated in a glacial acetic acid solution, glucose condenses with o-toluidine to form an equilibrium mixture of glucosylamine and the corresponding Schiff base. The condensation product has a greenish color whose absorbance is measured at 630 nm.

REAGENTS

1. *Glucose Stock Standard, 1000 mg/dl.* Dissolve 10.00 g reagent grade dextrose in 0.1% (w/v) benzoic acid solution and make up to 1 L volume.
2. *Glucose Working Standards, 50, 100, 150, and 200 mg/dl, respectively.* Transfer 5, 10, 15, and 20 ml each of Glucose Stock Standard to 100 ml volumetric flasks. Make each up to volume with the 0.1% benzoic acid solution.
3. *o-Toluidine Reagent, 60 ml/L in glacial acetic acid.* Dissolve 1.5 g thiourea in 940 ml glacial acetic acid. Then add 60 ml reagent grade o-toluidine. The thiourea acts as an antioxidant and reacts with any acetaldehyde impurity that may be present in the acetic acid.

PROCEDURE

1. Transfer 50 μl samples of standards, patient serums, control serum and water (for Reagent Blank) to appropriately labeled test tubes.
2. Add 3.0 ml o-toluidine reagent to each tube and mix.
3. Stopper the tubes with Teflon-lined screw caps or glass marbles and heat at 100° C for 12 minutes.
4. Remove from bath; cool in ice water for 5 minutes and at room temperature for 10 minutes.
5. Read the absorbance at 630 nm against the reagent blank.
6. Calculation:

$$\text{Glucose Concentration} = \frac{A_u}{A_s} \times C$$

Where A_u is absorbance of unknown, A_s is absorbance of standard, and C is concentration of standard.

An automated procedure for the determination of glucose by the o-toluidine procedure has been described by Snegoski and Freier.[10]

Enzymatic Methods

Clinical chemists turned to enzymatic methods in an attempt to obtain a "true" glucose determination because of the high specificity of an enzyme for a particular substrate. It was hoped by this approach that there would be fewer interfering substances when measuring the glucose concentration in serum. For the method to be practical, the procedure had to be simple, amenable to automation, and reasonably competitive in cost to the existing chemical procedures. The latter consideration is a serious one because, in general, purified enzyme solutions are costly; it thus became imperative to use minimum volumes of enzyme solution.

Glucose Oxidase Method

The first enzymatic method for glucose determination employed glucose oxidase, which is highly specific for the β-isomer of glucose. In the presence of oxygen, glucose oxidase converts β-glucose to gluconic acid and hydrogen peroxide, as shown in reaction (1). Solutions of glucose that have been standing contain an equilibrium mixture of the α and β forms. Some glucose oxidase mixtures contain the enzyme mutarotase, which rapidly converts α-glucose to the β-form as the latter is oxidized by glucose oxidase.

$$(1) \quad \beta\text{-glucose} + H_2O + O_2 \xrightarrow[\text{Oxidase}]{\text{Glucose}} \text{gluconic acid} + H_2O_2$$

Various procedures have been employed for the measurement of the hydrogen peroxide produced, which is equivalent to the amount of glucose oxidized, but many of these utilized a second enzyme, peroxidase, to decompose the hydrogen peroxide, as shown in reaction (2). This reaction requires a hydrogen donor, which becomes colored when it loses two of its hydrogen atoms to an oxygen atom coming from the decomposition of H_2O_2. Some commonly used hydrogen donors were benzidine derivatives (o-tolidine, o-dianisidine).

$$(2) \quad H_2O_2 + H_2N-\text{[o-tolidine]}-NH_2 \xrightarrow{\text{peroxidase}} \text{[o-tolidine blue (oxidized)]} + 2\,H_2O$$

o-tolidine   o-tolidine blue (oxidized)

It was soon found, however, that various serum components, such as uric acid, ascorbic acid, glutathione, and others competed with the chromogen as hydrogen donors and thus caused falsely low values by their interference with full color production. The interfering compounds could be removed by precipitation with the proteins by means of $Zn(OH)_2$ or a $Zn(OH)_2\text{-}BaSO_4$ mixture (Somogyi precipitation methods). This approach, however, required a great deal of handling and became an obstacle to automating the method. The method also posed a danger to the technologists, since benzidine and its derivatives are carcinogenic.

The problems of color inhibition by uric acid and of exposure to a carcinogen were eliminated by introducing a new type of chromogen to react with the H_2O_2-peroxidase mixture. Gochman and Schmitz[11] oxidatively coupled 3-methyl-2-benzothiazolinone hydrazone (MBTH) with N,N-dimethyl aniline (DMA) during the decomposition of H_2O_2 by peroxidase to produce a stable indamine dye. Trinder accomplished the same purpose by oxidatively coupling 4-amino phenazone with phenol in this reaction.[12]

The usually encountered serum levels of ascorbic acid (below 2 mg/dl), even in persons ingesting 2 g of this vitamin daily, do not interfere with glucose determination by this method.

BACK-UP METHOD USING GLUCOSE OXIDASE. A fast and simple method well-suited as a backup for an automated glucose oxidase method or for emergency work is the one employing the Beckman Glucose Analyzer.* Ten μl of serum or plasma are pipetted into a chamber to react with a glucose oxidase-catalase-ethanol solution. The H_2O_2 generated by the oxidation of glucose is decomposed by the catalase-ethanol mixture to form water and acetaldehyde. The rate of oxygen consumption during the oxidation of glucose is measured polarigraphically by an oxygen electrode cell; the glucose concentration is proportional to the rate of oxygen consumption and appears as a digital readout.

For those who do not have access to a special instrument dedicated to the analysis of glucose, the manual determination of glucose on a 1:20 protein-free filtrate by $Zn(OH)_2$ precipitation gives reliable values for "true" glucose.[8]

Hexokinase Method

The large number of potential interferences with the glucose oxidase method led to a search for other enzymatic methods that might be more specific. The hexokinase method turned out to have certain advantages over glucose oxidase and is widely used despite the higher cost of reagents; it is being considered in the USA as the reference method for glucose determination.

REFERENCE. A simple manual method is that employing the CalBiochem kit.

PRINCIPLE. Glucose is phosphorylated by hexokinase, in the presence of ATP, as shown in reaction (3). In reaction (4) the glucose-6-phosphate (G6P) is converted by a second enzyme, glucose-6-phosphate dehydrogenase (G6PD), in the presence of $NADP^+$, to 6-phosphogluconate; NADPH is produced in the reaction. The absorbance of NADPH is measured at 340 nm; the increase in absorbance is proportional to the amount of glucose originally present when the reaction goes to completion.

$$(3) \quad \text{Glucose} + \text{ATP} \xrightarrow[\text{Mg}^{2+}]{\text{Hexokinase}} \text{G6P} + \text{ADP}$$

$$(4) \quad \text{G6P} + \text{NADP}^+ \xrightarrow{\text{G6PD}} \text{6-phosphogluconate} + \text{NADPH} + \text{H}^+$$

---

*Beckman Instruments, Inc.

Uric acid, ascorbic acid, and the other reducing substances that affect some of the glucose oxidase methods have no effect upon the hexokinase method. The hexokinase preparation should be free from contamination with isomerases because they could slow down the hexokinase reaction. Although hexokinase can also phosphorylate mannose and fructose, these substances are normally not present in sufficiently high concentrations in serum to interfere. Protein precipitation is not required for the determination of serum or plasma glucose concentration.

REAGENTS

1. Vial B, containing $NADP^+$. Reconstitute by adding 15.5 ml water and gently swirling.
2. Vial A. Add the entire contents of vial B to vial A and dissolve by gentle inversion. According to the manufacturer, the reagent has the following composition:
   a. Tris buffer, pH 7.5, 50 mmol/L
   b. ATP, 0.5 mmol/L
   c. $NADP^+$, 0.45 mmol/L
   d. Mg^{2+}, 17 mmol/L
   e. Hexokinase, 666 U/L
   f. G6PD, 333 U/L
3. Stock Standard Glucose, 10.0 g/L. Dissolve 1.00 g pure anhydrous D-glucose in water containing 1.0 g benzoic acid per liter. Make up to 100 ml volume in the benzoic acid solution.
4. Working Glucose Standards. Prepare standards of 50, 100, 200, and 400 mg/dl by appropriate dilution of Stock Standard with benzoic acid solution.

PROCEDURE

1. Place 1.5 ml prepared reagent in a series of cuvets for standard, unknowns, and control serum, respectively.
2. Appropriate blanks are set up by placing 1.5 ml of 9 g/L NaCl in a series of cuvets.
3. Add 10 µl of standard, unknowns and control serum to the appropriate cuvets (tests and blanks). Cover with Parafilm and mix. Place in 37° incubator for a faster reaction. It is permissible to carry out the reaction at room temperature, even though it takes longer for the reaction to go to completion.
4. After incubating for 5 or 10 minutes, read the absorbance of each cuvet at 340 nm and check again a few minutes later to insure that an end point has been reached.
5. *Calculation.* Let ΔA be the difference in absorbance reading between each test and its blank.

$$\text{Glucose concentration (mg/dl)} = \frac{\Delta A_u}{\Delta A_s} \times C$$

where C = concentration of standard in mg/dl, and ΔA_u and ΔA_s are the respective differences in absorbance between unknown or standard and its blank.

Note:

1. When one is using a narrow bandpass spectrophotometer and 10 mm

cuvets, the above calculation can be checked against the theoretical value, as follows:

The absorbance of 1 μmole of NADPH per ml at 340 nm in a 10 mm light path = 6.22. In 1.5 ml of solution, the absorbance is 4.15.

$$\mu\text{moles of NADPH formed per 10 }\mu\text{l serum} = \frac{\Delta A}{4.15} = \mu\text{moles glucose oxidized}$$

$$\mu\text{moles of glucose oxidized per ml serum} = \frac{\Delta A}{4.15} \times \frac{1.0}{0.01} = \Delta A \times 24.1$$

$$\text{mg/dl glucose oxidized} = \mu\text{moles} \times \frac{180}{1000} \times 100 = \Delta A \times 434$$

2. If the ΔA exceeds 0.800, the blood glucose concentration exceeds 350 mg/dl and the analysis should be repeated with a smaller serum aliquot. The hexokinase method is also well-suited for analysis by automated, discrete and continuous flow instruments.

REFERENCE VALUES FOR SERUM GLUCOSE. The usually accepted reference values for "true" serum glucose are from 65 to 100 mg/dl (3.6 to 5.6 mmol/L).

The pathologic conditions associated with hyperglycemia and hypoglycemia were discussed earlier in this chapter.

Tests for the Detection of Hyperglycemia

Severe hyperglycemia is manifested by a fasting serum glucose concentration that is distinctly much higher than the upper limit of the reference values (at least 9 standard deviations above the upper limit). Milder hyperglycemia usually requires some type of glucose-loading test for its detection.

FASTING SERUM GLUCOSE LEVELS. A blatant hyperglycemia is revealed by a fasting serum "true" glucose level that exceeds 120 mg/dl (6.7 mmol/L). Many cases may be so mild, however, that the fasting glucose concentration may be within normal limits, ±2 standard deviations. A loading test should be given to those whose fasting serum glucose concentration is between 105 and 120 mg/dl (5.8 to 6.7 mmol/L) or for those whose histories warrant a further check despite being within normal limits.

POSTPRANDIAL SERUM GLUCOSE LEVELS. The simplest loading test is the giving of a 100 g carbohydrate breakfast to a patient after an overnight fast and measuring the serum glucose concentration 2 hours later. A glucose concentration in excess of 120 mg/dl is abnormal but should be confirmed by a glucose tolerance test. Values that lie between 110 and 120 mg/dl are suspicious and should be followed by a glucose tolerance test. A glucose concentration below 109 mg/dl (6.06 mmol/L) is considered to be within normal limits. Too much reliance must not be placed upon a single test, however, and all suspicious cases must be checked by an oral glucose tolerance test.

ORAL GLUCOSE TOLERANCE TEST (GTT). In order to avoid misleading responses, it is essential to place the patients upon a diet containing adequate calories, protein, and at least 150 g carbohydrate a day for 3 days before the test; this stimulates the production of inducible enzymes necessary for the conversion of glucose into glycogen and/or fat. Drugs that may influence the test should not be given for 3 days prior to the test.

PROCEDURE. A blood sample is drawn from a patient after an overnight fast. One hundred g of glucose as a 250 g/L solution are ingested over a 5-minute period. Blood samples are usually drawn for serum glucose analysis at 1, 2, and 3 hours after the glucose ingestion, but some systems also require a sample at 1.5 hours.

INTERPRETATION OF ORAL GTT. The American Diabetes Association called a conference to discuss the standardization of the oral GTT and the interpretation of the results. Their report[13] recommended the following criteria for establishing a diagnosis of diabetes mellitus:

1. One may use the Wilkerson point system as illustrated in Table 10.1 in which points are allocated for plasma glucose values that equal or exceed the concentrations shown in the table. A total of 2 points establishes the diagnosis of diabetes.
2. One may use the Fajans-Conn criteria in which plasma glucose concentrations equal to or exceeding the following are indicative of diabetes:
   a. A 1-hour value of 185 mg/dl, 1.5-hour value of 165 mg/dl, and a 2-hour value of 140 mg/dl.
   b. If the 1.5-hour sample is not taken for measurement, then the diagnosis is based upon the above values for the 1- and 2-hour samples.

In the University Group Diabetes Program a subject is considered to be diabetic if the sum of the serum glucose concentrations for the 0, 1, 2, and 3 hour samples exceeds 600 mg/dl.

If the serum glucose concentration obtained in an oral glucose tolerance test is plotted against time, different types of curves are obtained for diabetics and normals, as shown in Figure 10.7. The characteristic response of an overt diabetic is (1) an increased fasting serum glucose concentration; (2) a much greater peak rise in glucose concentration (exceeding 195 mg/dl); and (3) a delayed return to the fasting level. Some glucose is usually found in the urine collected during the test period.

A relatively flat curve in the glucose tolerance test, which may be encountered occasionally, is caused most frequently by impaired intestinal absorption; hyperinsulinism could also produce such a curve. The intravenous glucose tolerance test is sometimes given instead of the oral when there are significant gastrointestinal disturbances.

TABLE 10.1
Wilkerson Point System* for Interpretation of Oral GTT Test

TIME Hr. after Glucose	PLASMA GLUCOSE CONCENTRATION mg/dl	POINTS†
0	>130	1
1	>195	0.5
2	>140	0.5
3	>130	1

*Reference (13) and Wilkerson, H.L.C.: Diagnosis: oral glucose tolerance tests. *In* Diabetes Mellitus, edited by T.S. Danowski. New York, American Diabetes Association, 1964, p. 31.
†A total of 2 points is a positive test for diabetes mellitus.

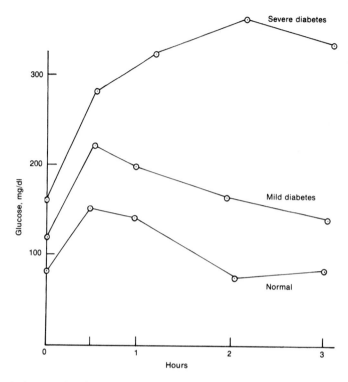

Fig. 10.7. Some typical types of oral glucose tolerance curves.

CORTISONE ORAL GLUCOSE TOLERANCE TEST. This test is sometimes used for the testing of subjects with a family history of diabetes who give a normal result with the conventional oral glucose tolerance test. The cortisone injection is a stress that accentuates the difficulty in handling a glucose load. Fifty to 62.5 mg of cortisone acetate are administered to subjects 8.5 and 2 hours before an oral glucose tolerance test. For subjects below the age of 50, serum glucose values greater than 185 mg/dl at 1 hour, 170 at 1.5 hours, and 160 at 2 hours indicate the presence of latent diabetes. Since patients over the age of 50 usually have higher glucose responses to the glucose tolerance test, the test becomes difficult to interpret in the elderly.

URINE GLUCOSE CONCENTRATION

Qualitative or Semiquantitative

COPPER REDUCTION. As described in the discussion of methodology for blood glucose testing, a hot, alkaline solution of cupric sulfate will oxidize all reducing sugars (glucose, fructose, galactose, maltose, lactose, xylulose, arabinose, ribose) and form a brick-red to yellow precipitate of Cu_2O. This is true whether Clinitest Tablets (Ames) or the Benedict method[14] is used as the source of the alkaline Cu reagent.

Five drops of urine plus 10 drops of water are placed in a test tube. When a Clinitest Tablet is added, the mixture begins to boil. The tube must *not* be agitated. About 15 seconds after the reaction ceases, check the color of the solution against a color chart. A blue color (of Cu^{2+}) is a negative result. The color change ranges from greenish (1$^+$) to yellow (4$^+$), depending upon the amount of Cu_2O produced. A 1+ reaction corresponds to about 0.5 g/dl of reducing sugar; the 4+ reaction indicates 2 or more g/dl. The test is adequate for making a semiquantitative estimate, but it is not specific for glucose.

GLUCOSE-OXIDASE IMPREGNATED STRIPS. There are dipsticks (Ames) and other impregnated strips (Bio-Dynamics, p. 123) that contain glucose oxidase, peroxidase, and a chromogen. The sticks or strips are dipped into the urine and checked for color at the appropriate time (10 seconds for Labstix, 30 seconds for Chemstrip). These glucose oxidase tests are more sensitive than the Cu-reduction tablets and are specific for glucose. False positive tests are rare, but these could occur if the glassware were contaminated with sodium hypochlorite (bleaching solution) or if the reaction were allowed to proceed too long before reading the result. False negative results occur more commonly because ascorbic acid and urates inhibit the reaction; they might be a real problem with subjects who are on a high vitamin C intake. It is advisable to check the sticks or strips again after 2 minutes to look for a delayed reaction. If present, confirm by Cu-reduction tablets or prepare a Somogyi $Zn(OH)_2$ filtrate of the urine and test again with the sticks. Because the $Zn(OH)_2$ removes the inhibitors, glucose, if present, should be detected easily.

Normal subjects excrete less than 150 mg of glucose in a 24 hour period (less than 10 mg/dl urine).

Quantitative

A quantitative estimation of the daily glucose excretion is frequently desired. The o-toluidine and hexokinase methods may be used for measuring urine glucose exactly as described for serum. Inhibitors, such as uric and ascorbic acids, must be removed before a glucose oxidase method can be employed. They can be removed by adsorbing the interferants with a mixed-bed resin or by preparing a $Zn(OH)_2$ filtrate.

No matter which glucose method is used for quantitation, the urine should be diluted with water if the glucose concentration exceeds 400 mg/dl. One of the semiquantitative methods for glucose determination should be employed first to determine whether dilution is required and, if so, to what extent.

GLUCOSE IN CEREBROSPINAL FLUID (CSF) OR OTHER BODY FLUIDS

The glucose concentration in cerebrospinal or other body fluids may be measured by the same methods and procedures as described for serum. See page 367 for a discussion of the clinical situations in which a knowledge of the glucose concentration in the cerebrospinal fluid is useful.

KETONE BODIES IN URINE

Ketones are not present in the urine of healthy individuals eating a mixed diet. They may be present in the urine of uncontrolled diabetics, of subjects who have been without food for several days, or of those who have been on a high fat, low carbohydrate diet. Infants have ketonuria much earlier than adults when deprived of food.

The most common method for the detection of ketones in urine makes use of a reaction between sodium nitroprusside and acetoacetate or acetone under alkaline conditions; a lavender color is produced. β-hydroxybutyrate does not react with the reagent.

IMPREGNATED STRIPS OR STICKS (p. 124). By comparison with a color chart, the concentration of acetoacetate + acetone is expressed as negative, small, moderate, or large. The color chart correlates approximately with the following concentrations of acetoacetate: small = 10 mg/dl, moderate = 30 mg/dl, large = 80 mg/dl.

REAGENT TABLETS. Acetest (Ames) tablets may be used instead of a dipstick or strip. Crush a tablet of Acetest and place 1 drop of urine on it. Compare the color with that in a color chart.

ROTHERA'S TEST. Grind separately to a fine powder each of the following: 10 g Na_2CO_3 (anhydrous), 20 g $(NH_4)_2SO_4$, and 0.5 g sodium nitroprusside. Mix the powders completely and stopper tightly.

Make a mound of about 0.5 g powder on a flat surface. Place 2 drops of urine on the powder and look for a lavender to purple color in about 30 seconds for a positive test.

IDENTIFICATION OF REDUCING SUGARS IN URINE

The reducing sugars that may be found in urine as a result of disease or a genetic defect are the disaccharide lactose, some common hexoses—glucose, fructose, galactose—and a few pentoses—xylulose and arabinose. Table sugar, sucrose, is not a reducing sugar. Reducing sugars in the urine are found by testing for copper reduction with a Clinitest tablet as described earlier in this chapter.

REFERENCE. Menzies, I.S., and Seakins, J.W.T.: Sugars. *In* Chromatographic and Electrophoretic Techniques, 3rd ed., Vol. 1, edited by I. Smith. New York, Interscience, 1969, p. 310.

PRINCIPLE. The separation of urinary sugars is accomplished by thin layer chromatography (p. 43) on plastic sheets or glass plates coated with fine cellulose particles. Identification of individual sugars is made from their relative migration distances and by the colors formed when heated with certain reagents.

MATERIALS AND REAGENTS

1. Chromatogram sheet. A plastic sheet 20 × 20 cm coated with cellulose (Distillation Products). Coated glass plates may also be used, but the coated plastic can be cut and used in smaller segments, as desired.
2. 10 μl pipets.
3. Standard Sugar Solutions as markers:
   a. General mixture, in 100 ml 10% (v/v) aqueous isopropanol, containing 200 mg each of xylose, glucose, galactose, fructose, and glucuronic acid and 400 mg of lactose.
   b. Individual solutions of the above sugars in 10% (v/v) aqueous isopropanol as individual markers, when necessary.
4. Solvent System (EtAc:Py). Ethyl acetate: pyridine: water in the volume ratio of 120:50:40.
5. Location Reagents:
   a. Aniline Phosphate. To 20 ml aniline, add with mixing after each addition, 200 ml H_2O, 180 ml glacial acetic acid, and 10 ml concentrated

phosphoric acid. Store at 4° C. When ready to use, dilute 2 volumes of this mixture with 3 volumes of acetone and mix.

b. Naphthoresorcinol. Dissolve 0.20 g naphthoresorcinol in 100 ml acetone. Add 18 ml H_2O and 2 ml concentrated H_3PO_4 and mix.

6. Glass atomizer for spraying reagents.
7. Chromatography chamber.
8. Fan or warm air blower (hair dryer).

PROCEDURE

1. It is usually best to apply the samples to the chromatography sheet as duplicates in two separate sections so that a different stain (location marker) can be applied to each section.
2. Using a spotting guide (ruler or commercial guide), apply the unknown urine samples and marker solutions in a line across the chromatography sheet, about 2 cm from the bottom. Leave a margin of 1 cm on each side. Apply the sample from a 10 μl pipet as a horizontal streak about 1 cm long. Dry with an airstream from a fan and reapply until the entire 10 μl sample has been applied to the one streak. Leave a space of 1 cm and apply the next sample.
3. When all samples and markers have been applied and the sheet has been dried in the airstream, place the sheet in a chromatography chamber. This may be a commercial "sandwich-type" chamber developed for this purpose or a battery jar closed with a glass plate. The EtAc:Py mixture is poured into a trough at the bottom of the sandwich chamber, and the bottom of the chromatography sheet is placed in it, vertically. The chamber is closed by clamping a glass plate to the front of it.
4. The sandwich chamber is kept in an upright position for 30 to 45 minutes until the solvent front has reached the top of the sheet. The chamber is opened and the sheet is air-dried.
5. The sheet is cut in half vertically, and each section is exposed to a different location marker.
   a. Spray the first section in a hood with the aniline phosphate reagent. Heat for 2 to 5 minutes at 100° C. This reagent reacts primarily with aldoses. Ketoses do not react unless present in high concentration.
   b. Traces of pyridine have to be removed before the second section is sprayed with the naphthoresorcinol reagent. Dry further by blowing hot air from a hair dryer across the surface. Then spray with the naphthoresorcinol reagent and heat at 100° for 5 to 8 minutes.

INTERPRETATION. The relative separation of the sugars on TLC cellulose is similar to that by conventional paper chromatography. In the solvent system described, glucose migrates about 0.28 as far as the solvent front. The other sugars, relative to glucose, migrate in the following ratios: xylose, 158%; fructose, 118%; galactose, 86%; lactose, 47%; glucuronic acid, 14%. The aniline phosphate location reagent gives red-brown colors with pentoses, brown with glucose, galactose, and lactose, and yellow-blue with glucuronic acid. The naphthoresorcinol location reagent turns red when fructose or ketoses are present. It does not react with xylose, glucose, galactose, or lactose. Glucuronic acid produces a blue spot. If sucrose should be present, it migrates very closely to galactose but turns red with the naphthoresorcinol reagent; it does not react with the aniline phosphate.

Glucose and galactose may not separate cleanly if both are present. In this event, add some glucose oxidase powder to a small portion of the urine, adjust pH to 6 to 7, let stand 20 minutes, and test with a glucose oxidase dipstick. When the glucose has been completely oxidized, repeat the chromatographic separation and test with the aniline phosphate location reagent. A brown spot with the Rf of glucose-galactose signifies the presence of galactose.

REFERENCES

1. Davidson, M.B.: Diabetes mellitus and hypoglycemia. *In* Endocrine Pathophysiology; A Patient Oriented Approach, 2nd ed., edited by J.M. Hershman. Philadelphia, Lea & Febiger, 1982, p. 171.
2. Porte, D., Jr., and Halter, J.B.: The endocrine pancreas and diabetes mellitus. *In* Textbook of Endocrinology, 6th ed., edited by R.H. Williams. Philadelphia, W.B. Saunders Co., 1981, p. 715.
3. Segal, S.: Disorder of galactose metabolism. *In* Metabolic Basis of Inherited Disease, 4th ed., edited by J.B. Stanbury, J.B. Wyngaarden, and D.S. Fredrickson. New York, McGraw-Hill, 1978, p. 160.
4. Froesch, E.R.: Essential fructosuria, hereditary fructose intolerance and fructose-1,6-diphosphatase deficiency. *In* Metabolic Basis of Inherited Disease, 4th ed., edited by J.B. Stanbury, J.B. Wyngaarden, and D.S. Fredrickson. New York, McGraw-Hill, 1978. p. 121.
5. Cooper, G.R.: Methods for determining the amount of glucose in blood. Crit. Rev. Clin. Lab. Sci. 4, 101, 1973.
6. Quam, E.F., Westgard, J.O., and Carey, R.N.: Selecting glucose methods that meet your laboratory requirements. Lab. Med. 6(2), 35, 1975.
7. Bunn, H.F.: Evaluation of glycosylated hemoglobin in diabetic patients. Diabetes 30, 613, 1981.
8. Reinhold, J.G.: Glucose. *In* Standard Methods in Clinical Chemistry, Vol. 1, edited by M. Reiner. New York, Academic Press, 1953, p. 65.
9. Bittner, D.L., and Manning, J.: Automated neocuproine glucose method: critical factors and normal values. Technicon Symposia, Automation in Analytical Chemistry 1, 33, 1966.
10. Snegoski, M.C., and Freier, E.F.: An automated o-toluidine glucose procedure without acetic acid. Am. J. Med. Tech. 39, 140, 1973.
11. Gochman, N., and Schmitz, J.M.: Application of a new peroxide indicator reaction to the specific, automated determination of glucose with glucose oxidase. Clin. Chem. 18, 943, 1972.
12. Trinder, P.: Determination of glucose in blood using glucose oxidase with an alternative oxygen acceptor. Ann. Clin. Biochem. 6, 24, 1969.
13. Committee on Statistics of the American Diabetes Association. Diabetes 18, 299, 1969.
14. Benedict, S.R.: The detection and estimation of glucose in urine. J. Am. Med. Assoc. 57, 1193, 1911.

Chapter 11

LIPID METABOLISM

LIPID METABOLISM AND COMPONENTS

Lipid is a term applied to naturally-occurring substances that are soluble in nonpolar solvents, such as hexane or chloroform, and insoluble in water. The main lipid classes of interest in clinical chemistry are fatty acids, triglycerides, phospholipids, sterols, sphingolipids, and lipoproteins.[1] Their functions are described in the following sections.

Lipid Constituents

Fatty Acids

Fatty acids are long, straight-chain hydrocarbons, with a terminal carboxyl group. They exist mostly as esters of glycerol in both triglycerides (Fig. 11.1) and some phospholipids (Fig. 11.2) and as esters of high molecular weight alcohols (cholesterol, Fig. 11.3, and sphingosine, Fig. 11.4). The fatty acids of triglycerides and phospholipids contain 16 to 20 carbon atoms in the chain, but those esterified with sphingosine (sphingomyelin, cerebrosides, gangliosides) usually have 24 carbons; human fatty acids always contain an even number of carbons. Some fatty acids, such as palmitic (C_{16}) and stearic (C_{18}) acids, are saturated (no double bonds); others are unsaturated, such as oleic acid (C_{18}, one double bond), linoleic (C_{18}, 2 double bonds), linolenic (C_{18}, 3 double bonds), arachidonic (C_{20}, 4 double bonds), and other acids. The body can synthesize the saturated fatty acids and oleic acid, but not linoleic and linolenic acids, which must be supplied in the diet. The two latter acids are called "essential" fatty acids because their presence in the body is necessary

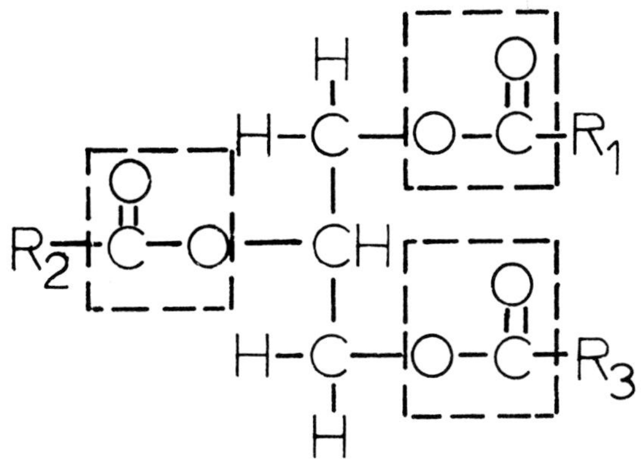

Fig. 11.1. Structure of triglyceride. R_1, R_2, and R_3 stand for long chain fatty acids. The broken lines enclose the ester bond.

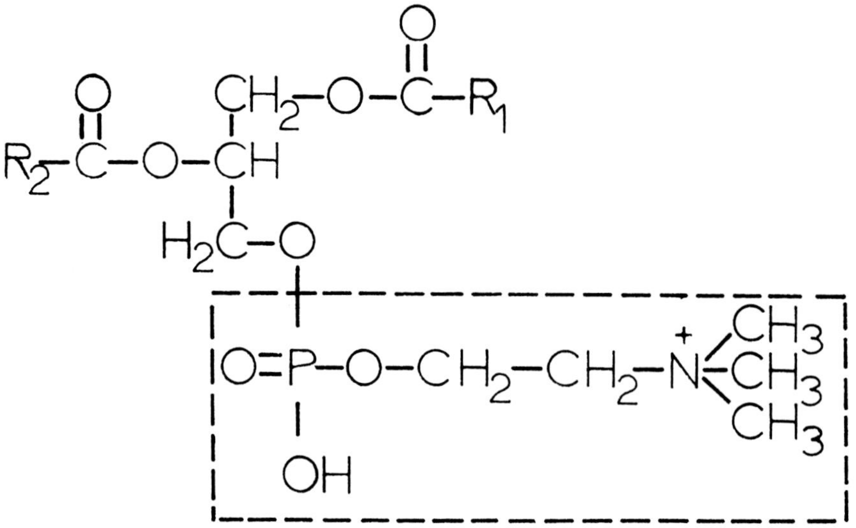

Fig. 11.2. Structure of lecithin. R_1 and R_2 stand for long chain fatty acids. The broken lines enclose phosphoryl choline.

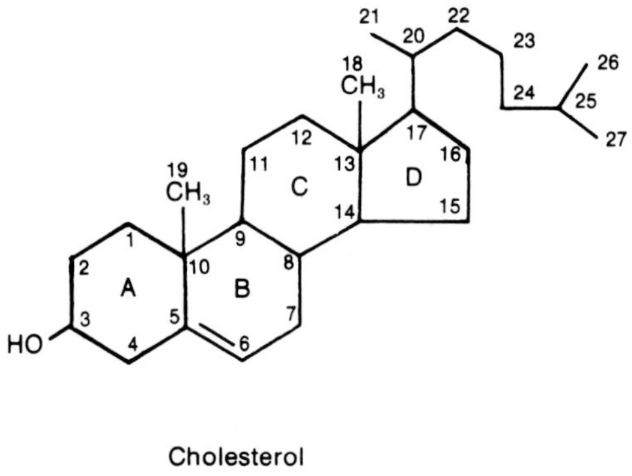

Cholesterol

Fig. 11.3. Structure of cholesterol. A cholesterol ester is produced when the hydroxyl group is esterified with an unsaturated fatty acid.

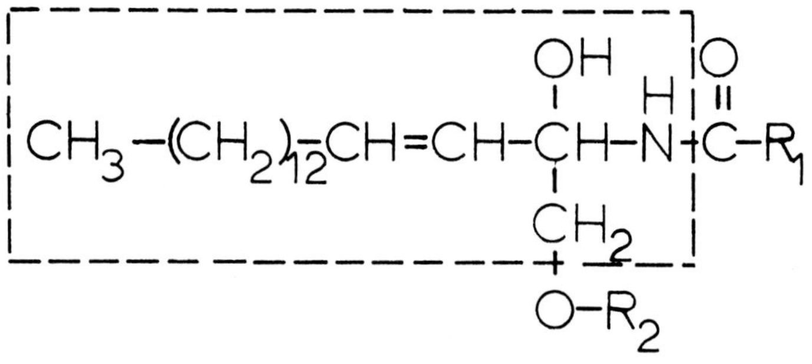

Fig. 11.4. Structure of sphingolipid. The broken lines enclose the sphingosine residue. In *sphingomyelin*, R_1 is a long chain fatty acid in amide linkage and R_2 is phosphoryl choline (see Fig. 11.2). In a cerebroside, R_1 is same as in *sphingomyelin* while R_2 is glucose or galactose in glycosidic linkage. A ganglioside is similar to a cerebroside except that the carbohydrate chain contains three molecules of hexose plus a hexosamine.

for the prevention of certain skin lesions and for the synthesis of arachidonic acid. Arachidonic acid is also essential for body function, but it can be formed from linoleic and linolenic acids by desaturation and chain elongation and need not be present in the diet. In general, the lipids derived from plant sources (seed oils) are less saturated and contain much more of the essential fatty acids than do animal fats.

Arachidonic acid may be released by hydrolysis of membrane phospholipids and enzymatically converted to a complex series of oxygenated fatty acids, the prostaglandins, that have powerful physiologic effects. Some of the prostaglandins are powerful stimulators of inflammatory reactions, potentiate the effect of histamine and bradykinin (a vasodilator) and frequently cause pain. Some prostaglandins initiate platelet aggregation, a precondition for blood clotting, while others oppose it. The release of arachidonic acid and formation of prostaglandins are stimulated by many factors, such as some hormones, inflammation, ultraviolet light, allergens and others; certain drugs (aspirin and others) inhibit their formation. Many of the effects of prostaglandins occur indirectly through their activation of adenyl cyclase. Prostaglandins and their metabolites are removed from the circulation within minutes because enzymes rapidly convert them to inactive compounds.

Triglycerides

Most of the fatty acids in the body are esterified with glycerol as triglycerides (Fig. 11.1) and are stored in the depots (adipose tissue) as fat. Triglycerides are formed in adipose tissue by esterification of fatty acids with glycerol-1-phosphate, which arises in the adipose cell from the catabolism of glucose via the glycolytic pathway (see p. 304). Glucose thus must be present in the cell for triglyceride formation; it is absent during periods of fasting, starvation, or uncontrolled diabetes mellitus, and in these conditions, hydrolysis of triglycerides and withdrawal of their fatty acids from the depots predominate. Excess carbohydrate ingested during a meal may be stored temporarily as triglycerides after conversion of glucose to fatty acids. The hormone, insulin, promotes the synthesis of triglycerides by adipose cells while its deficiency accelerates triglyceride hydrolysis.

When triglycerides are catabolized, a small portion of the total fatty acids appears in the plasma as nonesterified (free) fatty acids, bound to albumin as a carrier. Various tissues, but muscle cells in particular, take in the fatty acids and oxidize them to acetyl CoA and then to CO_2 and water; a great deal of energy in the form of ATP becomes available. As discussed on page 303, when the body is flooded by the catabolism of fatty acids, as in insulin deficiency (diabetes mellitus) or prolonged starvation, the acetyl CoA piles up and condenses into acetoacetyl CoA; this latter compound forms ketone bodies and is an intermediate in the synthesis of cholesterol (p. 304).

Phospholipids

The principal phospholipids are composed of a diglyceride esterified with phosphoric acid, which in turn is bound as an ester to a nitrogen-containing base (choline, ethanolamine) or to serine (Fig. 11.2). In lecithin, the base end, choline, carries a positive charge and is polar, while the fatty acid segments are nonpolar. The lecithins emulsify easily in water and help to form stable, colloidal emulsions of fat in water because the molecules orient so that the polar head (charged base) is in the water phase and the nonpolar end in lipid. A lecithin is the principal lipid component of a surfactant secreted by the lung alveoli and is essential for the proper distension of the alveoli upon breathing (see p. 374). Phospholipids are essential components of cell membranes because of their ability to align themselves between water and lipid phases. Phosphoethanolamine, a constituent of blood platelets, is a necessary participant of the clotting process. Phospholipids in lipoproteins also supply the fatty acids necessary for the esterification of cholesterol (p. 329). The phospholipids play a role in mitochondrial metabolism, blood coagulation, in lipid transport as part of lipoproteins, and are important structural components of membranes. Little useful information, however, can be gained by measuring the concentration of serum phospholipids in most instances.

Cholesterol

Cholesterol, the principal body sterol, is a complex alcohol formed of 4 fused rings and a side chain (Fig. 11.3); it is a solid at body temperature. It is present in all tissues and can be converted by the adrenals and the gonads into steroid hormones (see pp. 264 and 279). The cholesterol present in all membranes, including those of lipoproteins, is always *nonesterified (free)*; it is stored in cells and in the lipid core of lipoproteins as *cholesterol esters*.

Humans ingest cholesterol when the diet contains meat, dairy products, or eggs. Plants do not contain cholesterol, although they do have closely related sterols. Most cholesterol in the body is synthesized from acetyl CoA (p. 304), but this varies inversely to some extent with the cholesterol content of the diet. Excess dietary cholesterol over a period of time slowly increases the fasting plasma cholesterol concentration, but some of the excess is stored as cholesterol esters in the liver. Seventy to 75% of the cholesterol in plasma is esterified with long-chain, unsaturated fatty acids.

Cholesterol, a normal constituent of serum, may rise to very high levels in some pathologic states. An elevated serum cholesterol concentration has been implicated as one of several risk factors in coronary artery disease (atherosclerosis and myocardial infarction) so the measurement of serum cholesterol concentration is a fairly common laboratory procedure (see p. 330).

Sphingolipids

The sphingolipids are all compounds containing the long-chain, dihydroxy amino alcohol, sphingosine (Fig. 11.4). All of the sphingolipids bind a fatty acid in amide linkage to the amino group and are also known as ceramides because they are *cerebral* lipids containing an *amide* group. In the ceramide, sphingomyelin, phosphorylcholine is bound to a hydroxyl group (Fig. 11.4); sphingomyelin is also classified as a phospholipid since the molecule contains phosphate. Sphingomyelin is an essential component of cell membranes, particularly of red blood cells, and of the nerve sheath. Sphingomyelin accumulates in the liver and spleen of patients suffering from a rare lipid storage disease (Niemann-Pick disease); an enzyme capable of splitting off phosphorylcholine from sphingomyelin is lacking.

A ceramide becomes a cerebroside when a hydroxyl group binds a hexose (glucose or galactose) in a glycosidic linkage. The ceramides become gangliosides when the carbohydrate chain is composed of hexosamine or an acetylated amino sugar bound to 3 hexoses (Fig. 11.4). Cerebrosides and gangliosides are also glycolipids because of the carbohydrate moieties in a lipid structure. Cerebrosides are present as a high percentage of the lipid matter in brain and nerve tissue, particularly in the myelin sheath. Gangliosides are abundant in the gray matter of the brain and in nerve ganglia; they are also present in cell membranes in which the protruding carbohydrate chain serves as a determinant or recognition center for certain immunologic reactions (e.g., blood group reactions). A number of genetic lipid storage diseases have been discovered in which there is a deficiency of an enzyme that splits off a particular portion of the carbohydrate side chain of cerebrosides and gangliosides, with accumulation of these lipids in various tissues. These diseases (lipidoses) are quite rare. Little clinical information is obtained from measurement of cerebrosides or gangliosides in tissues except in research studies or to confirm a diagnosis at autopsy; identification of the enzyme defect is more convincing. For pregnant patients at risk for lipidoses, cell cultures are grown from amniotic fluid samples and a search is made for a specific enzyme deficiency in splitting the carbohydrate chain of glycolipids (p. 377). Enzyme assays of biopsy samples may be carried out for diagnostic purposes at some large hospital centers.

Absorption of Lipids

The body lipids are derived partially from fats and oils (mostly triglycerides) in the diet that have been digested (hydrolyzed), absorbed and reassembled, while the rest are synthesized in the body from simple precursors. Ingested lipids are emulsified in the small intestine by means of bile salts and phospholipids in bile coming from the gallbladder and liver. They are emulsified into microscopic droplets, micelles in which the hydrophilic portions (base ends of the phospholipids) orient toward the outer, water phase while the hydrophobic portions (fatty acids) face toward the central oil phase; the hydrophobic core dissolves the triglycerides, cholesterol, and other lipid substances. This emulsification is a precondition for hydrolysis of the various lipid esters (triglycerides, phospholipids, cholesterol esters) by pancreatic lipase, phospholipase and esterases, respectively, as well as for absorption of the other lipids that require no hydrolysis, such as sterols (cholesterol) and the fat-soluble vitamins (Vitamins A, D, E, and K).

The digested lipids are absorbed by the intestinal mucosa cells. The long-chain fatty acids (C_{12} or greater) are recombined as triglycerides in the mucosa, but not necessarily in the same combination as in the food. The newly formed triglycerides, together with the absorbed cholesterol, are secreted into lymph as chylomicrons (microscopic droplets of triglycerides enclosed by an envelope of protein, phospholipids, and nonesterified cholesterol). The chylomicrons are carried in lymph by way of the thoracic duct to enter the venous blood circulation.

The chylomicron, as well as all lipoproteins, functions as a micellar membrane with a lipid core. The lipoproteins contain the hydrophilic portions of proteins interlaced with the polar head of phospholipids and the hydroxyl group of cholesterol in the outer layer that is in contact with water (plasma); the hydrophobic, nonpolar portions of the above molecules form an inner layer that dissolves a core of nonpolar lipids (triglycerides, cholesterol esters). Chylomicrons consist mostly of a core of triglycerides surrounded by a thin, micellar envelope. The other lipoproteins transport lesser amounts of triglycerides and more of cholesterol, cholesterol esters, and phospholipids (Table 11.1). The proteins in the outer membrane are called apolipoproteins (apoproteins). Each type of lipoprotein contains a particular combination of apolipoproteins (Table 11.1) that is described in the following sections.

LIPOPROTEINS. (See reviews by Levy[2] and by Brewer and Bronzert.[3])

All lipids, whether derived from dietary sources or from synthesis by the liver, must be transported in plasma to the various body tissues for storage or use. Likewise, the lipids in depots must also be carried in the plasma to sites of catabolism when the need arises. The solubilization and transportation of these water-insoluble lipids is accomplished by their incorporation into lipoproteins (see above).

There are four classes of lipoproteins: chylomicrons, which are described above; very low density lipoproteins (VLDL); low density lipoproteins (LDL); and high density lipoproteins (HDL). All lipoproteins contain apolipoproteins, phospholipid, cholesterol, and triglyceride, although in different proportions.

Classification

The classification of the lipoproteins arises from two different sets of properties: the density of the lipoproteins upon being centrifuged in salt solutions of certain densities and the charge carried by the lipoprotein, which enables separation into four zones or species of lipoproteins when subjected to electrophoresis.

Classification by Density

The lipoproteins containing the least protein and the most triglycerides (chylomicrons and VLDL) have the lowest relative density. When plasma or serum is subjected to ultracentrifugation for 1 hour, the chylomicrons float to the top as a cream layer and are removed. Continuation of the centrifugation for 24 hours produces a pellet of the plasma (serum) proteins, reduction in density of the protein-free plasma to 1.006, and flotation of the VLDL as a top layer. The LDL and HDL can likewise be separated by flotation after increasing the density to 1.063 and 1.21, respectively, by the addition of salt solutions and ultracentrifuging.

Classification by Electrophoresis

This technique is more practical for a hospital laboratory than ultracentrifugation. A serum sample is subjected to electrophoresis, with agarose being preferred as a support medium. At the end of the run, the support medium is dyed with a lipid stain, washed, and dried, and the stain quantified. If chylomicrons are present, they remain at the origin because they carry no charge. The lipoprotein that migrates the furthest toward the anode in the position usually taken by the α_1-proteins in serum protein electrophoresis is called an α-lipoprotein; it corresponds to the HDL fraction by centrifugation. The next lipoprotein band appears in a position shortly before the β band of protein electrophoresis. It corresponds to the VLDL and is called pre-beta. The slower band that migrates in the beta protein position corresponds to the LDL fraction (see Fig. 11.5).

The electrophoretic separation of lipoproteins is described on page 344.

Composition

All of the proteins of lipoproteins are in the outer surface layer of the lipoprotein; they are called apolipoproteins (apoproteins) and have been studied by removing the lipids from isolated lipoprotein classes. There are at least 8 different types of these apoproteins; most are of relatively low molecular weight (from 6600 to 33,000 daltons), but the Apo B group ranges from 145,000 to 550,000.[14] The individual lipoprotein classes contain differing amounts and combinations of the apoproteins, which have different functions (Table 11.1). Apo A-I and A-II comprise 90% of the apoproteins of HDL and also are found in chylomicrons. Apo B is almost the sole apoprotein in LDL, and is also present in chylomicrons and VLDL but is absent from HDL. Apo D occurs to a small extent only in HDL while Apo E is found in both HDL and VLDL.

Apo A-I and C-I activate the plasma enzyme, lecithin-cholesterol acyltransferase (LCAT) while Apo C-I and C-II activate lipoprotein lipase; these apoproteins are transferred to HDL as the two lightest lipoproteins, VLDL and chylomicrons are catabolized. LCAT transforms unesterified cholesterol into a cholesterol ester by transferring an unsaturated fatty acid to it from lecithin. Lipoprotein lipase is attached to the capillary walls of adipose and muscle cells. Upon activation by Apo C-II or C-I, the lipoprotein lipase cleaves the triglycerides in chylomicrons and VLDL into fatty acids and glycerol that then enter adipose cells for storage or muscle cells for utilization.

Function

The lipoproteins have different functions. The chylomicrons, which initially are composed of 85 to 95% triglycerides, with small amounts of cholesterol esters and phospholipids, are the vehicles for transporting dietary fat from the intestines to adipose and muscle cells. VLDL is the prime transporter of endogenous (liver-synthesized) triglycerides to adipose and muscle cells and participates to a small extent in carrying triglycerides from the intestinal mucosa to the periphery. VLDL has a higher content of apoproteins, cholesterol, and

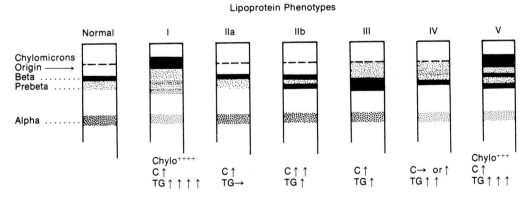

Fig. 11.5. Schematic representation of lipoprotein electrophoretic patterns; classification of Fredrickson and Levy.[8a] Chylo = chylomicrons; C = cholesterol; TG = triglycerides.

phospholipids and less triglycerides than chylomicrons. LDL is the principal vehicle for transporting plasma cholesterol esters to tissues while HDL is thought to transfer cholesterol esters from tissues to the liver for conversion to bile salts (see p. 222). VLDL and HDL are synthesized mainly in the liver and to a lesser extent in intestine while LDL arises from VLDL remnants as the triglycerides in this lipoprotein are catabolized. VLDL passes through a transitory, intermediate lipoprotein form, IDL, before becoming LDL.

Genetic diseases affecting the plasma lipoproteins[4] became known in 1963 when a patient was found to have no beta-lipoproteins in the plasma. The disease was named *abetalipoproteinemia* and it is characterized by the absence of chylomicrons, VLDL, and LDL in plasma; only HDL is present. Individuals with the disease cannot absorb ingested fat or fat-soluble vitamins and usually die by age 10 or 20 years.

A second type of inherited lipoprotein defect is *Tangier disease*, in which there is no Apo A-I, and thus, less than 5% of the normal amount of HDL. In this disease, there are heavy deposits of cholesterol esters in all tissues of the reticuloendothelial system, including the blood vessels, and there is a greatly increased risk of cardiovascular disease.

Other familial genetic disorders of lipoproteins were soon discovered, some of which are manifested by cholesterol and triglyceride abnormalities. Typical lipoprotein electrophoretic patterns and plasma lipid concentrations characteristic of these disorders are discussed on page 342.

SERUM LIPIDS

The serum lipid tests most commonly requested are total cholesterol, triglycerides, and HDL cholesterol; LDL cholesterol may be obtained by calculation; lipoprotein electrophoresis is requested less frequently. An elevated concentration of serum cholesterol is considered to be a factor that increases the risk of atherosclerosis (deposition of plaques containing cholesterol esters

TABLE 11.1

Lipoproteins: Composition, Origin and Function

LIPOPROTEIN	ELECTRO-PHORETIC MOBILITY	APOPROTEIN CONTENT	COMPOSITION		TISSUE OF ORIGIN	FUNCTION
			MAJOR LIPID	% APOPROTEIN		
Chylomicron	At origin	A-I, -II, C-I, -II, -III, B	TG-95%	1	Intestine	Transports dietary lipids to body tissues
VLDL	Pre-β	B, C-I, -II, -III, E	TG-55%	10	Liver (primarily), Intestine	Transports endogenous TG from liver to adipose tissues
LDL	β	B	CE-40%	20		Carries cholesterol to tissues
HDL	α₁	A-I, -II, C-I, -II, -III, D, E	PL-25%	50	Liver	Carries cholesterol from tissues to liver. Increases catabolism of chylomicrons and VLDL.*

TG = triglycerides; CE = cholesterol esters; PL = phospholipids
*Provides C-apoproteins that activate lipoprotein lipase and lecithin-cholesterol acyl transferase.

and calcium in arterial walls) and myocardial infarcts[2]; a simultaneous increase in the serum triglyceride concentration increases the risk. Other risk factors are hypertension (high blood pressure), obesity, smoking, a personality that cannot cope with stress, and lack of exercise. The nicotine in tobacco is a vasoconstrictor and thereby decreases the blood supply to the heart muscle. It is possible to reduce the concentration of plasma cholesterol and triglycerides by reduction in the dietary intake of cholesterol and fat, by reduction in the total caloric intake, by partial substitution of unsaturated oils for animal fat, or by a combination of dietary measures. In severe cases, the ingestion of a certain resin or drugs is effective in lowering the plasma cholesterol concentration. Periodic monitoring is necessary for the success of these measures because of the temptations to eat rich and tasty foods.

Few clinical situations require the analysis of plasma phospholipids or non-esterified fatty acids (NEFA) so references only will be supplied. Phospholipid phosphorus can be measured in a trichloroacetic acid precipitate of serum or in an isopropanol extract of serum by the method of Zilversmit and Davis.[5] The method of Trout, Estes, and Friedberg[6] for the estimation of serum NEFA concentration is satisfactory.

Serum Total Cholesterol

Approximately 60% of the total cholesterol in male plasma is carried by the LDL, 22% by HDL, 13% by VLDL, and 5% by chylomicrons. Methods for the measurement of total serum cholesterol have been in use in clinical laboratories for many decades but information concerning HDL- and LDL cholesterol is relatively new; methods sufficiently practical for adoption by smaller laboratories became available only since 1978. Measurement of HDL- and LDL cholesterol is discussed in the section following *total* cholesterol.

Enzymatic Methods

Enzymatic methods for the hydrolysis of cholesterol esters and oxidation of cholesterol are simpler, faster to perform, and use less corrosive chemicals than the more tedious chemical analyses. The serum can be used directly, with little interference from serum constituents. The required volume of serum varies from 5 to 100 μl, depending upon the system. Enzymes and reagents are available in commercial kit form and the tests can be performed manually or adapted to various types of discrete or continuous flow automated instruments. Oxidation of cholesterol by cholesterol oxidase yields H_2O_2, which can be quantitated in a manner similar to that in the glucose assay (p. 313).

REFERENCES. Allain, C.C., Poon, L.S., Chan, C.S.G., Richmond, W., and Fu, P.C.: Enzymatic determination of total serum cholesterol. Clin. Chem. *20*, 470, 1974.
Roeschlau, P., Bernt, E., and Gruber, W.: Enzymatische Bestimmung des Gesamt-Cholesterins im Serum. Zeit. Klin. Chem. Klin. Biochem. *12*, 226, 1974.

PRINCIPLE. Cholesterol esters in serum are hydrolyzed by cholesterol ester hydrolase (EC 3.1.1.13). The total cholesterol is then oxidized by cholesterol oxidase (cholesterol:oxygen oxidoreductase EC 1.1.3.6) to the corresponding ketone, with a shift in the location of a double bond. The H_2O_2 generated by the oxidation is decomposed by horseradish peroxidase (horseradish source; donor: H_2O_2 oxidoreductase EC 1.11.1.7) in the presence of 4-aminoantipyrine and phenol to yield a quinoneimine dye. The absorbance of the dye, measured at 500 nm, is proportional to the cholesterol concentration. All of the enzymes and chromogen are contained in a single, buffered reagent.

REAGENTS AND MATERIALS. Several companies now prepare the reagents for the enzymatic determination of cholesterol in kit form for use in either manual or automated methods.

1. Cholesterol Ester Hydrolase (Ames)
2. Cholesterol Oxidase (Whatman)
3. Horseradish Peroxidase (Worthington)
4. Sodium Cholate (General Biochemicals)
5. Phenol (Mallinckrodt)
6. 4-Aminoantipyrine (J. T. Baker)
7. Triton X-100 (Rohm & Haas)
8. Cholesterol (Pfanstiehl)
9. Carbowax 6000 (Schwarz/Mann)
10. Phosphate Buffer, 0.10 mol/L, pH 6.7, containing per liter: 3.0 mmol sodium cholate, 0.8 mmol 4-aminoantipyrine, 14 mmol phenol, 67,000 U horseradish peroxidase, 120 U cholesterol oxidase, 40 U cholesterol ester hydrolase, and 0.17 mmol Carbowax-6000.
11. Cholesterol Standard Solutions:

    A. 600 mg/dl cholesterol. Dissolve 300 mg cholesterol in isopropanol and make up to 50 ml volume in this solvent.

    B. 300 mg/dl. Transfer 25 ml of Standard A to a 50 ml volumetric flask and make up to volume with isopropanol.

    C. 150 mg/dl. Transfer 25 ml of Standard B to 50 ml volumetric flask and make up to volume with isopropanol.

PROCEDURE

1. Transfer 3.0 ml of reagent to a series of tubes labeled "Sample," "Control," "Standard," and "Blank" and place in 37° C water bath for 5 minutes. Add 30 µl of appropriate serum to the sample and control tubes, 30 µl of standards (150, 300, and 600 mg/dl, respectively) to standard tubes and 30 µl water to the blank.
2. After exactly 10 minutes of incubation for each tube, read the absorbance against the blank at 500 nm.
3. Construct a standard curve from the standard readings and obtain the cholesterol concentration of the samples and controls from the standard curve.

REFERENCE VALUES. Serum cholesterol values are low at birth (about 90 mg/dl) but increase about 40% by the third day of life.[7] The mean serum concentration increases with age; the upper limit of normal is 195 mg/dl at 1 year of age and 220 mg/dl at 14 years.[8] Goldstein et al.[9] report the mean and 90% limits for serum cholesterol in various age groups to be the following:

Age Group years	Mean Serum Cholesterol Concentration	
	mg/dl	mmol/L (mean)
0–19	175 (120–230)*	4.55
20–29	180 (120–240)	4.68
30–39	205 (140–270)	5.33
40–49	225 (150–310)	5.85
50–59	245 (160–330)	6.37

*90% limits

The serum cholesterol concentration of women before menopause does not differ significantly from that of men but may become higher after the menopause. The serum cholesterol concentration, however, varies with lifestyle (amounts of cholesterol and saturated fat in the diet, amount of exercise, stress and other factors). The serum cholesterol concentration of people in the United States and Scandinavia is much higher than for people living in China or India.

INCREASED CONCENTRATION. Total serum cholesterol concentration is increased in hypothyroidism, uncontrolled diabetes mellitus, nephrotic syndrome, extrahepatic obstruction of the bile ducts, and in various types of hyperlipidemias (see p. 343), especially in xanthomatosis. Cholesterol rises in late pregnancy but returns to normal levels within a month after delivery. Moderately elevated cholesterol is frequently but not always present in individuals with atherosclerosis and coronary artery disease and is considered a risk factor.

DECREASED CONCENTRATION. A hypocholesterolemia usually is present in hyperthyroidism, hepatocellular disease, anemias, starvation, and in certain genetic defects. In a rare genetic disease, abetalipoproteinemia, affected individuals have no beta lipoproteins (LDL), and the serum cholesterol concentration is very low.

Chemical Method

Chemical colorimetric methods have been used for many years in the analysis of serum cholesterol. The reagents and materials are less costly than in the enzymatic methods but much more labor time is required, especially for those methods requiring extraction.

REFERENCES. Leffler, H.H., and McDougald, C.H.: Estimation of cholesterol in serum by means of improved technics. Am. J. Clin. Path. 39, 311, 1963.
Kessler, G.: Automated techniques in lipid chemistry. In Advances in Clinical Chemistry. Vol. 10, edited by O. Bodansky and C.P. Stewart. New York, Academic Press, 1967, p. 45.
Edwards, L., Falkowski, C., and Chilcote, M.E.: Semiautomated fluorometric measurement of triglycerides. In Standard Methods of Clinical Chemistry, Vol. 7, edited by G.R. Cooper. New York, Academic Press, 1972, p. 69.

PRINCIPLE. Since concentrations of triglycerides and of cholesterol are frequently ordered on the same blood specimen, the extraction procedure selected permits the performance of both tests upon the same extract. Proteins are precipitated, and all major interfering compounds are removed: bilirubin, which interferes with the cholesterol method, and phospholipids, monoglycerides, diglycerides, glucose, and other substances capable of forming formaldehyde during the triglyceride method. Cholesterol in the isopropanol extract is reacted with $FeCl_3$ and concentrated sulfuric acid to produce a color whose absorbance is measured at 540 nm.

REAGENTS AND MATERIALS
1. Isopropanol, reagent grade (Mallinckrodt).
2. Adsorbent Mixture* for the removal of bilirubin, phospholipids, monoglycerides, diglycerides, glucose, and other chromogenic material. Mix well the following materials:
   a. Alumina (Merck), 50–100 mesh, 900 g

---

*A filter column packed with the adsorbent materials listed in #2, but in a different ratio is available from Analytical Products, Inc. as Lipo-Frax.

    b. Zeolite (Taylor), ground and sifted to 20–80 mesh, activated by heating at 110° C overnight, 50 g.

    c. Lloyd's reagent, 50 g.

    d. $CuSO_4$ anhydrous powder, 10 g.

    e. $Ca(OH)_2$ anhydrous, 20 g.

3. Sulfuric Acid, concentrated, reagent grade.

4. Ferric Chloride Color Reagent. Place 500 mg $FeCl_3 \cdot 6H_2O$ in a 500 ml volumetric flask, add glacial acetic acid to the mark, and mix. The reagent is stable in the dark for 1 year at room temperature.

5. Cholesterol Standard, 200 mg/dl. Dissolve 200.0 mg cholesterol in isopropanol and make up to 100 ml volume.

6. 20 × 150 mm screw-capped culture tubes with Teflon-lined caps.

PROCEDURE. The development and measurement of the color after preparation of the extract has been automated by the method of Kessler (reference, p. 333).

1. *Extraction.* Pipet 9.5 ml isopropanol into all culture tubes to be used for samples and controls and 9.0 ml isopropanol plus 0.5 ml water to tubes for standards.

2. Pipet 0.5 ml of serum into the appropriate sample and control tubes and 0.50 ml of cholesterol standard into the standard tubes. Tightly stopper and shake vigorously or mix on a vortex-type mixer for 20 seconds.

3. Allow to stand about 20 minutes, add about 2 g of adsorbent mixture to each tube, and mix thoroughly for 20 seconds. Let stand for 30 minutes, and shake vigorously for 5 seconds every 10 minutes.

4. Centrifuge for 10 minutes at 1100 to 1200 g. Aliquots of the extract can be used for the determination of both cholesterol and triglycerides. Samples with grossly elevated concentrations can be diluted with isopropanol and re-assayed.

5. *Color reaction.* Prepare a blank by pipetting 1.0 ml of isopropanol into a tube. Transfer 1.0 ml of sample, control, and standard extracts, respectively, into appropriately marked tubes.

6. To each tube, add 2 ml $FeCl_3$ reagent and mix.

7. Add 2 ml concentrated H_2SO_4 to a tube by allowing the acid to run down the side of the slanted tube, tightly stopper, and mix by inversion 6 times. Then proceed to the next tube.

8. Let color develop for 10 minutes, transfer to a cuvet, and read the absorbance against the blank at 540 nm.

9. *Calculations.* Let A_u be absorbance of sample and A_s the absorbance of standard; read against the blank.

$$\text{Cholesterol, mg/dl} = \frac{A_u}{A_s} \times 200 \text{ mg/dl}$$

In SI units, cholesterol, mmol/L = mg/dl × 0.026.

HDL Cholesterol

The serum concentration of *total* cholesterol is of limited value in assessing the risk of coronary artery disease since cholesterol is transported primarily by two lipoprotein classes with different functions.[2,3] HDL transports cholesterol *from tissues to the liver for catabolism* (conversion to bile salts) while LDL transports cholesterol *from sites of origin to deposition in tissues,* including blood vessels. The HDL competes with LDL for binding to tissue receptors and may thus reduce cholesterol accumulation in blood vessel walls. The concentration of HDL cholesterol appears to be *inversely* related to the risk of cardiovascular disease[13] but a word of caution must be inserted. The normal range of HDL cholesterol is narrow and the precision of the analysis is poorer than that for total cholesterol, a combination that provides a high relative error in the estimation of risk of cardiovascular disease. Commercial kits are available for the HDL cholesterol test.

PRINCIPLE. Serum is centrifuged to remove any chylomicrons that may be present; they float as a cream layer. All of the remaining lipoproteins contain Apo B except the HDL. The VLDL and LDL are precipitated by the addition of polyvalent anions (heparin, phosphotungstate or dextran sulfate) and either Mn^{2+} or Mg^{2+}. The polyvalent anions complex with the lipoproteins containing Apo B and precipitate. After centrifugation, HDL remains in the supernatant and its cholesterol may be measured by the same enzymatic methods as for total cholesterol.

REFERENCE. Kostner, G.M.: Enzymatic determination of cholesterol in high-density lipoprotein fractions prepared by polyanion precipitation. Clin. Chem. *22,* 695, 1976.

The Beckman HDL-Cholesterol reagent is described below. Other companies also supply kits for performing the test.

REAGENTS AND MATERIALS

1. Precipitant tube containing 1 mg dextran sulfate tablet and 14.2 mg magnesium sulfate tablet.
2. Cholesterol calibrator solution, 500 mg cholesterol/L.
3. Reagent tube containing the following ingredients when reconstituted:

a. 4-aminoantipyrine	1.6 mmol/L
b. Phenol	21.2 mmol/L
c. Peroxidase (horseradish)	50,000 U/L
d. Cholesterol oxidase	500 U/L
e. Cholesterol esterase	440 U/L
f. Phosphate buffer salts	pH 7.50

PROCEDURE

1. Reconstitute the cholesterol reagent by adding the appropriate amount of water to the tube and mixing gently. Do not shake.
2. Pipet 1.0 ml each of water, controls and samples to be assayed into appropriately marked glass test tubes.
3. Add one precipitant tablet to each tube and let stand for 5 minutes.
4. Mix all tubes with a vortex mixer for 20 to 30 seconds and then let stand at room temperature for 5 minutes.
5. Mix briefly with a vortex mixer and then centrifuge all tubes for at least 5 minutes at 1000 rpm.
6. Carefully transfer 50 μl of the clear supernatant and place in appropriate cuvets or test tubes for assay. The assay may be carried out as described for total cholesterol on page 331, or the use of the Beckman kit may be continued, as follows:
7. Transfer 50 μl of the cholesterol calibrator solution (#2) to a cuvet or test tube.

8. To each cuvet or test tube, add 1.0 ml of reconstituted cholesterol reagent (#3) and mix well.
9. Incubate all tubes or cuvets at 37° C for 10 minutes.
10. Measure absorbance of calibrator, controls, and samples against the reagent blank at 500 nm, using a 1.0 cm light path.
11. Calculation:

HDL Cholesterol (mg/dl)

$$= \frac{A}{13.78} \times \frac{1}{1000} \times 2 \times 387 \times \frac{1.05}{0.050} \times 100 = A \times 118$$

Where A = absorbance reading; 13.78 = millimolar absorptivity/liter of quinoneimine; 1/1000 changes it to absorptivity/ml; 2 × 387 changes millimoles of quinoneimine produced to mg of cholesterol oxidized; 1.05/0.05 = total volume/sample volume; and 100 is factor for changing per ml serum to per dl.

HDL- and LDL cholesterol may be obtained directly from a single lipoprotein electrophoretic separation, followed by enzymatic oxidation of cholesterol, decomposition of the liberated H_2O_2, and oxidation of a chromogen to produce colored bands in the HDL-, LDL-, and VLDL fractions; the intensity of the color is proportional to the cholesterol concentration. Some electrophoretic methods for HDL cholesterol are too variable to be of clinical use but the method of Cobb and Sanders[12] (kit by Helena) has been deemed adequate for the purpose; electrophoretic separation on polyacrylamide gradient gels is also effective.

LDL Cholesterol

The risk of cardiovascular disease is correlated directly with a high serum concentration of LDL cholesterol but the predictive value of this test appears to be only slightly better than that of total cholesterol and less than the value of HDL cholesterol. The highest correlations have been obtained between disease risk and the ratio, $\frac{\text{LDL cholesterol}}{\text{HDL cholesterol}}$.

The value for LDL cholesterol may be obtained by calculation, however, if the concentrations of total and HDL cholesterol and serum triglycerides are measured.

LDL Cholesterol = Total Cholesterol − (HDL Cholesterol + $\frac{1}{5}$ × triglycerides).

The concentrations of all constituents should be expressed in the same units, mg/dl or mg/L. The factor, $\frac{1}{5}$ × serum triglycerides, corrects for the average cholesterol concentration in serum VLDL.

REFERENCE VALUES. The reference ranges (and mean values in parentheses) for various age groups appear below in mg/dl for both HDL- and LDL cholesterol. For HDL cholesterol, the values for both males and females are given but they are grouped together in LDL cholesterol. Note the higher HDL cholesterol in women after puberty.

Age Group Years	HDL Cholesterol Male	HDL Cholesterol Female	LDL Cholesterol
0–19	30–65(48)	30–70(50)	50–170(110)
20–29	30–65(48)	36–78(57)	60–170(115)
30–39	30–59(45)	33–77(55)	70–190(130)
40–49	25–61(43)	40–81(61)	80–190(135)
50–75	29–72(50)	38–91(65)	80–210(145)

INCREASED CONCENTRATION. Present evidence suggests that an elevated HDL cholesterol reduces the risk of cardiovascular disease. Results of the Framingham Study[13] indicate that a rise in HDL cholesterol of 10 mg/dl above the mean may decrease the risk of an MI by one-third. The serum HDL cholesterol concentration may be increased by moderate to vigorous exercise, estrogens, and the moderate consumption of alcohol.

DECREASED CONCENTRATION. A lowered serum concentration of HDL cholesterol increases the risk of cardiovascular disease. HDL cholesterol is very low to nonexistent in the genetic disorder, Tangier disease (see p. 330). It is lowered by heavy cigarette smoking, obesity, very high carbohydrate diets, uncontrolled diabetes mellitus, and by male sex hormones.

Serum Triglycerides

Serum triglycerides are usually quantitated by measurement of the glycerol moiety, by colorimetric, fluorometric, or enzymatic methods. The enzymatic method is preferred because it is simpler and quicker to perform, readily automated, and as precise as the more laborious colorimetric or fluorometric techniques. The serum triglyceride concentration determined enzymatically, however, is about 10 mg/dl higher than that measured by an extraction method that eliminates free glycerol. Serum contains about 1.0 mg/dl of glycerol, which is equivalent to approximately 10 mg/dl of triglyceride.

Enzymatic Method

Several methods are readily available in kit form in which all reactions, including the hydrolysis of the triglycerides, are carried out by a series of enzymes. The assay method produced by Boehringer-Mannheim is described below because of the stable shelf life and reliability of the reagents.

REFERENCE. Wahlefeld, A.W.: Triglyceride: determination after enzymatic hydrolysis. In Bergmeyer, H.U. (editor): Methods of Enzymatic Analysis, 2nd English ed. New York, Academic Press, 1974, p. 1831.

PRINCIPLE. Triglycerides are hydrolyzed in three steps to glycerol and fatty acids by a lipase-esterase combination that does not hydrolyze phospholipids. The glycerol is converted to glycerol-1-P by glycerol kinase and ATP. This reaction is linked with that of pyruvate kinase (PK), which transfers the phosphate of phosphoenolpyruvate to the ADP formed in the above reaction, leaving pyruvate as a product. Another enzyme, lactate dehydrogenase, converts pyruvate to lactate, oxidizing NADH to NAD^+ in the process. The decrease in absorbance at 340 nm is proportional to the amount of glycerol and hence, triglycerides. The reactions follow:*

Reaction 1: $TG \xrightarrow{\text{lipase}} DG + FA + glycerol$

$\xrightarrow{\text{lipase}} MG + FA + glycerol$

$\xrightarrow[\text{esterase}]{\text{lipase}} glycerol + FA$

Reaction 2: $Glycerol + ATP \xrightarrow{\text{GK}} glycerol\text{-}1\text{-}P + ADP$

Reaction 3: $ADP + phosphoenolpyruvate \xrightarrow{\text{PK}} ATP + pyruvate$

Reaction 4: $Pyruvate + NADH + H^+ \xrightarrow{\text{LD}} lactate + NAD^+$

REAGENTS AND MATERIALS

1. The reagent set consists of four bottles, but aliquots of three are combined to form a Working Solution; the enzyme in the fourth bottle is added to start Reaction 2 after preliminary incubation of sample with Working Solution.
   a. Buffer, 20 mmol/L of sodium phosphate, pH 7.0 containing 4 mmol/L of $MgSO_4$, 0.35 mmol/L of sodium dodecylsulfate, and nonreactive preservatives.
   b. NADH/ATP/PEP. When reconstituted with the appropriate volume of water, the concentrations in mmol/L are 10, 22, and 18 for NADH, ATP, and PEP, respectively.
   c. LD/PK/Lipase/Esterase. The solutions contain the following U/ml (25° C) of enzyme activity: ≥300 LD, ≥50 PK, ≥4000 lipase, ≥30 esterase.
   d. GK. >150 U/ml (25° C).
   e. Working Solution. The Working Solution is stable for 8 hours at 20 to 25° C or 30 hours at 2 to 8° C, so prepare a volume sufficient for 1-day's work, according to the manufacturer's directions. For eight assays, this requires 20 ml of bottle #1, 0.4 ml of reconstituted bottle #2 and 0.4 ml of the suspension in bottle #3.
2. Water bath at 30° C or automated instrument with heating unit set at 30° C.
3. Pipetting devices suitable for delivering 2.50 ml of Working Solution and 50 μl of sample.

PROCEDURE

Serum must be obtained from a fasting patient (no food for 12–16 hours); the serum triglycerides are stable for 3 days at 4 to 8° C, or for 1 to 2 weeks with an antibiotic/azide mixture added as a preservative. Lipemic serum should be diluted 1 + 9 with isotonic saline because of the high triglyceride concentration; multiply the test result × 10. Plasma with either EDTA or heparin as anticoagulants may be substituted for the serum.

---

*The following abbreviations are used: TG = triglyceride; DG = diglyceride; MG = monoglyceride; FA = fatty acid; GK = glycerol kinase; ATP = adenosine triphosphate; ADP = adenosine diphosphate; PK = pyruvate kinase; PEP = phosphoenolpyruvate; NADH and NAD^+ are the reduced and oxidized forms, respectively, of nicotinamide adenine dinucleotide.

Note: A reagent blank (A_{RB}) should be determined once for each lot number of reagents. The procedure is carried out in the same manner as described below for specimens, except that water is substituted for the serum sample.

1. Pipet 2.50 ml of Working Solution (#1e) into a series of test tubes or cuvets for the specimens and control serum.
2. Add 50 µl of sample, mix, and incubate for 20 minutes at 30° C.
3. Measure the initial absorbance, A_1, at 340 nm.
4. Start Reaction 2 by adding 10 µl of solution #1d, the GK enzyme. Incubate for 5 minutes at 30° C.
5. Read the final absorbance, A_2, at 340 nm. The net absorbance = $A_1 - A_2$ = ΔA_s.
6. Correct for reagent blank: $\Delta A = \Delta A_s - \Delta A_{RB}$
7. Calculation:

$$\text{Triglyceride, mmol/L} = \frac{\Delta A}{6.22 \times 10^3} \times \frac{2.56}{0.050} = \Delta A \times 0.00823, \text{ where}$$

6.22 is the absorbance of 1.0 mmol/L NADH, $\dfrac{2.56}{0.050}$ are the volumes in ml of the total reaction mixture and specimen, respectively. The factor 10^3 converts the absorbance values from per ml to per liter. Triglyceride, mg/

$$\text{dl} = \text{mmol/L} \times \frac{1}{10} \times 885 = \Delta A \times 0.728, \text{ where } \frac{1}{10} \text{ converts liter to dl,}$$

and 885 is the molecular weight of triolein.

REFERENCE VALUES. The serum triglyceride concentration of newborns is low (about 38% of the adult level) but rises to the adult level by the third day of life.[7] The range for healthy adults and children in the United States is 40 to 150 mg/dl or 0.45 to 1.7 mmol/L. Mean values rise slowly with age after the third decade.

INCREASED CONCENTRATION. The concentration of serum triglycerides is moderately elevated following a meal containing fat and may rise as high as 260 mg/dl (3 mmol/L) after a meal containing 50 g fat. The peak of the triglyceride elevation occurs in 4 to 5 hours postprandial. Since a high triglyceride concentration is one of the risk factors in ischemic heart disease, elevated serum values are not unusual in individuals with atherosclerosis or with a history of myocardial infarction.

The serum triglyceride concentration is greatly elevated in hyperlipoproteinemia types 1 and 5 (Table 11.2), less so in type 4 and moderately increased in types 2b and 3. A very high triglyceridemia may induce pancreatitis. The above hyperlipoproteinemias refer to those of genetic origin but hypertriglyceridemia secondary to the following pathologic conditions commonly occurs: hypothyroidism, nephrotic syndrome, acute alcoholism, obstructive liver disease, acute pancreatitis, uncontrolled diabetes, glycogen storage disease (type I), and sometimes insulin-treated diabetes. The serum or plasma is usually turbid or milky when the triglycerides are elevated. The plasma triglycerides are mildly increased in Tangier disease (deficiency of high density lipoproteins).

DECREASED CONCENTRATION. The plasma triglyceride concentration is low in the rare disease, abetalipoproteinemia (absence of low density lipoproteins), and seldom exceeds 15 mg/dl.

Colorimetric Method

REFERENCE. Modification of Soloni, F.G.: Simplified manual micromethod for determination of serum triglycerides. Clin. Chem. *17*, 529, 1971.

PRINCIPLE. Serum triglycerides are extracted with isopropanol in the presence of an alumina adsorbent mixture that removes phospholipids, monoglycerides and diglycerides, glucose, bilirubin, and other interfering substances. Triglycerides in the extract are saponified to glycerol and soaps of the fatty acids or transesterified to yield glycerol and ethyl esters of the fatty acids as shown in the reactions 5a and 5b. The glycerol is oxidized to formaldehyde by means of periodate, yielding 2 moles of formaldehyde per mole of glycerol (Reaction 6). The formaldehyde is determined by the Hantzsch condensation of formaldehyde with NH_3 and acetylacetone (Reaction 7). The resulting diacetyl dihydrolutidine is colored yellow and is also fluorescent when activated with light at 400 nm. It may be determined colorimetrically as well as fluorometrically; the latter method is more sensitive.

Reaction 5a: Triglyceride + 3 KOH → glycerol
+ 3 K-soaps (K-salt of fatty acid)

Reaction 5b: Triglyceride + 3 Na ethoxide → glycerol
+ 3 ethyl esters of FA

Reaction 6:

$$H-\underset{\underset{OH}{|}}{\overset{\overset{H}{|}}{C}}-\underset{\underset{OH}{|}}{\overset{\overset{H}{|}}{C}}-\underset{\underset{OH}{|}}{\overset{\overset{H}{|}}{C}}-H + 2\ HIO_4 \rightarrow 2\ H-\overset{\overset{H}{|}}{C}{=}O + H-COOH + 2\ HIO_3 + H_2O$$

Glycerol　　Periodate　Formaldehyde　Formic Acid　　Iodate

Reaction 7: Hantzsch Reaction:

$$H-\overset{\overset{H}{|}}{C}{=}O + 2\ CH_3-\overset{\overset{O}{\|}}{C}-\underset{\underset{H}{|}}{\overset{\overset{H}{|}}{C}}-\overset{\overset{O}{\|}}{C}-CH_3 + NH_3 \rightarrow CH_3-\overset{\overset{O}{\|}}{C} \underset{CH_3 \ \underset{H}{N} \ CH_3}{\diagup \overset{H_2}{\diagup} \diagdown} \overset{\overset{O}{\|}}{C}-CH_3 \quad + 3\ H_2O$$

Formaldehyde　　Acetylacetone　　Ammonia

3,5-diacetyl-1,4-dihydrolutidine

The dihydrolutidine compound may be determined from either its absorbance or fluorescence.

REAGENTS AND MATERIALS

1. Isopropanol. Same as Reagent 1 in colorimetric cholesterol method (p. 333).
2. Adsorbent. Same as Reagent 2 in colorimetric cholesterol method (p. 333).
3. H_2SO_4, 0.8 mol/L. Dilute 44 ml of concentrated H_2SO_4 to 1 L with water.

4. Sodium Ethoxide, 0.1 mol/L in isopropanol. Let fine sediment settle overnight and use clear supernatant solution.
5. Sodium Metaperiodate, 0.02 mol/L. Dissolve 4.28 g $NaIO_4$ in water and make up to 1 L volume.
6. Sodium Arsenite, 0.2 mol/L. Dissolve 25.8 g $NaAsO_2$ in water and make up to 1 L volume.
7. Ammonium Acetate, 3 mol/L. Dissolve 231 g ammonium acetate in 800 ml water, adjust pH to 6.00 ± 0.05 with glacial acetic acid and make up to 1 L volume.
8. Acetylacetone Reagent. One to 5 drops of 2,4-pentanedione (Eastman) are added to 2.5 ml of ammonium acetate solution, mixed, and used within 1 hour.
9. Standard Triolein Solution, 100 mg/dl. Dissolve 100 mg of triolein (Applied Science) in isopropanol and make up to 100 ml volume.

PROCEDURE

Extraction

1. Prepare an isopropanol extract as described in step 1 of the Leffler Chemical method for the determination of cholesterol (p. 333) or use an aliquot of that extract.
2. 0.5 ml of sodium ethoxide are transferred to a series of tubes labeled "Sample," "Control," "Standard," and "Blank." To the sample and control tubes, add 0.1 ml isopropanol extract; add 0.1 ml standard triolein to the standard tube. Pipet 0.1 ml isopropanol into the blank tube.
3. Mix and incubate for 15 minutes at 60° C. The triglycerides are converted to glycerol and ethyl esters of fatty acids.
4. *Oxidation.* Add 0.1 ml periodate solution to each tube, mix, and let stand for 10 minutes.
5. Add 0.1 ml arsenite solution, mix, and let stand for 5 minutes.
6. *Condensation.* Add 2 ml of freshly prepared acetylacetone reagent to each tube. Mix with a vortex-type mixer for 2 to 3 seconds and incubate for 10 minutes at 60° C.
7. *Quantitation.* Read the absorbance at 415 nm against the blank.
8. *Calculation:*

$$\text{Triglyceride, mg/dl} = \frac{A_u}{A_s} \times 100 \text{ mg/dl}$$

where A_u is the absorbance of the unknown, A_s the absorbance of the standard, and 100 mg/dl is the concentration of the standard.

In SI units, triglyceride, mmol/L = mg/dl × 0.0113

Note: The factor, 0.0113 is correct only when triolein (mol. wt 885) is used as a standard; it is 0.0124 for tripalmitin (mol. wt 807).

Fluorometric Method.

The triglyceride fluorometric method has been automated after the first step of preparing the isopropanol extract.[10,11]

REFERENCE. Same as for colorimetic method (p. 340).

PRINCIPLE. The principles involved in the fluorometric determination of triglycerides are exactly the same as those described for the colorimetric method. They include extraction of triglycerides into isopropanol, with removal of interfering compounds by adsorption, hydrolysis of the triglycerides to glycerol, oxidation of glycerol to formaldehyde, and condensation of the formaldehyde with acetylacetone and NH_3 to form a diacetyl dihydrolutidine. The latter compound fluoresces when irradiated with light of 400 nm. The fluorometric method is several times more sensitive than the colorimetric method.

REAGENTS AND MATERIALS. Same as for colorimetric method; in addition a fluorometer is necessary.

PROCEDURE. Same as for colorimetric method, steps 1 through 6, except that 25 μl of extract instead of 100 μl should be used in step 2 because the fluorometric method is more sensitive.

7. Read directly in a fluorometer, using a 405 filter for the activating light and a Wratten #4 for the emitted light. Adjust the instrument to zero with a water blank.
8. Calculation: Let F_u and F_s stand for the difference in fluorescence between the unknown and its blank and between the standard and its blank, respectively.

$$\frac{F_u}{F_s} \times 100 \text{ mg/dl} = \text{mg/dl triglyceride}$$

In SI units, triglycerides, mmol/L = mg/dl × 0.0113.

HYPERLIPOPROTEINEMIAS

A hyperlipoproteinemia is defined as an elevation in the plasma concentration of one or more of the lipoprotein classes. This disorder may be familial (inherited) because of a single gene defect or secondary to a number of different metabolic disturbances that may or may not be of genetic origin. The increased plasma levels of triglycerides and/or cholesterol of the vast majority of individuals may be ascribed to the latter category of disturbances secondary to metabolic dysfunction or to life-style; the hyperlipoproteinemias caused by a single gene defect are rare. The classification of hyperlipoproteinemias serves a useful purpose, nevertheless, because it calls attention to the particular lipoprotein and lipid abnormalities.

Familial Hyperlipoproteinemias[3,9]

Five distinct familial hyperlipoproteinemias have been described; each is believed to be caused by a defect in a single gene. Each of these five genetic diseases gives a typical pattern or type when the serum is subjected to lipoprotein electrophoresis. The five phenotypes are the following:

Type 1, which is familial lipoprotein lipase deficiency. Chylomicrons in plasma circulate for a very long time because there is a deficiency of lipoprotein lipase for hydrolysis of triglycerides in chylomicrons.

Type 2, in which there is a deficiency of LDL receptors on tissue membranes. LDL is elevated in Type 2a (familial hypercholesterolemia), LDL and VLDL in Type 2b.

Type 3, which is found in familial dysbetalipoproteinemia. The defect is lack of Apo E-III, which is believed to cause defective HDL's. It is manifested as tuberous xanthelasma (deposition of lipids in skin of elbows and knees). VLDL is increased and LDL has an abnormal lipid composition which give a broad beta band merging with the prebeta upon lipoprotein electrophoresis.

Type 4, which is expressed in familial hypertriglyceridemia. The cause is unknown but plasma VLDL and triglyceride concentrations are elevated. It is commonly secondary to many other conditions.

Type 5, the cause of which is unknown. The plasma concentration of VLDL and chylomicrons is high and imparts a turbidity to the plasma even after standing overnight in a refrigerator; a cream layer denotes the presence of chylomicrons.

The serum concentrations of cholesterol and triglycerides typical of these familial hyperlipoproteinemias are summarized in Table 11.2. Visualization of a lipoprotein pattern by electrophoresis assists in the evaluation of patients with hypercholesterolemia. In many instances, however, a hyperlipoproteinemia can be classified without electrophoresis merely by having analytical values for serum cholesterol and triglycerides and by noting whether there is a cream layer or turbid serum after the serum has stood overnight in a refrigerator. Inspection of Table 11.2 shows that a type 1 hyperlipoproteinemia has a serum with greatly elevated triglycerides, a cream layer on standing, and clear serum underneath. This combination could be produced only by a high concentration of chylomicrons. The only other hyperlipoproteinemia to yield a cream layer is type 5, but here the serum is turbid; the

TABLE 11.2

Common Findings in the Familial Hyperlipoproteinemia Types

FINDINGS	TYPE 1	TYPE 2 a	TYPE 2 b	TYPE 3	TYPE 4	TYPE 5
		By electrophoresis:				
Chylomicrons	+ + + +	0	0	0 or ±	0	+ + +
VLDL (prebeta)	→ or ↓	→	↑	↑ Broad	↑	↑
LDL (beta)	→ or ↓	↑	↑	band†	→	→
HDL (alpha)	→ or ↓	→	→	→	→	→
Serum appearance:*						
Cream layer	+ + + +	0	0	0 to ±	0	+ +
Turbidity	0	0	±	+	+ +	+ + +
Serum analysis:						
Cholesterol	↑	↑	↑↑	↑	→	↑
Triglycerides	↑↑↑↑	→	↑	↑	↑↑	↑↑↑
Prevalence	Rare	Common		Fairly common	Common	Uncommon
Characteristic abnormalities	CL TG ↑	Beta ↑ Chol ↑	Prebeta ↑ Beta ↑ Chol ↑ TG ↑	Broad beta, ↑ Prebeta Chol ↑ ST when TG ↑	Prebeta ↑ ST TG ↑ Chol →	Prebeta ↑ CL, ST TG ↑

Symbols: + means present; 0 means absent; → means no change, ↑ increased, and ↓ decreased; TG = triglycerides; Chol = cholesterol; ST = serum turbid; CL = cream layer.
*After standing for at least 18 hours at 4° C.
†"Floating beta" LDL of abnormal lipid composition.

responsible lipoprotein can only by VLDL (prebeta by electrophoretic classification). In like manner, most hyperlipoproteinemias can be classified by chemical analysis and the application of logic and observation.

Secondary Hyperlipoproteinemias[9]

Classification of the hyperlipoproteinemias in those patients with lipidemias secondary to other factors is useful in prescribing treatment for reduction of plasma lipids where appropriate. For example, the hypertriglyceridemia in a type 4 hyperlipoproteinemia can be reduced by decreasing the patient's carbohydrate or ethanol intake; in this type, the liver converts excess carbohydrate to triglycerides, which are transported by VLDL. The typing is also useful in devising the best ways to reduce plasma cholesterol where necessary in high risk patients, whether by diet or by the use of drugs or resins. Some of the common hyperlipoproteinemias secondary to various pathologic states are of the following types:

1. Hypothyroidism: 2a or 3
2. Nephrotic syndrome: 2a, 2b, or 5
3. Alcoholic hyperlipemia: 4 or 5
4. Obstructive liver disease: lipoprotein X
5. Pancreatitis: 1, 4, or 5
6. Diabetes, uncontrolled, with severe hyperlipidemia: 5
7. Systemic lupus erythematosus: 1 or 3
8. Glycogen storage disease (type I): 4 or 5
9. Uremia, renal dialysis: 4

Lipoprotein Electrophoresis

The principal reason for requesting a lipoprotein electrophoresis is to ascertain the type of a hyperlipoproteinemia in patients with hyperlipidemia. As pointed out previously, in many cases, a classification can be made by correlating the analytical values for total cholesterol and triglycerides with the appearance of the serum after standing overnight in a refrigerator. An electrophoresis pattern does help in difficult situations.

REFERENCE. Noble, R.P.: Electrophoretic separation of plasma proteins in agarose gel. J. Lipid Res. 9, 693, 1968.

PRINCIPLE. Lipoproteins carry an electric charge at pH 8.6 and can be separated by electrophoresis in a manner similar to serum proteins. Chylomicrons are uncharged and remain at the origin. After electrophoretic separation, the lipoproteins are stained with a fat stain; the support medium is destained and then dried. The lipoprotein bands and chylomicrons are visible because of the fat stain dissolved in them. Abnormal bands may be seen by visual inspection or may be quantitated from a densitometric scan of the membrane.

REAGENTS AND MATERIALS

1. Agarose film, Pol-E-film (Pfizer).
2. Barbital buffer, pH 8.6, 0.05 mol/L containing 350 mg disodium-EDTA per liter. Dissolve 20.6 g sodium barbital in 1500 ml water, add 700 mg Na_2-EDTA, adjust pH to 8.6 with molar HCl, and make up to 2-L volume with water.

3. Fat Red 7B stain, stock solution. Dissolve 1 vial (about 0.9 g) Fat Red 7B stain (Pfizer) in 4 L methanol and let stand to ripen for 2 weeks; shake solution occasionally.
4. Working Solution, Fat Red B stain. Prepare freshly as needed 6 ml per membrane. To 5 ml stock Fat Red B add 1 ml 0.1 mol/L NaOH and 10 μl Triton X-100 (Rohm & Haas).
5. Glycerol, 2% (v/v). Dilute 20 ml glycerol to 1000 ml with water.
6. Destaining solution. Dilute 750 ml methanol with 250 ml water.
7. Electrophoresis apparatus that can be set for constant voltage. It is convenient but not essential to use the Pfizer apparatus.

PROCEDURE. Serum samples should be obtained after an overnight fast in order to avoid a postprandial hyperlipidemia. The lipid-protein bond is labile and easily ruptured by freezing. Samples should be run as soon as possible and kept in a *refrigerator*, not a freezer. On removal from the refrigerator, samples should be allowed to stand at room temperature for at least 30 minutes to allow for warm-up and reformation of any lipid-protein bond. Observe for turbidity and cream layer. Mix by inversion before taking sample for electrophoresis.

1. *Electrophoresis*. Apply 1 μl serum samples of patients and controls (preferably a normal and an abnormal) to the slots in the agarose film.
2. Set instrument for constant voltage, turn on, and adjust the voltage to 100 v. Check the current, which should be about 10 ma.
3. Continue the electrophoresis for 60 minutes. Turn off current, remove the film, and dry it at 72° C for approximately 30 minutes or until clear.
4. *Staining*. Cover the film with working solution of Fat Red 7B stain and let stand for 15 minutes.
5. *Destaining*. Place 200 ml of the methanol destaining solution in a dish and immerse the film for 20 seconds. Then gently agitate the dish until the background is clear.
6. Immerse the film for 2 minutes in a 2% glycerol solution.
7. Remove and dry in an oven at 72° C for 10 minutes.
8. Evaluation of electrophoresis. Inspect visually and look for abnormal bands, or scan with a densitometer and look for abnormal peaks.

INTERPRETATION. The interpretation is made from a combination of three sets of data: abnormal bands on the electrophoresis strip; the appearance of the serum, i.e., whether clear, turbid, or milky, or whether there is a cream layer upon standing; analytical values for serum cholesterol and triglycerides. The interpretation is made from the correlation of these data with the criteria outlined in Table 11.2.

REFERENCES

1. Martin, D.W., Jr., Mayes, P.A., and Rodwell, V.W.: Harper's Review of Biochemistry, 18th ed. Los Altos, Lange Publications, 1981, p. 186.
2. Levy, Robert I.: Cholesterol, lipoproteins, apoproteins and heart disease: Present status and future prospects. Clin. Chem. *27*, 653, 1981.
3. Brewer, H.B., Jr., and Bronzert, T.J.: Human plasma lipoproteins, Fractions #1. Palo Alto, Spinco Div., Beckman Instruments, 1977, p. 1.
4. Herbert, P.N., Gotto, A.M., and Fredrickson, D.S.: Familial lipoprotein deficiency. *In* Metabolic Basis of Inherited Disease, 4th ed., edited by J.B. Stanbury, J.B. Wyngaarden, and D.S. Fredrickson. New York, McGraw-Hill, 1978, p. 544.
5. Zilversmit, D.B., and Davis, A.K.: Microdetermination of plasma phospholipids by trichloroacetic acid precipitation. J. Lab. Clin. Med. *35*, 155, 1950.

6. Trout, D.L., Estes, E.H., Jr., and Friedberg, S.J.: Titration of free fatty acids of plasma: a study of current methods and a new modification. J. Lipid Res. *1*, 199, 1960.

7. Kaplan, A., and Lee, V.F.: Serum lipid levels in infants and mothers at parturition. Clin. Chim. Acta *12*, 258, 1965.

8. Pediatric Clinical Chemistry, edited by S. Meites. Washington D.C., American Association for Clinical Chemistry, 1981, p. 148.

8a.Fredrickson, D.S., and Levy, R.I.: Familial hyperlipoproteinemia. *In* Metabolic Basis of Inherited Disease, 3rd ed., edited by J.B. Stanbury, J.B. Wyngaarden, and D.S. Fredrickson. New York, McGraw-Hilll, 1972, p. 545.

9. Goldstein, J.L., and Brown, M.S.: The familial hyperlipoproteinemias. *In* Metabolic Basis of Inherited Disease, 4th ed., edited by J.B. Stanbury, J.B. Wyngaarden, and D.S. Fredrickson. New York, McGraw-Hill, 1978, p. 604.

10. Noble, R.P., and Campbell, F.M.: Improved accuracy in automated fluorometric determination of plasma triglycerides. Clin. Chem. *16*, 166, 1970.

11. Edwards, L., Falkowski, C., and Chilcote, M.E.: Semiautomated fluorometric measurement of triglycerides. *In* Standard Methods of Clinical Chemistry. Vol. 7, edited by G.R. Cooper. New York, Academic Press, 1972, p. 69.

12. Cobb, S.A., and Sanders, J.L.: Enzymic determination of cholesterol in serum lipoproteins separated by electrophoresis. Clin. Chem. *24*, 1116, 1978.

13. Kannel, W.R., Castelli, W.P., and Gordon, T.: Cholesterol in the prediction of atherosclerotic disease: New perspective on the Framingham Study. Ann. Intern. Med. *90*, 85, 1979.

14. Kane, J.P., Hardman, D.A., and Paulus, H.E.: Heterogeneity of apolipoprotein B: Isolation of a new species from human chylomicrons. Proc. Nat'l. Acad. Sci. *77*, 2465, 1980.

Chapter 12

MINERAL METABOLISM

For proper growth and development, a balanced diet must provide adequate amounts of minerals in addition to sufficient calories, high quality proteins, and vitamins. The roles of Na, K, and Cl are discussed in Chapter 3; among the other minerals, Ca and P (as phosphate) are essential for the skeletal components, bones, and teeth. Various minerals play a vital role in essential physiologic processes such as blood coagulation (Ca), hemoglobin production (Fe), action of ATP-ase (Mg), formation of thyroid hormones (I), formation of metalloenzymes (Ca, Mg, Zn, Mn, Cu, Mo). The seven principal minerals required by the body are Na, K, Cl, Ca, P, Mg, and organically-bound S (as R-SH or R_1-S-S-R_2). At least nine other elements, Fe, I, Zn, Cu, Mn, Se, Cr, Mo, and Co (as a component of vitamin B_{12}) are required in trace amounts; they are toxic when taken in excess. Fluoride strengthens bone and makes teeth more resistant to decay, but its absolute requirement for humans has not been demonstrated unequivocally.

CALCIUM

Calcium is the mineral present to the largest extent in the body. Approximately 99% of the total body calcium is deposited in the skeleton as a mixture of amorphous calcium phosphate and crystalline hydroxyapatite, a hydrated calcium phosphate crystal. Small amounts of fluoride are incorporated in the calcium phosphate deposits in both bone and teeth. The calcium phosphate laid down in bone is by no means an inert component of the skeleton; it is in dynamic equilibrium with the Ca^{2+} and HPO_4^{--} of body fluids and is constantly turning over by the process of dissolution (resorption) and deposition.

A higher proportion of the nonskeletal calcium is present within cells than in extracellular fluids but most of the intracellular calcium is bound to proteins in the cell membrane, mitochondria, and nucleus. The concentration of calcium ion in extracellular water is several thousand times as large as Ca^{2+} in cytoplasmic water because of this binding.[2] Only the ion is biologically active.

Ca^{2+} passively diffuses from extracellular to intracellular fluids but energy in the form of ATP is necessary for passage from cytoplasmic water through *any* membrane, including cellular, mitochondrial, or nuclear membranes. Ca^{2+} is physiologically active and regulates a variety of cellular functions, such as contraction of fibers, secretion of fluids or active substances (hormones), activation of enzymes, and transfer of ions across membranes. Ca^{2+} decreases neuromuscular excitability and is essential for blood coagulation.

Most individuals are in a state of calcium balance, ingesting about 500 to 1000 mg daily in the food and excreting a like amount in the urine and feces.

Dairy products (milk, cheese) are the best food sources of calcium. A large portion of the dietary calcium is not absorbed because of the formation of insoluble calcium compounds (phosphate, phytate, oxalate, soaps) in the intestines that are excreted in the feces. The calcium balance is maintained by a complex control system that regulates the absorption of calcium from the intestine, its excretion by the kidneys, and the movement of calcium in and out of bone.

Control of Calcium Balance and Plasma Ion Concentration

Calcium balance and a relatively constant concentration of plasma Ca^{2+} are maintained by the interplay of two hormones and a body-modified vitamin acting upon the above organs. Together, these factors control the input (absorption) of calcium, its output (renal excretion) and its movement in and out of the great calcium phosphate storage pool, bone. The absorption of calcium is facilitated by the biologically active form of vitamin D_3. Vitamin D_3 (cholecalciferol) is a sterol that is hydroxylated in the 25-position by the liver; the kidney inserts another hydroxyl group to produce the active vitamin, 1,25-dihydroxy-D_3. The active vitamin is also a hormone since it is a physiologically active molecule produced by one organ and carried by the blood to target organs where its effects are exerted. The 1,25-dihydroxy-D_3 not only promotes the intestinal absorption of calcium by stimulating an active transport mechanism,[12] but acts upon bone in cooperation with PTH to dissolve bone mineral and thereby increase plasma Ca^{2+} concentration. It may also increase the reabsorption of Ca^{2+} from the glomerular filtrate.

The plasma concentration of Ca^{2+} is maintained within close limits by the actions of two hormones: PTH from the parathyroid gland and calcitonin from the thyroid (p. 296). A minor decrease in plasma Ca^{2+} concentration stimulates the secretion of PTH. PTH enables the kidney to reduce the excretion of Ca^{2+} by increasing its reabsorption in the renal tubules, promotes the dissolving (resorption) of bone to increase the plasma levels of Ca^{2+} and phosphate, stimulates the kidney to produce more 1,25-dihydroxy-D_3 from its precursor. This latter hormone increases the intestinal absorption of calcium.

A high plasma Ca^{2+} concentration inhibits the secretion of PTH and stimulates the secretion of calcitonin. Calcitonin acts only upon the bone osteoclasts and inhibits their dissolution of bone. The effects of the two hormones are transitory because of their short half-lives (20–30 minutes) but more of the appropriate hormone will be secreted as long as the plasma concentration of Ca^{2+} is beyond the normal limits.

Plasma calcium exists in three forms: ionic or free calcium, Ca^{2+}, the physiologically active ion; unionized calcium bound primarily as a complex of citrate; and a nondiffusible form bound to plasma proteins. Ca^{2+} and unionized calcium pass through the membrane of the glomerulus and appear in the glomerular filtrate, but protein-bound Ca cannot do this. About 43 to 47% of the total plasma calcium is protein-bound, primarily to albumin, but also to some extent to the α-, β-, and γ-globulins. The amount of calcium attached to plasma proteins varies with the protein concentration. Ionized calcium constitutes 48 to 52% of the total calcium, and the unionized, diffusible form amounts to about 5%.

Serum is the fluid of choice for measurement of total calcium concentration.

The Ca^{2+} is altered by the following factors: blood pH (alkalosis lowers Ca^{2+} and acidosis increases it); diseases of the parathyroid gland or of the calcitonin-producing cells of the thyroid; intestinal disorders of malabsorption; vitamin D deficiency; severe renal disease; and disorders of bone turnover, some of which may be genetic while others may be caused by malignancies. A high concentration of Ca^{2+} presents the danger of deposition of calcium salts in soft tissues, including the kidney, and frequently is accompanied by symptoms of fatigue, weakness, and confusion. An elevated Ca^{2+} concentration triggers the secretion of the hormone, calcitonin, which inhibits the dissolution of bone, but this is insufficient to lower the plasma Ca^{2+} in the face of excess PTH and/or 1,25-dihydroxy-D_3. The three hormones must operate synchronously in order to attain a balanced calcium metabolism.

A low concentration of Ca^{2+} increases neuromuscular excitability to the point where tetany or convulsions may occur. The parathyroid gland is stimulated to secrete PTH but this alone may not suffice to restore a normal plasma Ca^{2+} concentration if the intake of calcium is inhibited by a malabsorption syndrome or by a deficiency of 1,25-dihydroxy-D_3.

Pregnancy and lactation increase the dietary requirement for calcium because of its utilization for fetal skeleton formation in the former case and its loss to the body as a component of milk in the latter. If not provided by the diet, the necessary calcium and phosphate are obtained by the resorption of maternal bone, but there is little change in the plasma Ca^{2+} concentration.

Total Serum Calcium

Innumerable methods have been proposed for the determination of total serum calcium, mute testimony to the fact that past methods were not entirely satisfactory. The Clark and Collip method (calcium precipitation with oxalate and titration with permanganate)[3] was the standard procedure for three or more decades, but it is cumbersome and time-consuming. Because its inherent errors tend to be compensating, results obtained by it agree reasonably well with the best methods available today.

Total Serum Calcium by Atomic Absorption Spectrophotometry

The most accurate method today for total serum calcium is by atomic absorption spectrophotometry (AAS), and a reference method by atomic absorption spectrophotometry has been proposed by Cali, Bowers, and Young.[4] Details of a simpler atomic absorption spectrophotometry method for everyday use are described.

REFERENCE. Trudeau, D.L., and Freier, E.F.: Determination of calcium in urine and serum by atomic absorption spectrophotometry (AAS). Clin. Chem. *13*, 101, 1967.

PRINCIPLE. When a dilute solution of serum is introduced into a hot flame, the elements present dissociate from their chemical bonds and appear as free atoms in the ground state (not activated—see p. 37). If a light beam from a calcium hollow cathode lamp is passed through the flame containing the vaporized sample, the calcium atoms arising from the serum sample will absorb some of the light emitted by the calcium hollow cathode lamp. The effect is specific for the 422.7 nm resonance line of calcium emitted by the lamp, and only calcium atoms in the ground state will absorb a portion of it. A photomultiplier tube measures the decrease in light intensity, which is proportional to the serum calcium concentration. The serum diluting fluid contains lanthanum chloride ($LaCl_3$) in order to form lanthanum phosphate instead of refractory calcium phosphate in the aerosol. It is more difficult to generate calcium atoms in the flame from calcium phosphate than from calcium chloride.

REAGENTS AND MATERIALS

1. Stock Lanthanum, 50 g/L La (0.36 mol/L) in 3 mol/L HCl. Place 58.7 g high purity lanthanum oxide and 300 ml water in a liter volumetric flask. Add 252 ml concentrated HCl and swirl until the solution is clear. Cool and dilute to the mark with water.
2. Working La Diluent, 1.0 g/L La (7.2 mmol/L) in 60 mmol/L HCl. Dilute 20 ml Stock La solution to 1 L with water.
3. Stock Ca Standard, 100 mg/dl Ca (25 mmol/L) in 60 mmol/L HCl. Dry some National Bureau of Standards $CaCO_3$ overnight at 120° C and cool in a desiccator. Transfer 2.50 g $CaCO_3$ to a liter volumetric flask; add 300 ml water and 5 ml concentrated HCl. Mix until the solution is clear, cool, and make up to volume with water.
4. Working Ca Standard, 10.0 mg Ca/dl (2.50 mmol/L). Dilute 10.0 ml stock standard to 100 ml volume with blank reagent.
5. Blank Reagent. Place 9.0 g NaCl, 0.373 g KCl, and 0.072 g Na_2HPO_4 in a liter volumetric flask, dissolve in water, and make up to volume. The reagent concentrations are 154 mmol/L Na, 5.0 mmol/L K, and 1.4 mg/dl P.
6. Diluter-Pipetter. Set diluter-pipetter to pick up 0.1 ml sample and to deliver with 9.9 ml working La diluent. Use for diluting specimens, standard and blank.
7. Atomic Absorption Spectrophotometer.

PROCEDURE

1. Follow manufacturer's directions for lighting and adjusting the AAS instrument.
2. Set the wavelength to 422.7 nm.
3. Prepare a blank by diluting 0.1 ml blank reagent with 9.9 ml working La diluent. Adjust the instrument to zero while aspirating the blank.
4. Dilute 0.1 ml working standard, serum, and controls, respectively, with 9.9 ml of diluent.
5. Aspirate the diluted standard and adjust the controls so that the instrument reads 10.0 mg Ca/dl (2.50 mmol Ca/L).
6. Aspirate controls and samples, and record results. No calculation is necessary with an instrument having a direct readout.

Total Serum Calcium by Colorimetric Methods

A number of laboratories analyze for serum calcium by automating a colorimetric method.[5-7] Calcium complexes with several different compounds to form highly colored chromophores whose absorbance can be measured. One of the reagents used most widely for this purpose is o-cresolphthalein complexone in alkaline solution. The addition of a little 8-hydroxyquinoline to the reagent prevents the binding of magnesium. The absorbance of the colored complex is measured at 580 nm and is proportional to the total calcium concentration.

REFERENCE VALUES. The reference values for total serum calcium have sharpened as methods and quality control have improved. The range is from 8.7 to 10.5 mg/dl (2.18 to 2.63 mmol/L).

INCREASED CONCENTRATION. The serum calcium concentration is increased in hyperparathyroidism, neoplasms with metastasis to bone, some malignancies (lung, kidney, bladder) without bone involvement, hypervitaminosis D, sarcoidosis, multiple myeloma when the plasma proteins are elevated, in patients taking large amounts of milk and alkali as treatment for peptic ulcer, acromegaly (excess growth hormone), some cases of hyperthyroidism, and in Paget's disease (a bone disorder). It is not unusual to find hypercalcemia in patients who had chronic renal disease and recently received a kidney transplant. This type of hypercalcemia is ascribed to parathyroid hyperplasia induced by the chronically low serum calcium concentration during the long period before the kidney transplant; the hyperplastic parathyroids secrete too much PTH because the new, normal kidney after the transplant is able to convert 25-hydroxy-D_3 to the 1,25-dihydroxy molecule, which improves the intestinal absorption of calcium and helps to elevate the serum calcium concentration. Prolonged elevation of plasma calcium can cause the deposition of insoluble calcium salts in soft tissues and is a factor in the production of renal calculi (stones).

DECREASED CONCENTRATION. A decrease in the serum ionized calcium concentration below a critical level causes tetany. It is important to notify the physician immediately if low serum calcium levels are found because it can be stopped or prevented by the intravenous injection of calcium gluconate solution. The most common causes of a low serum calcium concentration are hypoparathyroidism, pseudohypoparathyroidism (failure of the target organs to respond to PTH), a deficiency of vitamin D, gastrointestinal disease that interferes with the absorption of vitamin D and/or calcium (sprue, steatorrhea), nephrosis or other conditions in which the serum protein concentration is low, chronic renal disease (failure of the kidney to hydroxylate 25-hydroxy-D_3), acute pancreatitis, and magnesium deficiency.

The differential diagnosis of hypercalcemia and hypocalcemia is made easier by simultaneous determination of serum calcium, phosphate, alkaline phosphatase, and, in special instances, PTH.

Ionized Calcium

Calcium has to be in the ionized or free form (Ca^{2+}) in order to exert its physiologic effects upon the neuromuscular junction, membranes, and bone deposition. Hence, the most valuable clinical information[9] is provided by a knowledge of the concentration of Ca^{2+} rather than that of total Ca. Unfortunately, the state of the art is such today that the measurement of [Ca^{2+}] in serum or heparinized blood is a cumbersome procedure and not well-suited to a routine operation. The sample has to be treated anaerobically, processed, and measured without delay; the measurement is temperature-sensitive and varies with pH. A detailed procedure for measuring Ca^{2+} by means of a flow-through, ion-selective electrode is given by Ladenson and Bowers.[8] Their reference values for Ca^{2+} are 1.175 to 1.375 mmol/L (4.70 to 5.50 mg/dl); those for total Ca are 2.275 to 2.575 mmol/L (9.10 to 10.30 mg/dl). The ratio of ionized Ca to total Ca is 48 to 56%. In a series of patients with various types of hyperparathyroid disease, Ladenson and Bowers[9] found a better correlation of the disease state with Ca²⁺ than with total Ca. Although the instruments for measuring Ca²⁺ have improved since the above work was done, the analysis for ionized calcium is still not widely requested for two reasons: the precision of measurement has to be increased because the normal range is so narrow, and the test is still not well adapted for routine performance or mass production.

PHOSPHATE

The phosphorus in the body exists only as inorganic phosphate or as organic phosphate esters. Eighty percent of it is laid down in bone matrix as insoluble calcium salts. The organic phosphate esters are primarily confined within cells, associated with nucleoproteins, hexoses (glucose-6-phosphate), deoxygenated hemoglobin in red blood cells (2,3-diphosphoglycerate), and purines (ATP, GTP); phosphate forms high energy bonds in ATP, GTP, and creatine phosphate. The phospholipids are also present in cells, particularly as a component of membranes, but they also circulate in plasma in fairly high concentration (see Chap. 11). Inorganic phosphate ions ($H_2PO_4^-$ and HPO_4^{--}) are mostly confined to the extracellular fluid where they are part of the buffer system. At pH 7.4, 80% of the inorganic phosphate is in the form of HPO_4^{--}.

Phosphate is ubiquitous in food, and there is usually no problem in ingesting sufficient amounts. The organic phosphates in food are hydrolyzed in the gastrointestinal tract, and inorganic phosphate is liberated. A large proportion of the ingested phosphate is precipitated as insoluble salts, however, and not absorbed. The principal route of excretion for the absorbed phosphate is in the urine.

The control of plasma phosphate concentration is closely linked with that for calcium because they are deposited together in bone as an amorphous or hydrated calcium phosphate; bone resorption increases the plasma concentrations of both calcium and phosphate ions while bone deposition decreases the concentrations of both. Thus, the three hormones (PTH, calcitonin, and 1,25-dihydroxy-D₃), which regulate bone resorption or deposition, affect the phosphate concentration at the same time. PTH, however, stimulates the kidney to *excrete* phosphate while *conserving* calcium. This is a protective mechanism because the solubility product of calcium phosphate in body fluids is readily exceeded if the concentrations of both ions are elevated; the concentrations of Ca²⁺ and phosphate tend to vary inversely because of the PTH effect upon the kidney.

The phosphate ion concentration of the plasma is influenced by the calcium ion concentration, by parathyrin concentration, which promotes phosphaturia by decreasing the renal tubular reabsorption of phosphate, by calcitonin, which inhibits bone resorption, and by growth hormone, which increases the renal tubular reabsorption of phosphate; in chronic renal disease there is phosphate retention because of impaired glomerular filtration.

Serum Phosphate

In most of the determinations of serum phosphate, the inorganic phosphate is converted into an ammonium phosphomolybdate complex. A mild reducing agent is added to reduce the Mo^VI and convert the complex to highly colored "molybdenum blue." This is similar to the blue arsenomolybdate complex formed in the Nelson-Somogyi method. Because the reducing agent is added in excess, phosphate concentration is the limiting factor for color development.

Many different reducing agents have been employed in the determination of phosphate. Aminonaphtholsulfonic acid was introduced for this purpose more than 50 years ago and is still widely used, but p-methylaminophenol is more stable and easier to prepare.

It has been customary to report the serum phosphate concentration in terms of mg P per 100 ml rather than mg HPO_4^-. In the SI system, however, mol/L of P are identical with mol/L of HPO_4^-.

REFERENCE. A modification of Power, M.H.: Inorganic phosphate. *In* Standard Methods of Clinical Chemistry, Vol. 1, edited by M. Reiner. New York, Academic Press, 1953, p. 84.

PRINCIPLE. Serum proteins are precipitated by means of trichloroacetic acid, and the phosphate is converted to a phosphomolybdate (Mo^{VI}) complex by the addition of sodium molybdate. The addition of p-methylaminophenol reduces the Mo^{VI} in the complex to yield an intensely blue-colored phosphomolybdate complex (Mo^V). The absorbance of the solution at 700 nm is proportional to the serum phosphate concentration.

REAGENTS AND MATERIALS

1. Trichloroacetic Acid (TCA), 50 g/L. Transfer 50 g TCA to a flask and dissolve in a liter of water.
2. Molybdate Reagent. Dissolve 25 g sodium molybdate ($Na_2MoO_4 \cdot 2H_2O$) in about 500 ml water in a 1-L volumetric flask; stir while adding 250 ml 5 mol/L sulfuric acid, and make up to volume with water.
3. Reducing Agent. Dissolve 1.0 g p-methylaminophenol sulfate (Elon, Eastman) and 3 g sodium bisulfite ($NaHSO_3$) in 100 ml water. Filter through an ashless filter paper (Whatman #40 or #42, Reeve Angel) and store in a brown bottle. The solution is stable for at least 2 months.
4. H_2SO_4, 5 mol/L. Add slowly with stirring 278 ml of concentrated sulfuric acid to 700 ml cold water in an Erlenmeyer flask. When cool, make up to 1000 ml volume with water.
5. Stock Standard P, 1.0 mgP/ml (32.2 mmol/L). Dissolve 4.910 g KH_2PO_4 in water, add 10 ml 5 molar H_2SO_4, and make up to 1000 ml volume with water.
6. Working P Standard, 4.0 mgP/dl (1.29 mmol/L). Dilute 4.0 ml of stock standard P to 100.0 ml volume with water.

PROCEDURE

Protein Precipitation

1. Place 0.5 ml of unknown serum, controls, and working standard, respectively, in 13 × 125 mm test tubes.
2. Rapidly pipet 9.5 ml 50 g/L TCA into each, cover tubes with Parafilm, and shake.
3. Let stand for 5 minutes and centrifuge or filter through Whatman #40 filter paper.
4. Transfer a 5.0 ml aliquot from each tube to a clean test tube. Use 5.0 ml TCA for a blank.

Color Development

5. Add 1 ml molybdate reagent to each tube.

6. Add 0.25 ml reducing reagent to each tube, mix, and let stand for 5 minutes.
7. Read absorbance of all solutions against the blank at 700 nm.

Calculation

$$\frac{A_u}{A_s} \times 4.0 = mgP/dl$$

To convert to mmol/L: $mgP/dl \times \dfrac{10 \, dl/L}{31 \, mg/mmol} = mgP/dl \times 0.32$

REFERENCE VALUES
At birth: 4.2 to 9.5 mg/dl (1.34 to 3.36 mmol/L)
Children: 4.0 to 7.0 mg/dl (1.28 to 2.24 mmol/L)
Adults: 3.0 to 4.5 mg/dl (0.96 to 1.44 mmol/L)

INCREASED CONCENTRATION. The serum phosphate concentration is increased in advanced renal insufficiency, true and pseudohypoparathyroidism, hypervitaminosis D, and in patients with hypersecretion of growth hormone (GH). The higher phosphate levels in infants and children are associated with an increased concentration of growth hormone.

DECREASED CONCENTRATION. The serum phosphate concentration is temporarily lowered when carbohydrate is absorbed; glucose and other sugars are phosphorylated when they enter cells. A lowered concentration of serum phosphate typically appears in hyperparathyroidism, rickets (vitamin D deficiency), steatorrhea, and in some renal diseases when there is impaired tubular absorption of phosphate (Fanconi syndrome). The prolonged ingestion of antacids containing $Mg(OH)_2$ or $Al(OH)_3$ lowers the serum phosphate because of precipitation of insoluble phosphates in the gastrointestinal tract. Parenteral hyperalimentation causes a movement of phosphate from plasma into muscle and adipose cells, which lowers plasma phosphate concentration. A dangerously low hypophosphatemia may be produced unless the fluids contain sufficient phosphate.

Urine Phosphate

The procedure for determination of urine phosphate is exactly the same as that for serum phosphate except that a smaller aliquot of the TCA filtrate is taken. Steps 1, 2, and 3 are the same as for serum, but in Step 4 take a 0.5 and a 2.5 ml aliquot for analysis. Add sufficient TCA to bring the volume to 5.0 ml in each tube. Proceed with steps 5 through 7 exactly as for serum. Use the result for the aliquot that gives an absorbance closer to that of the standard.

Calculation

A. For the 0.5 ml aliquot:

$$mg \, P/dl = \frac{A_u}{A_s} \times 4.0 \times \frac{5.0 \, ml}{0.5 \, ml} = \frac{A_u}{A_s} \times 40$$

B. For the 2.5 ml aliquot:

$$mg \, P/dl = \frac{A_u}{A_s} \times 4.0 \times \frac{5.0}{2.5} = \frac{A_u}{A_s} \times 8$$

SI units: P as mmol/L = mg/dl × 0.32

MAGNESIUM

Magnesium is the fourth most abundant cation in the body; much of it is present in bone, associated with calcium and phosphate. The magnesium concentration within cells is second only to potassium; its level in extracellular fluids is much lower. About 30% of the plasma magnesium is bound to albumin, and 70% exists as the ion Mg^{2+}.

The main dietary sources of magnesium are meat and green vegetables. Like calcium, magnesium is absorbed in the upper intestines, but vitamin D is not essential for magnesium absorption. Magnesium compounds have solubility properties similar to those of calcium, and the major portion of the ingested magnesium is not absorbed because of the formation of insoluble phosphates and soaps in the gut. The kidney is the organ responsible for maintaining magnesium balance; it can conserve magnesium when the intake is low and excrete the excess when the intake is high. The exact control mechanism is unknown, but the hormone aldosterone may play a role; aldosterone promotes the excretion of Mg^{2+}, together with K^+, and the retention of Na^+.

Mg^{2+} is an activator for a number of enzymes, particularly for those involved in phosphorylation, protein synthesis, and DNA metabolism; it also affects neuromuscular excitability. Like calcium, a low serum Mg^{2+} concentration is associated at times with tetany. A high concentration of Mg^{2+} in plasma has a depressing effect upon the central nervous system and alters the conducting mechanism of the heart.

Serum Magnesium

There are some colorimetric and fluorometric methods for the determination of the serum magnesium concentration, but the best one employs atomic absorption spectrophotometry (AAS). The measurement of serum magnesium by AAS is very similar to that of calcium except that a different hollow cathode lamp is used as a light source and a different wavelength is employed. Some AAS instruments are designed to measure magnesium and calcium simultaneously from a single serum sample.

REFERENCES. Hansen, J.L., and Freier, E.F.: The measurement of serum magnesium by atomic absorption spectrophotometry. Am. J. Med. Tech. *33*, 158, 1967.
Trudeau, D.L., and Freier, E.F.: Determination of calcium in urine and serum by atomic absorption spectrophotometry (AAS). Clin. Chem. *13*, 101, 1967.

PRINCIPLE. The principle is the same as that described for the calcium AAS method. When the light from the magnesium hollow cathode lamp is passed through the flame, the vaporized magnesium atoms in the ground state coming from the aspirated serum sample absorb some of the light. The effect is specific for the 285.2 nm line of the lamp and is proportional to the total magnesium concentration in serum. The serum is diluted with a lanthanum solution to minimize the effects of interfering substances.

REAGENTS AND MATERIALS

1. Stock Lanthanum, 50 g/L. Same as for calcium method.
2. Working La Diluent, 1.0 g/L. Same as for calcium method.
3. Stock Standard Mg, 100 meq/L (50 mmol/L). Dissolve 1.216 g pure Mg metal in 10 ml concentrated HCl and dilute to 1 L with water.
4. Working Standard Mg Solutions:

    a. 1.0 meq/L (0.5 mmol/L). Dilute 1.0 ml stock standard to 100 ml with water.

    b. 2.0 meq/L (1.0 mmol/L). Dilute 2.0 ml stock standard to 100 ml with water.

    c. 3.0 meq/L (1.5 mmol/L). Dilute 3.0 ml stock standard to 100 ml with water.

5. Blank Reagent. Same as for calcium method.
6. Diluter-Pipetter. Same as for calcium method.
7. Atomic Absorption Spectrophotometer. Same as for calcium method.

PROCEDURE. All steps are identical to those described for calcium except:

1. Wavelength of 285.2 nm is used.
2. Working standards of Mg are substituted for the Ca standards.

If the AAS instrument is designed to measure calcium and magnesium simultaneously, working standards containing both magnesium and calcium are prepared.

REFERENCE VALUES. The concentration of total serum magnesium in normal individuals ranges from 1.3 to 2.1 meq/L or 0.65 to 1.05 mmol/L.

INCREASED CONCENTRATION. Some elevation in the serum magnesium concentration occurs in chronic renal disease, severe dehydration, and adrenal insufficiency in which there is a deficiency of aldosterone. The oral intake of Epsom salt ($MgSO_4$) for constipation may produce hypermagnesemia in persons with impaired renal function.

DECREASED CONCENTRATION. Hypomagnesemia is frequently associated with gastrointestinal disorders (malabsorption, prolonged diarrhea, bowel or kidney fistulas, acute pancreatitis), acute alcoholism, prolonged parenteral fluid therapy without magnesium supplementation, and the use of some diuretics. Tetany caused by a low serum magnesium concentration, with a normal serum calcium, has been reported; the parenteral administration of magnesium eliminates the tetany.

LITHIUM

Lithium is an element in the same periodic series as sodium and potassium. Its salts are widely distributed on earth but in small quantities; even though lithium salts are ubiquitous, their concentrations in river and well water are usually low. Consequently, the concentration of lithium compounds in plant or animal tissue is also very small.

Lithium carbonate (Li_2CO_3) is of value in the treatment of manic-depressive psychosis; it has a calming effect upon the manic stage and is frequently given as a therapeutic measure to forestall possible attacks. The use of lithium in psychiatry has been reviewed by Maletzky and Blachly.[10]

Since the concentration of Li^+ is toxic at a plasma concentration of about 2 mmol/L or greater, the serum Li^+ level of patients receiving this drug is usually monitored weekly during the early phase of treatment. A serum concentration of 5 mmol/L can be lethal. Therapeutic levels are usually considered to be from 0.5 to 1.5 meq/L or mmol/L, and the maintenance concentration is 0.5 to 1.2 mmol/L.

Serum Lithium

The concentration of serum Li^+ is readily measured by flame emission photometry or by atomic absorption spectrophotometry. The flame emission method will be described because the use of flame photometers is more widespread than that of atomic absorption spectrophotometers.

REFERENCE. Levy, A.H., and Katz, E.M.: Comparison of serum lithium determinations by flame photometry and atomic absorption spectrophotometry. Clin. Chem. *16*, 840, 1970.

PRINCIPLE. Serum is diluted 1:50 with a diluent containing potassium as an internal standard. When aspirated into a hot flame, some of the lithium atoms are forced into an activated state by thermal collisions; upon returning to the ground state, these atoms emit light at the characteristic wavelength for lithium, 671 nm. The emitted light of lithium is electronically compared to that given off by the internal potassium standard.

REAGENTS AND MATERIALS
1. Serum Diluent consisting of 1.5 mmol/L of K and 1 g/L of Triton X-100, a nonionic detergent (Rohm and Haas). Transfer 111.8 mg KCl to a liter volumetric flask containing about 500 ml water. Add 1 g Triton X-100 and make up to volume with water.
2. Blank Reagent, 140/5.0 mmol/L of Na/K. Dissolve 8.182 g NaCl and 0.373 g KCl in water and make up to 1 L volume.
3. Li Stock Standard, 10.0 mmol/L of Li. Dissolve 424.0 mg LiCl, reagent grade, in 140/5 mmol/L of Na/K and make up to 1 liter volume.
4. Li Working Standard, 1.0 mmol/L of Li and 140/5 mmol/L in Na/K. Dilute 10.0 ml stock standard Li to 100.0 ml volume with solution 140/5 mmol/L in Na/K.

PROCEDURE
1. Dilute 0.1 ml standard, unknowns, controls, and blank, each with 5.0 ml serum diluent.
2. Perform analysis as described by the manufacturer, using compressed air and propane for the flame.
3. Record Li concentration from the readout.
   Note: Since blank and standards contain 140/5 mmol/L of Na/K before the 1:50 dilution, the concentrations of Na and K are similar to that of serum.

REFERENCE VALUES
Therapeutic levels: 0.5 to 1.5 mmol/L
Maintenance levels: 0.5 to 1.2 mmol/L

Li^+ is not measurable in the serum of individuals who are not taking any lithium medication.

INCREASED CONCENTRATION. Toxic levels have been reported as low as 2.0 mmol/L. Telephone the physician if serum levels exceed 1.6 mmol/L.

IRON[11]

Iron is an essential component of a group of heme proteins that function in oxygen transport, or as enzymes in oxidation-reduction systems. The bulk of the body iron is contained in hemoglobin, the protein of the red blood cell that transports oxygen from the lungs to the tissues, but a portion is also present in myoglobin, a protein in muscle that is capable of binding and releasing oxygen. Iron is also a component of the cytochromes, enzymes that act as electron transfer agents in oxidation-reduction reactions, and in the enzymes, catalase and peroxidase, which decompose H_2O_2 or other peroxides. The iron component in all of these active compounds is heme (ferroprotoporphyrin, Fig.

7.3, Chap. 7); the iron is covalently bound to the porphyrin. The protein moiety of the different heme proteins confers the properties and specificity peculiar to each. A small amount of iron is contained in a number of nonheme metalloenzymes.

The total amount of iron in a 75 kg male is approximately 4 g but is less in females during the reproductive years. As shown in Figure 12.1, about 70 to 75% of the iron is incorporated in the following heme compounds: hemoglobin, myoglobin, and heme enzymes. Approximately 25% of body iron stores in the male are complexed to a protein, apoferritin, to form soluble ferritin and an insoluble product, hemosiderin (denatured ferritin?). The soluble ferritin comprises the major portion of the iron stores and recycles its iron as needed for heme products; it is present in all tissues but primarily in cells of the reticuloendothelial system (bone marrow, liver, and spleen) while hemosiderin is deposited only in the latter group of cells. The serum concentration of ferritin reflects the amount of *iron stored* in reticuloendothelial cells. The amount of ferritin and stored iron is greater after puberty in males than in females; children have less of each than adults. Bleeding experiments in humans have revealed that the iron stores in mg approximately equal the serum ferritin concentration (μg/L) $\times$ 8.

Two to 4 mg of iron circulate in plasma bound to a β-globulin, transferrin; two atoms of Fe^{3+} are bound per mole of transferrin. The transferrin serves to bind the iron very tightly and make it unavailable to nonspecific complexing agents in plasma; it releases the iron only to specific receptor sites on cell membranes. The HCO_3^- ion in plasma is essential for the binding of Fe^{3+} by transferrin. The total amount of iron in plasma is small and does not exceed 1.5 mg/L (150 μg/dl).

Iron is unique among the mineral elements in that it is recycled so effectively that there is virtually no loss of body iron through excretion into the urine; the

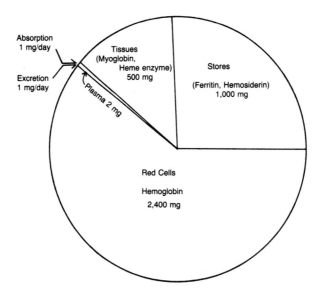

Fig. 12.1. Approximate distribution of iron in average-size male. Females have less storage iron and fewer red blood cells.

daily loss of iron in urine and sweat in an adult male is less than 1 mg. Since the kidney does not regulate the concentration of plasma iron by excreting excessive intake, the control must reside elsewhere. Apparently the intestinal mucosa contains special proteins necessary for the absorption of iron, which, to some degree, regulate its absorption.

The average daily diet contains from 5 to 15 mg of iron, but only a portion of this is absorbed, since many iron salts have a low solubility. Some iron is absorbed in the upper bowel as Fe^{2+}; as it traverses the intestinal mucosa, it is oxidized to Fe^{3+} and stored temporarily in ferritin. Upon release from the mucosal cell into plasma, the Fe^{3+} becomes tightly bound to transferrin, the iron transport protein. Each molecule of transferrin is capable of binding two atoms of Fe^{3+}, but there is usually an excess of this compound in the plasma of normal individuals so that the transferrin is only about 30% saturated with iron.

During the formation of erythrocytes in the bone marrow, storage iron from ferritin is transferred to normoblasts (nucleated red blood cells). Hemoglobin is synthesized in the normoblasts; as part of the process, the iron becomes inserted into a protoporphyrin molecule to become heme, which then binds to a protein (globin) to form hemoglobin. The iron in hemoglobin is Fe^{2+}; when oxidized to Fe^{3+}, the compound becomes methemoglobin, which cannot transport oxygen.

On page 224, it was pointed out that approximately 0.8% of the red blood cells are destroyed daily as they become senescent; formation of new cells keeps pace with destruction. The iron in the phagocytized, old cells is recycled and appears in newly synthesized hemoglobin.

Serum Iron and Iron-Binding Capacity

REFERENCE. Orynich, R.E., Tietz, N.W., and Fiereck, E.A.: *In* Fundamentals of Clinical Chemistry, edited by N.W. Tietz. Philadelphia, W.B. Saunders, 1976, p. 926.

PRINCIPLE—SERUM IRON. The iron in serum is dissociated from its Fe^{3+}-transferrin complex by the addition of an acid-buffer reagent containing ascorbic acid. The ascorbic acid reduces the Fe^{3+} to Fe^{2+}, which no longer binds to transferrin in acid solution. A chromogen, ferrozine, in the buffer solution, forms a highly colored Fe^{2+}-complex whose absorbance is measured at 562 nm. Sodium lauryl sulfate is a component of the buffer reagent in order to prevent the precipitation of serum proteins.

TOTAL IRON-BINDING CAPACITY (TIBC). A known amount of Fe^{3+}, which is more than sufficient to fully saturate the serum transferrin with iron, is added to a serum sample. The excess Fe^{3+} not bound to transferrin is removed by passing the solution through a $MgCO_3$ column. The total iron in the serum filtrate is measured in the same manner as for serum iron.

UNSATURATED IRON-BINDING CAPACITY (UIBC). The UIBC is a calculated value and equals TIBC minus serum Fe.

$$\% \text{ Saturation. The } \% \text{ saturation} = \frac{\text{serum Fe}}{\text{TIBC}} \times 100.$$

REAGENTS AND MATERIALS. Iron is ubiquitous, so all glassware must be washed in HCl, 6 mol/L or HNO_3, 5 mol/L and rinsed thoroughly with deionized water. Use only reagent grade chemicals and deionized water for all solutions.

1. NaOH 12.5 mol/L. Dissolve 100 g NaOH in about 150 ml water in an Erlenmeyer flask. When cool, transfer to a volumetric cylinder or flask and make up to 200 ml volume. Store in a polyethylene bottle.
2. Acetate Buffer, pH 4.5, 1 mol/L. Place 500 ml water in a liter volumetric flask; add 60 ml glacial acetic acid and 28 ml NaOH 12.5 mol/L. Cool, add water to the mark, and adjust pH to 4.5 with either NaOH or acetic acid.
3. Ferrozine Color Reagent. Place 400 mg ferrozine (Hach) and 2.5 g thiourea in a 100 ml volumetric flask. Dissolve in water and make up to volume. Store in a dark bottle. The thiourea complexes with serum copper and prevents it from forming a color with ferrozine.
4. Iron Reagent A. Dissolve 30 g sodium lauryl sulfate in about 300 ml water in a 500 ml volumetric flask. Add 50 ml acetate buffer and dilute to the mark with water. The reagent is stable for at least 1 month.
5. Iron Reagent B. Transfer 50 ml acetate buffer and about 200 ml water to a 500 ml volumetric flask. Add 30 g ascorbic acid and 1.0 g sodium metabisulfite ($Na_2S_2O_5$). When dissolved, make up to the mark with water. The reagent is stable for 2 weeks.
6. Iron Buffer Reagent. Prepare fresh daily by mixing equal volumes of iron reagents A and B.
7. $MgCO_3$ Filter Columns. Fill Quik-Sep columns containing filter paper discs (#QS-P, Isolab) with 0.15 to 0.20 g $MgCO_3$ powder (#MO-125, Sigma) and gently tap to pack the powder.
8. Stock Standard Fe, 1.0 g/L. Weigh out 1.000 g iron wire (analytical grade) and place in a flask containing 25 ml 6 mol/L HCl. When dissolved, transfer quantitatively to a 1000 ml volumetric flask with water and make up to volume. The reagent is stable.
9. Working Standard Fe, 200 μg Fe/dl. Transfer 1.0 ml of stock standard Fe to a 500 ml volumetric flask and make up to volume with water.

PROCEDURE FOR SERUM IRON

1. Pipet 0.5 ml portions of unknown sera, control serum, working standard, and water into appropriate tubes for "Unknowns," "Control," "Standard," and "Reagent Blank," respectively. Set up in duplicate.
2. To each tube, add 4.5 ml iron buffer reagent and mix well.
3. Heat all tubes for 15 minutes in a 37° C water bath.
4. Read the absorbance (A_1) of each tube against the reagent blank at 562 nm.
5. Add 0.1 ml ferrozine color reagent to each tube, mix, and heat at 37° C in water bath for 15 minutes.
6. Read the absorbance (A_2) of each tube against reagent blank.
   Note: The absorbance difference ($A_2 - A_1$) corrects for possible serum turbidity and bilirubin color. Let ΔA_u and ΔA_s stand for $A_2 - A_1$ of unknowns and standard, respectively.
7. Calculation:

$$\mu g \ Fe/dl = \frac{\Delta A_u}{\Delta A_s} \times 200$$

$$\mu mol \ Fe/L = \mu g \ Fe/dl \times 0.179$$

PROCEDURE FOR TIBC

1. Pipet 0.5 ml portions of unknown sera and controls into appropriately marked tubes. Add 1.0 ml of working standard Fe to each tube and mix gently.
2. Let tubes stand at room temperature for 10 minutes.
3. Carefully transfer the contents of each tube into a filter column that is seated on a 12 × 75 mm collection tube. Do not disturb the $MgCO_3$ packing during the transfer.
4. Centrifuge the filter column-collection tube assembly for 10 minutes at approximately 500 g.
5. Discard the columns. Remove the collection tubes from the centrifuge and recentrifuge those tubes in which particles of $MgCO_3$ have leaked into the filtrate.
6. Pipet 0.5 ml of filtrate into appropriately labeled, dry tubes.
7. Pipet 0.5 ml of standard and water, respectively, into standard and reagent blank tubes.
8. Add 4.5 ml iron buffer reagent to each tube, mix, and follow steps 3 through 6 as described for serum iron.
9. Calculation:

$$\text{TIBC, } \mu g \text{ Fe/dl} = \frac{\Delta A_u}{\Delta A_s} \times 200 \times 3 \text{ (dilution factor)} = \frac{\Delta A_u}{\Delta A_s} \times 600$$

$$\mu mol/L = \mu g/dl \times 0.179$$

$$\% \text{ saturation} = \frac{\text{serum Fe}}{\text{TIBC}} \times 100$$

A faster method and one requiring less serum (10–100 μl for serum Fe and 50–100 μl for TIBC) may be performed coulometrically with the Ferrochem Iron Analyzer (esa). There are few interfering substances with this method.

A serum sample is injected into the apparatus, which is flushed into an electrode cell with an acid-alcohol solution. The instrument measures the summation of electrons to convert all of the Fe^{3+} to Fe^{2+} and records the serum iron concentration.

To measure TIBC, a serum sample is shaken for 5 minutes with an ion exchange resin saturated with Fe^{3+}. Transferrin binds and removes Fe^{3+} from the resin until the transferrin in the serum sample becomes saturated. An aliquot of the supernate is measured for Fe^{3+} as for serum iron; the TIBC value is obtained directly.

REFERENCE VALUES

SERUM IRON. The values range from 65 to 165 μg/dl or from 11.6 to 29.5 μmol/L. The serum iron concentration is higher at birth but falls within 3 or 4 days. The concentration of serum iron in a geriatric population is lower than the above reference values.

The diurnal variation in serum iron concentration is considerable; the levels are high in the early morning and much lower in the evening.

TIBC. The total iron binding capacity varies from about 300 to 400 μg/dl or from 53.7 to 71.6 μmol/L. For persons over the age of 70 years, the TIBC may normally be about 70 μg/dl lower.

% SATURATION. Under normal circumstances, the serum transferrin is about 20 to 50% saturated with iron.

INCREASED CONCENTRATION. See Table 12.1.

SERUM IRON. The concentration of serum iron is elevated in conditions of (1) increased red blood cell destruction (hemolytic anemia); (2) ineffective or decreased erythrocyte formation (pernicious anemia in relapse, aplastic anemia, marrow damage by toxins); (3) blocks in heme synthesis (lead poisoning, pyridoxine deficiency); (4) increased release of storage iron (acute hepatic cell necrosis); (5) increased intake or impaired control of iron absorption (ingestion of large amounts of iron, hemochromatosis, hemosiderosis); and (6) in megaloblastic anemia.

TIBC. An increase in the plasma concentration of transferrin elevates the TIBC. This may occur during late pregnancy; in iron deficiency anemia; after acute hemorrhage; and after destruction of liver cells.

% SATURATION. This is increased in iron overload states (hemochromatosis, hemosiderosis); hemolytic anemias; acute hepatitis; and pernicious anemia in relapse.

DECREASED CONCENTRATION. See Table 12.1.

SERUM IRON. The concentration of serum iron is decreased when the dietary intake of iron is insufficient (iron deficiency anemia, malabsorption of iron); loss of iron is accelerated (acute or chronic blood loss, late pregnancy); and release of stored iron from reticuloendothelial cells is impaired (infection, neoplasia, rheumatoid arthritis).

TIBC. The TIBC is decreased when there is decreased synthesis of transferrin (infection, neoplasia, uremia) or increased loss of this protein in the urine (nephrosis).

% SATURATION. The % saturation is decreased in iron deficiency anemia, late pregnancy, infection, neoplasia, and after acute hemorrhage.

SERUM FERRITIN. The protein apoferritin is synthesized in the liver but rapidly scavenges and stores the iron of broken down red cells to become ferritin; iron may constitute up to 20% of the weight of this protein, which carries out its storage function mostly in the reticuloendothelial system. A low concentration

TABLE 12.1

Changes in Serum Iron, TIBC, Percent Saturation and Ferritin Concentration in Various Conditions

CONDITION	SERUM IRON	TIBC	PERCENT SATURATION	SERUM FERRITIN
Iron deficiency	↓↓	↑↑	↓↓	↓↓
Inflammatory states (infection, inflammation, neoplasias)	↓	↓	↓→	↑↑
Reduced transferrin concentration (nephrosis, Kwashiorkor)	↓	↓↓	↑	→↓
Ineffective erythropoiesis (B$_6$, folate, or B$_{12}$ deficiencies, Pb poisoning, thalassemia)	↑	→	↑	↑
Iron overload (excessive intake, absorption, hemochromatosis)	↑↑	↓	↑↑	↑↑
Liver disease (acute necrosis)	↑	↑	↑	↑↑

→ signifies no changes from normal, ↑ an elevation, and ↓ a decrease. Double arrows indicate large increases or decreases.

of circulating ferritin reflects decreasing iron stores before any other laboratory test.

METHOD OF MEASUREMENT. The concentration of ferritin in serum is so low that it must be measured by sensitive radioimmunoassay methods (RIA) or by some other radiometric method. Commercial kits are available for the measurement of serum ferritin and manufacturer's directions must be followed. See page 253 for the principle involved in immunochemical methods.

REFERENCE VALUES. There is a broad spread in the serum concentration of ferritin because of variation in iron stores. The mean value for normal females is about 35 µg/L, with a range of 12 to 125; the mean value for males is 95 µg/L, with a range from 30 to about 250.

INCREASED CONCENTRATION. Elevated concentrations of serum ferritin occur in inflammation produced by various disease states, endotoxins, neoplasms, surgical procedures, and rheumatoid arthritis. Acute and chronic liver disease, iron toxicity, and hemochromatosis also increase the ferritin concentration (see Table 12.1).

DECREASED CONCENTRATION. A ferritin concentration below 12 or 10 µg/L is indicative of iron deficiency. This finding appears before the anemia.

SERUM TRANSFERRIN. Although serum transferrin can be easily measured by immunonephelometric methods, the same information may be obtained by measuring the serum iron, the total iron binding capacity and obtaining the % saturation by calculation.

REFERENCES

1. Sanders, H.J.: Nutrition and health. Chemical and Engineering News *57*, 27, 1979.
2. Knox, F.G. (editor): Physiology of Calcium and Phosphate Metabolism. 1980 Refresher Course Syllabus. Bethesda, American Physiological Society, 1980, p. 1.
3. Clark, E.P., and Collip, J.B.: A study of Tisdall method for determination of blood serum calcium with a suggested modification. J. Biol. Chem. *63*, 461, 1925.
4. Cali, J.P., Bowers, G.N., Jr., and Young, D.S.: A referee method for the determination of total calcium in serum. Clin. Chem. *19*, 1208, 1973.
5. Kessler, G., and Wolfman, M.: An automated procedure for the simultaneous determination of calcium and phosphorus. Clin. Chem. *10*, 686, 1964.
6. Fu, P.C., Yokas, T., Gonzales, R., and Lubran, M.: A rapid biochromatic analysis of serum calcium. Clin. Chem. *20*, 908, 1974.
7. Sage, G.W., LeFever, D., and Henry, J.B.: Evaluation of the revised serum calcium procedure used with the DuPont automatic clinical analyzer. Clin. Chem. *21*, 850, 1975.
8. Ladenson, J.H., and Bowers, G.N., Jr.: Free calcium in serum. I. Determination with the ion-specific electrode, and factors affecting the result. Clin. Chem. *19*, 565, 1973.
9. Ladenson, J.H., and Bowers, G.N., Jr.: Free calcium in serum. II. Rigor of homeostatic control, correlations with total serum calcium, and review of data on patients with disturbed calcium metabolism. Clin. Chem. *19*, 575, 1973.
10. Maletzky, B., and Blachly, P.H.: The use of lithium in psychiatry. Crit. Rev. Clin. Lab. Sci. *2*, 279, 1971.
11. Bothwell, T.H., Charlton, R.W., Cook, J.D., and Finch, C.A.: Iron Metabolism in Man. Oxford, Blackwell Scientific Publications, 1979.
12. DeLuca, H.F.: The Vitamin D hormonal system: Implications in bone disease. Hosp. Pract. *15*, 57, 1980.

Chapter 13

CEREBROSPINAL FLUID AND OTHER BODY FLUIDS AND SECRETIONS

Serum is the fluid used most frequently for analysis in a clinical chemistry laboratory when the concentration of a blood constituent is requested. It is usually a matter of convenience or training whether plasma is used instead of serum because for many constituents the fluids are interchangeable. There are a number of other body fluids or secretions, however, that occasionally may be sent to the laboratory for analysis. In most instances, the analytical method will be similar to the one employed for measuring that same constituent in serum. The one major exception is the determination of protein in cerebrospinal fluid (CSF) because of the very much lower concentration of protein in cerebrospinal fluid than in blood.

CEREBROSPINAL FLUID (CSF)

Cerebrospinal fluid is a clear fluid that bathes the brain and spinal cord. The cerebrospinal fluid originates partially as an ultrafiltrate of plasma and partially as a secretion from the choroid plexus, a tuft of blood capillaries that protrudes into the fourth ventricle of the brain. The cerebrospinal fluid flows through the various brain cavities (ventricles) and subarachnoid space, the space between the outer, tough protective membrane (dura mater) and the inner, delicate membrane that carries the blood vessels to the brain and spinal cord. A portion of the cerebrospinal fluid is absorbed by some of the villi that it bathes and enters the venous blood system.

Measurement of the pressure of cerebrospinal fluid and examination of the fluid may be requested in various types of damage or diseases involving the central nervous system. The majority of requests is concerned with the confirmation or ruling out of infection (bacterial meningitis, encephalitis) and involves both the microbiology and chemistry laboratories. Since many of the samples of cerebrospinal fluid require microscopic examination, with staining and culture for microorganisms, the specimens are usually sent to the microbiology laboratory before the chemistry laboratory. Since the amount of cerebrospinal fluid taken from adults for laboratory examination usually does not exceed 5 to 10 ml because of the danger of headache, the amount available to the clinical chemistry laboratory is usually very much less. The samples are precious because of the difficulty and inconvenience of performing a lumbar puncture and must be used sparingly; surplus fluid should be saved in the refrigerator for at least 2 weeks in case further work should be required.

CSF Protein Concentration

Total Protein

The protein concentration of cerebrospinal fluid is quite low and resembles that of an ultrafiltrate to which a small amount of some proteins produced locally has been added. In normal individuals, the protein concentration varies from 15 to 45 mg/dl or about 0.3 to 0.5% of that in serum. The protein concentration rises when there is damage to the membranes covering the brain and spinal cord. Any type of damage or inflammation, whether caused by bacteria, viruses, neoplasms, or trauma, is accompanied by an increase in cerebrospinal fluid protein concentration as plasma proteins cross the blood brain barrier. The increase varies with the degree and extent of the damage; in a severe meningitis, protein concentrations of over 500 mg/dl have been found. The presence of blood or pus in a CSF sample is certain to elevate the protein concentration of the fluid because of the presence of blood and plasma proteins in the former case and cell proteins and an exudation of plasma from the inflamed surfaces in the latter instance. Inspection of the sample will reveal this and protein quantitation provides no additional information. The concentrations of total protein and gamma globulin usually rise in the following non-purulent conditions: encephalitis, tuberculous meningitis, syphilitic meningitis, and multiple sclerosis.

The method for measuring CSF total protein concentration and its interpretation is described on page 167, in the chapter dealing with proteins in body fluids.

Gamma-Globulin

The most common abnormalities of CSF proteins are due to injury to the blood brain barrier (inflammation or trauma) that allows plasma proteins to enter the CSF, or abnormal immunoglobulins produced within the blood brain barrier by lymphocytes and plasma cells in response to a chronic inflammation, such as multiple sclerosis or subacute sclerosing panencephalitis.

The determination of γ-globulin in the above cases provides helpful diagnostic information. The analysis of IgG in CSF may be performed directly by immunonephelometry[1] (p. 253) or by protein electrophoresis.

Protein Electrophoresis

It is usually necessary to concentrate CSF about 100-fold to bring the protein concentration into the range necessary for successful electrophoretic separation (p. 168). Protein electrophoresis of cerebrospinal fluid is most useful in the diagnosis of multiple sclerosis, a disease in which the gamma-globulin fraction of the cerebrospinal fluid may be elevated even though the total protein concentration may be within normal limits. Multiple myeloma patients who produce light chains of the myeloma protein in excess (Bence Jones protein) also exhibit the light chain as a globulin spike upon CSF protein electrophoresis (p. 168). The CSF electrophoretic pattern sometimes displays a gamma-globulin abnormality that cannot be picked up by an immunonephelometric method; small quantities of abnormal bands of IgG may be visible.

The diagnostic information provided by quantitation of CSF IgG may be improved by comparing the ratio of $\dfrac{CSF}{Serum}$ IgG to that of $\dfrac{CSF}{Serum}$ albumin. If the ratio of the IgG has increased more than that of albumin, it means that the increase in IgG has come from within the blood brain barrier and has not diffused from plasma.

CSF Glucose Concentration

The concentration of glucose in cerebrospinal fluid is about 60 to 75% of that in plasma and is in equilibrium with it; equilibration is slow and requires about 2 hours for changes in plasma glucose to show up in the cerebrospinal fluid. Serum glucose should be measured at the same time as the glucose concentration of the cerebrospinal fluid.

The determination of the glucose concentration is useful in the diagnosis of meningitis. The large number of bacteria and leukocytes in the infected cerebrospinal fluid rapidly metabolizes glucose and lowers its concentration. The concentration of glucose in normal individuals varies from 45 to 70 mg/dl but usually is less than 30 in various types of bacterial meningitis. A viral meningitis usually does not affect the CSF glucose concentration.

The method for cerebrospinal fluid glucose determination is the same as that for serum glucose (p. 311).

CSF Chloride Concentration

The concentration of chloride in the cerebrospinal fluid is sometimes requested, but this test yields little clinical information. The CSF Cl⁻ concentration is higher than that of plasma because there are so few negatively charged proteinate ions in CSF. The Cl⁻ concentration in CSF increases to balance the positive charges. The Cl⁻ concentration decreases as CSF protein concentration increases in pathologic states.

The method for the determination of the concentration of chloride is the same as that for serum Cl⁻ (p. 89).

CSF Electrolytes, pH, and Enzymes

Sometimes there are requests for measurement of cerebrospinal fluid electrolytes, pH, enzymes (lactate dehydrogenase), or other tests. These determinations are usually carried out in the same manner as for the serum constituents.

GASTRIC JUICE

Another fluid that is occasionally collected and sent to the clinical chemistry laboratory for analysis is gastric juice. Gastric juice is a secretion by specialized cells of the stomach; it is composed of hydrochloric acid, an inactive protease, pepsinogen, which becomes activated in the acid environment, a little lipase, some electrolytes, and mucus (mucoproteins and mucopolysaccharides). Among the mucoproteins is the intrinsic factor, a compound necessary for the absorption of vitamin B_{12}.

The ingestion of food or the anticipation of the ingestion of pleasing food items stimulates the outpouring of gastric juice. Nevertheless, some fluid usually remains in the stomach 12 to 16 hours after a meal. This is known as the gastric residue, which varies in volume from 20 to 100 ml.

The principal reason for requesting a gastric analysis is to measure the amount of HCl contained in it. Hydrochloric acid is absent from the gastric juice in a defect known as achlorhydria, in pernicious anemia and in some types of gastric carcinoma. The amount of gastric HCl is frequently increased in patients with the Zollinger-Ellison syndrome, overproduction of the hormone gastrin, by a pancreatic neoplasm (p. 297) and in some patients with duodenal ulcers.

COLLECTION. The gastric juice may be collected under the following conditions for measurement of its HCl:

1. After an overnight fast. The gastric residue or juice present in the stomach 12 to 16 hours after a meal is collected.
2. As the basal secretory output. Specimens are collected at specified time intervals while the patient is fasting and completely at rest.
3. As a 12-hour overnight secretory test. This is similar to situation 1, but the time is exactly 12 hours.
4. After stimulation of maximal gastric secretion by the infusion of histamine or injection of a histamine analog (Histalog, Lilly). Samples are collected every 15 minutes for several hours during the infusion (or after the injection) until a steady, maximal flow is achieved.
5. During hypoglycemic stimulation of gastric secretion (Hollander test) as a check for completion of vagotomy (severance of vagus nerve fibers to the stomach to decrease HCl output).

In all of these tests, the free HCl in gastric juice is determined by titrating an aliquot of the juice with 0.1 mol/L NaOH to pH 4. The millimoles of HCl per liter or per specimen are calculated.

TABLE 13.1
The Basal Secretory Output of Gastric Juice*

CONDITION	SEX	VOL. (ml)	CONCENTRATION HCl mmol/Lt	TOTAL HCl mmolt
Reference values	M	79	26	2.1
	F	65	21	1.4
Duodenal ulcer	M	100	51	5.6
	F	89	29	2.6
Gastric ulcer	M	81	28	2.3
	F	77	4	0.3

*Modified from J. B. Kirsner and W. L. Palmer. The problem of peptic ulcer. Am. J. Med. *13*, 615, 1952.
tmmol = meq

TABLE 13.2
12-Hour Nocturnal Gastric Secretory Test*

	AVERAGE 12-HOUR NOCTURNAL SECRETION		
CONDITION	VOL. (ml)	CONCENTRATION HCl mmol/Lt	TOTAL HCl mmolt
Reference values	580	29	16.8
Duodenal ulcer	1000	61	61.0
Gastric ulcer	600	21	12.6
Gastrin-producing tumor			100–500

*Modified from J. B. Kirsner and W. L. Palmer, The problem of peptic ulcer. Am. J. Med. *13*, 615, 1952.
tmmol = meq

INTERPRETATION

Gastric Residue. The gastric residue (no stimulation) normally contains from 0 to 40 mmol/L of HCl.

Basal Secretory Output. Table 13.1 illustrates the reference values of the basal secretory output in normal males and females; results are given in concentration of HCl (mmol/L) and in mmoles per total specimen.

In ulcer patients: the volume of gastric juice, the concentration and total output of HCl are increased in patients with duodenal ulcer. Patients with gastric ulcer do not secrete excess HCl.

12-hour Overnight Gastric Secretory Test. Table 13.2 illustrates the customary values that are usually obtained in a 12-hour nocturnal secretion test. Patients with duodenal ulcers have a moderately increased secretion of gastric HCl; those with gastric ulcers do not have a secretory rate different from that of normal individuals. Patients with a gastrin-producing tumor (Zollinger-Ellison syndrome) usually have an enormous increase in HCl secretion.

Histamine Stimulation Test. Originally, histamine was infused at a constant rate[2] to obtain maximal secretion of gastric juice. Later, a histamine analog, Histalog,[3] was substituted because of fewer side reactions. The gastric juice is collected every 15 minutes and analyzed for HCl until a steady state is achieved. Normal individuals secrete from 20 to 50 mmol/hr of HCl in the gastric juice under these conditions, with a mean being 33.

Insulin Hypoglycemic Stimulation of Gastric Secretion (Hollander test).[4] Sometimes a vagotomy (severing of vagus nerve connections to the stomach) is performed upon individuals with duodenal ulcers in an attempt to reduce the gastric secretion of HCl by cutting the nerve pathways. Because the operation is not always successful, the patient is subjected to insulin hypoglycemia to check for possible intact vagal pathways. Hypoglycemia stimulates the vagus nerve centrally to send stimuli to the stomach by vagal fibers for gastric secretion. If the vagotomy is complete, insulin hypoglycemia has no effect upon gastric secretion of HCl (it remains at basal levels); an increase in HCl secretion during hypoglycemia indicates that the vagotomy was incomplete.

HCl in Gastric Juice

REAGENTS AND MATERIALS

1. Concentrated NaOH, carbonate-free. Place 500 g NaOH pellets in a beaker and cautiously add 500 ml water. Stir until dissolved. When cool, transfer to a polyethylene bottle and protect against CO_2 with a tube filled with soda lime. Caution: Wear goggles when preparing this reagent because a great deal of heat is generated and splashing may occur. The concentration is approximately 18 to 20 mol/L.

2. NaOH, 0.10 mol/L. When the carbonate precipitate in the concentrated NaOH has settled, dilute about 5.5 ml of concentrated NaOH to 1000 ml with water. Transfer to a polyethylene bottle protected by a soda lime tube. Check the molarity by titration with standard acid. Adjust molarity to 0.100 mol/L.

3. Töpfer's Reagent* (0.5% p-dimethylaminoazobenzene in alcohol). Dissolve 0.5 g p-dimethylaminoazobenzene in 100 ml of 95% ethanol.

PROCEDURE

1. *Treatment of specimen*
   a. Record the number, time, and volume of each specimen.
   b. Filter each specimen through glass wool if solid food particles are present.
2. *Analytical procedure*
   a. Transfer a 10 ml aliquot of each specimen into a 25 or 50 ml beaker. If less than 10 ml are available, pipet as much as can be measured accurately and modify calculation accordingly.
   b. Add 3 drops of Töpfer's reagent.* If the color turns red, free HCl is present.
   c. Record the pH of sample and titrate to pH 4 with a glass electrode or to the point where the Töpfer's reagent* turns from red to yellow.
3. *Calculation*

HCl, mmol/L

$$= \text{ml of 0.100 molar NaOH} \times \frac{100 \text{ mmol}}{1000 \text{ ml}} \times \frac{1000 \text{ ml}}{10 \text{ ml}} = \text{ml NaOH} \times 10$$

$$\text{HCl, mmol/volume gastric fluid} = \text{mmol/L} \times \frac{\text{volume (ml)}}{1000}$$

Notes:

1. Report the gastric HCl both as mmol/L and as mmol per volume of sample. For HCl, mmol/L and meq/L are identical.
2. The procedure is carried out as above on each individual sample, whether single or multiple samples are sent to the laboratory (1 sample for gastric residue, 4 samples for basal secretory output, 10 samples for Hollander test, 4 samples for histamine stimulation test).
3. In the Hollander test, blood samples are drawn at the same time as the gastric samples for determination of serum glucose.

TRANSUDATES AND EXUDATES

Sometimes fluids from surfaces between membranes or from body cavities are brought in as specimens for analysis. This includes pleural, ascitic, pericardial, and peritoneal fluids. Amniotic fluid will be treated in a separate section.

A fluid that accumulates as an ultrafiltrate is called a transudate. The capillary pores of the membranes usually pass protein molecules of 200,000 to 300,000 molecular weight but hold back the larger ones. The fluid is low in protein relative to plasma, but the protein concentration still amounts to 2 or 3 g/dl. Transudates accumulate because of increased hydrostatic pressure in the capillaries, as in congestive heart failure, or because of a decreased oncotic pres-

---

*Töpfer's reagent is carcinogenic (see p. 79). Substitute an aqueous solution containing 100 mg methyl orange per 100 ml or use a pH meter. Methyl orange changes color from red at pH 3.1 to yellow at pH 4.4.

sure, as in the nephrotic syndrome. Ascitic fluid accumulates in the peritoneal cavity in chronic, severe liver disease because of both factors: an increase in portal vein pressure caused by fibrotic scarring (cirrhosis) and a low plasma osmotic pressure accompanying the hypoalbuminemia.

Exudates are the fluids that accumulate when the membranes are injured by inflammatory or infectious processes. The affected membranes allow larger molecules to pass through, and hence the protein concentration of exudates is higher than in transudates.

The most frequent clinical chemistry analysis performed upon transudates and exudates is the total protein assay ordered to differentiate between the two processes. If the total protein concentration is below 3.0 g/dl, the fluid is considered to be a transudate; if the total protein concentration exceeds 3.0 g/dl, it is considered to be an exudate, and an explanation for the inflammation (infection, trauma, neoplasm, or other cause) must be found. Other analytical requests, such as glucose or electrolytes, may be made much less frequently. The glucose concentration of transudates and exudates reflect the plasma concentration unless the fluid contains pus and bacteria; these cells utilize glucose and lower the concentration. In rare instances, an amylase assay may be requested upon peritoneal fluid; the presence of amylase indicates the leakage of pancreatic juice into the peritoneal cavity by way of a fistula or trauma to the pancreas. Tests for these body fluid constituents are performed exactly as for serum.

AMNIOTIC FLUID[5]

Amniotic fluid is the fluid contained in the amniotic sac that envelops the fetus. A portion of the fluid arises from the fetal respiratory tract, urine, the amniotic membrane, and the umbilical cord. The volume of the amniotic fluid at term[6] usually ranges from 500 to 1100 ml. There is significant water transfer between the three intrauterine compartments: the fetus, the placenta, and the amniotic fluid.

The amniotic fluid has become increasingly important in the clinical chemistry laboratory as a fluid for analysis. It is obtained by amniocentesis (aspiration by hypodermic needle and syringe) and is employed for three general purposes: as a guide to action in erythroblastosis (Rh-positive babies in Rh-negative, sensitized mothers, p. 229); as a guide to fetal maturity; and in genetic counseling. Amniocentesis is illustrated in Figure 13.1.

In Assessment of Erythroblastosis

In the section on hemolytic disease of the newborn (HDN, p. 229), it was pointed out that an Rh-positive fetus was at risk in an Rh-negative mother who had become sensitized to the Rh antigen (Rh-positive red blood cells). The immunoglobulin IgG is sufficiently small to cross the placenta; hence, the danger is great in sensitized mothers that some maternal antibodies to the Rh antigen will cross the placenta and enter the fetal circulation. Such antibodies coat the Rh-positive red cells of the fetus and cause their destruction; bilirubin is formed from the degradation of hemoglobin as the coated red cells are destroyed in the reticuloendothelial system.

In 1961, Liley[8] reported that the analysis of the amniotic fluid for bilirubin by spectrophotometric scanning during the latter weeks of pregnancy provided

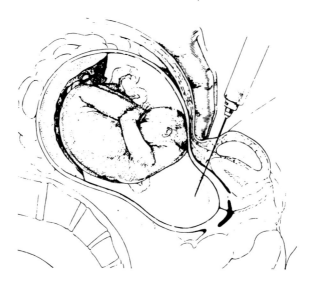

Fig. 13.1. Illustration of amniocentesis, the technique for obtaining amniotic fluid. (From V.J. Freda, The control of Rh disease. Hosp. Pract. *2*, no. 1, 54, 1967; and R.A. Good and D.W. Fisher, Immunobiology. Sunderland, MA, Sinauer Associates.)

an indication of the severity of hemolytic disease of the newborn. Liley plotted the net absorbance at 450 nm (the absorbance peak for bilirubin) against the gestational age and correlated the graph with clinical findings. Figure 13.2 is modified from Liley and shows three zones where (A) is the area indicating very great fetal risk (severe hemolysis), (B) an area ranging from mild to severe fetal risk (mild to severe hemolysis), and (C) a zone of little or no fetal danger (little or no hemolysis). The presence of oxyhemoglobin in the sample elevates the absorbance somewhat at 450 nm. Figure 13.3 illustrates normal and abnormal bilirubin absorbance curves.

REFERENCE. Liley, A.W.: See Reference #8, end of this chapter.

MATERIALS. A good recording spectrophotometer with a narrow bandpass.

PRINCIPLE AND PROCEDURE. A centrifuged specimen of amniotic fluid is scanned in a recording spectrophotometer from 350 to 700 nm. The absorbance is plotted on semilog paper as the ordinate versus wavelength as the abscissa. A straight line (baseline) is drawn on the curve from about 365 nm to 550 or 600 nm. The difference in absorbance, ΔA, is obtained by measuring the difference between the baseline and the peak at 450 nm (See Fig. 13.3.). The fetal risk is estimated by plotting the ΔA_{450} versus weeks of gestation (Fig. 13.2) and noting into which zone the point falls.

Note: The wavelength accuracy of the spectrophotometer should be checked in the vicinity of 450 nm, and adjustments should be made, if necessary.

Attempts have been made to improve the accuracy of the measurement of bilirubin in amniotic fluid. Burnett[9] recommends filtering the amniotic fluid and measuring the absorbance at 350, 455, and 550 nm or scanning from 350 to 550 nm. The baseline is drawn from 350 nm (instead of Liley's 365 nm) to 550 nm when the scan is made on semilog paper.

Gambino and Freda[10] measured the bilirubin directly in amniotic fluid by the Jendrassik and Grof method[11], plotted the values against gestational age,

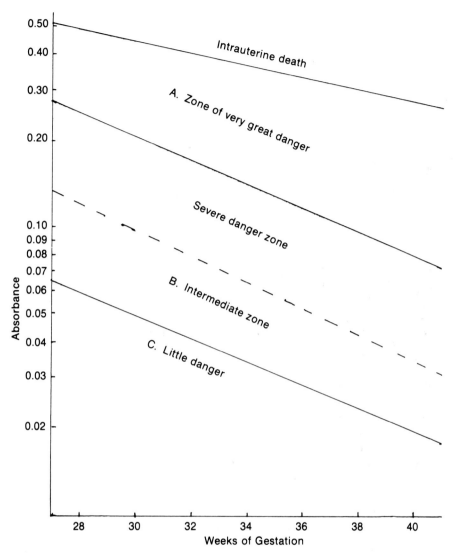

Fig. 13.2. Chart for estimating the danger of hemolytic disease of the newborn from the bile pigment concentration in amniotic fluid. (Modified from A.W. Liley.[8])

and established the various danger zones by correlation with clinical data. Dubin[12] reported an elegant way to accurately measure the bilirubin in amniotic fluid containing hemoglobin by using a Perkin-Elmer Model 156 Digital Two-Wavelength Spectrophotometer. He measured ΔA as $A_{451.8 \text{ nm}}$ minus $A_{578 \text{ nm}}$ because the absorbance of oxyhemoglobin was the same at the two wavelengths, 451.8 and 578 nm. The instrument was calibrated with a pure bilirubin solution and gave the same ΔA when hemoglobin was added to it.

In Assessment of Fetal Maturity

In certain high-risk pregnancies (Rh disease, diabetes, preeclampsia, premature rupture of membrane, premature labor), it is advantageous to terminate the pregnancy early, provided the fetus has attained a sufficient degree of

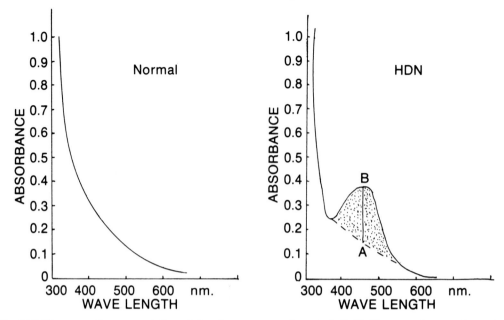

Fig. 13.3. Diagrammatic representation of absorbance curves of amniotic fluid. The normal curve shows the absence of bilirubin. The HDN curve shows a distinct bilirubin peak (the stippled area) at 450 nm and indicates that the fetus has a hemolytic problem. The line AB is the absorbance at 450 nm (A_{450}) ascribed to bilirubin. (Modified from A.W. Liley.[8])

physiologic maturity to cope with life in the outside world. If the fetal lungs have not matured sufficiently, the newborn fetus may have respiratory difficulties (the respiratory distress syndrome, RDS) or may even die from hyaline membrane disease.

An important factor in fetal lung maturity is the ability of the lung to synthesize in sufficient quantity the phospholipid lecithin (p. 325), which coats the alveolar sac linings.[7] Lecithin is a surfactant that is essential for respiration because it lowers the alveolar surface tension and permits the lungs to retain some residual air upon expiration, instead of collapsing.

The pulmonary synthesis and release of surfactant usually becomes well established after 32 to 36 weeks of gestation. It is manifested by a sharp increase in the lecithin concentration of amniotic fluid as some of the alveolar surfactant is regurgitated during normal respiratory movements of the fetus. Measurement of the surfactant concentration in amniotic fluid is the best method we have of assessing fetal lung maturity before birth. Infants born after there is evidence of sufficient surfactant usually have no respiratory problem; infants born before a critical amount of amniotic fluid surfactant is reached have a high risk for the respiratory distress syndrome. In the respiratory distress syndrome, many of the alveolar sacs cannot remain open for gaseous exchange; in some infants, the sacs may become filled with a proteinaceous fluid, the first step in the formation of a hyaline membrane.

Gluck and Kulovich[13] in 1971 proposed that the ratio of two phospholipids in amniotic fluid, the lecithin/sphingomyelin (L/S) ratio could serve as a suitable means for estimating fetal lung maturity. Their rationale was based upon the large increase in amniotic fluid lecithin between the 32nd week of gestation and term, the slight decrease in amniotic fluid sphingomyelin during the same period, and the assumption that lecithin changes in amniotic fluid reflected those taking place in the lungs. Sphingomyelin is not a surfactant and has no effect upon lung maturity, but it serves as an internal standard that corrects for volume variability in the amniotic fluid and variability in extraction of the lipids. Since both lecithin and sphingomyelin are affected to the same extent by variations in amniotic fluid volume and by extraction irregularities, use of the L/S ratio is a self-correcting procedure; expressing the surfactant in absolute terms as mg/dl of amniotic fluid is meaningless when the total volume of fluid and completeness of the extraction are unknown. The Gluck and Kulovich method[13] consisted of extraction of the amniotic fluid lipids, partial purification of the phospholipids by acetone precipitation, separation by thin layer chromatography (TLC) and quantitation of the lecithin and sphingomyelin spots by densitometry. They followed a number of pregnancies by measuring the L/S ratio in amniotic fluid and found that few newborns had serious respiratory problems when the L/S ratio was greater than 2.0 before birth. The morbidity was high among those whose L/S ratio was less than 2.0 but there were still many infants in this group who did not develop RDS. The predictive value of the L/S ratio for pulmonary *immaturity* was not high enough; premature labor was induced in some women who did not need it. Although the predictive value of the L/S ratio for pulmonary *maturity* was quite good in uncomplicated pregnancies, there was a high degree of false predictions of lung *maturity* in children of high risk pregnancies (diabetes, hypertension). In cases of maternal diabetes, an L/S ratio greater than 2.0 is not too dependable for assuring fetal lung maturity. Each laboratory should establish its own critical L/S ratio for estimating fetal maturity because of methodologic variability; the predictability of outcome is also dependent upon the prevalence of RDS in the patient population.

There have been many attempts to improve the discriminant capabilities of a surfactant test for fetal lung maturity.[7] Method modifications in the separation and measurement of lecithin and sphingomyelin have been proposed. Others have concentrated upon measuring the concentration of lecithin only in amniotic fluid because sphingomyelin has no surfactant properties. Since there are many different lecithins (the fatty acid composition of phosphatidylcholine can vary widely), the question has been raised: What is the composition of the surfactant? It is now known that a disaturated phosphatidylcholine (dipalmitoyl phosphatidylcholine) is the primary surfactant and not the lecithins containing an unsaturated fatty acid. There may be other phospholipids closely related to lecithin, such as phosphatidyl glycerol and phosphatidyl inositol that provide information concerning the lung maturation process; these two acidic phospholipids help to stabilize the surfactant properties of lecithin. The amniotic fluid concentrations of phosphatidyl inositol decrease sharply with lung maturity while phosphatidyl glycerol begins to be synthesized after lung lecithin synthesis has reached its peak. Some laboratories prefer to separate and measure the individual phospholipids rather than the L/S ratio.

Other laboratories utilize the properties of surfactants and make biophysical measurements. The common ones are measurement of the surface tension-lowering ability of different aliquots of amniotic fluid extract; the foam stability test of Clements et al.[15]; and the use of fluorescence polarization to assess the characteristics of the phospholipid component of amniotic fluid. The foam stability test is the easiest to perform and the most rapid. Equal volumes of 95% ethanol and amniotic fluid are shaken vigorously for 15 seconds; 15 minutes later the meniscus is examined for the presence of a complete ring of bubbles, a finding that indicates the presence of surfactant (disaturated phosphatidylcholine). The interpretation of the endpoint is subjective, however, and there are many equivocal results. Fluorescence polarization requires a very special instrument. A fluorescent hydrocarbon is mixed with amniotic fluid and the fluorescence polarization measured; the greater the concentration of surfactant, the lower is the viscosity of the solution and the lower the fluorescence polarization.

All of the above methods perform comparably with the results of the L/S ratio and all have the same shortcomings: good predictability of fetal lung maturity, except in cases of maternal diabetes and only fair predictability of fetal immaturity. The methods do vary, however, with the speed and convenience with which they can be carried out.

In their excellent review, Freer and Statland[14] critically discuss the various methods for assessing fetal lung maturity. They also point out many factors in collecting and handling the amniotic fluid sample that may affect surfactant concentration or L/S ratio found by analysis. Some factors are beyond the laboratorians' control, such as maternal volume of amniotic fluid (may vary from 250 to 1500 ml), unequal intrauterine mixing of amniotic fluid (surfactant comes from fetal mouth by regurgitation), or contamination of fluid aspirate by blood or meconium. Differences in preparation of the specimen may produce major analytic changes, of which the speed and time of centrifugation are the most important. The surfactant is associated with lamellar bodies that tend to stratify or layer at low speed centrifugation and to be spun down at higher speeds. Wagstaff[16] reported that 71% of the lecithin is lost from the supernate after centrifugation at 750 × g. After centrifugation, it is essential to *decant* the *supernate completely* and to *mix gently* before use because of the tendency of the lecithin fraction to form layers. There are no phospholipases in amniotic fluid that act upon surfactant so it is permissible to store the specimens at room temperature for a few days or at 4 to 8° C for several weeks. It is *imperative*, however, to *gently mix* the fluid *before sampling*.

Analysis of fluid collected after rupture of the amniotic sac is frequently requested. Since vaginal mucus contains phosphodiesterases (enzymes capable of hydrolysing lecithin but not sphingomyelin), the least contaminated samples will be obtained from freely flowing fluid collected immediately at the time of membrane rupture. Even so, the interpretation of a low L/S ratio based upon fluid collected from the vaginal vault has a great degree of uncertainty because of possible enzymatic destruction of surfactant; a high L/S ratio is evidence of fetal lung maturity.

In view of the many uncertainties involved in assessing fetal lung maturity, perhaps the recommendation of Gluck[20] to assess the components of the surfactant system is best. This can be approximated by using the one-dimensional chromatographic kit, Fetal-Tek 200 (Helena Laboratories), and following the manufacturer's directions. In this system, phosphatidylglycerol and phosphatidylinositol are separated, in addition to lecithin and sphingomyelin. The amniotic fluid concentration of phosphatidylglycerol becomes appreciable only after fetal lung maturity, while the concentration of phosphatidylinositol falls sharply at maturity. This provides two independent bits of evidence for cross-checking the results of the L/S ratio. This represents a compromise over Gluck's two-dimensional chromatographic system; the only information lost is the spot for disaturated phosphatidylcholine.

The concentration of amniotic fluid creatinine has also been used as an indicator of fetal maturity; a concentration of more than 2.0 mg/dl is usually associated with a fetal age of 37 weeks or greater, whereas a concentration of 1.5 mg/dl or less is associated with a high fetal risk for respiratory distress syndrome. There are two problems with the use of amniotic fluid creatinine for assessing fetal maturity: the test is not reliable if the mother has an elevated concentration of serum creatinine, as in preeclampsia and renal disease; and some mature, near-term infants have an amniotic creatinine level in the "high risk zone."

In Genetic Counseling

Various types of examination of the amniotic fluid have proven to be useful in ascertaining *in utero* whether a fetus is afflicted with or is a carrier of certain genetic diseases. This information is particularly useful for patients whose previous pregnancies have produced at least one infant with a genetic defect or who are over 35 years of age. Some information may be obtained by direct examination and analysis of the cells in the amniotic fluid, but most of the techniques require cell culture. This is a specialized area of investigation that is taking place principally in large medical centers.

The information obtained by study of the cells from the amniotic fluid (either directly or after cell culture) falls into three general classes:

1. Determination of the sex of the fetus. Since some genetic diseases are sex-linked, it is important to know the sex and hence, the degree of risk, if any.

2. Identification of fetal karyotypes. Tegenkamp and Tegenkamp[17] list 15 chromosomal abnormalities amenable to prenatal detection by chromosomal analysis of cultured amniotic fluid cells. Chromosomal abnormalities appear with relatively greater frequency when the mother is over 35 years old; the abnormality rate rises to 1 in 40 births for mothers older than 40 years.

3. Specific enzyme defects in cultured cells. Seegmiller[18] and Milunsky[19] list many genetic diseases that have been predicted successfully in utero by the lack of a specific fetal enzyme in cultured amniotic fluid cells and identify many more that are potentially possible to confirm by the same technique; more than 50 defects are tabulated. The list includes lipidoses (e.g., Tay-Sachs disease, with its absence of hexosaminidase A, p. 327), mucopolysaccharidoses (e.g., Hurler's syndrome, with abnormal deposits of mucopolysaccharides), amino acid disorders (e.g., urine maple syrup disease, with a lack of branched chain, keto acid decarboxylase), disorders of carbohydrate metabolism (e.g., galactosemia, with a deficiency of galactose-1-P-uridyl transferase, p. 217), and a large miscellaneous group. The list of defects amenable to prenatal diagnosis will increase as our knowledge of the biochemical lesions in genetic diseases becomes larger.

REFERENCES

1. Savory, J.: Nephelometric immunoassay techniques. *In* Ligand Assay, edited by J. Langan and J.J. Clapp. New York, Masson Publishing, 1981, p. 257.
2. Lawrie, J.H., Smith, G.M.R., and Forrest, A.P.M.: The histamine-infusion test. Lancet 2, 270, 1964.
3. Laudano, O.M., and Roncoroni, E.C.: Determination of the dose of Histalog that provokes maximal gastric secretory response. Gastroenterology 49, 372, 1965.
4. Hollander, F.: Laboratory procedures in the study of vagotomy (with particular reference to the insulin test). Gastroenterology 11, 419, 1948.
5. Fairweather, D.V.I., and Eskes, T.K.A.B. (editors): Amniotic Fluid: Research and Clinical Application, 2nd Ed. New York, Excerpta Medica, 1978.
6. Abramovich, D.R.: The volume of amniotic fluid and its regulatory factors. *In* Amniotic Fluid: Research and Clinical Application, 2nd Ed., edited by D.V.I. Fairweather and T.K.A.B. Eskes. New York, Excerpta Medica, 1978, p. 31.
7. Whitfield, C.R.: Prediction of fetal pulmonary maturity—clinical applications. *In* Amniotic Fluid: Research and Clinical Application, 2nd Ed., edited by D.V.I. Fairweather and T.K.A.B. Eskes. New York, Excerpta Medica, 1978, p. 393.
8. Liley, A.W.: Liquor amnii analysis in the management of the pregnancy complicated by rhesus sensitization. Am. J. Obstet. Gynecol. 82, 1359, 1961.
9. Burnett, R.W.: Instrumental and procedural sources of error in determination of bile pigments in amniotic fluid. Clin. Chem. 18, 150, 1972.
10. Gambino, S.R., and Freda, V.J.: The measurement of amniotic fluid bilirubin by the method of Jendrassik and Grof. Am. J. Clin. Pathol. 46, 198, 1966.
11. Jendrassik, L., and Grof, P.: Vereinfachte photometrische Methoden zur Bestimmung des Blutbilirubins. Biochem. Zeit. 297, 81, 1938.
12. Dubin, A.: Application of multi-wavelength spectroscopy to analysis of amniotic fluid bilirubin. *In* Amniotic Fluid, edited by S. Natelson, A. Scommegna, and M.B. Epstein. New York, John Wiley & Sons, 1974, p. 191.
13. Gluck, L., and Kulovich, M.: Lecithin/sphingomyelin ratio in normal and abnormal pregnancy. Am. J. Obstet. Gynecol. 115, 539, 1973.
14. Freer, D.E., and Statland, B.E.: Measurement of amniotic fluid surfactant. Clin. Chem. 27, 1629, 1981.
15. Clements, J.A., et al.: Assessment of risk of respiratory distress syndrome by a rapid test for surfactant in amniotic fluid. N. Engl. J. Med. 286, 1077, 1972.
16. Wagstaff, T.I., Whyley, G.A., and Freeman, G.: Factors influencing the measurement of the lecithin/sphingomyelin ratio in amniotic fluid. J. Obstet., Gynaecol. Br. Commonw. 81, 264, 1974.
17. Tegenkamp, T.R., and Tegenkamp, I.E.: Cytogenetic studies of amniotic fluid as a basis for genetic counseling. *In* Amniotic Fluid, edited by S. Natelson, A. Scommegna, and M.B. Epstein. New York, John Wiley & Sons, 1974, p. 281.
18. Seegmiller, J.E.: Amniotic fluid and cells in the diagnosis of genetic disorders. *In* Amniotic Fluid, edited by S. Natelson, A. Scommegna, and M.B. Epstein. New York, John Wiley & Sons, 1974, p. 291.
19. Milunsky, A.: Prenatal diagnosis of hereditary biochemical disorders of metabolism. *In* Genetic Disorders and the Fetus, edited by A. Milunsky. New York, Plenum Press, 1979, p. 209.
20. Merritt, T.A., Saunders, B.S., and Gluck, L.: Antenatal assessment of fetal maturity: The lung profile. *In* Clinical Perinatology, 2nd Ed., edited by S. Aladjem, A.K. Brown, and C. Sureau. St. Louis, C.V. Mosby, 1980, p. 213.

Chapter 14

TOXICOLOGY

We are surrounded in our daily lives by a large variety of chemicals that have toxic effects if they enter our bodies in excess of certain limiting amounts. The chemicals may be in the medicine chest (prescription drugs or over-the-counter remedies), kitchen (detergents and cleaning compounds), garden area (insecticides), or in the place of work (industrial chemicals and solvents). When we are sick, physicians prescribe and administer many different medications that could have adverse effects under certain conditions; many people uncritically take over-the-counter drugs without any type of medical supervision. Even in our social lives, we may consume large amounts of alcoholic drinks, while some experiment with or become addicted to various drugs of abuse. Accidental or purposeful exposure to toxic levels of drugs or chemicals is a frequent occurrence.

FUNCTIONS OF A TOXICOLOGY LABORATORY

In the past, the toxicology section of a clinical chemistry laboratory was relatively small and was devoted primarily to the identification and measurement, where possible, of offending chemicals or drugs in toxic or comatose patients. This still has to be done today but the emphasis and volume of work has changed considerably. The main efforts of a current toxicology section are devoted to therapeutic drug monitoring, the measurement of the serum concentration of a wide spectrum of medications in order to achieve optimum concentrations and benefits for the patients. Advances in analytical technology and the commercial availability of reliable and convenient kits for the rapid assay of many therapeutic agents have made this possible. The list of monitored drugs includes such groups as sedatives, tranquilizers, anticonvulsants, antihypertensives, antibiotics, antiarrhythmics, cardiac stimulators, antidepressants, antiasthmatics and others; the list is growing steadily. Thus, the clinical toxicology laboratory has two main functions:

1. *Therapeutic Drug Monitoring (TDM)*. The laboratory measurement of serum drug concentrations enables the physician to adjust and optimize the dosage upon an individual basis. Refer to the review of Pippenger[1] for a discussion of TDM. Toxicology books by Sunshine[2] and Baselt[3] should be consulted for methodologies and Gibaldi[4] for pharmacokinetics.

2. *Identification of drugs in acute intoxication*. When poisoning is suspected, the laboratory attempts to identify the offending drug or drugs. This helps to establish the diagnosis, assess the level of intoxication, and, in a few instances when antidotes or therapeutic guides are established, suggest the course of therapy. In general, there are few effective antidotes, and clinical treatment relies heavily upon supportive measures.

Drug identification in the comatose patient can be a very difficult problem when there are no clues; the number of possible poisons is enormous and identification may be tenuous. The laboratory can "play the odds," however, by setting up analyses for the eight or ten drugs that are involved in 80 to 90% of the overdosage cases. These include serum ethanol, salicylate, acetaminophen, barbiturate, chlordiazepoxide, diazepam, and tricyclic antidepressants. The employment of thin layer chromatography (TLC) for identifying classes of drugs or for screening for drugs of abuse is still the most cost effective way for tentative drug identification; the cost of equipment is low and the technique is relatively simple.

Pharmacokinetics[4]

The aim of drug administration is the attainment of a rapid therapeutic effect without deleterious side reactions. In general, the therapeutic effects of a drug, as well as its toxicity, correlate better with the serum drug concentration than with the drug dosage. Hence, it is necessary to monitor serum drug levels in order to accommodate individual variation and achieve the optimum effect.

The important physiologic factors that affect the blood level of an administered drug are *absorption* of the drug into the blood; *distribution* or equilibration of the drug with body tissues; *metabolism* (biotransformation) of the circulating drug, primarily by the liver; and *excretion* of the drug, usually by the kidney. Genetic factors are often responsible for individual differences, but for any person, the processes change with time due to maturation or aging, with organ disease, diurnal variation and drug interactions. In addition, there are nonphysiological factors that influence the processes, such as differences in dosage regimen, differences in drug formulation, medication errors, and patient compliance. Many patients do not follow directions and may take the drug erratically or take too much or too little. The immaturity of a young, pediatric population also must be borne in mind. Patients with liver or renal disease present problems because of the impairment of drug metabolism and excretion; the usual drug dose may reach toxic levels when there is malfunction of the liver or kidney.

Therapeutic drug monitoring (TDM) is necessary for drugs that have a low therapeutic index (the ratio between the minimum toxic and maximum therapeutic serum concentrations) and cannot be monitored by clinical observation or other laboratory tests. For these drugs, the therapeutic range should be well defined and a reliable assay must be available. *TDM* should be considered under the following circumstances:

1. The serum concentration should be determined after initiation of therapy or when the dosage regimen or drug formulation has been changed. This is usually done after the drug has reached a steady-state concentration (see below).
2. In neonates and infants, more frequent monitoring is necessary because renal and hepatic functions change rapidly during development.
3. Drug levels should be monitored during illness, especially when there is impairment of renal, hepatic, gastrointestinal, or cardiovascular function, since significant changes in serum concentration may occur.
4. Monitoring is indicated whenever the medication is unexpectedly ineffective or toxic.

5. Dosage adjustment may be required when a new drug is added to the patient's regimen because of possible drug interactions.

The concept of the plasma half-life ($t_{1/2}$, the time necessary for the serum concentration of a drug to decrease by $^1/_2$) helps to explain the peaks and troughs obtained in the serum concentration of a drug during a steady-state period. The trough occurs just prior to the administration of a dose, while the peak occurs during the period of maximal absorption. The steady-state is achieved when the drug intake balances the drug inactivation and excretion so that an average constant level between peaks and troughs is maintained. It has been shown that the administration of a maintenance drug dose over a time period of 5 times the $t_{1/2}$ is required to reach the steady-state level. In therapeutic monitoring of drug levels, the optimal time for obtaining samples is in the steady-state at the end of a $t_{1/2}$ period, just before the administration of a new dose; this provides the minimum concentration of the drug in the plasma. An example of how to use the $t_{1/2}$ in overdosage cases is given in the discussion of acetaminophen on page 392.

Therapeutic Drug Monitoring (TDM)

The administration of all drugs has a potential for hazards and side reactions; optimal drug concentration in blood can be assured only by TDM. Furthermore, the margin between therapeutic and toxic levels is small with some medications (see, digoxin, Table 14.1) and requires monitoring upon a regular basis. TDM helps to provide an estimate of patient compliance in taking the medication as directed. Therapeutic and toxic concentrations of a number of drugs are listed in Table 14.1.

Optimal Time for Blood Specimens

Drug monitoring is of maximum benefit when blood samples are drawn from a patient who is at the steady-state concentration, that is, one who has received regular maintenance doses of the drug for about five half-lives ($5 \times t_{1/2}$) of the drug. For most orally administered drugs, the most consistent results are obtained if the sample is taken during a trough (pre-dose) period. Blood may be drawn at peak levels (post-dose), however, in those patients who exhibit toxic symptoms shortly after receiving the drug. When patients receive their drug by intravenous infusion or intramuscular injection, generally the best times to draw blood samples are 30 minutes after the infusion has stopped and about 60 minutes after the intramuscular injection.

Hundreds of different drugs are administered to patients daily but the bulk of the monitoring (80–90%) is confined to fewer than 20 drugs. Table 14.2 lists the drugs most commonly assayed in 1980 by most hospitals. The number will increase, of course, as new methods become available to perform the assays rapidly.

A particular drug can usually be assayed by several of the techniques described in the following section but time is a factor in drug monitoring. The physician wants the results on hospitalized patients, sometimes within a few hours, in order to modify the regimen; in outpatient clinics, the results may be desired within 30 minutes while the patient is still present. Hence, the most widely used technique is that of enzyme immunoassay, which can be performed within several minutes. Kits are available for all of the drugs listed in Table

TABLE 14.1
Therapeutic Serum Concentrations of Some Common Drugs

Concentrations are expressed as µg/ml (mg/L) except where otherwise indicated.

DRUG NAME		SERUM CONCENTRATIONS	
GENERIC	PROPRIETARY	THERAPEUTIC	TOXIC
Acetaminophen	Tylenol		>300 or when $t_{1/2}$ > 4 h
Aminophylline*		10–20	25
Barbiturates			
Amobarbital	Amytal	7–15	30
Pentobarbital	Nembutal	4–6	20
Phenobarbital	Luminal	10–30	55
Secobarbital	Seconal	3–5	10
Carbamazepine	Tegretol	6–12	
Chloramphenicol	Chloromycetin	5–20	50
Chlordiazepoxide	Librium	1–3	
Chlorpromazine	Thorazine	0.5	2
Diazepam	Valium	>0.5	
Digitoxin		14–30 ng/ml	30 ng/ml
Digoxin	Davoxin	0.9–2.0 ng/ml	2.0 ng/ml
Diphenylhydantoin	Dilantin	10–20	
Ethchlorvynol	Placidyl	0.2–5.0	
Ethosuximide	Zarontin	40–100	
Glutethimide	Doriden	2	10–80
Lidocaine	Xylocaine	1.2–5.0	
Lithium		0.5–1.5 mmol/L	2.0 mmol/L
Meprobamate	Miltown	5–20	100
Primidone	Mysoline	8–12	50–80
Procainamide +	Pronestyl	5–25	
N-Acetylprocainamide			
Quinidine		2–5.6	10
Salicylate	Aspirin	50–300	>350
Theophylline		10–20	25
Tricyclics			
Amitriptyline	Elavil	>120 ng/ml	
Nortriptyline	Aventil	50–150 ng/ml	
Imipramine	Tofranil	>180 ng/ml	
Desipramine	Norpramin	50–160 ng/ml	>500 ng/ml
Doxepin +	Sinequan	>110 ng/ml	
Desmethyldoxepin			
Protriptyline		>70 ng/ml	

*Aminophylline is a mixture of theophylline and ethylenediamine; only theophylline is measured.

TABLE 14.2
Drugs Most Commonly Assayed in U.S.A. Hospitals in 1980

Phenytoin	Ethosuximide	Lidocaine
Phenobarbital	Digoxin	Gentamicin
Carbamazepine	Valproic acid	Methotrexate
Theophylline	Procainamide	Tobramycin
Primidone	N-Acetylprocainamide	Quinidine
	Lithium	Tricyclic antidepressants

14.1 except lithium and tricyclic antidepressants. Fluorescent immunoassay can compete with enzyme immunoassay for rapidity of performance and can be expected to increase in usage in the future. The principles for both of these techniques are outlined in the following section. No details will be given because the manufacturers' directions have to be followed for each different drug.

Theophylline in Serum by HPLC

The use of HPLC for TDM is also widely used so a method for the determination of serum theophylline by this technique is presented. Theophylline is administered for the treatment of newborn apnea and for asthma. The customary therapeutic concentrations in serum are 6 to 11 μg/ml in newborn apnea and 10 to 20 μg/ml in asthma; the toxic level is above 25 μg/ml.

REFERENCE. Butrimovitz, G.P., and Raisys, V.A.: An improved micromethod for theophylline determination by reversed-phase liquid chromatography. Clin. Chem. *25*, 1461, 1979. As modified by V. Ainardi and K. Opheim, Children's Orthopedic Hospital and Medical Center, Seattle, WA 98105.

PRINCIPLE. Serum proteins are precipitated by mixing two volumes of methanol containing an internal standard with one volume of serum. After centrifugation, an aliquot of the supernate is injected into a liquid chromatograph. Theophylline and internal standard are eluted from the column by means of a liquid phase (phosphate buffer/methanol) pumped at high pressure. The absorbances of the components in the eluate are measured at 273 and 254 nm. Theophylline is quantitated by comparing its peak height/internal standard ratio at 273 nm with a standard curve prepared from pure theophylline solutions of known concentration. The ratio of theophylline absorbance at 273 nm to 254 nm can be used as a check for purity of the eluting peak; the ratio should be constant for all samples and identical to that of the standards.

REAGENTS AND MATERIALS

1. High Performance Liquid Chromatograph (Waters Associates), with sample injector, solvent delivery system, UV absorbance detector (variable wavelength), C_{18} column, dual pen recorder and integrator. Flow rate: 3.0 ml/min.
2. Millipore Filtration system with 0.22 μm filters.
3. High speed centrifuge with Eppendorf polypropylene centrifuge tubes.
4. *SMI pipets*, 25 and 50 μl.
5. *Methanol*, HPLC grade (Burdick and Jackson).
6. *Stock potassium phosphate buffer, 1.0 mol/L.* Dissolve 34.0 g KH_2PO_4 in water and make up to 250 ml volume.
7. *Working phosphate buffer, 0.01 mol/L, pH 6.0.* Dilute 20 ml of Stock buffer to 2.0 L with water. Adjust pH to 6.0 with 10 mol/L potassium hydroxide. Prepare freshly each month.
8. *Mobile phase.* Combine 860 ml of Working Buffer with 140 ml methanol. Filter through Millipore system. Discard after 1 week.
9. *Stock Standards:* Prepare 1.00 mg/ml methanol solutions of:
   a. Theophylline (Regis Chemical Co.)
   b. β-hydroxyethyltheophylline (Pierce Chemical Co.). Solutions are stable for 1 year at −20° C.
10. *Working Standard,* theophylline, 30 μg/ml.
    Prepare by adding 300 μl of Stock Standard theophylline and 10 mg

sodium azide to calf serum and making up to a final volume of 10.0 ml with calf serum. Store in one ml aliquots at −20° C. Stable for 1 year frozen or 1 month if stored at 0° C after each use.

11. Internal Standard, β-hydroxyethyltheophylline, 16 μg/ml in methanol. Dilute 80 μl of Stock β-hydroxyethyltheophylline in methanol and make up to 5.0 ml volume with methanol.

PROCEDURE

1. Pipet 25 μl of working standard, controls and patient samples into 500 μl Eppendorf tubes with an SMI pipet.
2. Pipet 50 μl of Internal Standard/Methanol solution into each tube with an SMI pipet.
3. Vortex tubes for about 5 seconds.
4. Centrifuge at 7000 rpm for 2 minutes.
5. Inject 20 μl of supernate into the chromatograph. The retention times for theophylline and internal standard (I.S.) are approximately 6.7 and 7.4 minutes, respectively.
6. Calculation (from absorbance at 273 nm):

Theophylline, μg/ml

$$= \frac{\text{peak height sample}}{\text{peak height I.S.}} \times \mu\text{g/ml St} \times \frac{\text{peak height I.S.}}{\text{peak height St}}$$

where I.S. = Internal Standard and St = Standard Theophylline Solution. *Note:* Variations in the extraction or chromatography processes are automatically corrected by the use of the above two ratios. The first ratio of peak heights of Sample to Internal Standard adjusts for variations in the run. The second ratio of peak heights of Internal Standard to Standard Theophylline corrects for variation from run to run or day to day.

TECHNIQUES OF DRUG ANALYSIS

The techniques for identifying and measuring the concentration of small amounts of drugs in such a complex protein matrix as serum are varied and sometimes intricate. There are so many different types of drugs to be measured that no single instrument or technique can be utilized for all of them. A large medical center laboratory may employ many, if not all, of the following chromatographic, immunoassay, and spectrophotometric techniques in order to perform a large variety of drug analyses within a reasonable period of time.

Immunologic Techniques

The immunologic techniques (p. 249) include enzyme immunoassay (EIA), fluorescence immunoassay (FIA), and radioimmunoassay (RIA).

Enzyme Immunoassay Systems

There are two types of enzyme immunoassay systems frequently used for measuring serum drug concentrations, and the principles are described on page 256.

ENZYME MULTIPLIED IMMUNOASSAY TECHNIQUE (EMIT SYVA). This homogeneous system is the most widely used in the U.S.A.; in 1982, kits were available for about 18 or more drugs and more were in the development stage. The drug to be measured is the hapten part of an antigen to which antibody has been prepared. The pure drug is labeled by covalently linking it to an enzyme. This technique is more rapid than RIA and consists of first incubating the serum with a buffered mixture containing limiting amounts of antibody, a small amount of enzyme-labeled drug, substrate, and cofactors for the enzyme. Enzyme activity is measured kinetically in a spectrophotometer for less than a minute; the drug concentration is obtained from a standard curve in which enzyme activity is plotted against the drug concentration (see p. 258).

ENZYME-LINKED IMMUNOABSORBENT ASSAY, ELISA. The ELISA system has been used extensively for the identification of microbiological agents but is beginning to be applied to drug analyses. The ELISA system is a heterogeneous assay, for the antibody is bound to a solid state carrier (membrane, beads or some other surface) instead of being free in solution. The drug to be measured is also the hapten of the antigen against which the antibodies have been prepared. The principle is the same as described above for the EMIT system, but one extra step is necessary—the separation of the bound drug from the unbound. In the tests for digoxin and digitoxin (Immunotech), the antibody is bound to a membrane, so the separation step of bound from unbound merely consists of decanting the reaction mixture into another tube.

Fluorescence Immunoassay (FIA)

Three different approaches to the application of FIA to the measurement of serum analytes were described on page 258. To recapitulate briefly, all of the systems utilize antibodies to react with the antigen (drugs or other substances). The systems differ in their use of fluorophors, the molecules that fluoresce when irradiated with light of the proper wavelength.

In one system (Ames), the drug is covalently bound to an enzyme substrate (umbelliferyl-β-D-galactoside) which is nonfluorescent. The fluorogenic drug reagent (FDR) plus antibody plus β-galactosidase is incubated with a serum sample. The drug and FDR compete for binding; the amount of unbound FDR varies directly with the serum drug concentration. The enzyme splits the unbound FDR in the homogeneous reaction mixture and the fluorescence is measured. The system is capable of assaying several antibiotics and theophylline in serum and is bound to grow.

In a different approach (BioRad), antibody is bound to synthetic beads and the antigen is labeled with a fluorophor in a heterogeneous assay. The labeled antigen bound to antibody-coated beads is separated by centrifugation and the fluorescence measured. A different version of the same technique binds fluorophor to antibody. The antibody bound to antigen is separated by centrifugation and the fluorescence is measured.

The EMIT system is being modified to measure the fluorescence of NADH instead of its absorbance at 340 nm. This greatly increases the sensitivity of the assay and represents an extension of their enzyme immunoassay system. The system will be expanded to measure some constituents that are in sufficiently high concentration that the enzyme-multiplication effect is not a necessity; FIA will be used with a fluorophor instead of an enzyme as the label.

Radioimmunoassay (RIA)

This technique is scarcely used in toxicology. The drug to be measured is covalently bound as a hapten to a protein (e.g., bovine serum albumin) and antibodies to it are produced by repeated injections into an animal. The antibody is able to recognize and react with the hapten (drug), even in a complex mixture of proteins. The principle of a RIA test for a drug is the same as described for T_4 (p. 270) or for any of the other hormones, namely the following: (1) incubation of serum with a limiting amount of antibody plus radioactively-labeled drug; the serum and labeled drug compete for antibody-binding sites; (2) separation of the free from the antibody-bound drug; (3) counting the radioactivity in either the free or antibody-bound fraction; and (4) calculating the amount of drug in serum from a standard curve of % antibody-bound versus drug concentration. The method is cumbersome.

Chromatography

Three types of chromatography, high performance liquid chromatography (HPLC), gas chromatography (GC), and thin-layer chromatography (TLC), are commonly used in a toxicology laboratory. The first two techniques are employed primarily for quantitating serum drug levels in TDM but may also be used for confirming drug identification. TLC is primarily a screening technique for drug identification. All three techniques are based upon the adsorption of drug to a solid support and elution by means of a mobile, liquid phase.

High Performance Liquid Chromatography (HPLC, p. 46)

A sample or extract is injected onto a chromatographic column, eluted from the column at high pressure (up to 6000 lb/square inch) with an appropriate mobile phase, detected, recorded, and quantitated by peak height or by curve area integration. This has become a very versatile, useful technique for drug analysis.

Gas Chromatography (GC, p. 49)

This usually involves solvent extraction, concentration of extract, conversion to a volatile derivative (derivatization), injection into a gas chromatograph, elution from the column, and detection and quantitation. Positive identification of peaks may require the coupling of the gas chromatograph with a mass spectrograph (MS).

Thin Layer Chromatography (TLC, p. 43) for Drug Screening

This is the most efficient and useful screening procedure available for drug identification. The process requires a solvent extraction, concentration of the extract, application of concentrate to a TLC plate, separation of the components by means of a mobile phase, visualization of the separated drugs by chemical reactions, light absorbance or fluorescence, and identification by comparison of R_f values and colors with known standards.

Spectrophotometry

Tentative identification of some drugs may be made by means of a spectral scan. When the identity of the drug is known, quantitative analysis may be performed by spectrophotometry after protein precipitation or extraction. The measurement may be made by the following:

1. *In the visible spectrum,* after the conversion of the drug into a colored compound. The determination of salicylate concentration is an example.
2. *In the ultraviolet spectrum.* Serum barbiturates may be quantitated by measuring the differential ultraviolet absorption curve at pH 10.3 and 13.5 from 320 nm to 220 nm.
3. *By fluorescence.* Quinidine is an example of a drug whose concentration in a protein-free filtrate of serum can be measured readily by fluorescence.

Atomic Absorption Spectrophotometry (AAS)

The serum concentration of many elements can be determined readily by atomic absorption spectrophotometry. The AAS determination most frequently requested is for the TDM of serum lithium whose carbonate salt is used in the treatment of manic-depressive psychosis (p. 356); flame emission photometry could also be used for lithium analysis but the precision is a little greater with AAS. AAS is well-suited for the determination of some heavy elements in serum, particularly copper and zinc, but the determination of the concentration of lead in blood is best done with a graphite furnace or rod attachment to the atomic absorption spectrophotometer.

Most toxicology laboratories in hospitals are concerned with the care of patients and not with medicolegal aspects. This purpose greatly simplifies the problem of specimen handling because in cases that come to court, the laws of evidence are strict concerning the custody of the specimen from the moment it is obtained until the analysis is made. An analysis will not be admitted into evidence if an opposing counsel can successfully challenge the identity of the sample. Textbooks on forensic (used in court) toxicology or an experienced toxicologist should be consulted before engaging in toxicology work that has legal implications.

DRUG OVERDOSAGE

Procedures for the analysis of ethanol, salicylate, acetaminophen, and barbiturate are detailed because accidental or purposeful overdosage with these drugs is relatively common. The screening for various classes of drugs by thin layer chromatography (TLC) is also given.

Ethanol in Serum

Ethanol can be measured easily by gas chromatography, by diffusion into potassium dichromate in acid solution and oxidation to acetic acid, and by action of the enzyme alcohol dehydrogenase. Analysis by gas chromatography is a good method if one has a chromatograph dedicated to this purpose because the presence of other volatile alcohols (methanol, isopropanol) can be determined at the same time. The dichromate method is not specific for ethanol because it reacts with other volatile alcohols, if present. The enzymatic method is described because it is simple, rapid, and reasonably specific.

REFERENCE. Jones, D., Gerber, L.P., and Drell, W.: A rapid enzymatic method for estimating ethanol in body fluids. Clin. Chem. *16,* 402, 1970.

PRINCIPLE. Ethanol is selectively oxidized to acetaldehyde by yeast alcohol dehydrogenase while NAD^+ is reduced to NADH in the process. The reaction is forced to completion by chemically trapping the acetaldehyde by means of aminooxyacetic acid. The alcohol concentration is proportional to the absorbance of NADH at 340 nm.

REAGENTS. The reagents are available as kits from several manufacturers. The reaction mixture from CalBiochem contains the following after the two vials are reconstituted and mixed (pH 8.8 to 9.2):

1. Sodium pyrophosphate 0.03 mol/L.
2. Glycine, 0.04 mol/L.
3. Na_2CO_3, 0.11 mol/L.
4. Aminooxyacetic acid, 0.03 mol/L.
5. NAD^+ (alcohol-free), 0.0017 mol/L.
6. Yeast alcohol dehydrogenase, 25 U per assay.

Other solutions:

1. NaCl, 9 g/L. Dissolve 9 g NaCl in water and make up to 1000 ml volume.
2. Standard Ethanol Solution, 100 mg/dl (w/v).

PROCEDURE

1. Dissolve contents of vial B (NAD^+) in 13.5 ml water.
2. Transfer the entire contents of vial B to vial A, which contains buffer, salts, enzyme (alcohol dehydrogenase), and aminooxyacetic acid.
3. Dilute standard, unknown, and control sera 1:50 in 9 g/L NaCl by taking 0.5 ml and diluting to 25 ml volume.
4. Place 2.6 ml of reaction mixture in a series of cuvets.
5. Add 0.1 ml portions of diluted unknowns, control, and standard to the proper cuvets and mix by inversion. Set up a reagent blank with 0.1 ml water.
6. Incubate for 10 minutes at 37° C and read the absorbance against the reagent blank at 340 nm.
7. Incubate for an additional 5 minutes and record the absorbance reading. This is a check to see whether the reaction has been completed.
8. If the absorbance change is greater than 0.3, repeat the test on a greater sample dilution (1:100) and multiply result $\times$ 2.
9. *Calculation.* The volumes of sample and reagent have been selected such that the difference in absorbance between unknown and blank = g ethanol/dl when sample dilution = 1:50.

$$1000\ (A_u - A_B) = \text{mg ethanol/dl}$$

In SI units, mmol ethanol/L = mg/dl $\times$ 0.22

Notes:

1. The standard ethanol solution, when diluted 1:50 should give a ΔA of 0.100 ± 0.005 in the test.
2. Proteins must be precipitated if the serum is badly hemolyzed. Prepare perchloric acid, 0.34 mol/L by diluting 29 ml of 70% $HClO_4$ to 1 liter. To 4.9 ml of $HClO_4$ solution, add 0.1 ml hemolyzed serum. Mix and centrifuge for 5 minutes. Take 0.1 ml supernatant and incubate with 2.6 ml of reaction mixture as in procedure steps 4 to 8.
3. The test is sensitive for ethanol; therefore, no alcohol must be used for sterilizing the skin when the blood sample is drawn. Use aqueous antiseptics only.

4. Yeast alcohol dehydrogenase does not react at all with methanol and does so slightly with isopropanol; with 1-butanol and 1-propanol, however, absorbance values of approximately 45 and 17%, respectively, of that for equivalent amounts of ethanol, are obtained.

Interpretation. If no alcohol has been ingested, the serum alcohol levels should be 0 ± 10 mg/dl. For legal purposes, some states in the United States and many foreign countries consider a blood level of 80 mg/dl or greater as evidence of intoxication, i.e., slower reaction time and impairment of visual acuity. Other states have set the legal definition of intoxication as 100 mg of ethanol per dl of blood.

Concentrations between 100 and 150 mg/dl correlate with greater uncoordination, slower reaction time, slurring speech, perhaps difficulty in balance, and general central nervous system depression. Concentrations over 400 mg/dl may be lethal.

Salicylate in Serum

Salicylates are present in numerous over-the-counter medicines, particularly in cold remedies and analgesics; the most common salicylate is aspirin. As a consequence of the ready accessibility of salicylate-containing medicines around the home and in purses, salicylates are the main cause of accidental poisoning in children. Overdosage with salicylate causes hyperventilation at first as the respiratory center in the brain is stimulated; the hyperventilation produces a respiratory alkalosis that is characterized by a lowered serum total CO_2 concentration and a very low pCO_2 (p. 101). This later (2 to 4 or 5 hours) changes to a metabolic acidosis as salicylate anion begins to accumulate; the serum CO_2 content is still low, but the concentration of pCO_2 rises to nearly normal levels.

Aspirin is acetylsalicylic acid, the compound that is produced by acetylating the hydroxyl group of salicylic acid (Fig. 14.1). Because the acetyl group of aspirin is hydrolyzed by esterases shortly after the drug is ingested, partially in the gastrointestinal tract and the rest in tissues, the active form of aspirin in plasma and tissues is salicylate. Hence, the analysis is made for salicylate and not acetylsalicylate (aspirin).

The medical uses of aspirin are antipyretic (reduce fever), analgesic (reduce the feeling of pain), and antirheumatic (reduce the pain and inflammation of rheumatoid arthritis).

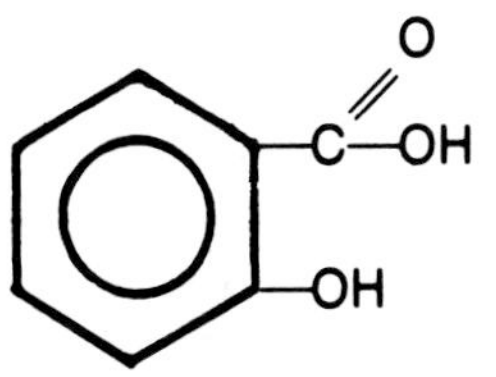

SALICYLIC ACID 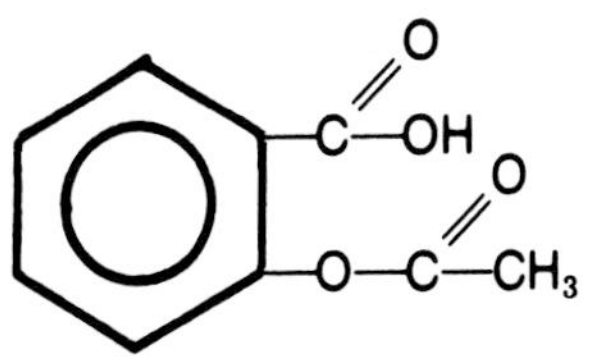   ASPIRIN

Fig. 14.1. Structural formulas of salicylic acid and aspirin. Aspirin is converted into salicylic acid by body esterases.

REFERENCE. Trinder, P.: Rapid determination of salicylate in biological fluids. Biochem. J. 57, 301, 1954. A modified version by R.P. MacDonald, entitled *Salicylate*, appears in Standard Methods of Clinical Chemistry, Vol. 5, edited by S. Meites, New York, Academic Press, 1965, p. 237.

PRINCIPLE. A serum sample is treated with an acid solution of $HgCl_2$ and $Fe(NO_3)_3$. The mercuric salt precipitates the serum proteins while the Fe^{3+} reacts with the phenolic group of salicylic acid to form a violet-colored complex. After centrifugation, the absorbance of the supernatant fluid is measured at 540 nm.

REAGENTS

1. Trinder's Reagent. Dissolve 40 g $HgCl_2$ in about 800 ml water. Add 120 ml 1 mol/L HCl and 40 g $Fe(NO_3)_3 \cdot 9H_2O$, and when completely dissolved, make up to 1 liter volume with water. The solution is stable indefinitely at room temperature.
2. Standard salicylate, 40.0 mg/dl. Dissolve 46.4 mg sodium salicylate (equivalent to 40.0 mg salicylic acid) in about 80 ml water and dilute to 100 ml volume with water. Add a few drops of chloroform as a preservative. The solution is stable for 6 months at 4° C. A low standard of 10 mg/dl is prepared by diluting one volume of above standard with 3 volumes of water.

PROCEDURE

1. Transfer 0.20 ml volumes of serum, standard, and water, respectively, to tubes labeled "Unknown," "Standard," and "Blank." Always run a control serum.
2. Add 1 ml Trinder's reagent to all tubes and mix.
3. Allow the tubes to stand for 5 minutes and then centrifuge for 10 minutes at high speed.
4. Transfer the supernatant fluid to a micro cuvet and read the absorbance against the blank at 540 nm.
5. Calculation:

$$\frac{A_u}{A_s} \times 40 = \text{mg/dl of salicylate}$$

In SI, mg/dl $\times$ 0.072 = mmol/L.

INTERPRETATION

1. There should be no salicylate in the serum of individuals who are not receiving the drug. Trinder's reagent does react slightly with some non-salicylate compounds in serum to give a serum blank that varies from about 0.5 to 1.2 mg/dl salicylate equivalent.
2. *Therapeutic levels.* A serum concentration of up to 30 mg/dl (2.2 mmol/L) may be reached in treating patients with rheumatoid arthritis. The concentration is usually lower in other conditions.
3. *Toxic levels.* A ringing sensation in the ears and hyperventilation may occur at serum concentrations of 30 to 40 mg/dl (2.2 to 2.9 mmol/L). In accidental poisoning or suicide attempts, the serum concentration may vary from 50 to 120 mg/dl, depending upon the amount of overdosage and the elapsed time since ingestion.

Acetaminophen in Serum

Acetaminophen (Fig. 14.2) is a common over-the-counter drug (Tylenol) that is present in many medications to relieve pain or reduce the extent of fever. It is of low toxicity when taken in the usual therapeutic dose but overdosage produces severe liver damage. The common causes for overdosage are accidental ingestion of large quantities, particularly by children, and by suicide attempts. The assay is of value for determining whether or not acetaminophen has been ingested, for assessing the level of overdose, and for deciding whether or not an antidote is required.

REFERENCE. Walberg, C.: Toxicology Laboratory, Los Angeles County, University of Southern California Medical Center and USC School of Medicine, Los Angeles, CA 90033.

PRINCIPLE. Acetaminophen in protein-free serum is converted into 2-nitro-5-acetaminophenol by mild nitration with nitrous acid ($NaNO_2$ + HCl). The nitro-derivative turns yellow in alkali and its absorbance at 430 nm is measured.

REAGENTS

1. Stock Standard acetaminophen, 1.00 mg/ml. Transfer 100.0 mg acetaminophen (Aldrich reagent grade) to a 100 ml volumetric flask, dissolve in water, and make up to volume.
2. High Working Standard, 150 μg/ml. Dilute 15.0 ml of Stock Standard to 100 ml volume with serum or plasma devoid of acetaminophen.
   Low Working Standard, 50 μg/ml. Dilute 5.0 ml of Stock Standard to 100 ml volume with negative serum or plasma.
3. Sodium nitrite, $NaNO_2$, 100 g/L. Place 10.0 g $NaNO_2$ in a 100 ml volumetric flask, dissolve in water, and make up to volume.
4. Ammonium sulfamate, $NH_4NH_2SO_3$, 150 g/L. Place 15.0 g of ammonium sulfamate in a 100 ml volumetric flask, dissolve in water, and make up to volume.
5. NaOH, 500 g/L. Place 50 g NaOH in a 250 ml beaker. Dissolve in 50 ml water while stirring. (Caution: solution becomes hot). When cool, transfer to a 100 ml volumetric cylinder and make up to volume with water.
6. HCl, 6 mol/L. In a hood, add 255 ml concentrated HCl to 245 ml water in a liter bottle. Mix and stopper.
7. Trichloroacetic acid (TCA), 100 g/L. Place 10 g TCA in a beaker containing 50 ml water. When dissolved, make up to 100 ml volume with water.

PROCEDURE

1. Transfer 1.0 ml of High and Low Working Standard, patient serum, and water, respectively, into 125 × 25 mm test tubes, appropriately labeled.
2. Add 2.0 ml TCA solution to each.

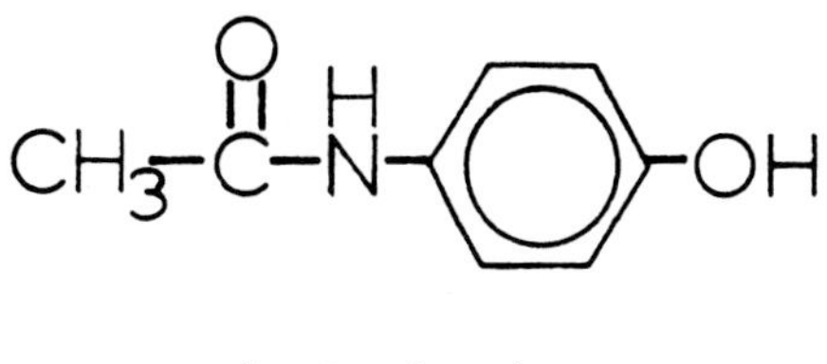

Acetaminophen

Fig. 14.2. Structure of acetaminophen.

3. Vortex and let stand 5 minutes. Centrifuge at high speed.
4. Quantitatively transfer the supernatant to clean, labeled test tubes.
5. Add 0.5 ml 6 mol/L HCl to each tube.
6. Add 0.5 ml $NaNO_2$ to each tube. Agitate by hand and wait 2 minutes. Tubes containing acetaminophen turn a faint yellow color.
7. Slowly add 0.5 ml ammonium sulfamate solution to each tube. The mixture fizzes as excess HNO_2 is decomposed.
8. Add 0.5 ml NaOH solution and mix. The yellow color intensifies.
9. Centrifuge at high speed to remove bubbles.
10. Zero the spectrophotometer with the reagent blank at 430 nm. Read absorbance of samples. The cuvets must be bubble-free and read immediately after pouring. Do not use a flow-through cuvet.
11. Calculation:

$$\text{Acetaminophen, } \mu g/ml = \frac{A_u}{A_s} \times 150$$

where A_u and A_s are absorbances of Unknown and High Standard, respectively.
Notes:
1. Report values of 10 μg/ml or less as negative for acetaminophen.
2. If the patient value is greater than 200 μg/ml, dilute the colored specimen with an equal volume of Reagent Blank, read the absorbance and repeat the calculation after multiplying absorbance $\times$ 2.

INTERPRETATION

Toxic Level. The following serum concentrations are toxic and certain to produce severe liver damage even though the damage may not be evident for several days:
1. If the serum level exceeds 300 μg/ml 4 hours after the drug ingestion.
2. If the serum half-life exceeds 4 hours.
3. If the half-life exceeds 10 hours, hepatic coma should be expected.

Antidote: The administration within 12 hours after drug exposure, of N-acetylcysteine, a sulfhydryl compound that replaces glutathione for detoxifying acetaminophen metabolites. The antidote causes gastrointestinal irritation, however, and neurological disturbances in some people, so it is not given unless clearly warranted. A recent review[9] discusses the pharmacokinetics of acetaminophen overdose.

Barbiturates in Serum

A barbiturate is a derivative of barbituric acid that has been developed for its sedative properties. Barbiturates are general depressants and are taken most frequently for preventing seizures (phenobarbital) or for inducing sleep (short- and intermediate-acting barbiturates). Barbiturates account for a large percentage of the adult patients treated in a general hospital for toxicity because of accidental or intentional (suicidal) overdosage; these drugs are responsible for many deaths each year.

The barbiturates are divided into four groups based upon their pharmacologic action: long-acting (phenobarbital, barbital, and others), intermediate-acting (amobarbital, butabarbital), short-acting (secobarbital, pentobarbital), and ultrashort-acting (thiopental and others). The toxicity (serum concentration at which coma is induced) of the different barbiturates varies with the type; the fastest acting barbiturates cause coma at the lowest serum level, and the critical serum concentration is highest for the longest acting compounds (Table 14.3). The ultrashort acting barbiturates are not prescribed for patients, since they are used almost exclusively for anesthesia; they are rarely involved in problems of overdosage. The serum concentrations causing coma are about 10, 30, and 55 μg/ml (10, 30, and 55 mg/L), respectively, for the short, intermediate, and long-acting barbiturates.

The barbiturates are inactivated in the liver by a series of enzymatic steps that may include oxidation of substituted radicals, removal of N-alkyl groups if present, the removal of sulfur from thiobarbiturates, and opening of the barbital ring. The drug metabolites and a portion of the unchanged barbiturate are excreted in the urine.

Barbiturate concentrations in serum may be measured by gas chromatography after extracting the drug into an organic solvent and preparing a volatile derivative. This is an excellent technique, since the type of barbiturate can be identified tentatively from its retention time and can be quantitated at the same time; it is tedious, however, and requires a gas chromatograph. Hence a spectrophotometric method that depends upon the differential ultraviolet absorption at two different pH levels is presented. See Figures 14.3 and 14.4.

REFERENCE. Williams, L.A., and Zak, B.: Determination of barbiturates by automatic differential spectrophotometry. Clin. Chim. Acta 4, 170, 1959.

PRINCIPLE. Barbiturates are quantitated by automatic differential spectrophotometry. Barbiturates have different spectra in 0.45 mol/L NaOH (approximately pH 13.5) from those at pH 10.3. Typical spectral curves, together with a differential spectral curve (automatically subtracting the absorbance at pH 10.3 from that of 13.5) are shown (Figs. 14.3 and 14.4). The differential spectrum of most barbiturates has a maximum absorbance at 260 ± 5 nm and a minimum

TABLE 14.3

Relation Between a Patient's Clinical Condition and the Barbiturate Concentration* in His Blood

| BARBITURATE | CLINICAL STAGES† | | | | |
	1	2	3	4	5
Amobarbital	7.0	15.0	30.0	52.0	66.0
Pentobarbital	4.0	6.0	15.0	20.0	30.0
Phenobarbital	10.0	34.0	55.0	80.0	150.0
Secobarbital	3.0	5.0	10.0	15.0	20.0

From Manual of Analytical Toxicology, edited by I. Sunshine. Copyright The Chemical Rubber Co., Cleveland, 1971.

*Concentration is expressed in μg per ml of blood, which is identical with mg/liter. Each figure represents the concentration of barbiturate at which the average nontolerant patient enters into a given clinical stage.

†The clinical conditions of the patients were grouped into five arbitrary categories on the basis of the following criteria:

Stage 1: awake, competent, and mildly sedated.
Stage 2: sedated, reflexes present, prefers sleep, answers questions when roused, does not cerebrate.
Stage 3: comatose, reflexes present.
Stage 4: comatose, reflexes absent.
Stage 5: comatose, circulatory difficulty, and/or depression of the respiratory center in the medulla.

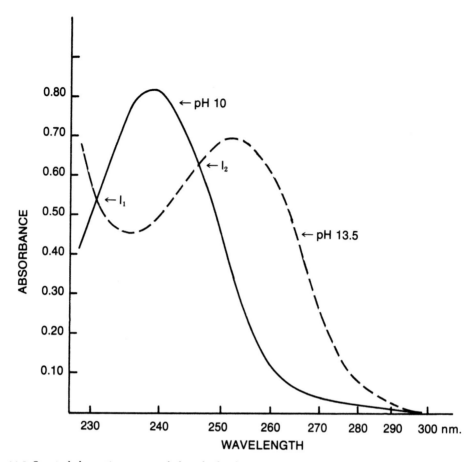

Fig. 14.3. Spectral absorption curves of phenobarbital at pH 13.5 and at pH 10. I_1 and I_2 are isobestic points, the places where the two curves intersect. (Modified from Manual of Analytical Toxicology, edited by I. Sunshine. Copyright The Chemical Rubber Co., Cleveland, 1971).

at 240 ± 5 nm; the difference between the two peaks is used for quantitation by comparing the differences between the unknown and a standard.

REAGENTS AND MATERIALS

1. KH_2PO_4, 1 mol/L. Dissolve 13.61 g anhydrous KH_2PO_4 crystals in water and make up to 100 ml volume.

2. NaOH, 1 mol/L. Dissolve 4.0 g NaOH pellets in water and make up to 100 ml volume.

3. Buffer, pH 7.4. Add 50.0 ml of the KH_2PO_4 solution (#1) to 39.1 ml of the NaOH solution (#2) and adjust to pH 7.4 if necessary. Make up to 100 ml volume with water.

4. NH_4Cl, 3 mol/L. Dissolve 16.0 g NH_4Cl in water and make up to 100 ml volume.

5. NaOH, 0.45 mol/L. Dissolve 1.8 g NaOH in 100 ml water. Adjust the solution so that when 0.5 ml NH_4Cl solution is added to 2.5 ml NaOH, the pH is 10.3 ± 0.2.

6. Extraction Solvent. Dichloromethane or chloroform (nanograde only).

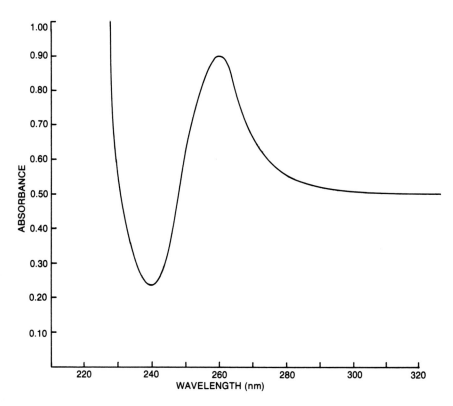

Fig. 14.4. Typical differential absorption spectrum between pH 10.3 and 13.5 for secobarbital.

7. Stock Phenobarbital Standard, 2.5 mg/ml in 95% ethanol. Dissolve 250 mg phenobarbital in 95% ethanol and make up to 100 ml volume.
8. Working Phenobarbital Standard, 25 μg/ml. Dilute 1.0 ml stock standard to 100 ml volume with water. Prepare fresh daily.
9. Screw cap tubes, 25 × 150 mm, with Teflon-lined caps.
10. Spectrophotometer. A double beam recording instrument capable of working in the ultraviolet spectrum down to 220 nm. Use quartz or fused silica cuvets only.

PROCEDURE

Extraction

1. Pipet 2.0 ml serum, control, and standard, respectively, to appropriately labeled screw cap tubes, and 0.5 ml of phosphate buffer (#3).
2. Add 25 ml of extraction solvent (CH_2Cl_2 or $CHCl_3$) to all tubes and mix on a rotator for 5 minutes or shake vigorously by hand.
3. Allow phases to separate and discard the upper, aqueous layer by aspiration.
4. Pour the organic layer through a rapid filtering filter paper (to remove traces of water) into a clean extraction tube.

Transfer to Alkaline Solution

5. Add 6.0 ml 0.45 mol/L NaOH (#5) and mix on rotator for 5 minutes or shake vigorously by hand.

6. After the phases separate, transfer as much as possible of the upper alkaline layer to 12 ml centrifuge tubes. Centrifuge for 3 minutes to clear the aqueous layer of organic solvent.
7. Pipet 2.5 ml NaOH layer to each of two matched cuvets.
8. Add 0.5 ml NH$_4$Cl (#4) to one cuvet, mix, and place in the spectrophotometer as the *reference* cuvet (pH 10.3).
9. Add 0.5 ml 0.45 mol/L NaOH (#5) to the other cuvet, mix, and place in the spectrophotometer as the *sample* cuvet (pH approximately 13.5).

Spectrophotometric Scan

10. Calibrate the instrument to zero on air.
11. Set the recorder pen to approximately the middle of the scale.
12. Set the recorder span at 0.50 absorbance units for full scale. If a peak or trough should go off the scale, change the span to 1.0 absorbance units and rerun.
13. Scan the absorbance from 325 nm to 200 nm.

Calculation and Quantitation

14. Qualitative identification of barbiturates is made if there is an absorption minimum at 240 nm and an absorption maximum at 260 nm (Fig. 14.4).
15. Quantitation. Calculate the concentration by measuring the distance between the 240 minimum peak and the 260 maximum peak for both unknown and standard. Let these differences be Δh_u and Δh_s, respectively.

$$\text{Barbiturate concentration, } \mu g/ml = \frac{\Delta h_u}{\Delta h_s} \times C = \frac{\Delta h_u}{\Delta h_s} \times 25$$

In SI units, phenobarbital, $\mu mol/L = \mu g/ml \times 4.3$

The concentration in $\mu g/ml$ is identical with that of mg/L.

Interpretation. No barbiturate should be present in serum unless the drug has been administered. The serum concentrations of several barbiturates at varying clinical stages are shown in Table 14.2.

Notes:

1. Be sure to use fused silica or quartz cuvets.
2. For barbiturate, the scan must have a maximum at 260 ± 5 nm and a minimum at 240 ± 5 nm.
3. A secondary peak at 300 nm may indicate the presence of diazepam; salicylate shows a slow decrease from 320 to 300 nm, followed by a rise.
4. Thiobarbiturates have a primary peak at 300 ± 5 nm, with a secondary peak at 240 ± 5 nm.
5. Several weakly acid drugs whose ultraviolet absorbance at 260 nm is significant can interfere, if present. Diphenylhydantoin is the most common and likely of these drugs because it is frequently administered with phenobarbital to control the seizures of epilepsy. Salicylate and a few other drugs are possible but not very likely interferents.
6. In case identification of the drug as a barbiturate is in doubt, run the UV absorption curve of the NaOH extract against its blank, and on the same chart paper, run the curve of the NH$_4$Cl-NaOH mixture (pH 10.3) against its blank. The curves should cross (isobestic points) at 247 to 250 nm and again at 227 to 230 nm (see Fig. 14.3).

Drug Screening (Identification)

Cases of suspected drug overdosage in comatose patients or of possible drug abuse are the most frequent situations that require the identification of drugs in body fluids. Screening by thin layer chromatogrphy (TLC) is the most efficient, rapid, and economical way of identifying the offending drug(s) in the above situations. Urine is the best single specimen for analysis; gastric fluid is good for drugs taken orally but should be diluted 1:5 prior to analysis.

There are many different TLC separation systems but they are all based upon the same general principles. The acidic drugs are separated from the basic drugs by proper pH manipulation of the extraction process, followed by TLC separation in two different solvent systems that also differ in pH. The neutral drugs tend to appear in both systems. The separated drugs on the chromatogram are visualized by several techniques: by their fluorescence when examined under ultraviolet light; by their colored derivatives when sprayed with or dipped into particular chemical solutions; and by their absorbance of light.

The TLC method described below is presented because it provides good separation and can be set up without dependence upon proprietary materials; the requisite materials are also available in kit form (Gelman Co., Quantum Assays Corp.). A faster but more costly TLC separation procedure is the Toxi-Lab Drug Identification system (Analytical Systems).

REFERENCES. Quantum Assays Corp., procedural bulletin QDS-3.
Gelman Drug Identification Systems. The following directions apply to the Gelman kit.

PRINCIPLE. Neutral, basic, and acidic drugs in fresh urine or gastric specimens are extracted into an organic solvent at pH 9.5. The solvent is evaporated, and the residue is dissolved in a small amount of methanol. Aliquots of the methanol solution are placed on each of two TLC plates and allowed to migrate in two separate solvent systems. Separation of the drugs is due to the polarity (degree of ionization) and solubility of the compounds spotted, as well as to the polarity of the solvent system. The degree of ionization of a drug is a function of the H^+ concentration. The more polar the substance, the less soluble it is in the organic solvent, the more strongly it binds to the silica gel, and the more slowly it migrates. Also, as the polarity of the solvent increases, the polar compound migrates further on the plate, because the solubility of the compound in the solvent increases. After migration of the compounds the plates are sprayed with several different solutions in order to develop typical colors for the different drugs. See Table 14.4 for relative migration distance (R_f) of various drugs.

REAGENTS AND MATERIALS
1. NH_4Cl Buffer, saturated, pH 9.5. Saturate 1000 ml water with NH_4Cl and adjust the pH to 9.5 with concentrated NH_4OH.
2. Extraction Solvent. Add 40 ml isopropanol to 960 ml dichloromethane (reagent grade).
3. Methanolic H_2SO_4. Add 0.5 ml concentrated H_2SO_4 to 100 ml anhydrous methanol.
4. NaOH, approximately 19 mol/L. Cautiously dissolve 50 g NaOH in 50 ml water.
5. Developing Solvents:
   a. *Solvent A (primarily for "acidic" drugs):*
      Ethyl Acetate   170 ml

TABLE 14.4

Relative Migration Distance of Common Therapeutic Drugs When Developed in Solvents* A and B on TLC Plates

TYPE	GENERIC NAME	PROPRIETARY NAME†	$R_f \times 100$ SOLVENT A	SOLVENT B
Acidic	Phenobarbital	Luminal	25–31	—
	Secobarbital	Seconal	56–64	—
	Amobarbital	Amytal	51–60	—
	Diphenylhydantoin	Dilantin	49–57	—
	Glutethimide	Doriden	91–95	—
Neutral	Meprobamate	Miltown	76–81	—
	Chlordiazepoxide	Librium	69–76	40–51
	Diazepam	Valium	92–96	85–89
	Methaqualone	Quaalude	92–95	85–91
Basic	Morphine		22–29	4–6
	Codeine		49–59	8–13
	Phenylpropanolamine	Norephedrine	95–100	9–30
	Quinidine/quinine		65–69	12–17
	d-Amphetamines	Dexedrine	—	16–22
	Meperidine	Demerol	86–92	32–42
	Phenothiazines	Phenergan	91–95	44–58
	Amitryptyline	Elavil	91–95	44–58
	Cocaine		90–96	72–87
	Dextropropoxyphene	Darvon	92–97	84–90

*Solvents described on page 397.
†Only one of the many proprietary names is given.

Methanol	20 ml
NH₄OH (conc.)	10 ml

b. *Solvent B (primarily for "basic" and "neutral" drugs):*

Ethyl Acetate	160 ml
Cyclohexane	25 ml
Methanol	13.5 ml
NH₄OH (conc.)	1 ml

6. Spray Reagents:
   a. Diphenylcarbazone. Dissolve 1 g diphenylcarbazone in 500 ml acetone and make up to 1000 ml volume with water.
   b. $HgSO_4$, 0.011 mol/L. Dissolve 2.5 g HgO in 100 ml concentrated H_2SO_4 and *cautiously* dilute to 1000 ml volume with water while cooling in an ice bath.
   c. Chloroplatinic acid, H_2PtCl_6, 0.19 mol/L. Dissolve the contents of a 3.5 g bottle of $H_2PtCl_6 \cdot 6H_2O$ in 35 ml water. Pipet 1 ml aliquots into test tubes, stopper, and freeze.
   d. Potassium Iodide, KI, 1.8 mol/L. Dissolve 300 g KI in water and dilute to 1 liter volume.
   e. Iodoplatinate. To 1 ml chloroplatinic acid, add 4 ml distilled H_2O, 20 ml 1.8 mol/L KI, and 25 ml methanol.
   f. Ninhydrin, 11 to 17 mmol/L. Prepare daily 0.1 to 0.15 g ninhydrin in 50 ml acetone.
7. Standard Controls. The Gelman Drug Control Set is the basic control used; however, other drugs are added to give a more complete identification system. Reconstitute each bottle in the control set with 50 ml water. Mix

gently and allow 5 to 10 minutes for complete solution. Reconstituted bottles should be refrigerated at 4 to 8° C and may be used for two weeks. These should be treated the same as unknown urine samples except use 15 ml rather than 20 ml.

C_1: glutethimide, secobarbital, diphenylhydantoin, phenobarbital, supplemented with meprobamate.

C_2: dextropropoxyphene, methadone, quinidine, methamphetamine, supplemented with ethchlorvynol and amitriptyline.

C_3: cocaine, meperidine, D-amphetamine, codeine, morphine, supplemented with methaqualone.

C_4: 12 ml saline (NaCl, 0.9 g/dl) supplemented with 3 ml chlorpromazine, chlordiazepoxide, diazepam, amobarbital, and pentazocine.

The supplements to Controls C_1, C_2, and C_3 consist of 2.0 ml of an aqueous solution containing 500 μg/ml of the supplemented drugs; the final concentration is 20 μg/ml. For C_4, add 3 ml of a 20 mg/dl solution of drugs to 12 ml saline solution to yield a 40 μg/ml concentration. See Tables 14.1 and 14.4 for a common trade name of each of the generic drugs listed above.

8. TLC plates and developing chamber (Gelman; Eastman).
9. Spray atomizer, glass.
10. Airjet blower or hair dryer.

PROCEDURE. PRECAUTIONS. Urine and gastric specimens should be fresh and refrigerated until analyzed. The only one of the above drugs that is stable in urine for months is morphine. Hydrolysis of the urine sample increases the sensitivity of detection of morphine because as much as 90% of the morphine in urine is excreted as the glucuronide, which requires hydrolysis for detection. Hydrolysis also improves the detection of diazepam and chlordiazepoxide.

Hydrolysis of Urine

1. Transfer 15 ml urine + 4.5 ml concentrated HCl to a test tube.
2. Heat at 100° C for 10 minutes.
3. Adjust pH to approximately 7 with approximately 19 molar NaOH (#4, about 3 ml). Check the pH.
4. Filter into an extraction tube and continue with the basic drug screen.

Extraction with pH 9.5 Buffer

5. Transfer 20 ml unhydrolyzed urine (or gastric juice) if hydrolysis is not necessary or 15 ml hydrolyzed urine to an extraction tube. Transfer 15 ml portions of control C_1, C_2, and C_3 and 12 ml of C_4 to appropriately labeled tubes.
6. Add 2 ml pH 9.5 NH_4Cl buffer (#1) and 25 ml extraction solvent (#2, isopropanol in dichloromethane) to each tube.
7. Mix on a rotator for 5 minutes and centrifuge if an emulsion forms.
8. Discard aqueous layer by aspiration. Filter the organic layer into large centrifuge tubes.

Concentration of Extract

9. Add 3 drops methanolic sulfuric acid (#3) to the extract and evaporate to dryness in a water bath. The amphetamines are volatile and easily lost by excessive or prolonged heating during evaporation; the addition of acid converts them to a less volatile salt.
10. After the centrifuge tubes are cool, reconstitute each tube with 6 drops

methanol and spot onto two predried TLC plates. Remove the solvent by blowing across the plates with an airjet blower or hair dryer.

11. After spots are dry, place plates into the TLC developing chambers, one of which contains the solvent for separating acidic drugs (solvent A) and the other for basic and neutral drugs (solvent B). Allow the solvent to rise approximately 12 to 16 cm (about 30 minutes).

12. Remove the plates from tanks, mark the solvent front, and dry the acid plate at 110° C in an oven for 10 minutes and air-dry plate B for at least 10 minutes. (High temperature or prolonged drying time of the basic plate will cause components other than amphetamines to turn pink with ninhydrin.)

Detection and Identification of Barbiturates on Plate A

13. Spray the cooled plate heavily with diphenylcarbazone and then lightly with mercuric sulfate.
    a. This step yields a blue to blue-purple response for compounds that have the typical barbital structure: pentobarbital, phenobarbital, secobarbital, amobarbital, diphenylhydantoin, and glutethimide.
    b. Diphenylhydantoin can very easily be misinterpreted as amobarbital; so careful observation of color and position related to the controls is necessary. Compare the R_f values of unknowns with the controls.
    c. Glutethimide may be missed because of incorrect (too heavy) spraying with mercuric sulfate.

14. Heat plate at 110° C for 2 or 3 minutes.
    a. Phenothiazines develop colors.
    b. At high levels of concentration, chlorpromazine responds before heating.

15. Examine plate under UV light. Quinidine and tranquilizers, such as chlordiazepoxide and diazepam, exhibit fluorescence.

16. Spray plate with iodoplatinate to (a) visualize methaqualone and (b) use as a cross check with plate B for the opiates.

Detection and Identification of Amphetamines and Alkaloids on Plate B

17. Examine dried plate under UV light.
    a. Compare R_f values of unknown fluorescent spots with that of controls.
    b. Quinidine fluoresces bright blue, codeine is violet, morphine bluish, and diazepam and chlordiazepoxide are greenish-blue.

18. Spray *very heavily* with ninhydrin. Phenylpropanolamine appears at this point.

19. Dry plate at 90° C for 5 minutes. High concentrations of amphetamines appear as brownish to pink spots.

20. Expose plate to UV light for 3 minutes to bring out the lower concentrations of amphetamines.

21. Spray plate with iodoplatinate.
    a. All basic drugs appear as purple to pink spots.
    b. The colors fade on standing and appear to be different shades, or even other colors. Colors reappear if the plate is resprayed.

22. Cross check with plate A for positive identification.

Notes:

1. Specimens stored without refrigeration for an extended period (4 to 5

days) have a tendency to exhibit moderate to severe background darkening and also retarded migration rates (lower R_f values).

2. The drier the extracted specimen is prior to evaporation, the lighter is the background of the plate. Small droplets of dispersed aqueous buffer retard the migration and distort the spot. These are the reasons for filtering the extract before evaporation and for drying the TLC plate directly before spotting.
3. When performing numerous determinations (over 12), reconstitute only half of the tubes at one time to avoid evaporation of the methanol.
4. Knowns must always be spotted on the same plate as unknowns. Small changes in conditions alter the R_f value (relative migration distance of sample and solvent front). These include chamber saturation, temperature, humidity, and solvent level.
5. When a plate is placed in solvent tank, care should be taken to obtain even, uniform contact with solvent surface across the bottom of the plate. The spots should be at least 1 cm above the solvent surface. Insure a tight seal of the chamber by using a light coat of stopcock grease on the top plate.
6. When removing plate B from tank, be sure to mark solvent front with pencil before drying, as it cannot be seen in the spraying process.
7. Solvents may be used 3 to 5 times during 1 day's testing; R_f values may drop slightly on each consecutive run, however.

Many of the drugs listed in Table 14.4 appear in the urine together with their metabolites, which usually separate on the chromatogram at a different spot (different R_f value). The colors of the metabolite spots may be similar to or different from those of the parent compound. The majority of the drugs listed in Table 14.4 are basic; they are best separated in solvent B.

The barbiturates are acidic drugs and are separated in solvent A. Barbiturates are taken as sleep inducers, to control seizures in epileptics in conjunction with anticonvulsants, and sometimes as a drug of abuse. Barbiturates are not found in the urine unless a large excess has been ingested; a positive urine test, therefore, is clinically significant. Addiction to barbiturate is dangerous because of the small margin of safety between effect and toxicity.

Phenothiazine tranquilizers are converted to several metabolites that are excreted very slowly by the kidney; it is possible to detect such a drug or its metabolites in the urine several days after its ingestion.

Some opiate drugs are metabolized and excreted in different forms. Codeine is partially metabolized (about 10%) to morphine prior to being cleared. Heroin, diacetylmorphine, is very rapidly hydrolyzed to morphine. Therapeutic levels of morphine and codeine can be detected in the urine. They usually are excreted as glucuronides and require hydrolysis prior to TLC assay.

Heavy Metals

Toxic metals suspected in a poisoning case (e.g., Cu, Zn, Pb, and Hg in body fluids) can be determined by atomic absorption spectrophotometry (p. 37). The Reinsch test[6] is sometimes used to screen for the presence of heavy metals (As, Sb, Hg, and Bi), but the test is not satisfactory when the concentrations are low. If the test should be positive, any one of the group of four heavy metals is present; further work is necessary for identification.

REFERENCES

1. Pippenger, C.E.: The rationale for therapeutic drug monitoring. *In* Therapeutic Drug Monitoring. Washington, D.C., American Association for Clinical Chemistry, 1979.
2. Sunshine, I. (editor): Manual of Analytical Toxicology. Cleveland, Chemical Rubber Co., 1971.
3. Baselt, R.C.: Analytical Procedures for Therapeutic Drug Monitoring and Emergency Toxicology. Davis, CA, Biomedical Pubns., 1980.
4. Gibaldi, M.: Biopharmaceutics and Clinical Pharmacokinetics, 2nd ed. Philadelphia, Lea & Febiger, 1977.
5. Jatlow, P.O.: Analytical toxicology in the clinical laboratory—an overview. Lab. Med. 6, no. 12, 10, 1975.
6. Gettler, A.O., and Kaye, S.: A simple and rapid analytical method for Hg, Bi, Sb, and As in biological material. J. Lab. Clin. Med. 35, 146, 1950.
7. Kabra, P.M., McDonald, D.M., and Marton, L.V.: A simultaneous high-performance liquid chromatographic analysis of the most common anticonvulsants and their metabolites in serum. J. Anal. Toxicol. 2, 127, 1978.
8. Opheim, K.: Lab News, Vol. 4, No. 4. Seattle, WA, Children's Orthopedic Hospital and Medical Center, 1978.
9. Rosenberg, J., Benowitz, N.L., and Pond, S.: Pharmacokinetics of drug overdose. Clin. Pharmacokinet. 6, 161, 1981.

APPENDIXES

TEST	SPECIMEN*	VALUE		FACTOR†
		CONVENTIONAL	SI UNITS	
Acetoacetic acid, qual	U	Negative		
Acetone, qual	U	Negative		
Adrenocorticotropic hormone (ACTH)	P	<50 pg/ml	<50 ng/L	1000
Alanine aminotransferase (ALT)	S	3–30 U/L	By method described	
Albumin	S	3.5–5.2 g/dl	35–52 g/L	10
Aldosterone, upright	P	5–20 ng/dl	140–560 pmol/L	27.8
recumbent	P	1–8 ng/dl	28–220 pmol/L	
Amylase	S	0.8–3.2 U/L	By method described	
	U	Up to 400 U/hr		
Aspartate aminotransferase (AST)	S	6–25 U/L	By method described	
Bicarbonate	P (Art)	21–28 meq/L	21–28 mmol/L	1.0
Bilirubin, total	S	0.1–1.0 mg/dl	1.7–17.1 μmol/L	17.1
Bilirubin, esterified	S	0.0–0.3 mg/dl	0–5 μmol/L	17.1
Bilirubin, qual	U	Negative		
Bromsulfophthalein (BSP)	S	<5% after 45 minutes		
Calcium, total	S	8.7–10.5 mg/dl	2.18–2.63 mmol/L	0.25
ionized	S	4.7–5.5 mg/dl	1.18–1.38 mmol/L	0.25
Catecholamines, total	U	<100 μg/24 hr	<0.59 μmol/24 hr	0.0059
Chloride	S	98–108 meq/L	98–108 mmol/L	1.0
	Sf	120–132 meq/L	120–132 mmol/L	1.0
	Sweat	5–40 meq/L	5–40 mmol/L	1.0
Cholesterol, total	S	140–250 mg/dl but varies with age	3.64–6.50 mmol/L	0.026
Cholesterol, HDL	S	Varies with age and		
LDL	S	sex. See table on page 337.		
CO_2 content	S	Ad 24–30 mmol/L	same	1.0
		Inf 18–26 mmol/L		
CO_2 pressure, pCO_2	B	M 34–45 mm Hg		
		F 31–42 mm Hg		
	P (Art)	M 24–30 mm Hg		
		F 21–30 mm Hg		

* S = serum	Af = amniotic fluid	M = adult male
P = plasma	qual = qualitative	F = adult female
B = whole blood	Art = arterial	Inf = Infant
U = urine	Ad = adult	
Sf = spinal fluid	Ch = children	

†Factor to convert from conventional to SI units.
‡Based on the molecular weight of dehydroepiandrosterone, 289.
**Therapeutic and toxic levels of various therapeutic agents are listed in Table 14.1, page 382.

| TEST | SPECIMEN* | VALUE | | FACTOR† |
		CONVENTIONAL	SI UNITS	
Cortisol	P	6–25 μg/dl	166–690 nmol/L	27.6
Creatine kinase (CK)	S	10–100 U/L	By method described	
Creatine kinase isoenzyme (CK–MB)	S	<6 U/L		
Creatinine	S	M 0.7–1.4 mg/dl	60–125 μmol/L	88.4
		F 0.6–1.3 mg/dl	53–115 μmol/L	
		Ch 0.4–1.2 mg/dl	36–107 μmol/L	
	U	Ad ~ 1 mg/min		
	Af	> 2.0 mg/dl	>177 μmol/L	
		taken as sign of maturity		
Creatinine clearance	S U	M 105 ± 20 ml/min		
		F 97 ± 20 ml/min		
Electrophoresis	S			
Albumin		3.5–5.2 g/dl	35–52 g/L	10
α₁-globulin		0.1–0.4 g/dl	1–4 g/L	10
α₂-globulin		0.5–1.0 g/dl	5–10 g/L	10
β-globulin		0.6–1.2 g/dl	6–12 g/L	10
γ-globulin		0.6–1.6 g//dl	6–16 g/L	10
Estriol (E₃), 3rd trimester,				
Unconjugated	S	5–40 ng/ml	17–140 nmol/L	3.5
Total	S	40–500 ng/ml	140–1750 nmol/L	
Estriol (pregnancy)	U	at term, 14–44 mg/24 hr	49–154 μmol/24 hr	3.5
Gamma glutamyl transferase (GGT)	S	M 3–35 U/L		
		F 3–30 U/L		
Globulins, total	S	2.3–3.5 g/dl	23–35 g/L	10
Glucose	S	65–100 mg/dl	3.6–5.6 mmol/L	0.055
	Sf	40–70 mg/dl	2.2–3.9 mmol/L	
Glucose, qual	U	<10 mg/dl	<0.55 mmol/L	
		Clinitest negative		
Growth hormone (GH) (during day)	P	~5 ng/ml	~5 μg/L	1000
Human chorionic gonadotropin (hCG)				
Nonpregnancy	S	F 0–10 mU/ml		
Pregnancy	S	F >30 mU/ml		
β-hCG (nonpregnancy)	S	F <3 mU/ml		
17-hydroxycorticosteroids (17-OHCS)	U	Ad 5–14 mg/24 hr	17–48 μmol/24 hr	3.46‡
Iodine, protein-bound (PBI)	S	4.0–8.0 μg/dl	0.32–0.63 μmol/L	0.079
Insulin	P	<860 pg/ml	<860 ng/L	1000
Iron, total	S	65–165 μg/dl	11.6–29.5 μmol/L	0.179
Iron binding capacity (TIBC)	S	300–400 μg/dl	53.7–71.6 μmol/L	0.179
% Saturation		20–50%		
17-Ketogenic steroids (17-KGS)	U	M 8–25 mg/24 hr	28–87 μmol/24 hr	
		F 5–18 mg/24 hr	17–62 μmol/24 hr	3.46‡
		Ch 2–4 mg/24 hr	6.9–13.8 μmol/24 hr	
17-Ketosteroids (17-KS)	U	M 8–20 mg/24 hr	28–69 μmol/24 hr	
		F 5–15 mg/24 hr	17–52 μmol/24 hr	3.46‡
		Ch <2 mg/24 hr	<6.9 μmol/24 hr	
Lactate dehydrogenase (LD)	S	P→L	By method described	
		125–290 U/L		
Lactate dehydrogenase isoenzymes	S			
LD₁		20–34% of total	<100 U/L	
LD₂		32–40% of total	<115 U/L	
LD₃		17–23% of total	<65 U/L	
LD₄		3–13% of total	<40 U/L	
LD₅		4–12% of total	<35 U/L	
Lecithin/sphingomyelin (L/S) ratio	Af	>2.0 indicates maturity		

APPENDIX 1, *continued*
Reference Values

TEST	SPECIMEN*	VALUE CONVENTIONAL	VALUE SI UNITS	FACTOR†
Lipase	S	28–200 U/L	By method described	
Magnesium	S	1.3–2.1 meq/L	0.65–1.05 mmol/L	0.5
Metanephrines	U	0.3–0.9 mg/24 hr	1.5–4.6 μmol/24 hr	5.1
5'-Nucleotidase (NTP)	S	0.4–7.0 U/L		
Ornithine carbamoyl transferase (OCT)	S	0–6 U/L		
Osmolality	S	278–305 mosm/kg		
Oxygen content	B	15–23 ml/dl		
Oxygen pressure, pO_2	B	85–95 mm Hg		
Oxygen saturation	B	94–97%		
pH	B (Art)	7.35–7.45		
Phenolsulfonphthalein (PSP)	U	15 min >25% excreted 30 min >40% excreted		
Phosphatase, acid (ACP), total tartrate inhibitable	S	0.2–1.8 U/L 0.2–0.6 U/L	By method described	
Parathyroid hormone (PTH)	S	0.10–0.65 ng/ml when Ca^{2+} is 1.02–1.16 mmol/L		
Phosphatase, alkaline (ALP)	S	Ad 20–105 U/L to 3 mo 70–220 U/L 3 mo–10 yr 60–150 U/L 10 yr–pub. 60–260 U/L	By method described By method described By method described By method described	
Phosphate, inorganic (as P)	S	Ch 4.0–7.0 mg/dl Ad 3.0–4.5 mg/dl	1.28–2.24 mmol/L 0.96–1.44 mmol/L	0.32
Potassium	S	3.8–5.4 meq/L	3.8–5.4 mmol/L	1.0
Protein, total	S	6.0–8.2 g/dl	60–82 g/L	10
	Sf	15–45 mg/dl	150–450 mg/L	10
	U	<10 mg/dl	<100 mg/L	10
Protein fractionation	See electrophoresis			
Sodium	S	136–145 meq/L	136–145 mmol/L	1.0
Testosterone	P	M 3.5–12 ng/ml	12–42 nmol/L	3.5
	P	F 0.3–0.9 ng/ml	1.0–3.1 nmol/L	
Thyroid hormones —by RIA	S	as thyroxine, 4.1–11.3 μg/dl	53–147 nmol/L	13
T_4—free	S	5 ng/dl	65 pmol/L	13
T_3—free	S	0.6 ng/dl	9 pmol/L	15
T_3 resin uptake	S	33–40%		
Thyroid stimulating hormone (TSH)	P	0–3 μU/ml		
Triglycerides	S	40–150 mg/dl	0.45–1.7 mmol/L	0.0113
Urea nitrogen	S	8–18 mg/dl	2.9–6.4 mmol/L	0.357
Uric acid	S	M 3.5–7.5 mg/dl	0.210–0.445 mmol/L	
		F 2.5–6.5 mg/dl	0.150–0.390 mmol/L	0.060
		Ch 2.0–5.5 mg/dl	0.120–0.330 mmol/L	
Urobilinogen 2 hr	U	0.1–1.0 Ehrlich units	Arbitrary unit	
24 hr		0.5–4 Ehrlich units		
VMA (4-hydroxy-3-methoxymandelic acid)	U	2–8 mg/24 hr	10–40 μmol/24 hr	5.05

APPENDIX 2
Greek Letters Commonly Used in Clinical Chemistry

α . . . . alpha
β . . . . beta
γ . . . . gamma
δ . . . . delta
κ . . . . kappa
λ . . . . lambda (obsolete term for wavelength, microliter)
μ . . . . mu (micro, 1/1,000,000)
Σ . . . . sigma (sum of)
Δ . . . . delta (difference)

APPENDIX 3
Concentrations of Common Reagents and Volumes Required to Dilute to 1 mol/L

REAGENT*	CONCENTRATION OF CONCENTRATED SOLUTION MOL/L	PURITY %	ML CONCENTRATED REAGENT TO PREPARE 1 MOL/L SOLUTION
Acetic acid	17.4	99–100†	57
Hydrochloric acid	11.6	36†	86
Nitric acid	16.4	69†	61
Phosphoric acid	14.6	85†	68
Sulfuric acid	17.8	95†	56
Ammonium hydroxide	14.8	28‡	67

*Reagent grade chemicals.
†% (w/v) of the acid.
‡% (w/v) of NH_3.

APPENDIX 4
Boiling Points of Commonly Used Solvents

NAME	ADDITIONAL NAMES	BOILING POINT (° C)
Acetone	2-Propanone	56.2
Benzene		80.1
Carbon tetrachloride	Tetrachloromethane	76.8
Chloroform	Trichloromethane	61.2
Ethanol	Ethyl alcohol	78.5
Ethyl acetate	Acetic acid ethyl ester	77.1
Ethyl ether	Diethyl ether	34.6
Ethylene dichloride	1,2-Dichloroethane	84
Heptane		98.4
Methanol	Methyl alcohol	65.0
Methylene chloride	Dichloromethane	40
Petroleum ether		40–120 but varies with fraction
Toluene	Methylbenzene	110.6
p-Xylene	1,4-Dimethylbenzene	138

APPENDIX 5
Acetic Acid-Sodium Acetate Buffer, 0.2 mol/L

pH 25° C‡	ACETIC ACID, 0.20 mol/L* ml	SODIUM ACETATE, 0.20 mol/L† ml
3.6	92.5	7.5
3.8	88.0	12.0
4.0	82.0	18.0
4.2	73.5	26.5
4.4	63.0	37.0
4.6	52.0	48.0
4.8	41.0	59.0
5.0	30.0	70.0
5.2	21.0	79.0
5.4	14.0	86.0
5.6	9.0	91.0
5.8	6.0	94.0

REFERENCE. Data calculated from Walpole, G.S.: Hydrogen potentials of mixtures of acetic acid and sodium acetate. J. Chem. Soc. *105*, 2501, 1914.

*Acetic acid, 0.20 mol/L. Dilute 11.5 ml glacial acetic acid to 1000 ml volume with water. Titrate with standard 0.10 molar NaOH to phenolphthalein end point and adjust to 0.20 mol/L.

†Sodium acetate, 0.20 mol/L. Dissolve 16.41 g sodium acetate (anhydrous) in water and make up to 1000 ml volume.

‡The pH is about 0.05 lower at 30° C.

APPENDIX 6

Phosphate Buffer of Constant Ionic Strength (0.05 and 0.1)
Volumes of 0.5 mol/L Solutions of KH_2PO_4 and Na_2HPO_4 Diluted to
1000 ml with Water

pH 25° C	IONIC STRENGTH			
	0.05		0.1	
	KH_2PO_4* ml	Na_2HPO_4† ml	KH_2PO_4* ml	Na_2HPO_4† ml
5.8	—	—	159	13.8
6.0	74.2	8.58	142	19.5
6.2	64.6	11.8	121	26.4
6.4	53.4	15.5	98.2	34.0
6.6	42.0	19.3	75.6	41.4
6.8	31.4	22.8	55.4	48.2
7.0	22.4	25.8	39.0	53.6
7.2	15.4	28.2	26.4	57.8
7.4	10.3	30.0	17.6	60.8
7.6	6.74	31.0	11.5	62.8
7.8	4.36	31.8	7.38	64.2
8.0	2.80	32.4	—	—

REFERENCE. Datta, S.P., and Grzbowski, A.K.: *In* Biochemist's Handbook, edited by C. Long. Princeton, NJ, Van Nostrand Co., 1961, p. 32.
*KH_2PO_4, 0.5 mol/L. The anhydrous salt is dried at 110° C for 2 hours; 68.05 g are dissolved in water and made up to 1000 ml volume.
†Na_2HPO_4, 0.5 mol/L. The anhydrous salt is dried at 110° C for 2 hours; 70.99 g are dissolved in water and made up to 1000 ml volume.

APPENDIX 7

Tris Buffer, 0.05 mol/L
Volumes of HCl to be added to 25 ml Tris*, 0.2 mol/L,
and diluted to 100 ml with water

HCl, 0.1 mol/L ml	pH† 23° C
5.0	9.10
7.5	8.92
10.0	8.74
12.5	8.62
15.0	8.50
17.5	8.40
20.0	8.32
22.5	8.23
25.0	8.14
27.5	8.05
30.0	7.96
32.5	7.87
35.0	7.77
37.5	7.66
40.0	7.54
42.5	7.36
45.0	7.20

REFERENCE. Gomori, G.: Buffers in the range of pH 6.5 to 9.6. Proc. Soc. Exp. Biol. Med. *62*, 33, 1946.
*Tris (hydroxymethyl) aminomethane, 0.2 mol/L. Dissolve 24.3 g Tris (Sigma or Schwarz/Mann) in water and make up to 1000 ml volume.
†At 30° C the pH values are 0.14 to 0.15 pH units lower.

APPENDIX 8
Atomic Weights of the Common Elements*

NAME	SYMBOL	INTERNATIONAL ATOMIC MASS	NAME	SYMBOL	INTERNATIONAL ATOMIC MASS
Aluminum	Al	26.98	Manganese	Mn	54.94
Antimony (stibium)	Sb	121.8	Mercury	Hg	200.6
Argon	Ar	39.95	Molybdenum	Mo	95.94
Arsenic	As	74.92	Neon	Ne	20.18
Barium	Ba	137.3	Nickel	Ni	58.71
Beryllium	Be	9.012	Nitrogen	N	14.01
Bismuth	Bi	209.0			
Boron	B	10.81	Oxygen	O	16.00
Bromine	Br	79.91	Palladium	Pd	106.4
Cadmium	Cd	112.4	Phosphorus	P	30.97
Calcium	Ca	40.08	Platinum	Pt	195.1
Carbon	C	12.01	Potassium (kalium)	K	39.10
Cerium	Ce	140.1	Selenium	Se	78.96
Cesium	Cs	132.9	Silicon	Si	28.09
Chlorine	Cl	35.45	Silver (argentum)	Ag	107.9
Chromium	Cr	52.00	Sodium (natrium)	Na	22.99
Cobalt	Co	58.93	Strontium	Sr	87.62
Copper	Cu	63.54	Sulfur	S	32.06
Fluorine	F	19.00	Tellurium	Te	127.6
Gold (aurum)	Au	197.0	Thallium	Tl	204.4
Helium	He	4.003	Thorium	Th	232.0
Hydrogen	H	1.008	Tin (stannum)	Sn	181.7
Iodine	I	126.9	Titanium	Ti	47.90
Iron (ferrum)	Fe	55.85	Tungsten (wolfram)	W	183.8
Lanthanum	La	138.9	Uranium	U	238.0
Lead (plumbum)	Pb	207.2	Vanadium	V	50.94
Lithium	Li	6.939	Xenon	Xe	131.3
Magnesium	Mg	24.31	Zinc	Zn	65.37

*Values as of 1963, based on carbon-12 and rounded off to four significant figures.

APPENDIX 9
Four-Place Logarithms
LOGARITHMS

Natural Numbers	0	1	2	3	4	5	6	7	8	9	PROPORTIONAL PARTS								
											1	2	3	4	5	6	7	8	9
10	0000	0043	0086	0128	0170	0212	0253	0294	0334	0374	4	8	12	17	21	25	29	33	37
11	0414	0453	0492	0531	0569	0607	0645	0682	0719	0755	4	8	11	15	19	23	26	30	34
12	0792	0828	0864	0899	0934	0969	1004	1038	1072	1106	3	7	10	14	17	21	24	28	31
13	1139	1173	1206	1239	1271	1303	1335	1367	1399	1430	3	6	10	13	16	19	23	26	29
14	1461	1492	1523	1553	1584	1614	1644	1673	1703	1732	3	6	9	12	15	18	21	24	27
15	1761	1790	1818	1847	1875	1903	1931	1959	1987	2014	3	6	8	11	14	17	20	22	25
16	2041	2068	2095	2122	2148	2175	2201	2227	2253	2279	3	5	8	11	13	16	18	21	24
17	2304	2330	2355	2380	2405	2430	2455	2480	2504	2529	2	5	7	10	12	15	17	20	22
18	2553	2577	2601	2625	2648	2672	2695	2718	2742	2765	2	5	7	9	12	14	16	19	21
19	2788	2810	2833	2856	2878	2900	2923	2945	2967	2989	2	4	7	9	11	13	16	18	20
20	3010	3032	3054	3075	3096	3118	3139	3160	3181	3201	2	4	6	8	11	13	15	17	19
21	3222	3243	3263	3284	3304	3324	3345	3365	3385	3404	2	4	6	8	10	12	14	16	18
22	3424	3444	3464	3483	3502	3522	3541	3560	3579	3598	2	4	6	8	10	12	14	15	17
23	3617	3636	3655	3674	3692	3711	3729	3747	3766	3784	2	4	6	7	9	11	13	15	17
24	3802	3820	3838	3856	3874	3892	3909	3927	3945	3962	2	4	5	7	9	11	12	14	16
25	3979	3997	4014	4031	4048	4065	4082	4099	4116	4133	2	3	5	7	9	10	12	14	15
26	4150	4166	4183	4200	4216	4232	4249	4265	4281	4298	2	3	5	7	8	10	11	13	15
27	4314	4330	4346	4362	4378	4393	4409	4425	4440	4456	2	3	5	6	8	9	11	13	14
28	4472	4487	4502	4518	4533	4548	4564	4579	4594	4609	2	3	5	6	8	9	11	12	14
29	4624	4639	4654	4669	4683	4698	4713	4728	4742	4757	1	3	4	6	7	9	10	12	13
30	4771	4786	4800	4814	4829	4843	4857	4871	4886	4900	1	3	4	6	7	9	10	11	13
31	4914	4928	4942	4955	4969	4983	4997	5011	5024	5038	1	3	4	6	7	8	10	11	12
32	5051	5065	5079	5092	5105	5119	5132	5145	5159	5172	1	3	4	5	7	8	9	11	12
33	5185	5198	5211	5224	5237	5250	5263	5276	5289	5302	1	3	4	5	6	8	9	10	12
34	5315	5328	5340	5353	5366	5378	5391	5403	5416	5428	1	3	4	5	6	8	9	10	11
35	5441	5453	5465	5478	5490	5502	5514	5527	5539	5551	1	2	4	5	6	7	9	10	11
36	5563	5575	5587	5599	5611	5623	5635	5647	5658	5670	1	2	4	5	6	7	8	10	11
37	5682	5694	5705	5717	5729	5740	5752	5763	5775	5786	1	2	3	5	6	7	8	9	10
38	5798	5809	5821	5832	5843	5855	5866	5877	5888	5899	1	2	3	5	6	7	8	9	10
39	5911	5922	5933	5944	5955	5966	5977	5988	5999	6010	1	2	3	4	5	7	8	9	10
40	6021	6031	6042	6053	6064	6075	6085	6096	6107	6117	1	2	3	4	5	6	8	9	10
41	6128	6138	6149	6160	6170	6180	6191	6201	6212	6222	1	2	3	4	5	6	7	8	9
42	6232	6243	6253	6263	6274	6284	6294	6304	6314	6325	1	2	3	4	5	6	7	8	9
43	6335	6345	6355	6365	6375	6385	6395	6405	6415	6425	1	2	3	4	5	6	7	8	9
44	6435	6444	6454	6464	6474	6484	6493	6503	6513	6522	1	2	3	4	5	6	7	8	9
45	6532	6542	6551	6561	6571	6580	6590	6599	6609	6618	1	2	3	4	5	6	7	8	9
46	6628	6637	6646	6656	6665	6675	6684	6693	6702	6712	1	2	3	4	5	6	7	7	8
47	6721	6730	6739	6749	6758	6767	6776	6785	6794	6803	1	2	3	4	5	5	6	7	8
48	6812	6821	6830	6839	6848	6857	6866	6875	6884	6893	1	2	3	4	4	5	6	7	8
49	6902	6911	6920	6928	6937	6946	6955	6964	6972	6981	1	2	3	4	4	5	6	7	8
50	6990	6998	7007	7016	7024	7033	7042	7050	7059	7067	1	2	3	3	4	5	6	7	8
51	7076	7084	7093	7101	7110	7118	7126	7135	7143	7152	1	2	3	3	4	5	6	7	8
52	7160	7168	7177	7185	7193	7202	7210	7218	7226	7235	1	2	2	3	4	5	6	7	7
53	7243	7251	7259	7267	7275	7284	7292	7300	7308	7316	1	2	2	3	4	5	6	6	7
54	7324	7332	7340	7348	7356	7364	7372	7380	7388	7396	1	2	2	3	4	5	6	6	7

LOGARITHMS

Natural Numbers	0	1	2	3	4	5	6	7	8	9	PROPORTIONAL PARTS								
											1	2	3	4	5	6	7	8	9
55	7404	7412	7419	7427	7435	7443	7451	7459	7466	7474	1	2	2	3	4	5	5	6	7
56	7482	7490	7497	7505	7513	7520	7528	7536	7543	7551	1	2	2	3	4	5	5	6	7
57	7559	7566	7574	7582	7589	7597	7604	7612	7619	7627	1	2	2	3	4	5	5	6	7
58	7634	7642	7649	7657	7664	7672	7679	7686	7694	7701	1	1	2	3	4	4	5	6	7
59	7709	7716	7723	7731	7738	7745	7752	7760	7767	7774	1	1	2	3	4	4	5	6	7
60	7782	7789	7796	7803	7810	7818	7825	7832	7839	7846	1	1	2	3	4	4	5	6	6
61	7853	7860	7868	7875	7882	7889	7896	7903	7910	7917	1	1	2	3	4	4	5	6	6
62	7924	7931	7938	7945	7952	7959	7966	7973	7980	7987	1	1	2	3	3	4	5	6	6
63	7993	8000	8007	8014	8021	8028	8035	8041	8048	8055	1	1	2	3	3	4	5	5	6
64	8062	8069	8075	8082	8089	8096	8102	8109	8116	8122	1	1	2	3	3	4	5	5	6
65	8129	8136	8142	8149	8156	8162	8169	8176	8182	8189	1	1	2	3	3	4	5	5	6
66	8195	8202	8209	8215	8222	8228	8235	8241	8248	8254	1	1	2	3	3	4	5	5	6
67	8261	8267	8274	8280	8287	8293	8299	8306	8312	8319	1	1	2	3	3	4	5	5	6
68	8325	8331	8338	8344	8351	8357	8363	8370	8376	8382	1	1	2	3	3	4	4	5	6
69	8388	8395	8401	8407	8414	8420	8426	8432	8439	8445	1	1	2	2	3	4	4	5	6
70	8451	8457	8463	8470	8476	8482	8488	8494	8500	8506	1	1	2	2	3	4	4	5	6
71	8513	8519	8525	8531	8537	8543	8549	8555	8561	8567	1	1	2	2	3	4	4	5	5
72	8573	8579	8585	8591	8597	8603	8609	8615	8621	8627	1	1	2	2	3	4	4	5	5
73	8633	8639	8645	8651	8657	8663	8669	8675	8681	8686	1	1	2	2	3	4	4	5	5
74	8692	8698	8704	8710	8716	8722	8727	8733	8739	8745	1	1	2	2	3	4	4	5	5
75	8751	8756	8762	8768	8774	8779	8785	8791	8797	8802	1	1	2	2	3	3	4	5	5
76	8808	8814	8820	8825	8831	8837	8842	8848	8854	8859	1	1	2	2	3	3	4	5	5
77	8865	8871	8876	8882	8887	8893	8899	8904	8910	8915	1	1	2	2	3	3	4	4	5
78	8921	8927	8932	8938	8943	8949	8954	8960	8965	8971	1	1	2	2	3	3	4	4	5
79	8976	8982	8987	8993	8998	9004	9009	9015	9020	9026	1	1	2	2	3	3	4	4	5
80	9031	9036	9042	9047	9053	9058	9063	9069	6074	9079	1	1	2	2	3	3	4	4	5
81	9085	9090	9096	9101	9106	9112	9117	9122	9128	9133	1	1	2	2	3	3	4	4	5
82	9138	9143	9149	9154	9159	9165	9170	9175	9180	9186	1	1	2	2	3	3	4	4	5
83	9191	9196	9201	9206	9212	9217	9222	9227	9232	9238	1	1	2	2	3	3	4	4	5
84	9243	9248	9253	9258	9263	9269	9274	9279	9284	9289	1	1	2	2	3	3	4	4	5
85	9294	9299	9304	9309	9315	9320	9325	9330	9335	9340	1	1	2	2	3	3	4	4	5
86	9345	9350	9355	9360	9365	9370	9375	9380	9385	9390	1	1	2	2	3	3	4	4	5
87	9395	9400	9405	9410	9415	9420	9425	9430	9435	9440	0	1	1	2	2	3	3	4	4
88	9445	9450	9455	9460	9465	9469	9474	9479	9484	9489	0	1	1	2	2	3	3	4	4
89	9494	9499	9504	9509	9513	9518	9523	9528	9533	9538	0	1	1	2	2	3	3	4	4
90	9542	9547	9552	9557	9562	9566	9571	9576	9581	9586	0	1	1	2	2	3	3	4	4
91	9590	9595	9600	9605	9609	9614	9619	9624	9628	9633	0	1	1	2	2	3	3	4	4
92	9638	9643	9647	9652	9657	9661	9666	9671	9675	9680	0	1	1	2	2	3	3	4	4
93	9685	9689	9694	9699	9703	9708	9713	9717	9722	9727	0	1	1	2	2	3	3	4	4
94	9731	9736	9741	9745	9750	9754	9759	9763	9768	9773	0	1	1	2	2	3	3	4	4
95	9777	9782	9786	9791	9795	9800	9805	9809	9814	9818	0	1	1	2	2	3	3	4	4
96	9823	9827	9832	9836	9841	9845	9850	9854	9859	9863	0	1	1	2	2	3	3	4	4
97	9868	9872	9877	9881	9886	9890	9894	9899	9903	9908	0	1	1	2	2	3	3	4	4
98	9912	9917	9921	9926	9930	9934	9939	9943	9948	9952	0	1	1	2	2	3	3	4	4
99	9956	9961	9965	9969	9974	9978	9983	9987	9991	9996	0	1	1	2	2	3	3	3	4

APPENDIX 10
Manufacturers and Suppliers

Advanced Instruments, Inc., Needham Heights, MA 02194.
Aldrich Chemical Co., Milwaukee, WI 53233.
Alpkem Corp., Clackamas, OR 97015.
American Can Co., Dixie/Marathon Division, Greenwich, CT 06830.
American Instrument Co., Silver Spring, MD 20910.
American Optical Corp., Scientific Instrument Div., P.O. Box 123, Buffalo, NY 14240.
Ames Division, Miles Laboratories, Inc., P.O. Box 70, Elkhart, IN 46515.
Analytical Products, Inc., P.O. Box 845, Belmont, CA 94002.
Analytical Systems, Inc., Laguna Hills, CA 92653
Applied Science Division, Milton Roy Laboratory Group, P.O. Box 440, State College, PA 16801.
J. T. Baker Chemical Co., Phillipsburg, NJ 08865.
Beckman Instruments, Inc., Health Care Products, Brea, CA 92621.
Bio-Dynamics/bmc, Div. Boehringer-Mannheim, Indianapolis, IN 46250.
Bio-Rad Laboratories, Richmond, CA 94804.
Boehringer-Mannheim Biochemicals, Indianapolis, IN 46250.
Brinkmann Instruments, Westbury, NY 11590
Buchler Instrument Div., Fort Lee, NJ 07024.
Burdick and Jackson Labs, Muskegon, MI 49442.
Calbiochem-Behring Corp., P.O. Box 12087, San Diego, CA 92112.
Corning Glass Works, Corning, NY 14830.
Corning Medical and Scientific, Medfield, MA 02052.
Distillation Products Industries, Div. Eastman Kodak Co., Rochester, NY 14650.
Dow Diagnostics, P.O. Box 68511, Indianapolis, IN 46240.
E. I. Du Pont de Nemours & Co., Inc., Wilmington, DE 19898.
Eastman Kodak Co., Rochester, NY 14650.
Environmental Science Associates (esa), Bedford, MA 01730.
Eppendorf. See *Brinkmann Instruments*.
Fisher Scientific Co., Pittsburgh, PA 15219.
Gelman Sciences, Inc., Ann Arbor, MI 48106.
General Biochemicals, Chagrin Falls, OH 44022.
Hach Co., P.O. Box 389, Loveland, CO 80537.
W. A. Hammond Drierite Co., Xenia, OH 45385.
Harleco, Div. American Hospital Supply Corp., Gibbstown, NJ 08027.
Helena Laboratories, P.O. Box 752, Beaumont, TX 77704.
Hynson, Westcott & Dunning, Inc., Baltimore, MD 21201.
Immunotech, Cambridge, MA 02139
Instrumentation Laboratory, Inc., Lexington, MA 02173.
Isolab, Inc., Akron, OH 44321.
Lexington Instruments Corp., Waltham, MA 02154.
Eli Lilly & Co., Indianapolis, IN 46206.
London Co., Cleveland, OH 44145 (agent in the USA for Radiometer).
Mallinckrodt, Inc., St. Louis, MO 63134.
MC/B Manufacturing Chemists, Cincinnati, OH 45212.
Merck & Co., Inc., Rahway, NJ 07065.
Millipore Corp., Bedford, MA 01730.
National Aniline Div., Allied Chemical Corp., New York, NY 10006.
National Bureau of Standards, Office of Standard Reference Materials, Washington, DC 20234.
Orion Research Inc., Cambridge, MA 02139.
Perkin-Elmer Corp., Norwalk, CT 06856.
Pfanstiehl Labs, Inc., Waukegan, IL 60085.
Pfizer Diagnostics Div., New York, NY 10017.
Pharmacia Laboratories, Piscataway, NJ 08854.
Pierce Chemical Co., Rockford, IL 61105
P-L Biochemicals, Inc., Milwaukee, WI 53205.
Quantum Industries, Fairfield, NJ 07006.

Radiometer, Copenhagen, Denmark. *See* London Co.
Reeve Angel, Clifton, NJ 07014.
Regis Chemical Co., Morton Grove, IL 60053
Rohm & Haas, Philadelphia, PA 19105.
Schleicher & Schuell, Keene, NH 03431.
Schwarz/Mann Research Laboratories, Orangeburg, NY 10962.
Scientific Industries, Inc., Bohemia, NY 11716.
Scientific Manufacturing Industries (SMI), Emeryville, CA 94608
Sclavo Diagnostics, Wayne, NJ 07470
Sigma Chemical Co., P.O. Box 14508, St. Louis, MO 63178.
E. R. Squibb & Sons, P.O. Box 4000, Princeton, NJ 08540.
Syva, P.O. Box 10058, Palo Alto, CA 94303.
Taylor Chemicals, Inc., Baltimore, MD 21204.
Technicon Industrial Systems, Tarrytown, NY 10591.
Turner Associates, Palo Alto, CA 94303.
Union Carbide, Carbon Products Div., New York, NY 10017.
Waters Associates, Milford, MA 01757.
Whale Scientific Co., Commerce City, CO 80239.
Whatman, Inc., Clifton, NJ 07014.
Worthington Diagnostics, Div. Millipore Corp., Freehold, NJ 07728.

INDEX

Page numbers in italics refer to illustrations; page numbers followed by a "t" refer to tables.